Imaging Life

Imaging Life

Biological Systems from Atoms to Tissues

EDITED BY GARY C. HOWARD

WILLIAM E. BROWN

AND

MANFRED AUER

OXFORD
UNIVERSITY PRESS

OXFORD
UNIVERSITY PRESS

Oxford University Press is a department of the University of Oxford. It furthers the University's objective of excellence in research, scholarship, and education by publishing worldwide.

Oxford New York
Auckland Cape Town Dar es Salaam Hong Kong Karachi
Kuala Lumpur Madrid Melbourne Mexico City Nairobi
New Delhi Shanghai Taipei Toronto

With offices in
Argentina Austria Brazil Chile Czech Republic France Greece
Guatemala Hungary Italy Japan Poland Portugal Singapore
South Korea Switzerland Thailand Turkey Ukraine Vietnam

Published in the United States of America by
Oxford University Press
198 Madison Avenue, New York, NY 10016

Library of Congress Cataloging-in-Publication Data
Imaging life : biological systems from atoms to tissues / edited by Gary C. Howard, William E. Brown, and Manfred Auer.
pages cm
Includes bibliographical references and index.
ISBN 978-0-19-531443-4 (alk. paper)
1. Microscopy. 2. Imaging systems in biology. I. Howard, Gary C. II. Brown, William E. III. Auer, Manfred.
QH205.2.I435 2014
570.28'2—dc23
2014001401

9 8 7 6 5 4 3 2 1
Printed in the United States of America
on acid-free paper

William E. Brown

During the preparation of this book and far too early, we lost Bill Brown. He was an extraordinary scientist, teacher, and leader, who influenced the lives and careers of scores of young scientists at Carnegie Mellon University for over thirty years. Bill looked for the very best in every experiment and every student. He was amazingly generous with his time, help, and encouragement. To me, he was a great mentor, role model, and friend.

Gary C. Howard

CONTENTS

LIST OF CONTRIBUTORS

Paul D. Adams, PhD
Senior Scientist
Molecular Foundary
Lawrence Berkeley National Laboratory

John R. Androsavich, PhD
Single Molecule Analysis Group
Department of Chemistry
University of Michigan
Current address: now Research Scientist,
Platform & Assay Development,
Regulus Therapeutics Inc.

Manfred Auer, PhD
Staff Scientist
Lawrence Berkeley National Laboratory

Chandrajit Bajaj, PhD
Computational Applied Mathematics Chair in Visualization
Professor of Computer Science
Director of Center for Computational Visualization
University of Texas, Austin

Liang Chen, PhD
Director, Berekeley Synchrotron Infrared Structural Biology Program
Lawrence Berkeley National Laboratory

Bertrand P. Cinquin, PhD
Postdoctoral Fellow
National Center for X-Ray Microscopy
Lawrence Berkeley National Laboratory

G. Marius Clore, MD, PhD
Head, Laboratory of Chemical Physics
National Institutes of Medicine

Steven Finkbeiner, MD, PhD
Associate Director and Senior Investigator
Gladstone Institute of Neurological Disease
Professor of Neurology and Physiology
University of California, San Francisco

Stefan P. Hau-Riege, PhD
Group Leader X-Ray Science and Technology Group
Physical and Life Sciences
Lawrence Livermore National Laboratory

Hoi-Ying N. Holman, PhD
Director, Berekeley Synchrotron Infrared Structural Biology Program
Lawrence Berkeley National Laboratory
hyholman@lbl.gov

Gary C. Howard, PhD
Manager, Scientific Editing
The Gladstone Institutes

Carolyn Larabell, PhD
Director, National Center for X-Ray Microscopy
Lawrence Berkeley National Laboratory
Professor of Anatomy
University of California, San Francisco

Mark A. Le Gros, PhD
Associate Director, National Center for X-Ray Microscopy
Lawrence Berkeley National Laboratory

Gerry McDermott, PhD
Senior Scientist
National Center for X-Ray Microscopy
Lawrence Berkeley National Laboratory

Daniel Müller, PhD
Professor of Biophysics
Department of Biosystems Science and Engineering
ETH Zurich

Valentin Nägerl, PhD
Professor of Neuroscience and Bioimaging
Interdisciplinary Institute for Neuroscience
University of Bordeaux

M. Gary Newton, PhD
Emeritus Professor of Chemistry
University of Georgia

Hanchuan Peng, PhD
Associate Investigator
Allen Institute for Brain Science

Pradeep N. Perera, PhD
Postdoctoral Fellow
Molecular Foundary
Lawrence Berkeley National Laboratory

Natalia Pinzon
Visiting Student
Lawrence Berkeley National Laboratory

Sethuramasundaram Pitchiaya, PhD
Postdoctoral Fellow
Single Molecule Analysis Group
Department of Chemistry
University of Michigan

Brendan Prideaux, PhD
Staff Scientist
Novartis, Inc.

John P. Rose, PhD
Associate Professor of Biochemistry and Molecular Biology
University of Georgia

K. Tanuj Sapra, PhD
Postdoctoral Fellow
Department of Biochemistry
University of Zurich

Martin Schmidt, PhD
Molecular Foundary
Lawrence Berkeley National Laboratory

James Schuck, PhD
Staff Scientist
Molecular Foundary
Lawrence Berkeley National Laboratory

Elizabeth A. Smith, PhD
Postdoctoral Fellow
National Center for X-Ray Microscopy
Lawrence Berkeley National Laboratory

Phoebe L. Stewart, PhD
Professor of Pharmacology and Director of Cleveland Center for Membrane and Structural Biology
Case Western Reserve University

Nils G. Walter, PhD
Professor of Chemistry
Director of Single Molecule
Analysis in Real-Time Center
University of Michigan

Bi-Cheng Wang, PhD
Professor and Ramsey/Georgia Research Alliance Eminent Scholar in
Structural Biology
University of Georgia

Alexander Weber-Bargioni, PhD
Staff Scientist
Molecular Foundary
Lawrence Berkeley National Laboratory

Imaging Life

1

Imaging Life

MANFRED AUER AND GARY C. HOWARD

> And in Physical Enquires, we must endeavor to follow nature in the more plain and easy the treads in the most simple and uncompounded bodies, to trace her steps, and be acquainted with her manner of walking there, before we venture ourselves into the multitude of meanders she has in bodies of a more complicated nature.
>
> —ROBERT HOOKE, *Micrographia or Some Physiological Descriptions of Minute Bodies Made by Magnifying Glasses with Observations and Inquiries thereupon*

Humans have been wondering about life for centuries. Mesmerized by the living world around them, early biologists were driven by the basic, yet still relevant question, "what is life?" Early scientists, such as Alcmaeon and Croton in the fifth century B.C. and Erasistratus and Herophilos in the third century B.C., began to describe the human body in remarkable detail, illuminating function through the description of form. Despite these early accomplishments, another two thousand years passed with little progress. Dissections of human and animal bodies were constrained by the visual abilities of our eyes, and our understanding was limited to the level of organs and tissues.

In the late seventeenth century, the invention of the optical microscope began a revolution in biology that merged our curiosity with our understanding of physics and light. When Antonie van Leeuwenhoek looked through his primitive microscope for the first time, he saw a world of microbes that no one even knew existed. Although his claims of single-cell organisms were greeted with great skepticism (as many important discoveries often are), it quickly became clear that "seeing was believing." Even today, few things are more powerful than an imagefor convincing skeptics.

The light microscope dominated biological discovery for a hundred years. Improvements to contrast-generating chemistry and better knowledge of the physical instrumentation improved biological imaging and allowed scientists to more fully exploit their potential. Clever selective chemical staining procedures enhanced imaging and laid the foundation for our understanding of life at the level of cell organelles, cells, and tissues.

At the turn of the twentieth century, scientists became more aware of the full range of the electromagnetic spectrum, including X-rays, electrons, neutrons and positrons, and gamma rays. Novel forms of imaging were based on the traditional optical, radio, and microwave window of the electromagnetic spectrum, but, more recently, were expanded to include approaches that provide chemical information, such as FTIR, Raman, and mass spectrometry imaging. In second part of the twentieth century, most of the modern imaging technologies were invented or substantially refined. Besides optical light microscopy, the use of X-ray and electrons in diffraction and microscopy, as well as radio waves and microwaves for NMR and fMRI, revolutionized biological imaging and led to exquisite insights into the sophisticated organization and biological functionality of tissues, cells, organelles, and proteins.

Through advances in imaging, the basic building blocks of life, such as proteins, nucleic acids, lipids, and carbohydrates, were described in increasing detail. The oldest discipline for structure determination is X-ray crystallography. Although it is not strictly an imaging method—a diffraction pattern is interpreted and refined by using an atomic model—it quickly became the "gold standard" for structural imaging. For example, it yielded molecular mechanisms that accurately explained protein function. The DNA double helix immediately suggested a mechanism of faithful replication of the genetic information, and enzymatic functions could be explained by ingenious yet simple stereochemical macromolecular organization.

Twentieth century biologists gathered around the idea that cells could be viewed as bags of enzymes, organelles as mere reaction vessels for enzymatic function. Hence, using the "divide and conquer" strategy of biochemistry made sense for determining the function of one protein at a time. As a result of this reductionist approach, we purified each component and used X-ray and electron crystallography or nuclear magnetic resonance (NMR) spectroscopy to determine its atomic or near-atomic structure. The success rate was substantially enhanced by genetic engineering to improve protein stability, reduce conformational heterogeneity, and introduce primary sequence amino acids that facilitate the process of subsequent X-ray structural analysis. Structural analysis combined with site-directed mutagenesis allowed mechanistic insights and prediction of protein function and into protein-protein interaction. Biochemistry was transformed from a discipline that merely sought to enrich and cleverly test enzyme activity to a powerhouse that, when combined with structural analysis, could explain in detail protein function, fueled by high-resolution structures of increasingly complex proteins and small (and not so small) protein complexes. The field of structural biology was born. System complexity was low (often one pure protein species), but mechanistic insight was extremely precise.

Biochemistry's sibling discipline was cell biology, in which the system was often messy and ill-defined, but the observations were tightly linked to actual cell biological phenomena. Within cell biology, electron microscopy has a unique place. Our understanding of cellular function is intimately linked to the advancement in electron microscopic imaging with a number of cellular macromolecular components, such as actin, microtubule, and intermediate filament networks, constituting the cytoskeleton, a variety of organelles and intracellular compartments, as well as multiprotein macromolecular complexes, such as ribosomes, proteasomes, nuclear pore, and various cell-cell and cell-matrix adhesion complexes. However, its shortcomings limited its potential to its early yet distinguished contributions to cell biology. These include an inability to study dynamic processes in real time as well as the very small part of the cell that could be imaged in great detail, along with artifacts of early sample preparation.

Once cell biologists decided they reasonably understood the basic ultrastructural framework of cells and the cell's inventory of basic organelle and macromolecular complexes, the emphasis changed to gaining temporal mechanistic insight, focusing on a small subset of macromolecules that could be easily detected and hence followed microscopically in real time. In recent years, advances in optics, histochemical staining and tag-based labeling have increased the sophistication of optical microscopic imaging and allowed the study of life at the subcellular and even the single macromolecular level. The discovery and subsequent engineering of genetically encoded fluorophores and the development of fluorescent protein-target protein fusion technology revolutionized optical microscopy. Now real-time in-vivo imaging of living cells is possible, adding dynamics to biological processes although only for the selected system component(s).

Cell biology has revealed that cells are much more sophisticated and cannot be assumed to be simple bags of enzymes. Even bacteria are not that. While the field of biology in the past five decades has been fixated on the reduction of life to its smallest constituents, there is a burning curiosity to understand life in all its complexity.

One solution to understanding the complexity of life is simply to acquire lots of data and to identify all key players. Hence, the era of "-omics" started with the human genome and expanded to the proteome and on to what is often called the structural genome. The last decades were spent compiling a phonebook of biological parts. However, imaging—in its various forms—will play a key role in making sense of the spatio-temporal relationship of the parts, turning the phone book into the biological equivalent of Google-Maps and, therefore, ultimately lead to an enhanced understanding of biological processes.

In preparing this book, our goal has been to describe the current state of the complex, interdisciplinary science of bioimaging. At the core of answering the fundamental questions of life, microscopy and related imaging techniques provide the visual-spatial framework for biological processes. Without imaging, the answers to the fundamental questions of life are not within our grasp. The mere presence of chemical reactions and physical forces alone is not sufficient to produce life. Spatio-temporal *organization* is the missing ingredient. Only the finely orchestrated coexistence of a mind-bogglingly large number of chemical processes and physical forces, rooted in billions of years of evolutionary experimentation and refinement, can begin to account for the mystery of life. Imaging allows us to uncover that organization.

The objective of this book is to provide an overview for the novice as well as sufficient detailed insight for the curious expert mind. To work together effectively, scientists from different but related fields must understand the nuances of each approach. At the same time, a new generation of scientists needs to be exposed to these approaches and what they can do for their area of interest. We asked each author to strike a balance between basic concepts and detailed expert knowledge. We are not looking for a specific solution, as we believe there is no single magic bullet. Instead, we want to examine frontiers. We also hope to contribute to an improved communication between scientists in different fields. The audience for this book will be scientists and others who wish to understand the potential role of imaging techniques in filling the knowledge gap, including those in the fields of structural and cell biology and also those in allied fields, such as computer science and mechanistic biology. While we deliberately did not design this book as a textbook, we certainly hope it will be useful as an important reference for teaching graduate students and advanced undergraduates, with the hope to stimulate students and researchers to take up the cause and bridge the currently existing divides.

We have included a list of what we consider the most powerful techniques for imaging-based holistic insight into the mechanisms of biological function. These methods will also allow us ultimately to predict biological behavior based on reality-based models for sustaining cellular function, both physiological and pathological, and thus allow us to approach again the age-old question of "What is life?"

PART I

Imaging the Macromolecular Inventory

Structure and Mechanistic Function of the Building Blocks of Life

Introduction to Part I

MANFRED AUER, NATALIA PINZÓN, AND GARY C. HOWARD

> Biologists work very close to the frontier between bewilderment and understanding.
>
> —Sir Peter B. Medawar, *1969*

What is life? While we cannot comprehensively answer this age-old question, we have come a long way in this almost philosophical quest. In the twentieth century, the basic building blocks of life that compartmentalize, catalyze, or otherwise enable the chemical reactions of biological metabolism were identified. Four classes of macromolecules lie at the heart of cellular function for all living organisms: proteins, nucleic acids, lipids, and complex carbohydrates. These four classes of macromolecules, all with clearly different roles, must work together to enable life as we know it and sustain metabolic processes. Their interplay ensures the formation of energy-rich molecules, such as sugars from water and carbon dioxide during photosynthesis, that can be metabolized to produce the even more valuable cellular currencies of ATP, NADH+, and proton gradients across membranes, allowing a variety of energy-dependant cellular processes, including nutrient uptake, signaling, protein biosynthesis, and hereditary material replication and transduction.

Proteins naturally dominate this carefully orchestrated metabolic dance. They have evolved to catalyze energetically chemical reactions favorable and sometimes even unfavorable and therefore to promote efficient (bio-)chemistry. They allow the synthesis of all small molecules and macromolecules and are crucial in cellular life. Not surprisingly, proteins are the most comprehensively studied macromolecules. Protein function is not limited to metabolism. It is also central to other aspects of cellular function, such as the recognition of molecules resulting in a cellular response, cell signaling, and the controlled entrance and/or selective uptake of nutrients, molecules, and ions. Often it is important to visualize proteins in various conformations to describe the reaction mechanism that underlies function in atomic and molecular detail and therefore to reduce biological function to stereochemistry considerations.

Nucleic acids have been studied in great detail to understand the nature and architecture of the hereditary material and the mechanisms of replication, transcription, and translation. However, DNAs of various sequences have similar structures, and many have been satisfied with the knowledge we have obtained from a few representative structures. RNAs fold back on themselves to create complex three-dimensional conformations, not unlike proteins. While such RNA complexes remain a focus of structure determination, we have a significant understanding of nucleic acid structure and function.

Lipids are the primary means of storing energy. They also arrange into bilayers and are used to form compartments and thus separate cellular processes that require a unique environment for biochemical function to occur most efficiently. The basic architectural concept of lipid bilayers forming flat membrane systems or highly curved membrane structures, including vesicles, is fairly well understood, hence currently little research is dedicated to lipid macromolecular structures, although questions remain regarding the nature of lipid membrane heterogeneities, such as the proposed lipid rafts.

Carbohydrate has long been the neglected family member of the four major macromolecules, in part due to the challenges of determining primary, secondary, and tertiary structure and the fact that primary structure isn't encoded genetically, at least not in a form we can easily identify, but rather relies on the presence and sequence of action of particular enzymes, all of which is somewhat difficult to address experimentally. Although clearly important to the evolution of life, the exact nature of the carbohydrate architecture is often not crucial for intracellular processes, including metabolism, DNA replication, RNA transduction, and protein biosynthesis, but often plays an important role for posttranslational modification and also cell-cell interactions.

When attempting to solve the riddle of life, one is tempted to follow the age-old approach of "divide and conquer." If a problem is too complex to be solved as a whole, we break it up into smaller pieces that can be solved and understood on their own. We then put the pieces together again to approximate the more complex system. In the twentieth century, this reductionist approach enlisted biochemistry as the central player for its ability to fractionate, enrich, and isolate enzymatic activities and to purify proteins to homogeneity. The ultimate test in organic chemistry of a molecule is whether pure crystals of the substance can be obtained, and hence early on, crystallization of proteins and other macromolecules were attempted and increasingly accomplished. X-ray radiation was used to study the order of the crystals, which diffract the X-rays into a pattern characteristic for each crystal type. Mineralogy extensively used this approach to study inorganic material and deduce the exact position of each type of atoms in a lattice. Such systems are relatively simple, and the respective patterns could be determined through clever mathematical Fourier analysis, only needing knowledge of the chemical composition and the stoicheometry of its elements, a diffraction pattern, and a pen and paper. The Fourier transformation of the intensities of the indices of the diffraction patterns, combined with some educated guesses about which atoms could occupy which position in the lattice, yielded density information and hence led to the determination of the three-dimensional organization of the mineral's atomic lattice structure.

The principles of crystallography had been in place for quite some time, but the complexity of proteins and the number of atoms prohibited a direct structure determination. It took Max Perutz almost two decades to solve the so-called "phase problem," which means to figure out how diffraction patterns of small and not so small proteins yield so-called electron density maps that indicate the position of the atoms. The breakthrough was possible through a combination of chemistry (the creation of heavy atom derivatives) and computational advances making the structure determination of the first two proteins, myoglobin and hemoglobin, a heroic undertaking, for which Max Perutz, like many structural biologists since then, was awarded the Nobel Prize. What is so appealing about high-resolution atomic structures is that one can literally see all atoms (excluding hydrogens) and therefore describe a mechanism (e.g., of an enzymatic reaction) in chemical detail. By examining the catalytic active site, it becomes obvious how the enzyme or protein functions in general.

Strictly speaking, the diffraction pattern doesn't yield an atomic structure directly. Solving the 3D structure of a protein requires the calculation of an initially noisy density map that itself is hard to interpret. However, with several rounds of model building to create density maps and refinement of the model—using molecular dynamics calculations—and subsequent iterative comparison of the experimental data to the structural factors calculated from the atomic model, one bootstraps oneself towards the right solution.

X-RAY DIFFRACTION/CRYSTALLOGRAPHY

X-ray diffraction is the oldest technique and still the "gold standard" for structural determination. Since inter-atom bonds and distances between neighboring atoms are 5–10 Å, the high-energy portion (X-rays) of the electromagnetic spectrum is needed to resolve neighboring atoms. While each macromolecule will diffract X-rays, the signal of individual proteins is too weak to allow determination of its structure. By arranging billions of proteins into a crystalline pattern, X-rays will systematically reinforce one another at certain points of the three-dimensional space, also known as diffraction spots, and therefore elevate the signal to a level well above the noise that can be measured quite accurately and thus allow the determination of structure factors. For the longest time the search for derivative crystals with adequate large electron beam–stable heavy atoms was problematic. However, various genetically encoded labeling tricks, as well as the clever use of the tuneability of synchrotron radiation, together with freezing and the brightness of synchrotron X-ray generators, have ensured that high-quality diffraction patterns can be readily obtained, even from rather small crystals. Also, the structure determination has been mainly automated, allowing rapid interpretation and model building for most proteins.

The downsides of X-ray crystallography are that the macromolecule of interest must be crystallized and that such crystals must be ordered to a high degree. Crystallization often requires the availability of tens of milligrams of highly purified and conformationally homogeneous proteins, concentrated to 1–10 mg/ml, which can be difficult for human proteins or notoriously challenging proteins, such as membrane proteins. Furthermore, posttranslational modifications may be important for protein function but may increase heterogeneity and interfere with the formation of highly ordered crystals. While the crystal structure of a protein is an exquisite starting point for determining its mechanism of function, the protein is locked into one particular conformation and the packing into the crystal lattice can lead to conformations, rarely adopted in solution. Hence, it is always a good idea to examine different crystal forms and ideally trap the protein into different conformations (e.g., by co-crystallization with substrate, activators, inhibitors, or macromolecule interaction partners). Inherently X-ray crystallography gives a static snapshot of a protein conformation. Nevertheless, some aspects of the structure are less well defined than others, suggesting a higher degree of such regions of conformational flexibility.

Leaving all potential difficulties aside, when a molecule can be crystallized, X-ray analysis provides the fastest route to high-resolution structure determination, and can be considered a high-throughput technique, particularly when synchrotron X-ray radiation is employed, an effort which is also known as structural genomics.

While not explicitly mentioned in the chapter, there is a way to study proteins in solution by X-rays, also known as small angle X-ray scattering (or SAXS). It provides a scattering curve for the macromolecule that allows one to build models that match the observed scattering curve. This approach, while not yet capable to determine *ab-initio* structures without prior knowledge about the macromolecular shape, excels in distinguishing between different conformational models and is therefore very valuable for proteins where some structural information is already available or where through homology modeling a particular coarse shape can be assumed. Among its big advantages is the throughput that one can accomplish with this approach, but independent validation of the proposed conformation (e.g., by biochemical, genetic, or other structural experiments) is advised.

NUCLEAR MAGNETIC RESONANCE SPECTROSCOPY

NMR does not require proteins to be in crystalline form, but still requires large amounts of proteins soluble at high concentration. For larger proteins, it requires selected isotope labeling that can render the biochemistry challenging and expensive. The technique exploits the fact that certain NMR active nuclei (such as ^{1}H or ^{13}C) absorb and resonate in the radiowave frequency range of the electromagnetic spectrum. The nuclei work essentially as small magnets that can be manipulated by an applied magnetic field with sophisticated pulse sequences. The resulting shift of the signal is linked to its precise chemical environment, allowing characteristic signatures that provide information about nearest interaction partners and thus result in constraints on the three-dimensional conformation of the macromolecule. Unlike X-ray crystallography, NMR does not yield a density map that can be interpreted by model building, but model building is the central theme of NMR, by keeping certain intermolecular distance constraints in consideration.

Using molecular dynamics approaches, such as simulated annealing, and applying the constraints observed by NMR, a set of structures can be found that are compatible with the experimental data. Often the spatial coordinates of some parts of macromolecules are in agreement, and others differ widely, which is interpreted as conformationally stable and flexible protein domains, respectively. Determination of the structure of large proteins requires very high magnetic fields. Proteins of greater molecular mass are not accessible to solution NMR. Solid-state NMR does not have such molecular mass limitations, but it suffers from a variety of other challenges so that it is not a routine technique for large protein structure determination and, like solution NMR spectroscopy, cannot be considered a high-throughput technique for atomic structure determination.

CRYO-ELECTRON MICROSCOPY

Electron microscopy (EM) imaging of macromolecular assemblies dates back to the 1960s, but the last two decades have seen remarkable success in determining macromolecular complexes, typically protein complexes that are larger than those studied by X-ray crystallography. Unlike most other imaging techniques, EM does not utilize a window in the electromagnetic spectrum, but instead uses charged particles (i.e., electrons) that interact with the nuclear field of the atoms and therefore provides a nuclear density map. In a first approximation, it is similar to the electron density map of X-ray crystallography: both detect atoms in a certain spatial location. Electrons interact with matter orders of magnitudes stronger than X-rays, limiting the specimen size that can be examined to about one micron for intermediate voltage microscopes typically employed in biological EM imaging. As a consequence of this strong interaction with matter, the specimen must be examined in a vacuum, and this can present a challenge for biological samples.

Since its invention in 1974 and its popularization in the 1980s, cryo-preservation of biological specimen has been a key ingredient of successful structure determination. The absence of staining and flattening artifacts typically encountered in the negative staining, combined with minimal dose imaging of frozen-hydrated samples, allows intermediate- to high-resolution structures to be determined. Cryo-EM imaging only relies on the phase contrast of the object and, due to the limited dose, is typically characterized by a low signal-to-noise ratio. Limiting the dose and imaging at liquid nitrogen temperature minimizes radiation damage, which is ultimately one of the fundamental limitations in cryo-EM. Such noisy data typically require either implicit or explicit data averaging. Implicit averaging exploits external symmetry (as in the case of two-dimensional crystals or helical symmetry in tubular crystals or along helical filaments) and internal symmetry (as is the case for viruses that often display icosahedral symmetry). Explicit averaging is the domain of single-particle analysis, where projection images of individual particles (proteins) are recorded, then classified into different classes of views (so-called class averages), to which angles are assigned that allow a three-dimensional reconstruction. Often several tens to hundreds of thousands of particles must be imaged, class-averaged, and reconstructed into a volume. None of the cryo-EM approaches is well suited for high-throughput: sample preparation and image data collection are often a bit of an art and hence require significant expert knowledge. One of the significant advantages of EM images is that they provide amplitude and phase information and hence yield a better starting point for structural determination than X-ray diffraction patterns. Hence, in electron crystallography, no heavy atom derivatives are necessary. While this is a theoretical advantage, practically speaking, X-ray analysis is often much faster than what is possible for cryo-EM. Few (but some) single-particle projects have yielded atomic-resolution insight, but more typical is the generation of an intermediate resolution envelope that can be readily interpreted, particularly if the protein is known to exist in a particular conformational state or if parts of the large complex have been determined by other structural approaches. Hence, often so-called hybrid-methods are employed where the high-resolution structure of known components are fitted into the envelope provided by cryo-EM.

ATOMIC FORCE MICROSCOPY

Atomic force microscopy (AFM) does not rely on a particle (photon) of the electromagnetic spectrum to penetrate or otherwise interact with the atoms of the biological macromolecule. Instead, AFM seeks out the topology of macromolecules, not unlike running your hand over the wall to find the light switch in a dark room. In AFM, a very fine probe moves over the sample and detects the topology of the macromolecule. The resolution obtained varies with the nature of the surface and the geometry of the tip. Thus, isolated well-exposed biological objects are best for this technique. In those cases, the resolution can be truly astounding.

However, biological samples introduce significant challenges. The sample must be firmly attached to the support stage so the specimen will not be pushed around by the tip. AFM works best with hard materials that can resist the interaction with the probe. Unfortunately, many biological samples are soft, and the probe can damage the sample. Alternatives modes of AFM, such as a tapping motion of the probe, can lessen the damage. A great advantage of AFM is that the specimen can be in an aqueous buffer, and conformational changes can be readily observed, more or less in real time. Of course, only the surface can be examined; any changes within the macromolecules or near its attachment site will likely be undetected. AFM can reveal valuable information about biological samples, including mechanical properties, electrostatics, binding energy profiles, and molecular interactions through the interaction of the probe with the sample or a macromolecule bound to the cantilever tip.

FREE-ELECTRON LASER DIFFRACTION

While currently more a concept than a reality, let alone a routine tool for structure determination, the free-electron laser promises to overcome a fundamental limitation of X-ray diffraction: the need for a crystallized sample. Instead, using powerful electron pulses from lasers, direct, although very weak, diffraction patterns of individual proteins can be detected, and the structure could be determined from a large number of pulsed diffraction patterns. While this approach would not prevent radiation damage, one could record diffraction patterns just before the complete destruction of the macromolecule. The ultimate goal of the free-electron laser diffraction is to obtain high-resolution three-dimensional structures of biological macromolecules and atomic resolution of single particles. Since it's a new technology, there are many challenges left to overcome. Most parameters have been defined by theoretical models; however, much experimentation on biological samples is needed before we will even know whether this approach can live up to its promise. It is not yet clear if the fragile biomolecule can survive the beam sufficiently long for a diffraction pattern to be collected. The short duration of XFEL pulses offers an opportunity for ultra-fast imaging for measuring static and time-dependent dynamic objects. If successful, it will be particularly interesting for proteins that are difficult to crystallize. Physics suggests such an approach should be possible, but whether it will ever become a routine or even a functional tool for *ab-initio* structure determination remains unclear and is somewhat controversial.

SUMMARY

This first section is dedicated to approaches that map the molecular inventory of single macromolecules or macromolecular machines. No single technique is better than any other one. Indeed, all have advantages and disadvantages, and often measurements by more than one technique are needed to gain biological insight. NMR and X-ray crystallography, as well as X-ray and single-particle EM, can be conducted on the same specimen, creating hybrid approaches that often yield complementary information about the biological system, such as the mechanism of function and its interaction with substrates, inhibitors, and potential binding partners. Thus, these will contribute to our understanding of macromolecular function, and may allow more specific drugs to be developed.

So while these techniques excel in providing understanding for macromolecular mechanisms, they do not tell us, however, how all these macromolecules work together to carry out cellular function, nor do they often shed much light on the exact physiology and pathogenesis.

	X-ray diffraction	Nuclear Magnetic Resonance	Cryo-Electron Microscopy	Atomic Force Microscopy	X-Ray Free Electron Lasers
Size range	Small molecules—supermolecular complexes	Small molecules to proteins <100 kDa*	>300 kDa **	Proteins—cells	Anticipated individual proteins
Resolution	~1–10 Å	Not applicable	~4.0 Å–20 Å	Lateral resolution 5 Å, vertical 1 Å	Unknown
Radiation type	X-rays (0.01–10 nm)	Magnetic field/radio waves	Electrons	A physical probe	High-intensity pulsed X-rays
Sample preparation	Requires crystallization	Normal purification to high concentration	Freezing at less than −135°C	A clean surface; sample immobilization to the matrix	Routine sample preparation; no crystallization
Dynamic or static	Static in crystal context	Both. Dynamic in solution	Static in frozen state	Static/dynamic in solution	Unknown
Type of Information obtained	3D architecture	3D architecture	3D architecture	Topology	3D architecture, anticipated
Advantages	Mature technology; straightforward for most systems; high degree of automation allowing high throughput	No crystallization needed, suitable for highly flexible macromolecules; examination in solution	Wide range of biological objects in their native frozen hydrated state	Direct study of 3D surface profile; single-molecule imaging; time-lapse imaging may reveal dynamic processes	Promises to allow the study of individual molecules in their native state in solution with high temporal and spatial resolution
Challenges	Obtaining suitable crystals; Determining the phase; Possible local distortions by crystal contacts	Isotope labeling required to larger molecules; Lengthy data collection; low throughput	Restricted to large macromolecules, *limited resolution due to minimal dose imaging due to radiation sensitivity*	Biological surfaces susceptible to mechanical damage; topology difficult to interpret	Unproven so far, in concept phase

* High molecular mass possible by solid state NMR.

** Lower molecular mass possible by phase plate EM.

2

Protein Crystallography and X-Ray Diffraction

JOHN P. ROSE, M. GARY NEWTON, AND BI-CHENG WANG

INTRODUCTION

X-ray crystallography, as the name implies, is the study of crystals (three-dimensional molecular assemblies) using X-rays. It is a powerful technique that, when successful, can offer a unique, unambiguous picture of the molecule in three dimensions. The technique, refined over the past century, works equally well on molecules ranging in size from simple ionic salts (1) to large assemblies of molecules such as the ribosome (2). X-ray crystallography together with nuclear magnetic resonance (NMR) spectroscopy constitutes the basis of structural biology. In combination, they form a powerful tool with crystallography providing detailed structural information and NMR providing information on chemical kinetics and dynamics of the system.

X-rays constitute the region of the electromagnetic spectrum that lies between the ultraviolet and gamma ray regions of the spectrum. X-rays are useful in imaging a molecule because their wavelengths, ranging from 0.01 to 100Å (1Å = 10^{-8} cm), are comparable to or smaller than the atoms making up the molecule. The ability of X-rays to resolve individual atoms is what is meant by the term atomic resolution. X-rays are scattered by the electrons of the atom or molecule producing a diffraction pattern. Since the amount of X-ray scattering is proportional to the number of electrons in a given atom, the diffraction pattern contains information about both the position and elemental composition of the atoms making up the molecule. However, the X-ray diffraction pattern produced by an isolated atom or molecule is too weak to be measured. Crystals, on the other hand, containing millions of copies of the atom or molecule in an ordered three-dimensional array can produce a measurable X-ray diffraction pattern. Thus, the X-ray diffraction experiment in its simplest terms requires (i) a crystal, (ii) an X-ray source, (iii) a means of recording the X-ray diffraction pattern, and (iv) a means of reconstructing the image of the molecule from the recorded diffraction pattern.

This chapter will introduce the reader to the X-ray diffraction experiment as applied to macromolecules and introduce some of the general basics of crystallography. The following is an outline of the topics covered:

1. Crystals and diffraction
2. Sample preparation and crystallization
3. X-ray sources and detectors
4. Data collection and processing
5. The phase problem
6. Model building
7. Refinement
8. Validation
9. Visualization and analysis
10. Chapter-related websites

The scope of the chapter is limited. The reader can obtain additional information on the topics covered above from the reference texts and web links provided at the end of the chapter, as well as the web resources listed below.

- Rupp's *Crystallography 101* (www.ruppweb.org/Xray/101index.html)
- Cowtan's *Book of Fourier Transforms* (www.ysbl.york.ac.uk/~cowtan/fourier/fourier.html)
- Sawaya's *Crystallography Tutorials* (www.doe-mbi.ucla.edu/~sawaya/tutorials/tutorials.html)
- IUCr's *Teaching Pamphlets* (www.iucr.org/education/pamphlets)

In addition, web links are provided at the end of this chapter for the suppliers of the crystallography hardware and software highlighted in the chapter.

CRYSTALS AND DIFFRACTION

A crystal is made up of atoms or molecules that form a pattern that is repeated periodically in three dimensions. The smallest repeat unit found within the crystal is called the unit cell. The unit cell can be described by three lengths termed a, b, and c, representing the edges or axes of the unit cell and three angles, denoted α, β, and γ, representing the angles between two of the corresponding axes. The angle between the b and c axes is called α, the angle between c and a axes is called β, and the angle between a and b axes is called γ. These six parameters are referred to as cell constants or lattice parameters. Based on the unit cell parameters and lattice symmetry (see below), crystals can be grouped into the seven crystal systems (Table 2.1).

Mathematics (3) has shown that there are only thirty-two combinations of symmetry operations (rotation, inversion, and reflection) that are consistent with a three-dimensional crystal lattice. The thirty-two-point groups, or crystal classes, can be grouped into one of the seven crystal systems given in Table 2.1. There are four types of crystal lattices: primitive (P), end-centered (C, B, A), face-centered (F), and body-centered (I). The primitive lattice contains a lattice point at each corner of the unit cell. The lattice point at each corner is also shared with neighboring unit cells

Table 2.1. The Seven Crystal Systems

System	Lattice Symmetry	Unit cell Constraints
Triclinic	$\bar{1}$	$a \neq b \neq c, \alpha \neq \beta \neq \gamma \neq 90^\circ$
Monoclinic	2/m	$a \neq b \neq c,\ \alpha = \gamma = 90^\circ, \beta > 90^\circ$
Orthorhombic	mmm	$a \neq b \neq c,\ \alpha = \beta = \gamma = 90^\circ$
Tetragonal	4/mmm	$a = b \neq c,\ \alpha = \beta = \gamma = 90^\circ$
Trigonal		
Rhombohedral lattice	$\bar{3}/m$	$a = b = c, \alpha = \beta = \gamma \neq 90^\circ$
Hexagonal lattice	6/mmm	$a = b \neq c, \alpha = \beta = 90^\circ, \gamma = 120^\circ$
Hexagonal	6/mmm	$a = b \neq c, \alpha = \beta = 90^\circ, \gamma = 120^\circ$
Cubic	m3m	$a = b = c, \alpha = \beta = \gamma = 90^\circ$

giving the primitive lattice a total of one lattice point. The end-centered lattice has lattice points at each corner of the unit cell plus an additional lattice point centered on one of the unit cell faces (shared by a neighboring unit cell) giving the end-centered lattice two lattice points. The face-centered lattice has lattice points at each corner of the unit cell plus an extra lattice centered on each of the unit cell faces (shared by a neighboring unit cell) giving the face-centered lattice four lattice points. Finally, the body-centered lattice has lattice points at each corner of the unit cell plus an extra lattice point at the center of the unit cell giving the body centered lattice two lattice points. By combining the seven crystal systems with the four lattice types (P, C, I, F), fourteen unique crystal lattices, also known as Bravais lattices (4), are produced.

In addition to the four (rotation, inversion, and reflection) symmetry operators that define the thirty-two point groups, the crystal lattice can also have translational symmetry (unit cell translations, screw axes, and glide planes). Unit cell translations are simply the translation of atoms or molecules at a point in one unit cell to the identical point in an adjacent unit cell (the addition of ± 1.0 to the atom's fractional coordinate). Screw axes combine a rotation with a unit cell translation. Screw axes are denoted by two integers: r defining the rotational operator (rotation = $360^\circ \div r$) and t the translation operator (translation = $t \div r$ unit cells). Glide planes combine a reflection (across a mirror plane) with a translation (usually ½ unit cell). Glide planes are denoted by a lower case letter (*a, b, c, n, d*) which defines the crystal axis (a, b, c, face, diagonal) that the glide is along. The additional screw axis and glide plane translation symmetry operators when combined with the fourteen Bravais lattices and the thirty-two crystallographic point groups gives the 230 possible crystallographic space groups (see Volume A, International Tables of Crystallography). Every crystal must belong to one of the 230 space groups. However, since proteins are chiral (they do not have internal mirror or inversion symmetry), their crystals belong to one of the sixty-five enantiomorphic (or non-centric) space groups. Space group symmetry when applied to the contents of the crystallographic asymmetric unit (the part of the unit cell having no symmetry) generates the contents of the entire unit cell. This simplifies the crystal structure determination, since once we determine the structure of the asymmetric unit, space group symmetry generates contents of the entire unit cell.

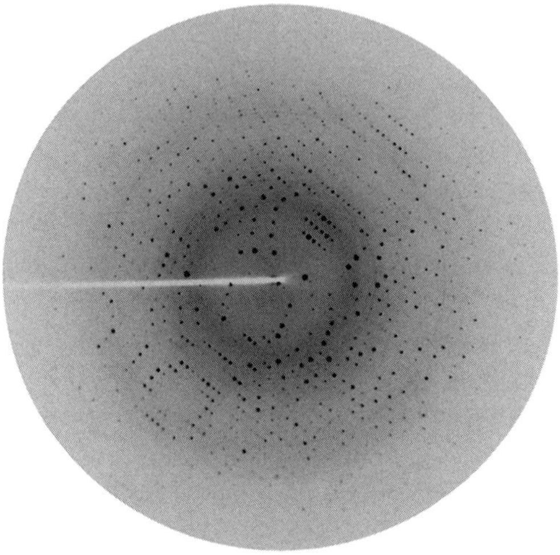

Figure 2.1 An image of an X-ray diffraction pattern recorded on a CCD detector. Each black dot (termed a reflection) can be assigned a three-integer index (called the Miller Index) that corresponds to the set of lattice planes in the crystal that produced the reflection. Photograph courtesy of Bruker-AXS.

The X-ray diffraction pattern (Figure 2.1) produced when a crystal is placed in the X-ray beam reflects the arrangement of atoms in the crystal (including space group symmetry) and is governed by Bragg's law (1).

$$n\lambda = 2d_{hkl} \sin \theta_{hkl} \quad \text{Bragg's Law}$$

where λ is the X-ray wavelength, θ_{hkl} is the diffraction angle, d_{hkl} is the lattice spacing, and n is the diffraction order, usually 1.

In Bragg's treatment, parallel planes of reflection passing through the crystal lattice points reflect the incoming X-ray beam much like a mirror reflects light (i.e., the angle of incidence is equal to the angle of reflection). The diffraction pattern then simply results from the interference of the reflections from sets of parallel planes within the crystal. The spacing of the lattice planes is determined by the lattice geometry (i.e., it is a function of the unit cell parameters). The orientation of the plane with respect to the axes of the unit cell is defined by three integers, h, k, and l (termed Miller Indices) that denote the fractional points (x, y, z) where the plane intersects the three unit cell axes. Miller indices are then defined as $h = (x/a)^{-1}$, $k = (y/b)^{-1}$, $l = (z/c)^{-1}$. Thus, the plane intersecting the unit cell at ½ a, ½ b, and ½ c would have Miller indices $h = (½\ a/a)^{-1} = (1/2)^{-1} = 2$, $k = (½\ b/b)^{-1} = (1/2)^{-1} = 2$ and $l = (½\ c/c)^{-1} = (1/2)^{-1} = 2$ or 222. Conversely, the plane intersecting the unit cell at -¼ a, -½ b, and running parallel to the c axis (intercept = ∞ c) would have Miller indices of -4 -2 0.

Knowing the unit cell constants and the orientation of the crystal in the X-ray beam, each spot or reflection observed in the diffraction pattern can be defined using the Miller indices corresponding to the set of lattice planes that produced the reflection. There is a reciprocal relationship between the diffraction angle θ_{hkl} and the spacing between the lattice planes d_{hkl} in the crystal. Thus, crystals with larger unit cells, such as proteins, will produce a more dense diffraction pattern (reflections are closer together) than that of salt crystals or other small molecules with much smaller unit cells.

Several good websites describing X-ray diffraction theory, crystal lattices, space groups, Miller indices, and the X-ray diffraction experiment are listed at the end of this chapter to provide further review.

SAMPLE PREPARATION AND CRYSTALLIZATION

It is always a good idea to do some upfront research on the protein that you wish to crystallize prior to going to the bench. A quick search for structures with similar sequences in the Protein Data Bank (PDB) (5) can save you a lot of time and effort, as well as providing valuable clues for crystallizing your protein. In addition, web resources such as XtalPred (6) can be used to analyze your sequence and give you an estimate of the crystallizability of your protein.

Sample Preparation

Successful crystallization often depends on the preparation of a pure and homogeneous protein sample. Thus, all factors that create chemical heterogeneity, such as proteolytic cleavage, denaturation, contamination, oxidation, etc., must be minimized or eliminated. In addition, the proper transport and storage of samples is crucial in preventing sample degradation. Ideally, several milligrams of pure soluble protein should be available. Sample purity should be assessed prior to setup by polyacrylamide gel electrophoresis (SDS-PAGE) (7) and native gel PAGE. Typically, the SDS-PAGE gels are overloaded with protein or stained with silver nitrate (8) or other highly sensitive stains to identify minor components in the preparation that may effect crystallization. Native gels are used to assess the protein's oligomerization state (e.g., monomer, dimer, trimer) and to identify the presence of protein aggregation.

Dynamic light scattering provides another powerful tool to characterize protein homology, purity, size, and molecular weight (9–11). Electrospray mass spectrometry, which can give the mass of the protein to within the mass of a single amino acid residue (12–14) can be used to detect protein cleavage and to monitor selenomethionine incorporation when preparing samples for multi-wavelength anomalous dispersion (MAD) experiments (15). The use of affinity tags, such as the histidine tag, for detection and purification (16, 17), can affect crystallization. These tags, if used, should be of minimal length or removed prior to crystallization. Insoluble proteins such as membrane proteins need to be solubilized, usually by adding detergent, before crystallization is attempted (18).

Crystallization

Protein crystallization can be viewed as a controlled precipitation of the protein from an aqueous medium. The solubility of proteins is dependent on the competition between solvent-solute interactions and intermolecular interactions between protein molecules. The balance of these interactions can be modified by several factors: (i) temperature (affects solubility), (ii) pH (affects both solute and solvent), (iii) salts (salt-in or salt-out effects), (iv) hydrogen-bond competitors (e.g., urea, guanidiuum salts), (v) hydrophobic additives (non-ionic detergents), and (vi) organic solvents (modification of dielectric constants) (19, 20). The goal of the crystallization process is to very slowly drive the protein-solvent system to a state of reduced solubility of the macromolecule by adjusting one or more of the six factors listed above. This is generally achieved by increasing the concentration of precipitating agents such as salts, polyethylene glycol (PEG) polymers, or organic solvents by a technique called vapor diffusion, the most common crystallization technique used today.

A simple phase diagram for the vapor diffusion process is shown in Figure 2.2. Initially the protein is in solution (A). Then, as either the protein or precipitant concentration increases, a point is reached (B) where crystal nucleation occurs. Nucleation and crystal growth depletes the local protein concentration and state C is reached where crystal growth occurs.

In practice, vapor diffusion crystallization trials are generally carried out using twenty-four-well crystallization plates with each well sealed by a microscope cover slip of appropriate size (see Figure 2.2). The plates, cover slips, crystallization screens, and a variety of other tools can be purchased from a number of vendors including Hampton Research, Molecular Dimensions, Jena Bioscience, Emerald Biosystems, Quiagen, and Microlytic (vendor websites are listed at the end of this chapter).

Each setup is produced by placing ~400 μl of the precipitant solution in one of the wells of the crystallization plate. A 2–4-μl drop, containing equal parts of protein solution (concentration ~10 mg/ml) and precipitant solution (taken from the well), is then placed on a siliconized microscope cover slip. The cover slip is then carefully inverted and placed over the appropriate well. A bead of high vacuum grease previously applied to the lip of each well provides the seal between the well and the cover slip. The X-ray diffraction experiment typically requires crystals whose size ranges from 10 to 300 microns depending on the X-ray source, optics, goniometer, and detector used.

Alternatively, one can use sitting drop vapor diffusion where the drop is placed in a raised depression within the well. Although the sitting drop vapor diffusion experiment requires special plates or devices, the setup and retrieval of crystals is much simpler. Once the plate is prepared, it is then incubated at 4 or 18°C for a period of time ranging from a few days to several weeks. During incubation, each well of the plate is checked periodically under a microscope for the presence of crystals or precipitate. Conditions derived from the initial screen are usually further refined by adjusting pH, additives, and concentrations to grow suitable crystals for data collection and structure analysis. Crystal seeding (21, 22) can also be explored to improve the crystal size and quality.

Several screening protocols have been developed to help quickly identify an initial set of crystallization conditions. These include footprint screening (21), sparse matrix sampling (23), and statistical methods (24). As noted above, premade crystallization screens based on these protocols are commercially available from a number

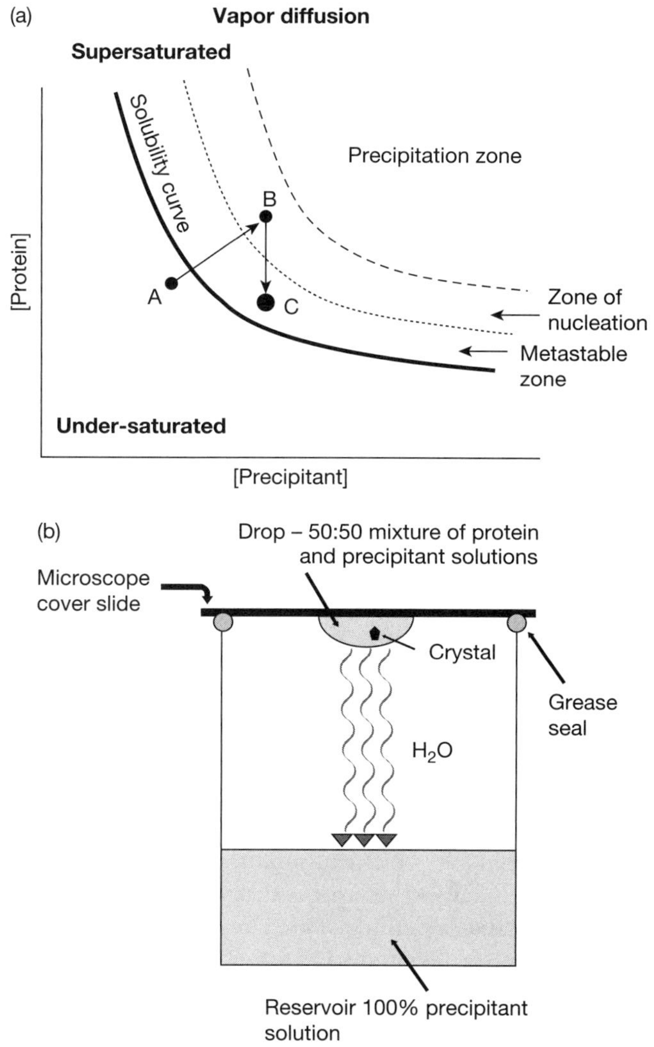

Figure 2.2 (a), a phase diagram of the vapor diffusion process showing the initial state (A), the nucleation point (B) and the final state (C) of the system. (b), an illustration showing a typical hanging drop vapor diffusion setup. Crystals grow as the concentration of protein and precipitants increase in the drop due to loss of water.

of vendors. In addition to the vapor diffusion method, other techniques such as the batch and micro-batch method, bulk and micro dialysis, free interface diffusion, liquid bridge and concentration dialysis have also been developed to produce crystals for X-ray diffraction analysis (19, 25–27).

Automated systems for vapor diffusion and micro-batch crystallization screening are now commercially available from a number of vendors including the CrystalMation system from Rigaku Americas and the Oryx systems from Douglas Instruments. These systems are particularly useful for rapid screening and optimization of new samples.

X-RAY SOURCES AND DETECTORS

X-Ray Sources

Today, most macromolecular X-ray diffraction experiments are carried out using X-rays produced either by a rotating anode X-ray generator or by electrons or positrons orbiting in a high-energy synchrotron storage ring. The most common laboratory X-ray source for macromolecular crystallography is the rotating anode generator. In the rotating anode generator, electrons produced by a tungsten filament (cathode) are accelerated to high velocities by a potential difference towards a metal target (anode) where their impact produces X-rays of a wavelength characteristic of the target material (1.5418 Å for the CuKα characteristic line). The anode is usually a sealed copper cup that is water cooled and rotated at a high speed (6500–9000 rpm) to dissipate the heat generated during X-ray production. The removal of heat from the target material limits the X-ray intensity to 3–5 kW for protein applications. Commercial rotating anode generators in a variety of anode/optic configurations are available from Rigaku and Bruker-AXS.

Synchrotron storage rings are very large and expensive facilities occupying a full city block. X-ray ports (called beamlines) are generally placed every 10° around the ring, allowing for a total of 30–35 user stations. The large number of user stations makes the synchrotron cost-effective. Synchrotron X-rays offer the crystallographer several advantages over in-house X-rays: (i) a small and highly intense X-ray beam, which allows data collection on microcrystals, and (ii) the ability to select (or tune) the X-ray wavelength. The use of intense synchrotron X-rays (100–1,000,000 times that of a rotating anode generator) generally produces structures of higher resolution (more structural detail) than structures obtained from rotating anode data. In addition, the higher resolution structures produced from synchrotron data have improved the overall accuracy of the protein structures deposited in the PDB and give researchers a deeper understanding of the biochemical behavior of the protein. The intense focused X-ray beams produced by third generation synchrotrons, such as the Advanced Photon Source at Argonne National Laboratory, also provide an excellent X-ray source for micro-crystallography experiments where the crystal size is generally less than 10 microns on edge and many facilities are equipping their beamlines with microdiffractometers to allow data collection on these small crystals.

The tunability of synchrotron X-rays offers a convenient means of recovering the phase information that is lost during the X-ray diffraction experiment but is needed to generate the electron density map of the protein on which the structure is based. Using techniques, such as MAD (28) and single-wavelength anomalous diffraction (SAD) (29), researchers are now able to solve a protein structure in a matter of minutes—a process which required months, or even years, a decade ago.

Since synchrotron light sources are expensive to build and operate, most are funded at the national level. In the United States, five major light sources are available to researchers for the collection of macromolecular single crystal diffraction data: (i) the Advanced Photon Source, Argonne National Laboratory, (ii) the Advanced Light Source, Lawrence Berkeley National Laboratory, (iii) the National Synchrotron Light Source, Brookhaven National Laboratory, (iv) Stanford Synchrotron Research Laboratory, Stanford University, and (v) the Macromolecular Diffraction Facility, Cornell High Energy Synchrotron Source (MacChess). Researchers can apply for time at these facilities by submitting proposals to their General User Programs via web-based portals such as (www.aps.anl.gov/Users/apply_for_beamtime.html).

Detectors

The introduction of commercial multi-wire X-ray area detectors, such as the Nicolet X100 in the mid-1980s (30), marked a new chapter in how data are collected from protein crystals. By recording many reflections simultaneously, these new detectors offered the ease of fast, automated data collection without the expense and effort associated with film data collection. Multi-wire detectors, although fast (with readout times under ten seconds), were limited by their active area and dynamic range (the ratio of the largest and smallest signal that can be measured). These shortcomings led to the development of storage-phosphor based imaging plate (IP) systems, such as the Rigaku and MarResearch IP systems of the early 1990s. These large, 30 cm IP systems allowed for the collection of higher resolution data during each exposure and were not as easily saturated compared to their multi-wire counterparts. A disadvantage of the IP systems was the relatively slow readout speed of one to three minutes, which makes IP's less efficient for data collection using the short (1–10 sec) exposure times commonly employed for synchrotron data collection. However, IP systems remain useful for collecting X-ray diffraction data on protein crystals in some home laboratory.

The demand for a large detector with high dynamic range and fast readout for synchrotron applications led to the development of the charge-coupled device (CCD)-based area detector system found today at most synchrotron beamlines for macromolecular crystallography. In the CCD detector, X-ray photons are converted to visible light by an X-ray-sensitive phosphor and focused onto the CCD chip by a reducing fiber-optic taper. The light photons induce a charge proportional to the number of X-ray photons striking the phosphor in the wells of the chip. CCD detectors are, however, limited by the size, cost, and dynamic range of the CCD chip and the light loss in the fiber-optic focusing system. CCD-based detector systems found in the home lab such as the Proteum (Bruker-AXS), SX-165 (Rayonix), and Saturn 944 (Rigaku) usually consist of a single, large (4k) CCD sensor coupled to a fiber optic taper. The synchrotron versions of these systems generally consist of four or more CCD-fiber optic modules placed in either a tiled 2 × 2 (the ADSC Quantum 210r and the Rigaku A200), 3 × 3 (the ADSC Quantum 315r and Rayonix MX225), and 4 × 4 (Rayonix MX300) arrays. In addition, next generation detectors having ultra-fast (~10 Hz) readout, improved dynamic range and lower noise are now becoming commercially available. The new detectors include the Pilatus 6M (Dectris), a photon counting detector based on pixel array technology and the Rayonix MX300HS based on frame shifting CCD technology.

DATA COLLECTION AND PROCESSING

X-ray diffraction data are generally collected by slowly rotating the crystal, in 0.2° to 1° steps, in a monochromatic X-ray beam while recording the resulting diffraction pattern at these positions by an X-ray area detector. Data collection can be broken down into seven basic steps: (1) crystal selection, (2) crystal mounting, (3) crystal cryoprotection, (4) crystal centering, (5) initial exposures and determination of experimental parameters, (6) determination of the data collection strategy, and (7) data collection.

Crystal Selection

Once crystals are obtained, they must be placed in the X-ray beam to record the diffraction pattern. The experimenter should carefully choose the crystal. First, it should have the same morphology or habit as the majority of crystals in the drop since crystals having similar habits should, in most cases, have the same unit cell and crystal symmetry. Choosing similar crystals also makes it easier to find isomorphous crystals for heavy atom derivatization, if needed. The crystal should be well formed with sharp faces and have no reentrant angles between its faces. The presence of reentrant angles between crystal faces usually indicates that the crystal is twinned. Twinned crystals contain two or more independent crystalline components (present in varying amounts) each capable of producing a diffraction pattern. Thus the diffraction pattern of a twinned crystal is a complicated combination of the diffraction patterns from its multiple crystalline components. Data from twinned crystals is usually difficult to process since the independent diffraction patterns from the various twin components must be identified and processed separately. Finally, it is always a good idea to survey the crystals using a polarizing microscope before choosing one for mounting. Single crystals will extinguish uniformly (every 90°) as they are rotated between the polarizer and analyzer of the polarizing microscope.

Crystal Mounting and Cryoprotection

Macromolecular crystals contain an appreciable amount of solvent, generally 40–60 percent. To prevent the crystal from drying out it must be either placed in a sealed glass capillary containing a small quantity of its mother liquor (31) or flash frozen (32, 33) into an amorphous glass and maintained at cryogenic temperatures during data collection. Although the latter technique using a small fiber loop (34) has become popular, capillary mounting is still employed for data collection at room temperature for crystals that cannot be flash frozen. It is generally good practice to first survey a new crystal at room temperature before trying to flash freeze it since the freezing process may affect or destroy the diffraction quality of the crystal. Flash freezing the crystal, however, offers certain advantages. The crystal is less susceptible to radiation damage, usually allowing for the collection of a complete data set from a single crystal. In addition, the amount of handling and the resulting mechanical stress on the crystal is minimized which may lead to better diffraction quality. Finally, the crystal can be recovered and stored at cryogenic temperatures for future experiments.

The commonly used loop freezing technique (35) is illustrated here. A small fiber loop (0.05–0.5 mm in diameter) is mounted at the end of a metal pin that is connected to a magnetic base for easy attachment to the goniometer (crystal positioner). Note: the loop diameter used should match the crystal's size to reduce error in the data. The pin length used should always be consistent. This will avoid problems when using remote or robotic systems for data collection (36). It is also important to make sure that the loop is properly centered in the cold stream for data collection. For proper centering, one generally mounts a mounting pin with an empty loop on the goniometer and centers the empty loop at the point where the rotational axes of the goniometer intersects the X-ray beam (this position is usually the center of the crosshairs of the alignment microscope). A nitrogen gas cold stream maintained at ~100K should have also been positioned such that the cold stream also intersects the position defined by the alignment microscope's crosshairs. However, it is always a good idea to ensure that the cold stream is positioned correctly.

To mount the crystal, the crystal is scooped up with the loop directly from the crystallization drop and, while the cold gas stream is momentarily blocked, the loop (and its attached mounting pin) is quickly transferred to the goniometer. The cold stream obstruction, usually an index card or a microscope slide, is then quickly removed, allowing the cold gas stream to flash freeze the crystal. If done properly, the flash frozen crystal should appear to be suspended in the loop surrounded by clear and colorless mother liquor. Once in place, the crystal is centered in the crosshairs of the alignment microscope to ensure that it is centered in both the cold stream and the X-ray beam.

A disadvantage of the flash-freezing technique is the increase in mosaicity that usually accompanies the process. Cryoprotectants (33), such as glycerol (13–25 percent), ethylene glycol (11–30 percent), PEG 400 (25–35 percent), 2-methyl-2,4-pentanediol (to 28 percent), and glucose (to 25 percent), can be used to minimize this increase in mosaicity. The first step in determining the correct cryoprotectant conditions is to simply flash freeze a loop full of cryoprotectant solution (cryoprotectant mixed with reservoir solution) in the cold stream. If the solution freezes as a clear colorless glass, the proper cryoprotectant concentration has been achieved. Next, a test crystal (i.e., not your best crystals) is transferred momentarily to a small drop of cryoprotectant solution to see how the cryoprotectant affects the crystal. If the crystal cracks or dissolves, the cryoprotectant or its concentration needs to be changed. The final step in the process is to record the diffraction pattern for the cryoprotected crystal. If the cryoprotectant is good, the diffraction pattern should show a significant reduction in crystal mosaicity (compared to a non-cryoprotected crystal) and no ice rings.

Cryo-loops, mounting pins, capillaries, and other crystal mounting supplies are available from several vendors including the Charles Supper Company, Hampton Research, Molecular Dimensions, and MiTeGen. The recently introduced MiTeGen MicroRT capillaries allows for data collection at both room and cryogenic temperatures. Crystals are harvested using a loop method but the pin has been modified to accept a special plastic sleeve, which seals the system much like a glass capillary for room temperature diffraction studies. The sleeve can then be removed and crystal flash frozen for data collection at cryogenic temperatures if desired.

Crystal Centering

Once the crystal is mounted on the goniometer, it must be precisely centered in the X-ray beam. It should be noted that a little time spent carefully centering the crystal in the X-ray beam could significantly improve the quality of the data. Note the following procedure requires that the crystal be placed such that X and Y translations of the goniometer head are perpendicular to the alignment microscope. The loop should have been approximately centered prior to mounting the crystal so the crystal should be visible in the alignment microscope. The crystal is then centered in the crosshairs of the alignment microscope ($\phi = 0°$) using the X and Y adjustments on the goniometer head. The crystal is then rotated 180° ($\phi = 180°$). The crystal should be centered at this new position if the crosshairs of the microscope are correctly centered. The crystal is then rotated to 90° ($\phi = 90°$) and centered. Again, rotation by 180° from this point ($\phi = 270°$) should result in a centered crystal at this new position. This process is repeated until the crystal is centered at every position. If the 180° crystal rotation suggests that the crosshairs of the alignment microscope are off, the user can account for this by applying half of the correction to the goniometer head and half of the correction to the position of the alignment microscope.

Initial Exposures and Determination of Experimental Parameters

Before collecting data, it is advisable to collect a series of images 90° apart that can be used to assess the crystal's diffraction quality in terms of diffraction resolution, crystal mosaicity, and proper cryoprotection. In addition, these images can often provide clues to the presence of crystal twinning or other crystal pathologies. This will ensure that data are collected on the best crystals. Robots for automated crystal diffraction characterization are available from Rigaku (ACTOR) and Bruker-AXS (BRUNO). In addition, many synchrotron facilities, such as the SER-CAT (www.ser-cat.org) facility at the Advanced Photon Source have automated crystal-screening systems capable of screening hundreds of crystals per day (37, 38).

Once a well diffracting crystal has been identified, several important parameters need to be determined. These include the direct beam position (at $2\theta = 0$) on the face of the detector, the crystal-to-detector distance, the oscillation step ($\Delta\phi$), the detector swing angle (2θ), and exposure time. The direct beam position is usually determined by the site manager following realignment of the X-ray optics and recorded in a computer file or notebook. The other parameters can usually be determined from the initial set of diffraction images described above.

The crystal-to-detector distance is dependent on crystal (unit cell), beam (size), and wavelength and should be chosen such that the reflections are well resolved (i.e., the reflection intensity profile goes to background between reflections). Once the minimum crystal-to-detector distance is identified, the detector should be moved away from the crystal such that the observed diffraction pattern covers 90–95 percent of the detector surface since this will increase the signal-to-noise ratio of the data set. If, on the other hand, the diffraction pattern is observed to extend beyond the edge of the detector at the minimum crystal-to-detector distance, the detector in some cases can be moved off center on its 2θ axis to allow collection of the higher-resolution data.

The choice of the oscillation step size used for data collection is both crystal and detector dependent and is usually chosen to match the crystal mosaicity and the readout speed of the detector. Image plate detector systems have slow readout times so data is usually collected using an oscillation step ranging from 0.5 to 1.5° while

0.1–0.5° oscillation steps are used for CCD detectors that have faster readout speeds. Note the use of smaller oscillation step increases the signal-to-noise ratio of the data set.

Once the crystal-to-detector distance, detector swing angle, and oscillation step have been decided, the exposure time can be chosen by taking a series of images with increasing exposure times. It is generally a good rule of thumb to select an exposure time such that the average intensity of the high-resolution data is at least three times the background on a single image. In addition, low-resolution reflections should also be checked for possible saturation of the detector.

Determination of the Data Collection Strategy

Generally, one should determine the lattice parameters and the orientation of the crystal with respect to the X-ray beam (called Indexing) before collecting the dataset. This will avoid problems later during data processing since, if you can't index the crystal, you can't process the data. In addition, if the crystal's orientation is known, a data collection strategy (39) can be determined to ensure that a complete as possible data set is obtained in the most efficient manner. Data collection strategies become increasingly important when beam time is limited or the detector is placed off axis. A good data collection strategy is also vitally important for crystals that suffer radiation damage at cryogenic temperatures or in the collection of highly redundant data sets for phasing applications.

Indexing is generally carried out by first locating the positions of a number of reflections harvested from the image in terms of their location in x and y on the detector surface and the rotation angle ϕ where the image was recorded. The spatial relationships (in x, y, and ϕ) between these reflections and autoindexing algorithms can be used to determine the unit cell constants, the crystal's orientation matrix, and the three-integer index (hkl) that describes the lattice plane that produced the reflection. Indexing also allows the estimation of the solvent content of the crystal and the number of molecules in the unit cell by Matthews's method (40).

$$Vs = 1 - \frac{1.23}{Vm} = 1 - 1.23MSN / Vcell$$

where V_m is the crystal volume per unit protein molecular weight, V_s is the fractional solvent content, M is the molecular weight of the protein, S is the number of asymmetric units in the unit cell, N is the number of molecules in the asymmetric unit, and V_{cell} is the volume of the unit cell.

Indexing that gives an unreasonable number of molecules (e.g., more than six molecules in the asymmetric unit) can be an indication that there may be problems with the crystal (e.g., cracks, twinning, presence of small satellite crystals, cryoprotection), reflection harvesting (e.g., split reflections) or indexing parameters. These problems should be addressed before data collection is initiated.

Most commercial detectors provide their own software to collect and process data. These packages include Proteum II (Bruker-AXS) and CrystalClear (Rigaku).

In addition, programs such as HKL2000/HKL3000 (41, 42), d*TREK (43), DPS (44), MOSFLM (45), and XDS (46) can also be used to process data from a variety of detectors. A brief synopsis of each package can be found at (www.ccp14.ac.uk/solution/ccd_software).

Data Collection

Once the user is satisfied that all data collection parameters have been determined and a good data collection strategy has been developed, data collection can begin. The user should monitor the collection of the first images to ensure that the parameters and strategy chosen are correct and that the instrument is running correctly. At this point the user can generally begin processing the first images. In most cases, the data reduction process can continue unattended since the data reduction program will wait until new images in the data set become available.

Data Processing

In general, the data reduction process consists of the following steps: (i) reflection harvesting, (ii) indexing, (iii) integration, (iv) scaling, and, in some cases, (v) post refinement.

Data processing begins with the harvesting of a sufficient number of strong reflections spaced widely apart in x, y, and ϕ from the collected images, which will be used for indexing. Reflections should be chosen carefully since the quality of the resulting orientation matrix has a direct impact on the quality of the resulting data set. Autoindexing, as described above, is then used to provide the crystal's orientation matrix, estimates of the unit cell parameters, and to assign a three-integer index to each reflection. The quality of the orientation matrix can be assessed by (i) assessing how well the predicted reflections matches the observed diffraction pattern, (ii) checking the R.M.S. differences between the observed and predicted spot centroids, and (iii) examining histograms showing the deviation from integer values of the observed and calculated h, k, l indices.

To integrate the data, positions for all reflections in the data set are predicted (in x, y, and ϕ) based on the orientation matrix and detector geometry. Integration of the X-ray intensity at these predicted positions is then carried out using either a two-dimensional summation or three-dimensional profiles. In both cases, model reflection profiles must be constructed to provide the program with information about the reflection's general size and shape so that the background measurements can be made correctly. Integration is usually monitored by tracking the average R.M.S. deviation or χ^2 values in x, y, and ϕ of the predicted peak position compared to the intensity maxima found at or near that position.

Finally, the integrated data must be merged and scaled together. Data merging and scaling are based on the space group symmetry of the crystal, which can be determined from the relationships of reflection intensities for certain classes of reflections observed in the complete data set. See Volume A of the International Tables for Crystallography. Most scaling programs provide detailed statistics about the data set including R_{merge}, completeness, redundancy, and $\langle I/\sigma \rangle$ as a function of resolution.

These values can be used to judge the quality of the data and to determine the effective resolution of the data set.

$$R_{merge(I)} = \sum_{hkl} \left| \langle I \rangle - I_{hkl} \right| \Big/ \sum_{hkl} \langle I \rangle$$

where I_{hkl} is the intensity of reflection hkl.

R_{merge} is an indication of the agreement between the intensities I of the scaled symmetry related reflections in the data set and is usually in the range of 0.03–0.08 for well-diffracting crystals. The R_{merge} value as a function of resolution can also be used to identify scaling problems and to estimate the effective resolution of the data set. Generally R_{merge} slowly increases with increasing resolution. If this trend is not observed, there may be some problems with the data reduction process. The resolution at which R_{merge} exceeds 0.25 is also a good indication of the true effective resolution of the data set which should match the resolution at which the $\langle I/\sigma_{(I)} \rangle$ equals 2 if the values for the error in the measurements are being estimated correctly. Data completeness should be very close to 100 percent if the crystal has not slipped or decayed during data collection, provided that the data collection strategy was designed correctly. Finally, carefully examine the statistics for each scaled image to identify problem images, which may affect the overall scaling process and the accuracy of the resulting scaled data set. This is especially true when scaling data from a sulfur-SAD experiment (47).

THE PHASE PROBLEM

The data recorded in the X-ray diffraction experiment are processed and eventually used as either intensities, I_{hkl}, or structure factor amplitudes, $|F_{hkl}|$. A structure factor amplitude, $|F_{hkl}|$, is simply the square root of the intensity, I_{hkl}, so that $|F_{hkl}| = \sqrt{I_{hkl}}$. A typical protein structure determination will measure thousands of intensities (reflections) with each data point having a unique set of Miller indices, *hkl*. However, the diffraction experiment cannot directly provide a vital piece of information, the phase of each reflection. This constitutes the so-called "phase problem" in X-ray diffraction. If the phase for each individual reflection could be measured, then an electron density map and, corresponding atomic positions, could be calculated in a direct and straightforward way. To restate the phase problem: in order to reconstruct the protein image from its diffraction pattern both the intensity and phase angle of each of the thousands of measured reflections must be known. The phase problem is further complicated by the fact that the phase of each of the thousands of reflections can, in most cases, assume any value from 0 to 2π. However, the needed phase information is lost in the diffraction experiment. Thus, considerable effort must be made to recover this lost information so that a reasonable and chemically meaningful (i.e., you can fit the amino acid sequence) electron density map can be produced. Once a viable map is available, the completion of the structure determination is merely a matter of refining and improving the original solution.

The phase problem can be best understood from a simple mathematical construct. The structure factors (F_{hkl}) are treated in diffraction theory as complex quantities,

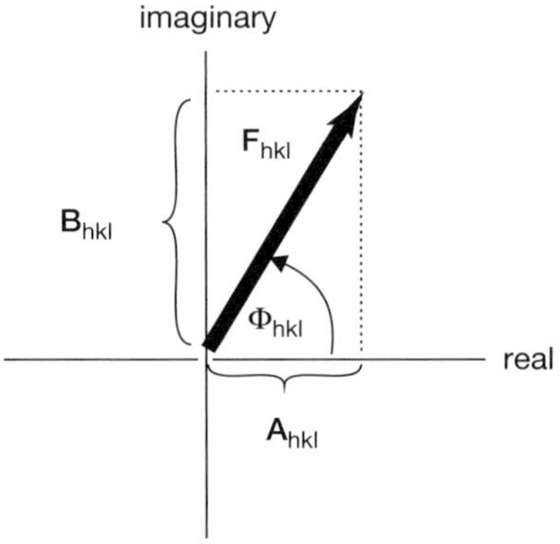

Figure 2.3 An Argand diagram of the structure factor $\mathbf{F}_{hkl}$ with phase Φ_{hkl}. The real ($\mathbf{A}_{hkl}$) and imaginary ($\mathbf{B}_{hkl}$) components are also shown.

i.e., they consist of a real part (A_{hkl}) and an imaginary part (B_{hkl}). If the phases, Φ_{hkl}, were available, the values of A_{hkl} and B_{hkl} could be calculated from very simple trigonometry:

$$A_{hkl} = |F_{hkl}| \cos(\Phi_{hkl})$$

$$B_{hkl} = |F_{hkl}| \sin(\Phi_{hkl})$$

this leads to the relationship:

$$(A_{hkl})^2 + (B_{hkl})^2 = |F_{hkl}|^2 = I_{hkl}$$

These relationships are often illustrated on an Argand diagram (see Figure 2.3). From the Argand diagram, it is obvious that A_{hkl} and B_{hkl} may be either positive or negative, depending on the value of the phase angle, Φ_{hkl}. Units of A_{hkl}, B_{hkl}, and F_{hkl} are in electrons.

With this information, the electron density, ρ_{xyz}, at a given position (x, y, z) can be calculated using the equation below where V is the volume of the crystals unit cell.

$$\rho_{xyz} = \frac{1}{V}\left\{\sum_{hkl} A_{hkl}\cos\left[2\pi(hx+ky+lz)\right] + \sum_{hkl} B_{hkl}\sin\left[2\pi(hx+ky+lz)\right]\right\}$$

Note that ρ_{xyz} has units of electrons/volume (i.e., electron density). In practice, the electron density for one three-dimensional unit cell is calculated by starting at x, y, z = 0, 0, 0 and stepping incrementally along each axis, summing the terms as shown in the equation above for all *hkl* (as limited by the resolution of the data) at each point in space. The electron density map will consist of regions of high ρ-values, corresponding to electron density around atomic positions, and regions of low ρ-values, where no atoms are found. In proteins, large regions of low electron density generally correspond to areas containing mostly non-ordered solvent (water) molecules. Electron density maps can be contoured at regular intervals to produce spheres or globular

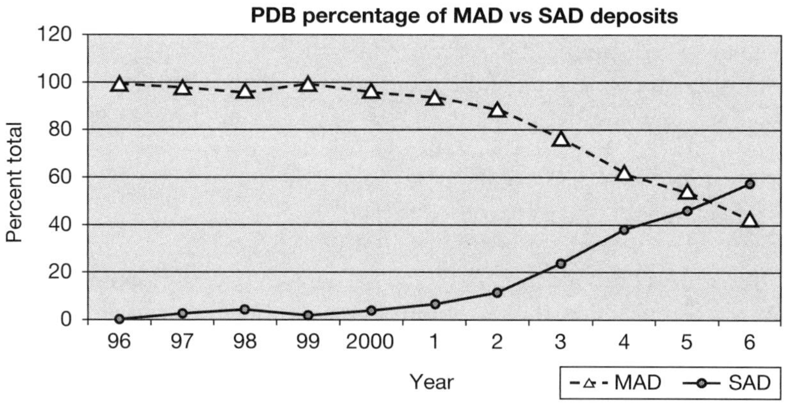

Figure 2.4 The number of MAD and SAD solved structures (percentage of total) deposited in the Protein Data Bank from 1996 to 2006.

regions of high electron density. When viewed in three-dimensions, regions of helical and/or β sheet can be easily identified in a well-phased electron density map.

One of the most common methods for solving the phase problem in macromolecular crystallography is the MAD phasing technique (15). MAD phasing became increasingly popular in the early 1990s and continues to find wide and successful application. MAD phasing exploits anomalous scattering information gained from diffraction experiments that are carried out at several different wavelengths close to the X-ray absorption edge of an anomalous scatterer, such as selenium or iron. Since a tunable X-ray source is needed, MAD experiments must be conducted using a synchrotron X-ray source.

The need for data collection at several wavelengths has now been shown to be unnecessary, since anomalous scattering data, collected at a single wavelength, is sufficient to routinely yield interpretable electron density maps. Although this means of structure determination was first demonstrated *via* simulation by Wang in 1985 by a process called iterative single-wavelength anomalous scattering (ISAS) (29) and later applied to the de novo structure determinations of a neurophysin-dipeptide complex (48) and ferrochelatase (49), the approach has gained general acceptance only recently. Today, the *ISAS* process or the SAD (single-wavelength anomalous diffraction) method, as some call it, is the most common method for de novo protein structure determination, with data from both home lab and synchrotron sources (see Figure 2.4).

In addition to the MAD and SAD methods there are the traditional isomorphous replacement methods, which include multiple isomorphous replacement (MIR), which uses several heavy atom derivatives, and single isomorphous replacement (SIR), which uses only one heavy atom derivative. The underlying principle to all these methods is the "phase-triangle relationship." To understand this relationship we shall begin our discussions with the isomorphous replacement method.

Isomorphous Replacement Methods

Until the introduction of MAD phasing in 1989, the isomorphous replacement method (50) was the primary technique used for solving the phase problem related to *de novo* structure determination. Here, a heavy atom (atom with a large number of electrons) must be introduced into the crystal lattice in such a way as not to disturb the conformation and packing of the protein in the crystal. Importantly, it is necessary that the crystals of the heavy atom derivative be isomorphous with the native protein crystals (i.e., have similar cell dimensions and symmetry (space group)). These derivatives are usually obtained by soaking protein crystals in a solution containing a small mount of heavy atom salts (Hg, Pt, Os, etc.).

There are two related isomorphous replacement-phasing methods, SIR and MIR. SIR requires the collection of the native data set and one heavy atom derivative data set. SIR produces an ambiguous phase doublet (two phase values, one of which is correct) which limited the approach until 1985 when Wang introduced the iterative single isomorphous replacement (ISIR) method (29). MIR used by Perutz to solve hemoglobin (50), the first protein crystal structure, requires the collection of a native data set plus two or more isomorphous heavy atom derivative data sets.

The large number of electrons associated with the heavy atom will produce large scattering amplitudes from this atom, and, as a result, many of the intensities will be dominated by scattering from this heavy atom. To locate the position of the heavy atom(s), a Patterson function is calculated which represents a three-dimensional vector space map, also known as an $|F|^2$ map. The Patterson function is shown below:

$$P(uvw) = \frac{1}{V}\sum\left|F_{hkl}\right|^2 \cos\left[2\pi(hu + kv + lw)\right]$$

This function is similar to the electron density function given earlier. Here $P(uvw)$ is the value of the Patterson function at Patterson coordinates u, v, w; these are the traditional coordinate symbols (instead of x, y, z) used for "squared" ($|F_{hkl}|^2$) space. All other symbols have their usual meaning. The Patterson function is a Fourier summation using the intensities as coefficients and setting all phase (Φ_{hkl}) values to 0. The resulting contoured map will have peaks corresponding to vector distances between all atoms in the structure. Vectors between an atom and itself is a zero vector; therefore, the Patterson functions always have a very large peak at the origin where $u, v, w = 0, 0, 0$.

In macromolecular crystallography, the Patterson coefficients used in the SIR and MIR techniques are the squared differences in the intensities between the native and heavy atom derivative data sets, $[\Delta F(hkl)]^2$. If the native data set is represented by $F_p(hkl)$ and the heavy atom derivative by $F_{PH}(hkl)$. The differences are $|\Delta F(hkl)| = ||F_{PH}(hkl)|-|F_p(hkl)||$, thus $|\Delta F(hkl)|^2$ are the Patterson coefficients. Because the intensity differences are primarily due to the heavy atoms, the resulting isomorphous difference Patterson map reveals the location of the heavy atoms. Programs, such as *SOLVE* (51), *SHELXD* (52), or *PHENIX* (53) will automatically search Patterson space for heavy atom positions.

The structure factor of the heavy-atom derivative may be expressed vectorially as $\mathbf{F}_{PH}$, where $\mathbf{F}_{PH} = \mathbf{F}_P + \mathbf{F}_H$. $\mathbf{F}_P$ and $\mathbf{F}_H$ are the structure factors of the native protein and the heavy atom, respectively, for the same reflection. Figure 2.5 illustrates this phase triangle on an Argand diagram.

A triangle has three sides. For this triangle, its sides represent the magnitudes of $\mathbf{F}_{PH}$, $\mathbf{F}_P$, and $\mathbf{F}_H$. If we know the orientation (phase or Φ) of one of the sides, then the orientation of the whole triangle may be fixed. If the orientation of the triangle is fixed, then the orientations of its three sides are also fixed. So, if we know the phase of one of the three quantities, F_{PH}, F_P, or F_H, then we will know the phases of the other two. F_{PH} and F_P are measured quantities, but F_H is calculated from the heavy atom parameters and the phase angle of F_H can also be calculated. This means that if we know the heavy atom positions in the crystals, then the isomorphous replacement method should give us the phase angle of F_P for all the reflections. In simplest terms, isomorphous replacement finds the orientation of the phase triangle from the orientation of one of its sides. It turns out, however, that there are two possible ways to orient the triangle if we fix the orientation of one of its sides (see Figure 2.6). Even with this ambiguity, we have reduced the number of possibilities of orienting the phase triangle from an infinite number to just two possibilities. This is a tremendous improvement from no phase information to two possible values. The situation of two possible SIR phases is called the "phase ambiguity" problem, since we obtain both a true and a false phase for each reflection. Both phase solutions are equally probable; that is, the phase probability distribution is bimodal.

The phase ambiguity problem can also be understood from the mathematical expression shown below. We can express the protein phase angle α_P from the following equation:

$$|\mathbf{F}_{PH}|^2 = |\mathbf{F}_P|^2 + |\mathbf{F}_H|^2 + 2|\mathbf{F}_P||\mathbf{F}_H| \cos(\alpha_P - \alpha_H)$$

which leads to

$$\alpha_P = \alpha_H + \cos^{-1}[(|\mathbf{F}_{PH}|^2 - |\mathbf{F}_P|^2 - |\mathbf{F}_H|^2)/2|\mathbf{F}_P||\mathbf{F}_H|] = \alpha_H \pm \beta$$

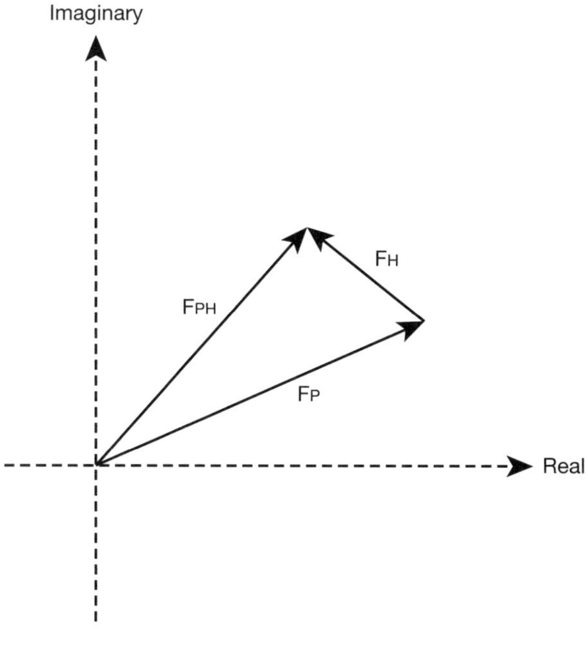

Figure 2.5 An Argand diagram showing the vector sum $\mathbf{F}_{PH} = \mathbf{F}_P + \mathbf{F}_H$.

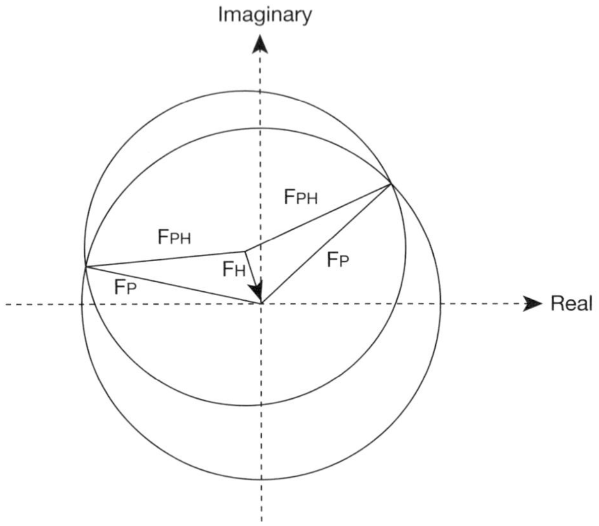

Figure 2.6 A Harker construction showing the SIR phase ambiguity.

Here we have two possible signs for β, because we have an arccosine function, which gives two possible solutions for a single argument. Thus, the phase ambiguity is actually a mathematical problem.

Since we get one true phase and one false phase, we do not know which phase is the correct one to choose. Thus, the next step in the isomorphous replacement method is to resolve the phase ambiguity. Although the phase ambiguity is a mathematical problem, most crystallographers routinely solve it by collecting more experimental data.

For example, if we add another heavy atom isomorphous derivative to the above analysis, where the heavy atom sites are different from those found in the first derivative, we again get two solutions, one true and one false for each reflection from the second derivative. The true phase should be consistent between the two derivatives while the false phase should show random variation. Thus, by comparing the solutions obtained from these two calculations, one can establish which solution represents the true phase angle. This is the principle of the MIR method. One can also utilize the anomalous scattering (AS) data of the first derivative to resolve the phase ambiguity. In this case, the technique is called the SIRAS approach. If two derivatives and anomalous data are used, then it is called the MIRAS approach.

Usually, isomorphous replacement methods are coupled with a density modification process called solvent flattening. The solvent flattening process was made practical by the introduction of the *ISIR/ISAS* program suite (29), and other phasing programs, such *DM* (54) and *PHASES* (55), are based on this approach. Wang pointed out that the phase ambiguity in both the SIR or SAS data can be resolved by an iterative noise filtering process using mathematics, not additional experimental data, to resolve the phase ambiguity problem. Using solvent flattening SIR data alone is sufficient for the determination of protein crystal structure, provided that the derivative crystal is isomorphous to the native crystal. Likewise, SAD data alone are sufficient for the protein structure determination. This will be described in more detail in the next section.

Table 2.2. PROTEIN RESIDUES AND THEIR AFFINITIES FOR HEAVY METALS

Residue:	**Affinity for:**	**Conditions:**
Histidine	K_2PtCl_4, $NaAuCl_4$, $EtHgPO_4H_2$	pH > 6
Tryptophan	$Hg(OAc)_2$, $EtHgPO_4H_2$	
Glutamic, Aspartic Acids	$UO_2(NO_3)_2$, rare earth cations	pH > 5
Cysteine	Hg, Ir, Pt, Pd, Au cations	Ph > 7
Methionine	$PtCl_4^{2-}$ anion	

Key to the MIR and SIR methods is the production of an isomorphous heavy atom derivative. This is traditionally done by soaking crystals in dilute (2–20 mM) solutions of heavy atom salts. Crystal cracking is generally a good indication that that heavy atom is interacting with the crystal lattice, and suggests that a good derivative can be obtained by soaking the crystal in a more dilute solution.

Table 2.2 lists some common heavy metals and their affinities to specific amino acid residues (56). This list can serve as an initial guide in preparing heavy atom derivatives. Once data on a potential isomorphous heavy atom derivative has been collected, the merging R factor between the two data sets is another indication of both heavy atom incorporation and the isomorphism between the native and derivative crystals. The merging R factor measures the agreement of identical reflections between the two data sets. Typically, R_{merge} values for isomorphous derivatives range from 0.05 to 0.15, with values below 0.05 indicating that there is little heavy atom incorporation and values above 0.15 indicating a lack of isomorphism between the two crystals. Finally, an isomorphous difference Patterson should be calculated and inspected for consistent heavy atom peaks.

Anomalous Dispersion Methods

Since the introduction of MAD phasing in 1990, anomalous dispersion (AD) effects have been used more and more frequently to solve the phase problem. All elements display an AD effect in X-ray diffraction. However, the elements in the first and second row of the periodic table (e.g., C, N, O) have negligible AD effects. For heavier elements, especially when the X-ray wavelength approaches an atomic absorption edge of the element, these AD effects can be very large. The scattering power of an atom exhibiting AD effects is:

$$f_{AD} = f_n + \Delta f' + i\Delta f''.$$

The f_n value is the normal scattering power of the atom in absence of AD effects. The $\Delta f'$ value arises from the AD effect and is a real factor added to f_n; it may be positive or negative. The $\Delta f''$ is an imaginary term which also arises from the AD effect and is always positive and 90° ahead of $(f_n + \Delta f')$ in phase angle. The values of $\Delta f'$ and $\Delta f''$ are highly dependent on the wavelength of the x-radiation. An excellent discussion of AD can be found on the University of Washington website (www.bmsc.washington.edu/scatter).

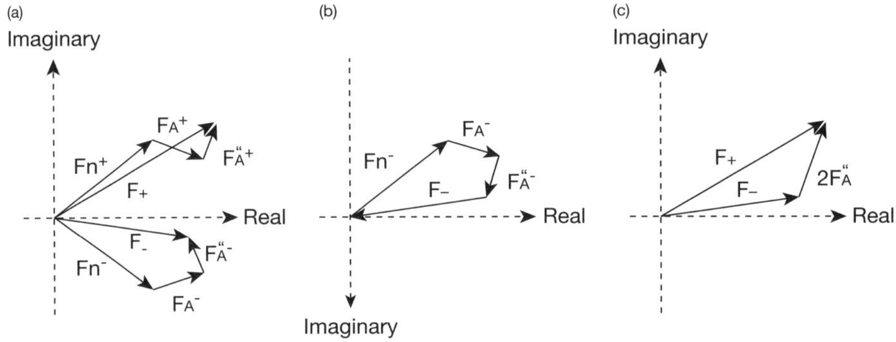

Figure 2.7 An illustration of the anomalous dispersion phase triangle relationships. (a) A diagram showing vector sums of $\mathbf{F}_{n+}$, $\mathbf{F}_{A+}$, and $\mathbf{F}_{A''+}$ which produce $\mathbf{F}_+$; corresponding vectors are shown for $\mathbf{F}_-$ at the bottom. (b) A diagram showing the $\mathbf{F}_-$ and associated vectors from (a) with the imaginary axis inverted. (c) A diagram composed of the $\mathbf{F}_+$ and $\mathbf{F}_-$ vectors from (a) and (b) respectively showing the phase triangle relationship with $2\mathbf{F}_{A''}$.

In the absence of AD effects, the intensity of I_{hkl} is the same as the intensity of its inverse, I_{-h-k-l} (Friedel's law). However, with AD effects the Friedel pairs, I_{hkl} and I_{-h-k-l} will be unequal. For many crystallographic applications, it is assumed that Friedel pairs are, at least approximately, equal. However, accurate measurement of Friedel pair differences can be used to extract starting phases if the AD effect is large enough.

Figure 2.7 illustrates how AD from a single heavy atom can cause the intensities of a Friedel pair to be different. In the left-hand diagram (a) the vectors F_{n+} or F_{n-} represent the total scattering by "normal" atoms without AD effects. The vectors F_{A+} or F_{A-} represent the sum of the normal and real AD scattering values ($f_n + \Delta f'$).

The $F_{A''+}$ or $F_{A''-}$ vectors are the imaginary AD components and appear 90° (at a right angle) ahead of the F_{A+} or F_{A-} vectors. The total scattering vector in the top of the diagram is F_+ and the bottom counterpart is the negative vector sum F_-. The resultant vector, F_- in this case, is obviously shorter than the F_+ vector. The F_- vector, however, is not always shorter than the F_+ vector, since about half of the time it is longer. What is interesting is that for each reflection we can have two measurable quantities, $|F_+|$ & $|F_-|$ with different magnitudes.

In Figure 2.7(b), the F_- and associated vectors are shown with the imaginary axis inverted. Figure 2.76(c) uses F_+ from (a) and F_- from (b) to construct the phase triangle relationship between F_+, F_- and $2F_{A''}$.

The above observation is also significant in terms of designing our experiments for solving the phase problem. In the isomorphous replacement method described previously, heavy atoms were introduced into the native crystals to obtain $|F_{PH}(hkl)|$ so that for each reflection there would be two quantities, $|F_{PH}(hkl)|$, and $|F_P(hkl)|$ with different magnitudes, which can be experimentally measured. Using anomalous scattering data there is no need to add heavy atoms in the crystals since both $|F_{+++}|$ and $|F_{---}|$ are measurable quantities.

With this introduction, one can immediately see that using the anomalous scattering data to obtain phase information can be carried out in a manner very similar to the one used for isomorphous replacement. So, the statement made earlier about the principle for the isomorphous replacement, "to find the orientation of a phase triangle from one of its sides," should also be applicable for the use of anomalous scattering data. From this point of view, the MAD method is conceptually the counterpart of the MIR method and SAD method is the counterpart of the SIR method. The major difference is only on how the two measurable quantities in the experiments are obtained; that is, are they measured in two separate experiments using two different crystals or are they obtained directly from a single experiment with one crystal? Compared with the isomorphous method, the anomalous scattering technique should save considerable time and effort in obtaining phase information.

Much has been said on the advantage of using anomalous scattering data for gaining phase information, but little has been said about the problems that one has to overcome in order to obtain the accurate $|F_{+++}|$ and $|F_{---}|$ values. The main problem in collecting anomalous scattering data is that the difference between $|F_{+++}|$ and $|F_{---}|$, or anomalous signal, is generally about one order of magnitude smaller than that observed between $|F_{PH}(hkl)|$ and $|F_P(hkl)|$. Thus, the signal-to-noise (S/N) level in the data plays a critical role in the success of AD experiments (i.e., the higher the S/N the greater the probability of success in producing an interpretable electron density map). Thanks to recent advances made in X-ray optics and detector technology, we can now measure X-ray diffraction data with much higher precision than ever before. This means that we can now consider anomalous scattering from elements having very weak anomalous scattering signals, such as sulfur, which were previously thought to be too weak for AD experiments. In fact, there have been several recent successful examples of direct determination of protein crystal structure from the native crystals using the anomalous scattering signal of sulfur (e.g., structure of obelin (47)). Although this approach is still under development, there are currently more than ninety sulfur SAD structures in the PDB. The use of the sulfur-SAD method is expected to grow as advances are made in X-ray detector and optics technology and with the development of dedicated long wavelength beamlines at a number of synchrotron facilities.

Since AD experiments are critically dependent on the data S/N, introducing a "strong" anomalous scatterer into the crystal can be used to increase the S/N of the data. Alternately, the S/N of the data can be increased by collecting the data with X-ray wavelengths at or near an anomalous scatterer's X-ray absorption edge. The MAD experiment takes advantage of both these points. A strong anomalous scatterer is introduced into the crystal and data are recorded at several wavelengths near the X-ray absorption edge of the anomalous scatterer. The phase ambiguity in each of the SAD experiments is then resolved by the use of additional SAD data collected at other wavelengths in a manner similar to the use of multiple derivatives in the MIR technique. Thus, the MAD method is geometrically similar to the MIR method but uses anomalous scattering information from different wavelengths instead of different derivatives.

The MAD method requires that the proteins under investigation contain anomalous scatterers whose X-ray absorption edges are in the tunable range (0.8–2.0 Å) of the synchrotron. Metalloproteins or selenomethionine-labeled proteins are good candidates for a MAD study (15, 17). However, the preparation of the protein samples is often the limitation of this method. Selenomethionine incorporation involves the expression of the protein in a bacterial cell line, which is incapable of producing methionine. Selenomethionine is then added to the minimal growth media and is incorporated into the expressed protein. Crystallization of the selenium labeled proteins usually produces crystals that are isomorphous with the native protein. Therefore it is generally a good idea to first crystallize and characterize the protein before going to the expense of producing the seleno-protein.

For those proteins containing metal ions or metal clusters, such as Fe, Ni, and Cu, the MAD experiment can be carried out directly on the native protein crystals without the need of seleno-derivatives. Crystals containing heavy atoms incorporated via chemical modification or by soaking (56) can also be used for MAD experiments. Finally, Dauter *et al.* (57) showed that ordered bromine in the protein's solvation shell could provide MAD phasing information.

The SAD method, which combines the use of SAD data and solvent flattening (29) to resolve phase ambiguity been used to phase several structures including a neurophysin *para-iodo*-Phe-Try-amide complex (I-SAD, λ=1.54Å, 46 kDa, 5 iodine atoms) (48) and ferrochelatase (Fe-SAD, λ=1.54Å, 86 KDa, 2[2Fe-2S] clusters) (49). It is important to note that data for these structures were collected on a normal laboratory X-ray generator using Cu radiation. Thus, structures containing atoms, such as iron, iodine, or xenon, with sufficiently large $\Delta f''$, for Cu radiation, can be very effectively solved with SAD with home source data.

The ultimate goal of the SAD method is the use of sulfur-SAD to phase protein data. Most proteins contain sulfur and thus it would be very convenient and efficient if data of sufficient accuracy could be measured to phase using the sulfur AD signal alone. Compare the anomalous scattering signal of sulfur ($\Delta f''$ = 0.56 electrons for Cu radiation) with that of iron ($\Delta f''$ = 3.20 electrons for Cu radiation) or xenon ($\Delta f''$ = 7.35 electrons for Cu radiation) and you begin see the nature of the problem: the sulfur AD signal is very weak. Despite this problem, Hendrickson and Teeter (58) were able to determine the structure of crambin (5 kDa, 6 sulfurs) using 1.5 Å resolution data collected with a copper home source. The technique used, termed resolved anomalous phasing, estimates phases based on the sulfur substructure in the protein. Later, Wang (29) showed that Bence Jones protein Rhe (12.5 kDa, 1 disulfide) could be solved at 3Å resolution from simulated sulfur SAD data using the ISAS method. The ISAS method differs theoretically from the RAS method in that it uses the filtered image of the protein molecule to compute the protein phases. Accuracy in phase prediction is improved because the filtered image of the molecule is an approximate structure of the protein. Thus, iterative image filters, also known as solvent flattening, can be used to resolve the phase ambiguity associated with SAD data.

To use sulfur as a phasing probe in S-SAD, several inherent problems must be addressed. First, since the sulfur K absorption edge is at 5.02 Å, well outside the tunable energy range of most current synchrotron facilities, the optimal wavelength for data collection must be identified. This wavelength should maximize the sulfur SAD signal while X-ray absorption effects are kept to a minimum. A strategy collecting S-SAD data has been developed in Wang's laboratory and seven *de novo* protein structures have been solved using this approach: the photoprotein obelin (PDB entry 1EL4) (47), Sso10A (PDB entry 1RJ7) (59), aequorin (PDB entry 1SL8) (60), and structural genomics targets PF1951 (PDB entry 1NNH) (61), Q15691 (PDB entry 1VKA) (61), PF0523 (PDB entry 1ZD0) (62), and AF1382 (PDB entry 3O3K) (63). In these studies, the *ISAS* program routinely resolved the phase ambiguity from S-SAD data collected using in-house or synchrotron X-ray sources with X-ray wavelengths ranging from 1.5418Å to 2.2909Å.

Molecular Replacement

Molecular replacement (MR) (64) has proved effective for solving macromolecular crystal structures. Most of the 90,000 plus structures in the PDB were solved by the molecular replacement method, which relies on the availability of a homologous protein structure. This method is straightforward and reduces the time and effort required for structure determination because there is no need to prepare heavy atom derivatives and collect their data. Model building is also simplified, since little or no chain tracing is required. This method is gaining increasing importance, as more structures become available. However, the three-dimensional structure of the search model must be very close (r.m.s.d Cα's less than 1.7Å) to that of the unknown structure for the technique to work. Sequence homology between the model and unknown protein is often helpful but not strictly required. Successful structure determinations have been carried out with search models with as little as 17 percent sequence similarity to the unknown protein. Several computer programs such as *AmoRe* (65), *X-PLOR/CNS* (66, 67), and *PHASER* (68) are available for MR calculations.

The MR method involves the following steps: (i) a rotational search to orient the structure of a homologous (known) model in the unit cell of the target (unknown) protein crystal using a Patterson correlation search and (ii) a translational search where the newly oriented model is correctly positioned in the unknown's unit cell as described below.

If M_1 represents the search model and M_2 represents the target structure, a simple relation can be described:

$$M_2 = M_1\,[\mathbf{R}] + \mathbf{T}$$

where [R] is the rotation matrix and T is the translation vector. A suitable search model (M_1) can be obtained by sequence or function homology searches (such as *BLAST* (69)) of known crystal and NMR structures. The rotation function [**R**] is first applied to M_1 to maximize the orientational agreement (highest Patterson correlation) between M_1 and M_2. Once the orientation of the model is known, a simple three-dimensional translation R-factor search over the asymmetric unit yields the solution. In a successful case, the Patterson correlation coefficient between M_1 and M_2 should be large and standout from the other solutions. The R factor from the translational search should generally be lower than 50 percent. A rigid body refinement of the oriented and translated search model against the unknown data is usually carried out in order to refine the initial solution and to generate the initial phases of M_2 for model rebuilding and further refinement. Since MR is dependent on a model structure, one must be careful not to introduce model bias into the unknown structure.

Finally, it should be noted that several popular software packages have been developed to aid in the phasing and/or refinement of protein structures. These include *CCP4* (70), *SHARP* (71), *OASIS* (72), *Auto-RICKSHAW* (73), *SGXPRO* (74), *PHENIX* (75), *HKL3000* (42). Links to these websites are provided at the end of this chapter.

MODEL BUILDING

For structures not determined by molecular replacement, the amino acid sequence of the protein must be fit into the experimental electron density map (Figure 2.8). This process is called model building or chain tracing. As one would expect, the success or failure of chain tracing is dependent upon the quality of the electron density map. A good (traceable) electron map should display most of the features listed below:

1. There should be a large contrast difference between the protein and solvent regions in the map resulting in a clear boundary.
2. The protein region of the map should have connected and clearly defined electron density. The density in the protein region should be consistent in terms of contour or sigma levels.
3. Secondary structure elements such as helices and β-sheets should be clearly visible.

In addition, one can look at the histogram of the electron density distribution. The electron density distribution is independent of the fold and space group, but is a function of resolution range and crystal solvent content. Thus, by comparing the histogram of the "experimental" electron density map with that of a correctly phased map, one can get an objective indication as to the "interpretability" of the map.

There are several steps in making the first trace of the electron density map. A molecular boundary needs to be defined and the asymmetric unit identified. If the

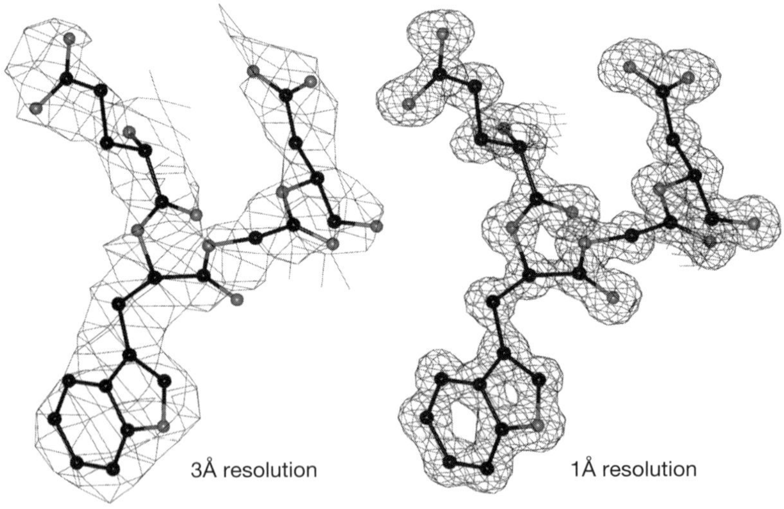

Figure 2.8 Left, a section of electron density calculated using 3Å resolution data. The overall main chain fold can be observed. In addition, side chain electron density is also apparent but lacks fine detail. Right, the same section of electron density calculated using 1Å resolution data. High-resolution electron density maps show much more detail including fine detail in the side chain electron density as observed for the arginine (top center) and tryptophan residues (bottom center) where electron density for the individual carbon and nitrogen atoms can be observed. Figure generated with CCP4mg (91).

asymmetric unit contains molecules related by non-crystallographic symmetry, only one molecule needs to be traced because the other molecules can be generated from the non-crystallographic symmetry operators. Non-crystallographic symmetry can be identified from a self-rotation function analysis (64) while the translation vector can be identified from the native Patterson or by analysis of the electron density map. Next, the initial Cα positions must be assigned either manually from a skeleton/bones map using programs such as *O* (76) or *COOT* (77), or if the resolution of the data permits automatically using *ARP/wARP* (78) or *PHENIX* (53). Once the initial Cα positions have been placed, the protein sequence is then fitted into the electron density using the Cα positions as guides and the size and shape of the side chain electron density to define the amino acid type. In the case of Se-MAD or S-SAS experiments, the selenium or sulfur atoms can serve as additional guide points during the fitting process. If non-crystallographic symmetry is present, this information should also aid in the interpretation of the electron density map. Again, if the resolution of the data and phases permits, automated fitting programs, such as *ARP/wARP* or *PHENIX*, can be used to fit the entire protein sequence of a moderate size protein in a matter of hours.

A good initial model should fit the electron density and make biochemical and biophysical sense. The model should have reasonable stereochemistry and have few clashes with symmetry mates. The model should fit all of the electron density; missing density is better than extra density that is unaccounted for.

REFINEMENT

Once the initial model has been built, it must be refined against the X-ray diffraction data. Each non-hydrogen atom in the structure can be described by four to nine parameters. Three parameters x, y, and z are used to describe the location of the atom in the unit cell plus one to six additional parameters, called thermal- or temperature-factors, which define the size, shape, and orientation of the electron density surrounding the atom. Based on these atomic parameters one can calculate the structure factor (F_{hkl}) associated with each reflection (*h k l*) from the following equation:

$$F_{(h.k.l)} = \sum_{j=1} f_{(j)} \exp\left[2\pi \cdot i\left(hx_{(j)} + ky_{(j)} + lz_{(j)}\right)\right]$$

here f_j is the atomic scattering factor for atom j. The *h k l* triplet describes the set of lattice planes which gives rise to the reflection and x_j, y_j, and z_j is the location of atom j in the unit cell.

Refinement of the model is carried out by adjusting the atomic parameters of each atom in the model to (i) minimize the difference between the observed and calculated structure factors and (ii) minimize the deviation of the geometry of the atoms defining an amino acid from the amino acid's ideal geometry.

Refinement progress is monitored by calculating the overall agreement (R factor) between the observed and calculated structure factor amplitudes as shown below.

$$R = \frac{\sum \left\| F_o \right| - \left| F_c \right\|}{F_o}$$

For the refinement to work (converge) properly, the number of observations (reflections) must exceed the number of parameters refined (usually four per atom) by a factor of six or more. Since the atomic positions for the atoms defining each amino acid have a well-defined geometry, each amino acid must also obey the stereochemical constraints (e.g., bond lengths, bond angles, peptide planarity) associated with the particular amino acid. These constraints can be viewed as additional observations (e.g., observed bond length versus ideal bond length, based on high-resolution structures (79)) and the refinement program can include additional minimization terms to reflect this.

The X-ray and stereochemistry terms in the refinement are usually given different weighting factors that can be adjusted during refinement to obtain the best-refined model (i.e., the model that agrees with both the X-ray data and the stereochemistry of the protein). The addition of stereochemical information to the refinement process is also important because it usually adds enough extra observations for least squares refinement to converge properly. Programs commonly used for model refinement include *X-PLOR/CNS* (67, 80), *REFMAC* (81), *SHELX* (82), and *PHENIX* (75).

The refinement of the model generally begins with a positional refinement where the positions (x, y, and z) of every atom in the model are refined against the X-ray data and stereochemistry. The selection of the resolution limit used in the refinement is based on the data quality and individual refinement requirements. Generally, one should start to refine the structure using medium resolution (3.0–2.0 Å) data and gradually include the higher resolution data. Free R factor (R_{free}) monitoring, introduced by Brunger in 1992 (83), during all stages of refinement is recommended in order to assess the accuracy of the model and prevent over-fitting of the data. The R_{free} calculation measures the agreement between observed and model-based structure factor amplitudes for a "test" set of randomly selected reflections (usually 5–10 percent of the data set), which have been excluded from the refinement process. The behavior of R_{free} should mimic the behavior observed for the R factor during the refinement and is generally 4–5 percent larger.

Once the positional refinement has reached convergence, manual revision of the model is carried out using residue deleted "OMIT" $2F_o-F_c$ electron density maps. "OMIT" maps remove model bias in the region under study since the investigated residues (usually 5 percent of the total) were omitted from refinement and phase calculation. The $2F_o-F_c$ coefficients used for the electron density map calculation further reduces model bias, since the observed data (F_os) are given more weight in the calculation. At this point, missing residues, bound substrates, or cofactors, ions, and other ligands can be identified from difference (F_o-F_c) electron density maps and added to the model. Finally, the model should be adjusted such that the main chain torsion angles Φ and Ψ fall into the allowed regions of the Ramachandran plot (84).

Subsequently, simulated annealing (SA) refinement can be carried out if desired followed by manual revision as described above. Simulated annealing refinement, introduced in 1989 by Brunger, serves to ensure that the refined model represents the true energy minimum of the system, since conventional positional refinement sometimes converges to false local minima. SA refinement is essentially a standard crystallographic refinement linked with molecular dynamics. During SA an energy term is added to the refinement targets and is used to simulate a rise in the temperature of the system to the point where all the atoms dissociate and randomly arrange themselves in the liquid state. As the system cools, the atoms can rearrange into their lowest energy state. Generally, SA refinement heats the system to 3000K–5000K and then slowly cools it in 25K steps to 300K. At each temperature step 0.5 femto seconds of energy minimization is carried out and the total energy of the system is monitored.

Once the SA refinement has converged and if the resolution (2.5Å or higher) of the data permits, solvent molecules identified from difference Fourier electron density maps can be included in the model. Generally, solvent molecules are accepted if (i) the identified peak in the difference F_o-F_c electron density correlates with a similar peak in the corresponding $2F_o-F_c$ electron density map, (ii) the peak height is well above the noise level of the map, and (iii) the potential solvent atom forms one or more hydrogen bonds (2.6–3.3 Å) with the protein or with other solvent molecules. After the addition of solvent, the model is further refined (positional and group or individual thermal factors) until convergence. Again, it is a good idea to visually check the model after refinement and make manual revisions of the model if required.

Refinement can be monitored observing the trends in the R and R_{free} values during the refinement process. Both values should decrease during the refinement process and plateau as the model is refined to convergence. If this trend is not observed, it could indicate problems with correctness of the refined model and/or an over-interpretation of the data. In general, a good model at 2.5Å resolution should refine to an R value in the neighborhood of 20 percent with an R_{free} value below 30 percent. However, these values are dependent on data resolution and data quality.

VALIDATION

The final structure should make biochemical and biophysical sense. The quality of the structure can be judged by how well the model fits the X-ray data (the R and R_{free} values) and its stereochemistry (the R.M.S. deviations from ideality in bond lengths and angles). In addition, a Luzatti plot (85) can be calculated to measure the general accuracy of the coordinate set as a function of resolution. Validation programs such as *MOLPROBITY* (86), *WHAT_CHECK* (87), and *PROCHECK* (88) have been developed to allow for a detailed analysis of the structure with regard to stereochemistry, hydrogen bonding, and other parameters. The PDB also validates all structures it receives and these reports are available online.

Typically, a "well" refined structure at 2.5 Å resolution should meet or exceed most of the following criteria: (i) the structure should make biochemical and biophysical sense, (ii) the structure should have an R-value in the low to mid-20 percent range, (iii) the structure should have an R_{free} value below 30 percent, (iv) the structure should have R.M.S. deviations from ideality (79) in bond lengths and angles of 0.01Å and 1.2°, respectively, and (v) the structure should produce a Ramachandran plot (88) with less than one percent of its non-glycine residues lying in unfavorable regions.

VISUALIZATION AND ANALYSIS

The crystal structure produced from the X-ray diffraction experiment is complex due to its three-dimensional nature and the large number of atoms, which make up the protein. Several visualization tools and analysis tools for protein structure analysis are available free to the academic user. A number of graphics-based visualization tools such as *CHIMERA* (89), *PYMOL* (90), *CCP4MG* (91), and *RASMOL* (92) have been developed to aid the researcher in the production of publication quality figures and in understanding the local environments of bound ligands or cofactors. The programs read the standard PDB formatted file and can produce a variety of structure representations including wire frame, ball and stick, ribbon, space filling, and molecular surfaces. The programs, in most cases, can also read the CCP4 formatted electron density map files. There are also a number of other structure analysis tools such *PDBsum* (93) and *PISA* (94) for analyzing the protein structure in terms of secondary structure, surface properties, possible oligomerization states, and the presence of voids or clefts which may be ligand binding sites. Links to these sites are given below.

CHAPTER-RELATED WEBSITES

Education-Related Websites:

Anomalous scattering	www.bmsc.washington.edu/scatter
Bragg diffraction	www.eserc.stonybrook.edu/ProjectJava/Bragg
Bravais lattice	en.wikipedia.org/wiki/Miller_index
Crystallography 101	www.ruppweb.org/Xray/101index.html
Crystallography tutorials	www.doe-mbi.ucla.edu/sawaya/tutorials/tutorials.html
Fourier transforms	www.ysbl.york.ac.uk/~cowtan/fourier/fourier.html
Miller Index	en.wikipedia.org/wiki/Miller_index
Space groups	www.uwgb.edu/dutchs/SYMMETRY/3dSpaceGrps/3dspgrp.htm
Symmetry	xrayweb.chem.ou.edu/notes/symmetry.html
Teaching pamphlets	www.iucr.org/education/pamphlets
X-ray diffraction Wiki	en.wikipedia.org/wiki/X-ray_crystallography
XrayView	phillips-lab.biochem.wisc.edu/software.html

Supplies-Related Websites:

Charles Supper Company	www.charles-supper.com
Emerald Biosystems	www.emeraldbiosystems.com
Hampton Research	hamptonresearch.com
Jena Bioscience	www.jenabioscience.com
MiTeGen	www.mitegen.com
Microlytic	www.microlytic.com
Molecular Dimensions	www.moleculardimensions.com
Quiagen	www.qiagen.com

Hardware-Related Websites:

ADSC	www.adsc-xray.com
Bruker-AXS	www.bruker-axs.com
Dectris	www.dectris.com

Douglas Instruments	www.douglas.co.uk
Mar Research	www.marresearch.com
Rayonix	www.rayonix.com
Rigaku Americas	www.rigaku.com

Synchrotron-Related Websites:

Advanced Photon Source	www.aps.anl.gov
Advanced Light Source	www.als.lbl.gov
National Synchrotron Light Source	www.nsls.bnl.gov
Stanford Synchrotron Research Laboratory	www.ssrl.slac.stanford.edu
MacChess	www.macchess.cornell.edu

Software-Related Websites:

AmoRe	www.ccp4.ac.uk
ARP/warp	www.embl-hamburg.de/ARP
Auto-RICKSHAW	www.embl-hamburg.de/Auto-Rickshaw
CCP4	www.ccp4.ac.uk
CHIMERA	www.cgl.ucsf.edu/chimera
COOT	www.biop.ox.ac.uk/coot
CRYSTALCLEAR	www.rigaku.com/software/crystalclear.html
DPS	staff.chess.cornell.edu/~szebenyi/DPS
*D*TREK*	www.rigaku.com/software/dtrek_news.html
HKL2000/HKL3000	www.hkl-xray.com
MOLPROBITY	molprobity.biochem.duke.edu
MOSFLM	www.mrc-lmb.cam.ac.uk/harry/mosflm
O	xray.bmc.uu.se/alwyn/TAJ/Home.html
PDB	www.rcsb.org/pdb
PDBSum	www.ebi.ac.uk/pdbsum/
PHASER	www.phaser.cimr.cam.ac.uk/index.php/ Phaser_Crystallographic_Software
PHENIX	www.phenix-online.org
PISA	www.ebi.ac.uk/msd-srv/prot_int/pistart.html
PROCHECK	www.ebi.ac.uk/thornton-srv/software/PROCHECK
PROTEUM	www.bruker-axs.com/software_proteum2.html
PYMOL	www.pymol.org
RASMOL	www.rasmol.org
REFMAC	www.ysbl.york.ac.uk/~garib/refmac
SHARP	www.globalphasing.com
SHELX	shelx.uni-ac.gwdg.de/SHELX
WHAT_CHECK	swift.cmbi.ru.nl/gv/whatcheck
XDS	xds.mpimf-heidelberg.mpg.de
X-PLOR/CNS	cns-online.org/v1.3
XTALPred	ffas.burnham.org/XtalPred

ACKNOWLEDGEMENT

After the first draft of this chapter was completed we lost Gary Newton. Gary was a thoughtful scientist and a resourceful mentor for many students, and participated in a variety of crystallographic activities at the University of Georgia that spanned 45 years. We would like to dedicate this chapter to him, as it would not be in its current form without his input – J.P.R. and B.C.W.

RECOMMENDED READING

Glusker, P. J., Lewis, M., and Rossi, M. *Crystal Structure Analysis for Chemists and Biologists.* VCH Publishers, Inc., New York, NY (1994).

Bergfors, T., ed. *Protein Crystallization, Second Edition.* International University Line, La Jolla, CA (2009).

Rupp, B. *Biomolecular Crystallography: Principles, Practice, and Application to Structural Biology.* Garland Science, Oxford United Kingdom (2009).

Doublié, S., ed. *Macromolecular Crystallography Protocols: Volume 1, Preparation and Crystallization of Macromolecules.* Methods in Molecular Biology 363, Springer, New York, NY (2007).

Doublié, S., ed. *Macromolecular Crystallography Protocols: Volume 2, Structure Determination.* Methods in Molecular Biology 363, Springer, New York, NY (2007).

REFERENCES

1. Bragg, W.L., *The structure of some crystals as indicated by their diffraction of X-rays.* Proc. Roy. Soc. (London), 1913. **A89**: p. 248–77.
2. Ban, N., et al., *The complete atomic structure of the large ribosomal subunit at 2.4 A resolution.* Science, 2000. **289**(5481): p. 905–20.
3. Hessel, J.F.C., *Krystallometric oder Kystallonomic und Krystallographie*, in *Physikalisches Worterbuch* J.S.T. Gehler, Editor, 1830, Schwickert: Leipzig. p. 1023–360.
4. Bravis, A., *Memorie sur les polyedres de forme symetrique.* J. de Math. (Louville), 1849. **14**: p. 137–80.
5. Berman, H.M., et al., *The Protein Data Bank.* Acta Crystallogr D Biol Crystallogr, 2002. **58**(Pt 6 No 1): p. 899–907.
6. Slabinski, L., et al., *XtalPred: a web server for prediction of protein crystallizability.* Bioinformatics, 2007. **23**(24): p. 3403–5.
7. Shapiro, A.L., E. Vinuela, and J.V. Maizel, Jr., *Molecular weight estimation of polypeptide chains by electrophoresis in SDS-polyacrylamide gels.* Biochem Biophys Res Commun, 1967. **28**(5): p. 815–20.
8. Chevallet, M., S. Luche, and T. Rabilloud, *Silver staining of proteins in polyacrylamide gels.* Nat Protoc, 2006. **1**(4): p. 1852–8.
9. Hitscherich, C., Jr., et al., *Static light scattering studies of OmpF porin: implications for integral membrane protein crystallization.* Protein Sci, 2000. **9**(8): p. 1559–66.
10. Georgalis, Y. and W. Saenger, *Light scattering studies on supersaturated protein solutions.* Sci Prog, 1999. **82 (Pt 4)**: p. 271–94.

11. Malkin, A.J. and A. McPherson, *Light-scattering investigations of nucleation processes and kinetics of crystallization in macromolecular systems.* Acta Crystallogr D Biol Crystallogr, 1994. **50**(Pt 4): p. 385–95.
12. Edmonds, C.G. and R.D. Smith, *Electrospray ionization mass spectrometry.* Methods Enzymol, 1990. **193**: p. 412–31.
13. Scoble, H.A., S.A. Martin, and K. Biemann, *Peptide sequencing by magnetic deflection tandem mass spectrometry.* Biochem J, 1987. **245**(2): p. 621–2.
14. Jardine, I., *Molecular weight analysis of proteins.* Methods Enzymol, 1990. **193**: p. 441–55.
15. Hendrickson, W.A., et al., *Multiwavelength anomalous diffraction as a direct phasing vehicle in macromolecular crystallography.* Basic Life Sci, 1989. **51**: p. 317–24.
16. Sambrook, J. and D.W. Russell, *Molecular Cloning: A Laboratory Manual.* Third Edition, 2001, Cold Spring Harbor, NY: Cold Spring Harbor Laboratory Press.
17. Hendrickson, W.A., J.R. Horton, and D.M. LeMaster, *Selenomethionyl proteins produced for analysis by multiwavelength anomalous diffraction (MAD): a vehicle for direct determination of three-dimensional structure.* Embo J, 1990. **9**(5): p. 1665–72.
18. Caffrey, M., *Membrane protein crystallization.* J Struct Biol, 2003. **142**(1): p. 108–32.
19. McPherson, A., *Introduction to protein crystallization.* Methods, 2004. **34**(3): p. 254–65.
20. Bergfors, T.M., *Protein crystallization.* 2nd ed. IUL biotechnology series, 2009, La Jolla, CA: International University Line. xxvii, 474 p.
21. Stura, E.A. and I.A. Wilson, *Applications of the streak seeding technique in protein crystallization.* J. Cryst. Growth, 1991. **110**: p. 270–82.
22. Walter, T.S., et al., *Semi-automated microseeding of nanolitre crystallization experiments.* Acta Crystallogr Sect F Struct Biol Cryst Commun, 2008. **64**(Pt 1): p. 14–18.
23. Jancarik, J. and S.H. Kim, *Sparce-matrix sampling—a screening method for crystallization of proteins.* J. Appl. Cryst., 1991. **24**: p. 409–11.
24. Carter, C.W.J.B., E.T. and Frick, L., *Statistical design of experiments for protein crystal growth and the use of a precrystallization assay.* J. Cryst. Growth, 1988. **90**: p. 60–73.
25. McPherson, A., *Preparation and analysis of protein crystals*, 1982: John Wiley & Sons, Inc.
26. Chayen, N.E., et al., *An Automated-System for Microbatch Protein Crystallization and Screening.* Journal of Applied Crystallography, 1990. **23**: p. 297–302.
27. Ng, J.D., J.A. Gavira, and J.M. Garcia-Ruiz, *Protein crystallization by capillary counterdiffusion for applied crystallographic structure determination.* J Struct Biol, 2003. **142**(1): p. 218–31.
28. Hendrickson, W.A., et al., *Crystal structure of core streptavidin determined from multiwavelength anomalous diffraction of synchrotron radiation.* Proc Natl Acad Sci U S A, 1989. **86**(7): p. 2190–4.
29. Wang, B.C., *Resolution of phase ambiguity in macromolecular crystallography.* Methods Enzymol, 1985. **115**: p. 90–112.
30. Rose, J.P. and B.C. Wang, *Factors Affecting the Data Reproducibility on an Area Detector.* J. Appl. Cryst., 1990. **23**: p. 234–40.
31. Glusker, J.P., M. Lewis, and M. Rossi, *Crystal structure analysis for chemists and biologists.* Methods in stereochemical analysis, 1994, New York: VCH. xvii, 854 p.
32. Hope, H., *Cryocrystallography of biological macromolecules: a generally applicable method.* Acta Crystallogr B, 1988. **44** (**Pt 1**): p. 22–6.

33. Garman, E., *Cool data: quantity AND quality.* Acta Crystallogr D Biol Crystallogr, 1999. **55** (Pt **10**): p. 1641–53.
34. Teng, T.Y., *Mounting of crystals for macromolecular crystallography in a freestanding thin-film.* Journal of Applied Crystallography, 1990. **23**: p. 387–91.
35. Rodgers, D.W., *Practical Cryocrystallography.* Methods in Enzymology, 1997. **276**: p. 183–203.
36. Rose, J.P., et al., *High throughput* de novo *structure determination on a home source using quick soaks, ACTOR and parameter space screening.* Rigaku J., 2006. **23**: p. 3–12.
37. Cork, C., et al., *Advanced beamline automation for biological crystallography experiments.* Acta Crystallogr D Biol Crystallogr, 2006. **62**(Pt 8): p. 852–8.
38. Cohen, A.E., et al., *Automation of high-throughput protein crystal screening at SSRL.* Sync. Rad. News, 2005. **18**: p. 28–35.
39. Minor, W., D. Tomchick, and Z. Otwinowski, *Strategies for macromolecular synchrotron crystallography.* Structure, 2000. **8**(5): p. R105–10.
40. Matthews, B.W., *Solvent content of protein crystals.* J Mol Biol, 1968. **33**(2): p. 491–7.
41. Otwinowski, Z. and W. Minor, *Processing of X-ray diffraction data collected in oscillation mode.* Methods in Enzymology, 1997. **A276**: p. 307–26.
42. Minor, W., et al., *HKL-3000: the integration of data reduction and structure solution—from diffraction images to an initial model in minutes.* Acta Crystallogr D Biol Crystallogr, 2006. **62**(Pt 8): p. 859–66.
43. Pflugrath, J.W., *The finer things in X-ray diffraction data collection.* Acta Crystallogr D Biol Crystallogr, 1999. **55** (Pt **10**): p. 1718–25.
44. Steller, I., B. Bolotovsky, and M.G. Rossmann, *An Algorithm for Automatic Indexing of Oscillation Images using Fourier Analysis.* J. Appl. Crystallogr., 1997. **30**: p. 1036–40.
45. Leslie, A.G., *The integration of macromolecular diffraction data.* Acta Crystallogr D Biol Crystallogr, 2006. **62**(Pt 1): p. 48–57.
46. Kabsch, W., *Xds.* Acta Crystallogr D Biol Crystallogr, 2010. **66**(Pt 2): p. 125–32.
47. Liu, Z.J., et al., *Structure of the Ca2+-regulated photoprotein obelin at 1.7 A resolution determined directly from its sulfur substructure.* Protein Sci, 2000. **9**(11): p. 2085–93.
48. Chen, L.Q., et al., *Crystal structure of a bovine neurophysin II dipeptide complex at 2.8 A determined from the single-wavelength anomalous scattering signal of an incorporated iodine atom.* Proc Natl Acad Sci U S A, 1991. **88**(10): p. 4240–4.
49. Wu, C.K., et al., *The 2.0 A structure of human ferrochelatase, the terminal enzyme of heme biosynthesis.* Nat Struct Biol, 2001. **8**(2): p. 156–60.
50. Green, D.W., V.M. Ingram, and M.F. Perutz, *The Structure of Haemoglobin. IV. Sign Determination by the Isomorphous Replacement Method.* Proceedings of the Royal Society of London. Series A. Mathematical and Physical Sciences, 1954. **225**(1162): p. 287–307.
51. Terwilliger, T.C. and J. Berendzen, *Automated MAD and MIR structure solution.* Acta Crystallogr D Biol Crystallogr, 1999. **55** (Pt **4**): p. 849–61.
52. Schneider, T.R. and G.M. Sheldrick, *Substructure solution with SHELXD.* Acta Crystallogr D Biol Crystallogr, 2002. **58**(Pt 10 Pt 2): p. 1772–9.
53. Zwart, P.H., et al., *Automated structure solution with the PHENIX suite.* Methods Mol Biol, 2008. **426**: p. 419–35.
54. Cowtan, K., *DMMULTI.* Joint CCp4 and ESF-EACBM Newsletter on Protein Crystallogrpahy, 1994. **31**: p. 34–38.

55. Furey, W. and S. Swaminathan, *PHASES-95: A Program Package for the Processing and Analysis of Diffraction Data from Macromolecules.* Methods in Enzymology, 1997. **277** B: p. 590–620.
56. Petsko, G.A., *Preparation of isomorphous heavy-atom derivatives.* Methods Enzymol, 1985. **114**: p. 147–56.
57. Dauter, Z. and M. Dauter, *Entering a new phase: using solvent halide ions in protein structure determination.* Structure (Camb), 2001. **9**(2): p. R21–6.
58. Hendrickson, W.A. and M.M. Teeter, *Structure of the Hydrophobic Protein Crambin Determined Directly from the Anomalous Scattering of Sulfur.* Nature, 1981. **290**(5802): p. 107–13.
59. Chen, L., et al., *The hyperthermophile protein Sso10a is a dimer of winged helix DNA-binding domains linked by an antiparallel coiled coil rod.* J Mol Biol, 2004. **341**(1): p. 73–91.
60. Deng, L., et al., *All three Ca2+-binding loops of photoproteins bind calcium ions: the crystal structures of calcium-loaded apo-aequorin and apo-obelin.* Protein Sci, 2005. **14**(3): p. 663–75.
61. Rose, J.P., et al., *Practical Aspects of SAS Structure Determination Using Chromium X-rays.* Rigaku, J., 2004. **21**: p. 1–9.
62. Habel, J.E., *Simulation Expansion and Structural Realization of the 1985 SAS Phasing Insight, in Department of Biochemistry and Molecular Biology*, 2005, University of Georgia: Athens, GA. p. 109.
63. Zhu, J. and University of Georgia., *Exploring molecular biological and structural aspects of non-Pfam proteins*, 2008. University of Georgia Electronic Thesis Service (https://getd.libs.uga.edu/pdfs/zhu_jinyi_200808_phd.pdf).
64. Rossmann, M.G., *The molecular replacement method; a collection of papers on the use of non-crystallographic symmetry.* International science review series, v. 13, 1972, New York: Gordon and Breach. viii, 267.
65. Navaza, J. and P. Saludjian, *AMoRe: An automated molecular replacement program package.* Macromolecular Crystallography, Pt A, 1997. **276**: p. 581–94.
66. Brunger, A., *Extension of Molecular Replacement: a New Search Strategy based on Patterson Correlation Refinement.* Acta Cryst., 1990. **A46**: p. 46–57.
67. Brunger, A.T., et al., *Crystallography & NMR system: A new software suite for macromolecular structure determination.* Acta Cryst,, 1998. **54** (Pt 5): p. 905–21.
68. McCoy, A.J., *Solving structures of protein complexes by molecular replacement with Phaser.* Acta Crystallogr D Biol Crystallogr, 2007. **63**(Pt 1): p. 32–41.
69. Altschul, S.F., et al., *Gapped BLAST and PSI-BLAST: a new generation of protein database search programs.* Nucleic Acids Res, 1997. **25**(17): p. 3389–402.
70. Winn, M.D., *An overview of the CCP4 project in protein crystallography: an example of a collaborative project.* J Synchrotron Radiat, 2003. **10**(Pt 1): p. 23–5.
71. Vonrhein, C. and G. Bricogne, *autoSHARP, an Automated Structure Determination System*, 2003, Global Phasing Ltd.: Cambridge, UK.
72. He, Y., et al., *OASIS and molecular-replacement model completion.* Acta Crystallogr D Biol Crystallogr, 2007. **63**(Pt 7): p. 793–9.
73. Panjikar, S., et al., *Auto-Rickshaw: an automated crystal structure determination platform as an efficient tool for the validation of an X-ray diffraction experiment.* Acta Crystallographica Section D-Biological Crystallography, 2005. **61**: p. 449–57.
74. Fu, Z.Q., J. Rose, and B.C. Wang, *SGXPro: a parallel workflow engine enabling optimization of program performance and automation of structure determination.* Acta Crystallogr D Biol Crystallogr, 2005. **61**(Pt 7): p. 951–9.

75. Adams, P.D., et al., *PHENIX: a comprehensive Python-based system for macromolecular structure solution.* Acta Crystallographica Section D, 2010. **66**(2): p. 213–21.
76. Jones, T.A., et al., *Improved methods for building protein models in electron density maps and the location of errors in these models.* Acta Crystallogr A, 1991. **47** (Pt **2**): p. 110–9.
77. Emsley, P. and K. Cowtan, *Coot: model-building tools for molecular graphics.* Acta Crystallogr D Biol Crystallogr, 2004. **60**(Pt 12 Pt 1): p. 2126–32.
78. Perrakis, A., R. Morris, and V.S. Lamzin, *Automated protein model building combined with iterative structure refinement.* Nat Struct Biol, 1999. **6**(5): p. 458–63.
79. Engh, R.A. and R. Huber, *Accurate Bond and Angle Parameters for X-Ray Protein-Structure Refinement.* Acta Crystallographica Section A, 1991. **47**: p. 392–400.
80. Brunger, A.T., M. Karplus, and G.A. Petsko, *Crystallographic refinement by simulated annealing: Application to a 1.5Å resolution structure of crambin.* Acta Cryst., 1989. **A45**: p. 50–61.
81. Murshudov, G.N., A.A. Vagin, and E.J. Dodson, *Refinement of macromolecular structures by the maximum-likelihood method.* Acta Crystallogr D Biol Crystallogr, 1997. **53**(Pt 3): p. 240–55.
82. Sheldrick, G.M. and T.K. Schneider, *SHELXL: high resolution refinement.* Methods Enzymol, 1997. **277**: p. 319–43.
83. Brunger, A.T., *Free R-Value—a Novel Statistical Quantity for Assessing the Accuracy of Crystal-Structures.* Nature, 1992. **355**(6359): p. 472–75.
84. Ramachandran, G.N. and V. Sasisekharan, *Conformation of polypeptides and proteins.* Adv Protein Chem, 1968. **23**: p. 283–438.
85. Luzzati, P.V., *Traitement Statistique des Erreurs dans la Determination des Structures Cristallines.* Acta Crystallogr, 1952. **5**: p. 802–10.
86. Davis, I.W., et al., *MOLPROBITY: structure validation and all-atom contact analysis for nucleic acids and their complexes.* Nucleic Acids Res, 2004. **32**(Web Server issue): p. W615–9.
87. Hooft, R.W., et al., *Errors in protein structures.* Nature, 1996. **381**(6580): p. 272.
88. Laskowski, R.a., et al., *Procheck—a Program to Check the Stereochemical Quality of Protein Structures.* Journal of Applied Crystallography, 1993. **26**: p. 283–91.
89. Pettersen, E.F., et al., *UCSF Chimera—a visualization system for exploratory research and analysis.* J Comput Chem, 2004. **25**(13): p. 1605–12.
90. DeLano, W.L., *The PyMOL molecular graphics system*, 2002, DeLano Scientific: San Carlos, CA, USA.
91. Potterton, L., et al., *Developments in the CCP4 molecular-graphics project.* Acta Crystallogr D Biol Crystallogr, 2004. **60**(Pt 12 Pt 1): p. 2288–94.
92. Goodsell, D.S., *Representing structural information with RasMol.* Curr Protoc Bioinformatics, 2005. Chapter 5: p. Unit 5 4.
93. Laskowski, R.A., *PDBsum new things.* Nucleic Acids Res, 2009. **37**(Database issue): p. D355–9.
94. Krissinel, E. and K. Henrick, *Inference of macromolecular assemblies from crystalline state.* J Mol Biol, 2007. **372**(3): p. 774–97.

3

Magnetic Resonance in Structural Biology

G. MARIUS CLORE

1. INTRODUCTION

Nuclear magnetic resonance (NMR) is a powerful spectroscopic technique that permits the detailed study at atomic resolution of the three-dimensional structures and dynamics of macromolecules and their complexes in solution (1–4). The major source of structural information resides in a large number of short (< 6 Å), approximate interproton distance restraints derived from nuclear Overhauser enhancement (NOE) measurements. These can be supplemented by torsion angle restraints derived from ^3J scalar couplings and backbone chemical shifts, orientational restraints in the form of residual dipolar couplings measured in weak alignment media, and long-range distance restraints (up to 35 Å) from paramagnetic relaxation enhancement (PRE) or paramagnetic pseudo-contact shift (PCS) measurements. Dynamical information can be derived over a wide range of time scales ranging from picoseconds to seconds (4). Techniques include relaxation spectroscopy to measure dynamics of bond vectors in the picosecond to low nanoscond regime (faster than the rotational correlation time), relaxation dispersion spectroscopy in the microsecond to millisecond regime, residual dipolar couplings potentially in the picosecond to millisecond regime, *z*-exchange spectroscopy for the millisecond to sub-second regime, and real-time spectroscopy for the second regime upwards. Recent interest in dynamics has largely focused on the application of relaxation dispersion spectroscopy to detect and characterize the kinetic properties of lowly populated species (5), and paramagnetic relaxation enhancement to detect and visualize such species (6).

2. NMR AND SOLUTION STRUCTURE DETERMINATION OF MACROMOLECULES: A BRIEF OVERVIEW

The development of two-dimensional ^{1}H-NMR spectroscopy (7) led to the first three-dimensional solution structure determinations of small proteins in the mid-1980s (8, 9). Subsequent work incorporating larger numbers of interproton distance and torsion angle restraints, together with stereospecific assignments, led to significant increases in both precision and accuracy (3, 10). Chemical shift overlap, however, limited the successful application of 2D ^{1}H-NMR techniques to proteins less than about 100 residues (~10 kDa). The late 1980s and early 1990s witnessed the development of 3D and 4D heteronuclear NMR spectroscopy which extended the range of applicability of the NMR method to significantly larger systems (10). The first structure determination of a protein larger than 150 residues using 3D and 4D NMR was interleukin-1β (~18 kDa), which at the time was ~50 percent larger than any previous NMR protein structure determination (11). Over the next few years, these methods were extended to a variety of protein-peptide and protein-DNA complexes (3). Hybrid approaches combining existing high-resolution structures determined either by crystallography or NMR with computational techniques involving the application of conjoined rigid body/torsion angle simulated annealing saw the structure determination of a number of significantly larger protein-protein complexes (up to ~70 kDa) (12, 13). At the same time, deuteration combined with transverse relaxation optimized spectroscopy further increased the molecular weight range by significantly reducing linewidths (14, 15). Very recently, novel techniques, based on the application of paramagnetic relaxation enhancement (PRE), have been developed to visualize lowly populated, highly transient species on the free energy landscape of macromolecular systems that are undetectable by conventional biophysical and structural techniques (6, 16–19).

3. FUNDAMENTALS OF NMR STRUCTURE DETERMINATION

Macromolecular structure determination by NMR is intrinsically a highly specialized, labor intensive, and time-consuming technique. In addition, for a system of any reasonable size (say greater than about 70 residues) isotopic labeling with ^{15}N and ^{13}C is required. For even larger systems, additional labeling schemes are also required, including site specific isotope-labeling, deuteration, and methyl protonation on a ^{13}C and ^{2}H background. Numerous reviews have been written on the subject detailing the experimental and computational methodologies involved (1–4, 10, 12).

Determining the structure of a single protein by NMR can be broken down into essentially four steps: (i) sequential resonance assignment making use of a number of experiments to identify through-bond connectivities along the backbone and side chains (usually 3D triple resonance experiments); (ii) assignment of cross-peaks in nuclear Overhauser enhancement spectra (usually 3D and 4D) to obtain short ($\leq$6 Å) interproton distance restraints which provide the main source of geometrical information; (iii) measurement of additional NMR observables that provide useful conformational information (these may include three-bond scalar couplings that are related to torsion angles by simple empirical equations; ^{13}Cα/^{13}Cβ chemical shifts which are related empirically to backbone ϕ/ψ torsion angles; long-range orientational restraints, such as residual dipolar couplings measured in dilute liquid crystalline media); (iv) calculation of the three-dimensional structure from the experimental NMR restraints using simulated annealing. Generally an iterative refinement strategy is employed (2, 10): calculations are initially carried out with a limited set of interproton distance restraints corresponding to NOE cross-peaks with unambiguous assignments; further interproton distance restraints from the remaining NOE cross-peaks are subsequently added in an iterative manner on the basis of a successively calculated series of structures. While improvements in spectrometer technology (e.g., the advent of cryoprobe technology that increases the signal-to-noise ratio 3–4 fold; higher field magnets that increase spectral resolution, thereby reducing spectral overlap) have reduced the measurement time to some extent, collecting all the data necessary to solve an NMR structure at high accuracy may still require several months. Similarly, improvements in both spectral analysis software (20–23) and structure calculation algorithms (24–28) has permitted the introduction of some degree of automation, but extensive human intervention is still necessary to fully and reliably interpret the data in all but the simplest of cases.

4. PLACE OF NMR SPECTROSCOPY IN STRUCTURAL BIOLOGY

In this light, what contribution can NMR make to structural biology and structural proteomics? There are two major methods for deriving high-resolution structural information at atomic resolution: NMR spectroscopy in solution and single crystal X-ray diffraction. In rare instances, electron microscopy is also capable of providing high-resolution information in the solid state. In addition, mass spectrometry in combination with cross-linking data is potentially capable of providing low-resolution structural information when combined with the computational techniques conventionally employed to derive structures from NMR data. If crystals can be rapidly obtained, there is little doubt that crystallography, particularly with the advent of synchrotron X-ray sources, offers the fastest route to high-resolution structure determination. However, complexes are generally more difficult to crystallize than isolated proteins, and it is usually the case that weak complexes (with K_D's in the 1 to 100 μM range) are extremely difficult to co-crystallize, while very weak complexes (K_D's > 1mM) are virtually impossible to crystallize. In the case of NMR, complexes are amenable to structural investigation providing exchange is either fast (weak binding) or slow (tight binding) on the chemical shift time scale. If exchange, however, is intermediate on the chemical shift time scale, the signals are broadened out precluding any detailed structural work.

A full structure determination of a protein-protein complex by NMR is extremely time consuming. For example, in the case of the 40 kDa EIN·HPr complex from the bacterial phosphotransferase system, the total NMR measurement time alone was ~3500 hours (or 4.8 months) (29). Clearly, therefore, the conventional approach is not suitable for high throughput. Fortunately, new developments have significantly shortened the amount of time required by making full use of prior knowledge in the form of existing high-resolution crystal structures of the free proteins (24, 30, 31). Measurement of residual dipolar couplings can quantitatively confirm that the structures of the components within the complex are either unchanged from that in the free state or exhibit specific regions with identifiable structural changes. Similarly, chemical shift perturbations can also be used qualitatively in this regard since small chemical shift perturbations do not entail any significant structural changes. With this information in hand it is then possible to derive high-resolution structures of complexes using limited intermolecular NOE data to provide translation (as well as orientational) information and, if measurable, residual dipolar coupling data (32, 33) to generate very accurate orientational information. In addition, strategies based on orientational information from residual dipolar couplings have been developed whereby translational information from NOE data can be entirely replaced in suitable cases by highly ambiguous intermolecular distance restraints derived from $^{15}N/^{1}H_N$ chemical shift perturbation mapping (34). Long-range distance restraints from a paramagnetic label attached to an engineered surface cysteine on one partner to nuclei of the other partner can also be very helpful and can replace the measurement of intermolecular NOE data completely if multiple paramagnetic labels are used.

5. EXPERIMENTAL AND COMPUTATIONAL CONSIDERATIONS

Given that structure determination of proteins has been extensively reviewed, we will briefly review the main experimental restraints and computational techniques used in the structure determination of complexes.

5.1 Intermolecular Distance Restraints

As noted above, the nuclear Overhauser effect (NOE) is the primary source of geometric information for NMR-based structure determination (1, 2). The NOE (in the initial rate approximation) is proportional to the sixth root of the distance between two protons. The upper limit for interproton distances that can be detected using the NOE is 5–6 Å. The key to deriving intermolecular NOE-derived interproton distance restraints lies in combining various isotope (^{15}N and ^{13}C) labeling strategies with isotope-filtering experiments that permit one to detect NOEs on protons attached to specific isotopes of nitrogen and carbon (i.e., NMR active, such as ^{15}N or ^{13}C, or NMR inactive such as ^{14}N and ^{12}C) (12). For example, in a complex comprising one protein labeled uniformly with ^{13}C and the other at natural isotopic abundance (i.e., ^{12}C), one can selectively detect NOEs from protons attached to ^{13}C to protons attached to ^{12}C.

5.2 Paramagnetic-Based Distance Restraints

It is possible to derive intermolecular distance restraints using another NMR-based approach which involves derivatizing (one at a time) suitable surface accessible cysteines (which may have to be introduced by site-directed mutagenesis) on one protein with either a nitroxide spin label or a metal binding site (such as EDTA) and measuring the resulting PRE or PCS on the other protein to yield long-range (15–35 Å) distance restraints (6, 35). Because in most cases the paramagnetic label is attached to the protein by a linker involving several rotatable bonds, it is essential to consider the conformational space sampled by the paramagnetic label in order to obtain accurate results. Two types of measurement can be made with paramagnetic labels, depending on the nature of the label. For paramagnetic labels with an isotropic *g*-tensor (e.g., Mn^{2+}, Gd^{3+}, nitroxide radical), pseudocontact shifts are not observed and PRE measurements can be carried out to determine the PRE rates for each paramagnetic-^1H interaction from the difference in relaxation rates (usually transverse) between the paramagnetic and diamagnetic samples. When the *g*-tensor is anisotropic (e.g., many lanthanide ions), pseudocontact shifts are observed.

Both PRE and PCS effects arise from the dipole-dipole interaction between the unpaired electron of the paramagnetic spin and a nuclear spin. Because of the large magnetic moment of the electron these effects are very large and measurable over much larger distances (up to 35 Å in suitable cases) than the dipolar interaction between two protons that gives rise to the NOE. The magnitude of the PRE is related to the sixth root of the distance between the paramagnetic center and a proton and can be directly used in refinement. The PCS has the same functional form as residual dipolar couplings and is dependent on both the cube root of the distance between the paramagnetic center and the nuclei of interest and the orientation that the paramagnetic center-nucleus vector makes relative to the χ paramagnetic tensor.

In general, paramagnetic effects arising from an extrinsic paramagnetic center (as opposed to one that is intrinsic as is the case for metalloproteins) can only be applied in a rational manner if one already has a good idea of the interaction surfaces involved in complex formation. Such information can be derived rather easily by either $^{15}N/^{1}H_{N}$ chemical shift perturbation mapping (36) or cross-saturation experiments (37). The latter experiment is far more challenging experimentally since it necessitates that one of the proteins is not only ^{15}N-labeled but fully deuterated as well.

Paramagnetic effects, however, do have to be used with some caution. In the case of very tightly binding complexes where exchange is slow on the relaxation time scale, the paramagnetic effects will arise solely from the stereospecific complex (6). However, when exchange is fast, the footprints from paramagnetic effects arising from minor species and configurations are apparent. This is more marked for PRE measurements than PCS ones owing to the respective sixth versus cube root distance dependencies. It is precisely these effects that have permitted states that are undetectable by conventional structural techniques to be studied.

5.3 Other Sources of Distance Information

NMR is not the only method that can be used to derive intermolecular distance restraints. It is also possible to derive distance restraints using a combination of cross-linking, proteolytic digestion, and mass spectrometry (38–40). In many cases, however, the data will not yield unique cross-linking partners but multiple possibilities.

Fluorescence energy transfer (FRET) through non-radiative dipolar-dipolar coupling from the fluorophore, the energy donor, to a second chromophore, the energy acceptor, scales as the sixth root of the distance between the two chromophores and can probe separations ranging from 10 to 100 Å (41). Likewise, double nitroxide spin-labeling coupled with pulsed EPR methods, such as double electron-electron resonance (DEER) based on the magnitude of the magnetic dipolar coupling of the unpaired nitroxide electrons which scales as the cube root of the separation between the two nitroxide labels, can yield remarkably accurate distances in the 20–60 Å range (42). FRET and EPR are not limited by the molecular weight of the system being studied, but suffer from a major drawback in so far that only a single pair-wise distance can be measured per sample (i.e., each distance requires a new double spin-labeled or double chromophore labeled sample, with the labels in different positions). Thus, although FRET and EPR can yield very specific information they do not afford a practical approach for solving three-dimensional structures of proteins or their complexes.

5.4 Orientational Restraints

Long-range orientational restraints can be derived from the measurement of residual dipolar couplings (32, 33) and chemical shift anisotropy (43, 44) in liquid crystalline media, and in suitable cases from heteronuclear T_1/T_2 data (45). The characteristic feature of these various parameters is that they yield direct geometric information on the orientation of an interatomic vector(s) with respect to an external axis system (e.g., the alignment tensor in liquid crystalline media, the diffusion tensor for relaxation measurements) expressed in terms of two angles: θ, the angle between the interatomic vector and the z axis of the tensor, and ϕ, the angle which describes the position of the projection of the interatomic vector on the x-y plane of the tensor.

For most practical purposes, residual dipolar couplings provide the easiest method for deriving orientational information. In an isotropic medium, the dipolar couplings average to zero. In the solid state, the maximum value of the N-H dipolar coupling is 20.7 kHz. To effectively measure dipolar couplings in solution, therefore, it is necessary to devise means of inducing only a small (ca. 10^{-3}) degree of order such that the N-H dipolar couplings lie in the ± 20 Hz range. Experimentally, this is achieved by dissolving the protein or protein complex of interest in a dilute, water soluble, liquid crystalline medium. Examples of such media include lipid bicelles, filamentous phages such as fd or pf1, rod-shaped viruses such as tobacco mosaic virus, polyethylene glycol/hexanol, and stretched polyacrylamide gels.

5.5 Some Computational Methods

In many instances, protein complex formation involves no significant changes in backbone conformation. Thus, if the structures of the individual proteins are already known at high resolution and it can be shown that the backbone conformation remains essentially unchanged upon complex formation (for example, by comparison of dipolar coupling data measured on the complex with the X-ray structures of the free proteins) one can then make use of conjoined rigid body/torsion angle dynamics to rapidly solve the structure of the complex on the basis of intermolecular NOE data and backbone NH dipolar couplings (24, 30). In this procedure, only the interfacial side chains are allowed to alter their conformation. The backbone and non-interfacial side chains of one protein are held fixed, while those of the second protein are only allowed to rotate and translate as a rigid body. This has been applied with considerable success in the case of several 30–70 kDa protein-protein complexes of the bacterial phosphotransferase system (13, 45–49), as well as to the 42 kDa ternary Oct1·Sox2·*Hoxb1*-DNA ternary transcription factor complex (50).

It should be emphasized that conjoined rigid body/torsion angle dynamics can readily be extended to cases where significant changes in backbone conformation are localized to specific regions of the protein, such as the binding interface. In such cases, both the interfacial side chains and the relevant portions of the protein backbone would be given torsional degrees of freedom, and the experimental data would also have to include intramolecular NMR restraints (e.g., NOE, dipolar coupling) relating to that portion of the backbone. This, for example, is the strategy that was employed to solve the structure of the IIA^{Mtl}·HPr complex (46). This was necessitated because the crystal structure of IIA^{Mtl} (51), which contains multiple copies of IIA^{Mtl} in the unit cell, revealed alternate conformations for four loops in relatively close proximity to the putative interaction surface with HPr.

Providing the complex under study can be aligned in a suitable liquid crystalline medium, the measurement of dipolar couplings is straightforward and permits one to determine the relative orientation of two proteins in a complex. Dipolar couplings, however, do not yield any translational information which is essential for docking. Clearly, NOE-derived intermolecular interproton distance restraints provide the most useful and reliable source of translational information. However, intermolecular NOEs are not always easy to observe and their unambiguous assignment is still difficult and time consuming, particularly for larger complexes. Backbone ^1H_N and ^{15}N chemical shifts, on the other hand, are highly sensitive to environment and have been extensively used to rapidly map interaction surfaces on proteins (36). Not surprisingly, examination of the NMR literature reveals hundreds of examples of chemical shift mapping studies; to date, however, only a handful of structures of macromolecular complexes have been determined by NMR. Recently, it has been shown that it is possible to convert chemical shift perturbation maps into highly ambiguous intermolecular distance restraints which, in combination with orientational restraints from dipolar couplings, can reliably and accurately dock the partner proteins in a complex by means of rigid body/torsion angle dynamics calculations (34). Clearly, this methodology provides a powerful tool for high throughput structural proteomics and, moreover, can greatly accelerate the determination of higher accuracy NMR structures of complexes (including the detailed placement of interfacial sidechains) by providing a good starting point for the assignment of intermolecular NOE data.

6. STRUCTURAL PROTEOMICS OF THE BACTERIAL PHOSPHOTRANSFERASE SYSTEM

In bacteria, carbohydrate transport across the membrane is mediated by the phosphoenolpyruvate:sugar phosphotransferase system (PTS), which provides tight coupling of translocation and phosphorylation (52). The PTS is a classical example of a signal transduction pathway involving phosphoryl transfer whereby a phosphoryl group originating on phosphoenolpyruvate is transferred to the translocated carbohydrate via a series of three bimolecular protein-protein complexes. The first two steps of the PTS are common to all sugars: enzyme I (EI) is autophosphorylated by phosphoenolpyruvate and subsequently donates the phosphoryl group to the histidine phosphocarrier protein HPr. The proteins downstream from HPr comprise the sugar specific enzymes II which fall into four distinct families: glucose

(Glc), mannitol (Mtl), mannose (Man), and lactose/chitobiose (Chb). Although the four families bear no sequence or structural similarity, they do possess similar organizations consisting of two cytoplasmic domains A and B, and one or two membrane-bound domains, C and D, which may or may not be covalently linked to one another. The active site residue of the A domains is always a histidine which accepts the phosphoryl group from HPr on its Nε2 atom and donates a phosphoryl group to either a cysteine residue (in the case of IIB^{Glc}, IIB^{Mtl}, and IIB^{Chb}) or to the Nδ1 atom of a histidine residue (in the case of IIB^{Man}). Subsequently the phosphoryl group is transferred onto the incoming sugar on the cytoplasmic side of the membrane bound C domain (also known as the sugar permease).

The complexes in this pathway are rather weak with K_D's ranging from 1 μM to 3–6 mM. The K_D's in the millimolar range relate to complexes involving isolated domains that are connected by 20–30 residue long flexible linkers in the intact protein (13, 49). Although binding in such instances is very weak, it is in fact perfectly tuned to the system. In particular, it can readily be calculated, based on the expected average end-to-end distance for the linkers, that these millimolar equilibrium dissociation constants correspond to 50–85 percent probabilities of the two linked domains interacting with one another at any given time. Although high-resolution crystal structures and NMR structures have been determined for many of the individual proteins of the PTS, crystallization of these protein-protein complexes has proven to be refractory, despite many years of trying. Thus, this system provides a showcase for the impact of NMR in structural proteomics.

Figure 3.1 shows ribbon diagrams of a number of structures of PTS complexes solved in our laboratory, including EIN-HPr (29); HPr complexed with IIA^{Glc} (45), IIA^{Mtl} (46), and IIA^{Man} (48); and IIA^{Glc}-IIB^{Glc} (47), IIA^{Mtl}-IIB^{Mtl} (48), and IIA^{Man}-IIB^{Man} (49). In the latter case both productive and non-productive complexes between the A and B domains are observed, and mutation of the active site histidine of IIA^{Man} to a glutamate, to mimic phosphorylation, shifts the equilibrium essentially entirely in favor of the productive complex. These complexes shed light on understanding fundamental aspects of protein-protein recognition, mechanisms for phosphoryl transfer between proteins, and the diversity of structural elements recognized by a single protein. Specificity of the protein-protein interaction surfaces is characterized by geometric and chemical complementarity, coupled with extensive redundancy to permit the effective recognition of multiple partners. There is little or no conformational change in the protein backbone before and after association. Some interfacial side chains, however, adopt different conformations (side chain conformational plasticity) depending on the interacting partner so as to achieve optimal intermolecular interactions. A consequence of these properties is increased velocity in signal transduction by eliminating any unnecessary time delay required for significant conformational change.

The interaction surfaces for HPr on EIN, IIA^{Glc}, IIA^{Mtl}, and IIA^{Man} are very similar despite the fact that their underlying structures are completely different in terms of linear sequence, secondary structure, and topological arrangement of structural elements. HPr makes use of essentially the same surface to interact with both its upstream partner EI and its downstream partners, the sugar-specific enzymes IIA. Concomitantly, the binding sites for the sugar-specific enzymes IIB and HPr on the corresponding sugar-specific enzymes IIA overlap extensively. One might therefore anticipate that the enzymes IIB could also interact with EIN. However,

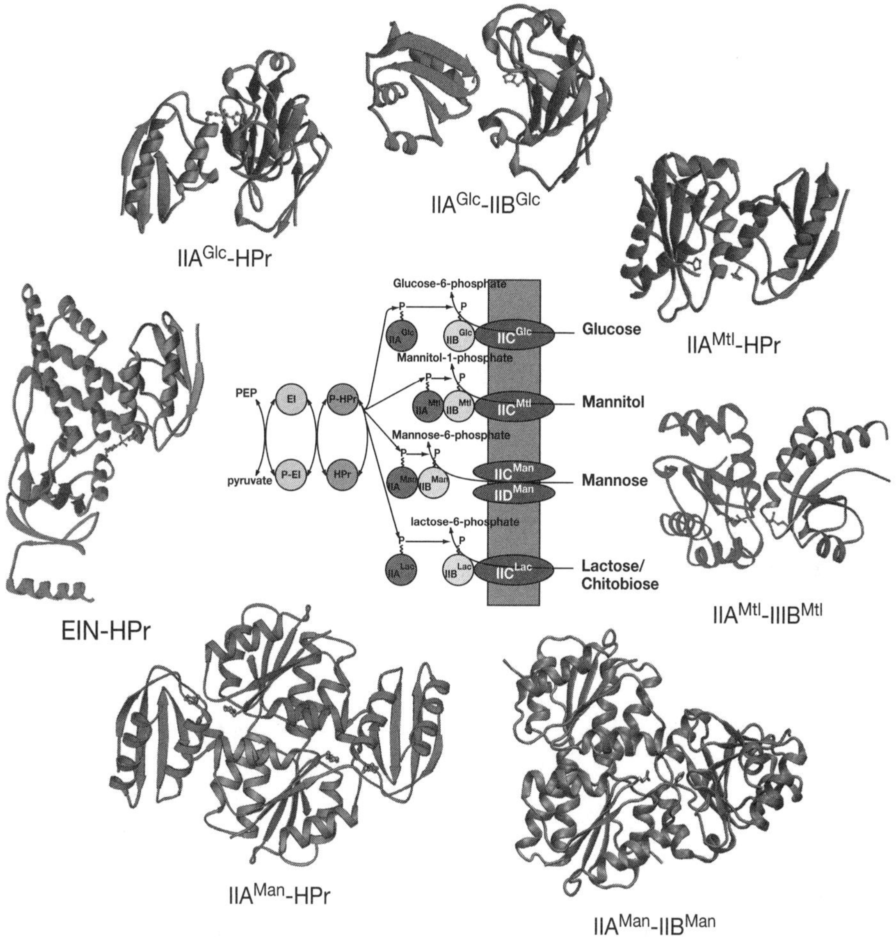

Figure 3.1 Complexes of the *E. coli* phosphotransferase system determined by NMR. EIN(light gray)-HPr(gray) (29); IIA^{Glc}(dark gray)-HPr(gray) (45); IIA^{Glc}(dark gray)-IIB^{Glc}(light gray) (47); IIA^{Mtl}(dark gray)-HPr(gray) (46); IIA^{Mtl}(dark gray)-IIB^{Mtl}(light gray) (49); IIA^{Man}(gray)-HPr(light gray) (48); IIA^{Man}(gray)-IIB^{Man}(light gray) (13). The IIA^{Mtl}-IIB^{Mtl} complex represents a post-transition state analog with the active site His554 of IIA^{Mtl} mutated to Gln to block phosphoryl transfer activity and the active site Cys384 of IIB^{Mtl} substituted by phosphoserine (49). IIA^{Man} is a dimer. In the case of the IIA^{Man}-HPr complex, two symmetry related molecules of HPr bound to IIA^{Man} are shown (48); for IIA^{Man}-IIB^{Man}, only one molecule of IIB^{Man} bound is displayed (13). The IIA^{Man}-IIB^{Man} complex represents an analogue of phosphorylated IIA^{Man} with the active site His10 of IIA^{Man} mutated to Glu; with this mutation, the complex is entirely in the productive configuration consistent with phosphoryl transfer from His10 of IIA^{Man} to His 175 of IIB^{Man} via a pentacoordinate phosphoryl transition state (13).

NMR data indicate that there is absolutely no interaction between EIN and any of the sugar-specific enzymes IIB at millimolar concentrations. From a functional perspective this is important since it ensures that the PTS cascade is not bypassed. In addition, prevention of any potential shortcuts circumventing HPr and the sugar-specific enzymes IIA is also necessary since these proteins in different states of phosphorylation also regulate the functions of proteins in other pathways (52). The structural basis for specificity and discrimination lies in the different charge distributions on the interaction surfaces of HPr and the sugar-specific enzymes IIB such that binding of a sugar-specific enzyme IIB to EIN is precluded by electrostatic repulsion.

7. VISUALIZING SPARSELY-POPULATED REGIONS OF THE FREE ENERGY LANDSCAPE OF MACROMOLECULAR COMPLEXES BY PRE

Conventional crystallography and NMR have yielded a wealth of information on the structures of highly populated static states of biological macromolecules and their complexes. However, little is known about the nature of transient sparsely-populated species that cannot be trapped and are therefore invisible to conventional structural and biophysical techniques. Yet many biological interactions proceed via the intermediary of such transient states and are also dependent on infrequent but rapid transitions between the global minimum (major configuration) and higher energy local minima within the free energy landscape sampled by macromolecular systems (53). Recent advances in NMR spectroscopy involving the application of the PRE (6, 54) have permitted such transient states to be detected and visualized (16–19). We illustrate the application of this methodology to the direct visualization of the distribution of nonspecific transient encounter complexes involved in the formation of stereospecific protein-protein complexes in section 7.2, and to the determination of the structure of a minor species for a multidomain protein where large interdomin motions are associated with ligand binding in section 7.3.

7.1 Underlying Phenomenon for the Detection of Minor Species by PRE

The observation of PRE effects for spectroscopically invisible states relies on rapid interconversion between the major and minor species (16). In this type of exchanging system, the observed PREs measured on the resonances of the major species will be modulated by the rate of exchange, k_{ex}, between the major and minor species, and the paramagnetic center-proton distances in the two species. Consider a two-site exchange system comprising a major species (denoted as A), populated at 99 percent, and a minor species (denoted as B) with an occupancy of only 1 percent, in which the paramagnetic center-proton distance is 30 Å for the major species and 8 Å for the minor one (Figure 3.2). For a 30 kDa system using Mn^{2+} as the paramagnetic label, the ^{1}H-Γ_2 rate will be ~2 s^{-1} for the major species and ~5600 s^{-1} for the minor one. When k_{ex} is slow (<50 s^{-1}), the presence of the minor species has no impact on the Γ_2 rate (Γ_2^{obs}) observed on the resonance of the major species; that is, Γ_2^{obs} is equal to the Γ_2 rate for the major species, Γ_2^A. As k_{ex} increases, Γ_2^{obs} increases, and in the fast exchange limit, defined as $k_{ex} >> (\Gamma_2^A - \Gamma_2^B)$, is given by the weighted population average of the two species: $\Gamma_2^{obs} = p\Gamma_2^A - (1 - p)\Gamma_2^B$, where p is the fractional population of the major species. In this example, Γ_2^{obs} in the fast exchange regime is ~30 times faster than Γ_2^A. Thus, in the fast exchange regime, providing the

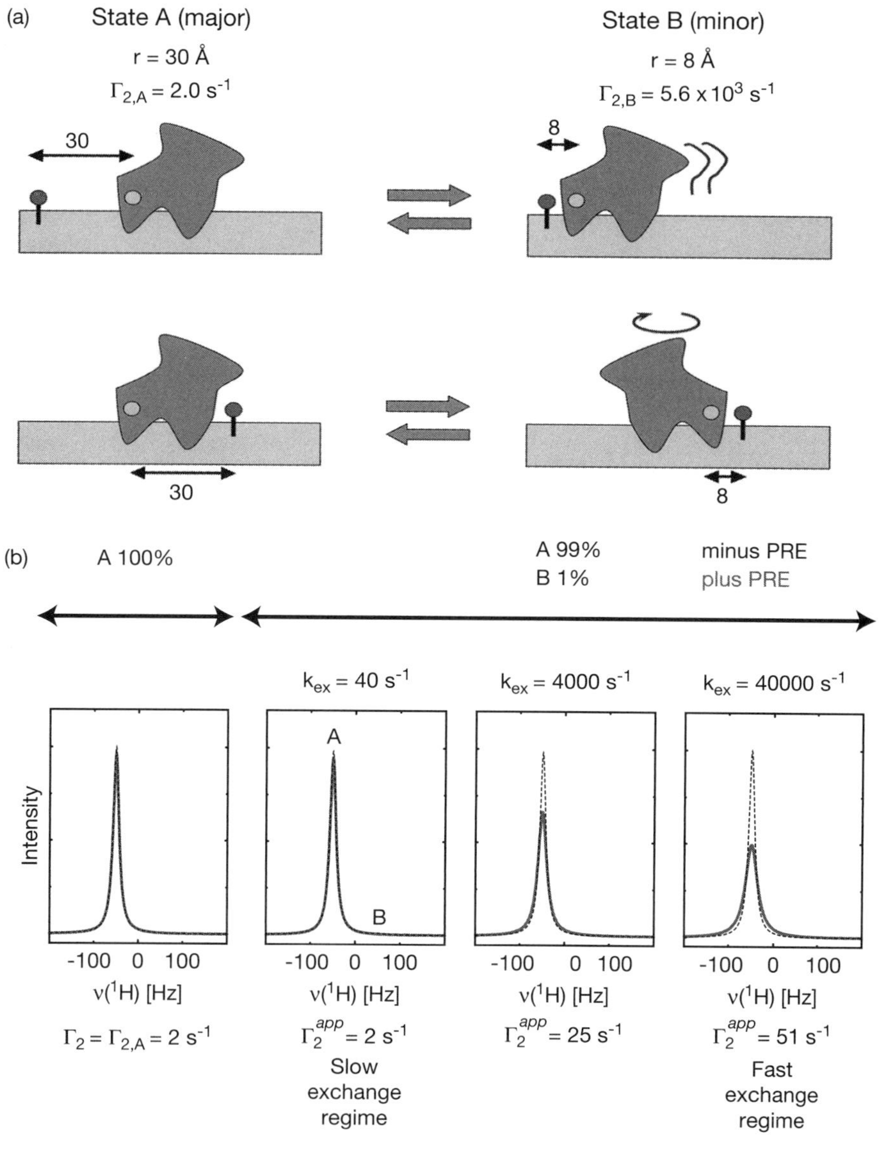

Figure 3.2 Intermolecular PRE in an exchanging system. (*a*) Diagrammatic depiction of a two site-exchange process involving major (99 percent) and minor (1 percent) species with paramagnetic-^{1}H distances of 30 and 8 Å, respectively. (*b*) Effect of increasing exchange rate on NMR line-shape with (dark gray, solid line) and without (light gray, dashed line) PRE. In the slow exchange regime the PRE is insensitive to the presence of a minor state; in the fast exchange regime, however, the PRE is sensitive to the presence of minor species and can be used to reveal the footprint of minor species. Adapted from (16).

corresponding paramagnetic center-proton distance is shorter in the minor species than the major one, Γ_2^{obs} will include a contribution from the minor species, allowing one to both infer the presence of the minor species and derive structural information on it (16–19). The latter is obtained not from a single PRE observed on a single proton, but from multiple PREs observed on the large number of protons in the system.

This is crucial since the footprint of the minor species can only be ascertained by discrepancies between the PREs that are not consistent with a single configuration corresponding to the spectroscopically visible state.

7.2 Transient Encounter Complexes in Protein-Protein Recognition

Kinetic data on a number of protein-protein associations have provided evidence for the initial formation of a pre-equilibrium encounter complex that subsequently relaxes to the final stereospecific complex. Further, site-directed mutagenesis (55, 56) and Brownian dynamics (57, 58) simulations have suggested that the rate of association can be modulated by perturbations in charge distribution outside the direct interaction surfaces. This suggests that nonspecific encounter complexes may play a role in protein-protein recognition by facilitating the formation of the stereospecific complex through a reduction in the dimensionality of the search process.

We set out to detect encounter complexes in protein-protein association by intermolecular PRE measurements using as a model system the relatively weak ($K_D \sim$ 1 μM) complex between the N-terminal domain of enzyme I (EIN) and the histidine phosphocarrier protein (HPr) of the bacterial phosphotransferase system (29). Exchange between free and bound states of the proteins is fast on the chemical shift time scale (29). The structure of EIN and HPr have been solved by both crystallography (59, 60) and NMR (61, 62), and the structure of the EIN-HPr complex has been solved by NMR on the basis of extensive NOE and dipolar coupling data (29). This complex catalyzes the reversible transfer of a phosphoryl group from the Nε2 atom of His189 of EIN to the Nδ1 atom of His15 of HPr, and the NMR structure of the complex (Figure 3.1a) is fully consistent with the formation of a pentacoordinate phosphoryl transition state intermediate without requiring any significant structural perturbation (29). Paramagnetic labels in the form of EDTA-Mn^{2+} were conjugated (one at a time) to three engineered surface cysteine residues on HPr (E5C, E25C, and E32C) located outside the binding surface with EIN (Figure 3.3) (17).

The intramolecular $^{1}H_N$-Γ_2 rates for HPr within the EIN-HPr complex are fully consistent with the structure of HPr, with a PRE Q-factor for all three sites combined of only 0.26 (Figure 3.3a). A comparison of the intermolecular PRE profile measured on EIN, however, shows that while features attributable to the stereospecific complex are clearly present, other features cannot be accounted for by either the stereospecific complex or any single alternative configuration, and reflect the presence of an ensemble of alternative binding modes involving nonspecific encounter complexes (Figure 3.3b (17). As a result, the overall PRE Q-factor calculated using the stereospecific complex is high with a value of 0.61 and the correlation between observed and calculated Γ_2 rates is poor (Figure 3.3c).

To derive a semi-quantitative structural description of the encounter complex ensemble from the intermolecular PRE data use was made of direct rigid body simulated annealing refinement against the PRE data. The representation employed comprised the stereospecific complex (whose structure is fixed) with population p in rapid exchange with an ensemble of encounter complexes comprising N states with population $(1 - p)$ (Figure 3.4a) Complete cross-validation indicated that the optimal ensemble size was 10–20 (Figure 3.4b) at a population of ~10 percent (Figure 3.4c). The resulting overall Q-factor is 0.21 (Figure 3.4d) (17).

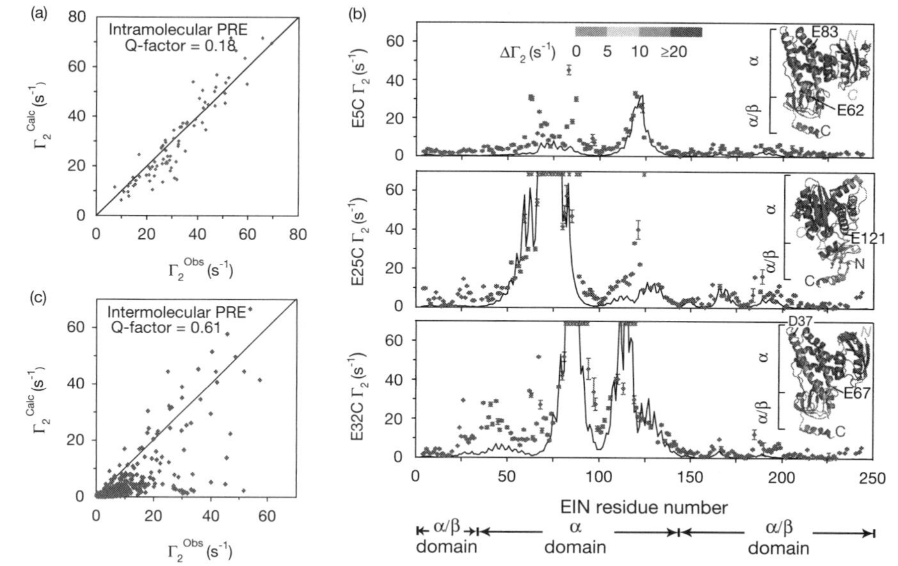

Figure 3.3 Intermolecular PREs for the EIN-HPr complex. EDTA-Mn^{2+} was conjugated to an engineered surface cysteine at three sites (E5C, E25C, and E32C). (*a*) Correlation between observed and calculated intramolecular Γ_2 rates for HPr. (*b*) Intermolecular PRE profiles observed for the three sites, with experimental Γ_2 rates denoted by the gray circles, and the theoretical Γ_2 rates calculated from the structure of the stereospecific complex by the black line. (*c*) Correlation between observed and calculated intermolecular Γ_2 rates measured on EIN and arising from paramagnetically labeled HPr. Adapted from (17).

The distribution of nonspecific encounter complexes is best visualized as an atomic probability map (Figure 3.4e) (63). Two features stand out. First, there is a qualitative correlation between the encounter complex distribution and the electrostatic surface potential, with the positively charged face of HPr populating regions of EIN with high negative electrostatic potentials. Second, the region occupied by the stereospecific complex is minimally populated by nonspecific encounter complexes, suggesting that once HPr reaches this region formation of the stereospecific complex occurs with high probability.

The relevance of weak nonspecific electrostatic interactions in the formation of nonspecific encounter complexes can be probed by examining the salt dependence of the intermolecular PREs (64). The magnitude of the intermolecular PREs that can be directly attributed to the stereospecific complex shows only a small dependence on salt concentration and can be accounted for by the dependence of the overall equilibrium dissociation constant on salt concentration. The magnitude of the PREs originating from the ensemble of nonspecific encounter complexes, however, shows a much larger salt dependence. Thus, the population of nonspecific encounter complexes is modulated to a significantly greater degree by ionic strength than the stereospecific complex. This finding is fully consistent with Debye-Hückel theory. Interfacial packing is on average less compact in the nonspecific encounter complexes than in the stereospecific complex (17). Hence the average intermolecular distance between oppositely charged residues is significantly longer in the nonspecific complexes than the stereospecific one, thereby permitting more effective screening of intermolecular electrostatic interactions by ions in solution.

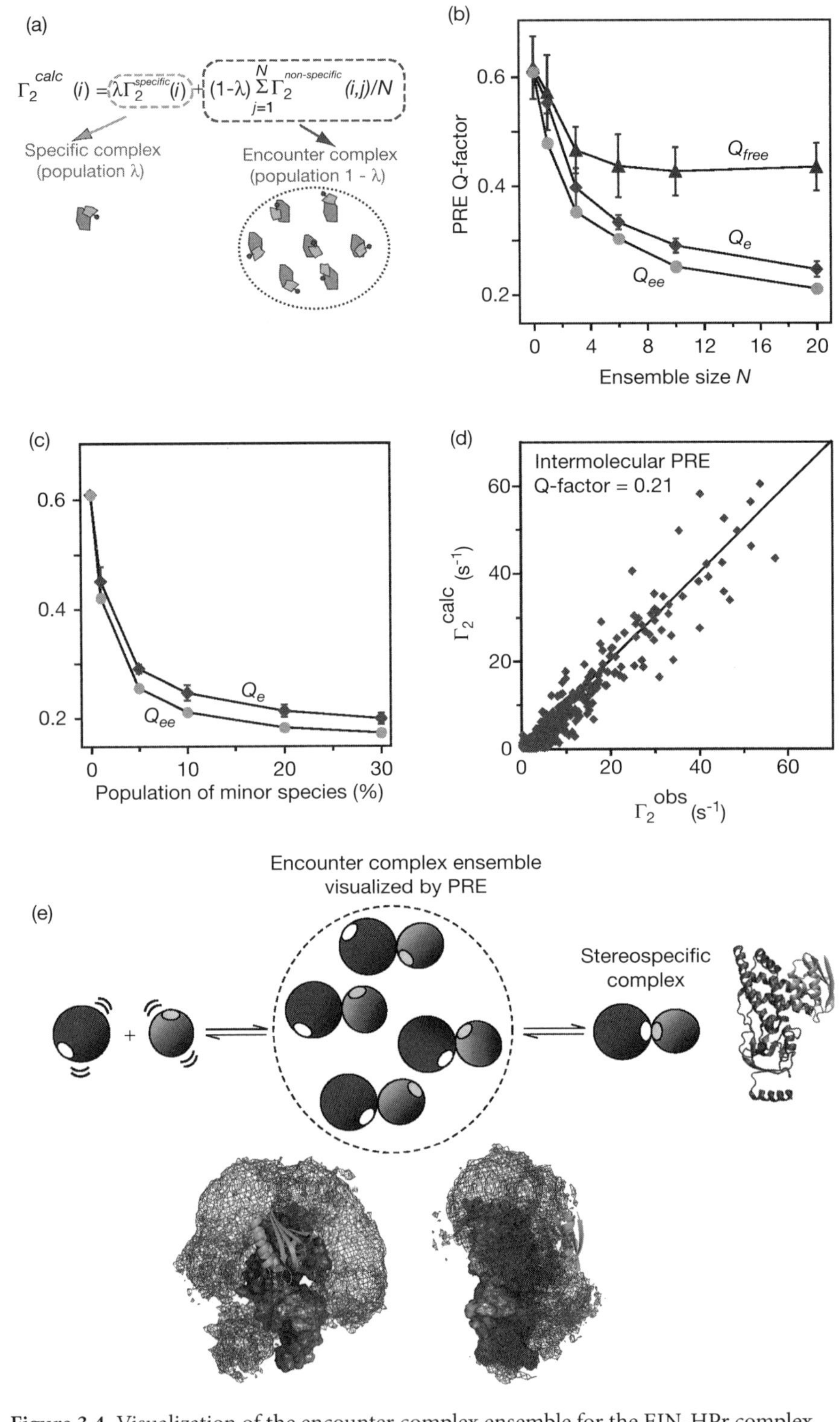

Figure 3.4 Visualization of the encounter complex ensemble for the EIN-HPr complex by PRE. (*a*) In ensemble refinement against the intermolecular PRE data the observed Γ_2 rates in the fast exchange regime are a weighted average of the Γ_2 rates for the specific

Thus, the picture that emerges from this study is one in which weak, highly transient nonspecific encounter complexes are initially formed by weak long-range electrostatic interactions, supplemented by short range van der Waals interactions. The life-times of the nonspecific encounter complexes are sufficiently long to permit a two-dimensional search on the surface of the proteins until the region of the specific interaction surfaces is reached and the complex falls down a narrow energy funnel (65). The stereospecific complex is located at the minimum of this energy funnel and is characterized by an array of complementary van der Waals and electrostatic interactions.

The findings on the EIN-HPr complex are quite general and direct detection of nonspecific encounter complexes by PRE has been observed for several other weak protein-protein complexes, including two other complexes involving HPr and the proteins IIA^{Mtl} and IIA^{Man} (17), and the redox complex of cytochrome c with cytochrome c peroxidase (66).

7.3 Transient, Sparsely-Populated States Sampled by Large-Scale Domain Motions

Large-scale domain rearrangements in proteins often play a critical function in ligand binding, recognition, catalysis, and regulation. Crystal and NMR structures have provided a static picture of the *apo* (usually open) and *holo* (usually closed) states, but the general question remains as to whether the *apo* state exists as a single species in which the *holo* conformation (in the absence of ligand) is energetically inaccessible and interdomain rearrangement is induced by ligand binding, or whether the predominantly *apo* conformation coexists in rapid equilibrium with an alternative, lowly populated, excited species which may correspond to a greater or lesser degree to the *holo* conformation. The PRE provides a powerful method to examine such phenomena (18).

---

complex and an encounter complex ensemble comprising *N* species. (*b*) Dependence of working (Q_e and Q_{ee}) and complete cross-validated (Q_{free}) *Q*-factors on ensemble size *N*. (Q_e is the average *Q*-factor <Q> for all 100 calculated ensembles, and Q_{ee} is the ensemble of ensembles average *Q*-factor). (*c*) Dependence of working *Q* factors on population of the encounter complex ensemble. (*d*) Correlation between observed and calculated Γ_2 rates obtained with a population of 10 percent for the encounter complex species represented by an ensemble of size $N = 20$. (*e*) Schematic of the reaction pathway involved in stereospecific protein-protein complex formation, with two views of a reweighted atomic probability density map (63) illustrating the distribution of HPr molecules on the surface of EIN that make up the ensemble of encounter complexes. The encounter complex probability map (light gray mesh plotted at a threshold of 20 percent maximum) is calculated from 100 independent calculations of ensemble size $N = 20$ at a population of 10 percent; the molecular surface of EIN is color coded by electrostatic potential (± 8 *kT*); and the location of HPr in the steterospecific complex is shown as a dark gray ribbon. A ribbon diagram of the stereospecific complex with EIN in gray and HPr in dark gray is also shown on the right hand side of the panel. Adapted from (17). See insert for color version.

Maltose binding protein (MBP) is a classic example of a protein undergoing large interdomain rearrangement which has been extensively studied by crystallography (67, 68), NMR (69, 70), and other biophysical techniques (71). Upon sugar binding a ~35° rigid body domain reorientation between the N- (NTD) and C- (CTD) terminal domains occurs, involving hinge-bending within the linker region connecting the two domains (67, 68). Residual dipolar couplings measured on the *apo* and *holo* states are fully consistent with the respective crystal structures (70). This is hardly surprising because the dipolar coupling observable is a linear weighted average of the species present in solution and is therefore insensitive to lowly populated states (18).

MBP was paramagnetically labeled with a nitroxide spin-label conjugated to surface engineered cysteine residues, D41C and S211C (one at a time), located in the NTD and CTD, respectively (18). The PRE data for the sugar-bound *holo* form are entirely consistent with the crystal structure of *holo* MBP (PRE *Q*-factor of 0.18). In the case of *apo* MBP, however, the intradomain PRE data are in agreement with the structure, but the interdomain PRE data arising from the nitroxide at D41C exhibit regions with large discrepancies between observed and calculated PRE values (with an interdomain PRE Q-factor of 0.49) (Figure 3.5a and Figure 3.5b, left panel). These discrepancies cannot be accounted for by the presence of a small amount of *holo* conformation in rapid equilibrium with the *apo* conformation since a linear combination of these two states results in a minimal decrease in the PRE *Q*-factor for the D41C data and an increase in the PRE Q-factor for the S211C data. This finding is fully consistent with an extrapolated population of ~0.002 percent for the *holo* conformation derived from thermodynamic data (71). Although the PRE data for *apo* MBP can be accounted for by a single alternative domain orientation of the NTD and CTD that is different from that in both the *apo* and *holo* structures, this alternative structure is completely inconsistent with the dipolar coupling data. Thus, *apo* MBP must exist as a rapidly exchanging mixture comprising the predominant *apo* (open) state and a minor species. The time scale for interconversion between the two states has an upper limit of ~20 μs from relaxation dispersion experiments and a lower limit of ~20 ns, corresponding to the rotational correlation time of MBP, from $\{^{15}N\}$-^{1}H heteronuclear NOE data (18).

The $<r^{-6}>$ average structure of the minor species of *apo* MBP can be determined by conjoined rigid-body/torsion angle simulated annealing refinement (24, 31), in which the PRE data originating from the D41C and S211C nitroxide labels are fitted simultaneously to a two-member ensemble of major open and minor species. The major species is held fixed in the *apo* crystal structure conformation whereas the domains of the minor species are allowed to move as rigid bodies by giving the linker residues torsional degrees of freedom. These calculations fully account for all the *apo* MBP PRE data with PRE *Q*-factors of 0.21 and 0.24 for the PREs originating from the D41C and S211C nitroxide labels (Figure 3.5b, right panel), respectively, and combined with RDC data, indicate that the population of the minor species is 5–7 percent (18).

The minor species of *apo* MBP represents a partially closed state (Figure 3.5c) (18). The transition between major (open) and minor (partially closed) forms of *apo* MBP involves a hinge rotation of ~33°, comparable to the ~35° rotation between open *apo* and closed *holo* MBP. However, the *apo* minor and closed *holo* states are not the same and are related by a domain reorientation of ~18° accompanied by a ~6 Å translation (Figure 3.5d). The interface between the NTD and CTD domains is lined by negatively charged residues that are responsible for an array of hydrogen bonds with the sugar substrate in *holo* MBP (67, 68). When the substrate is removed,

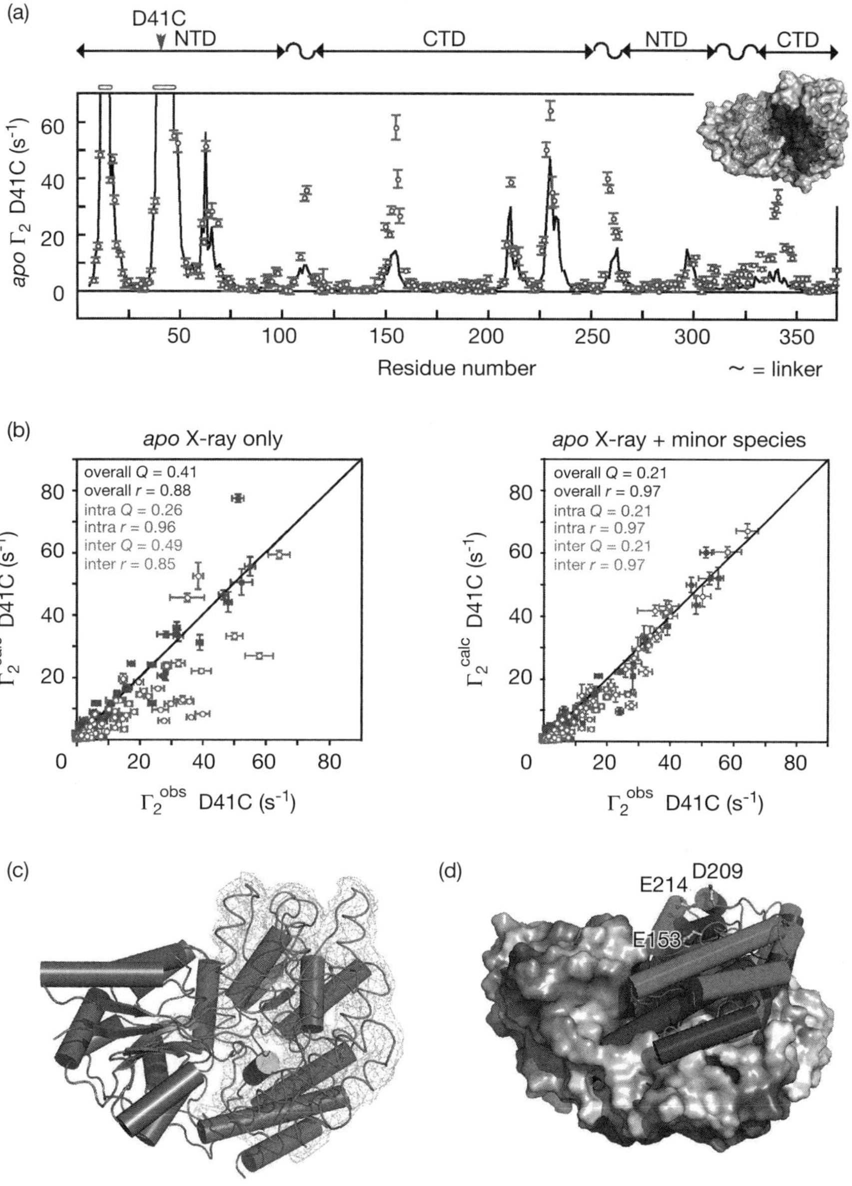

Figure 3.5 Open-to-partially closed transition in *apo* MBP. (*a*) Comparison of observed PRE profiles (gray circles) with back-calculated values obtained by ensemble paramagnetic probe refinement against the intermolecular PRE data for the nitroxide spin-label at D41C (black line). (*b*) Comparison of observed versus calculated Γ_2 rates for the D41C data obtained with the X-ray *apo* MBP structure alone (*left panel*) and upon inclusion of a minor species at an occupancy of 5 percent (*right panel*). (*c*) Equilibrium mixture of the major open (gray cylinder) and minor partially-closed (gray smoothed backbone trace with reweighted backbone atomic probability map shown as a light gray mesh) forms of the CTD of *apo* MBP with the NTDs of the two species superimposed. (*d*) Structural comparison of the CTD of the minor partially closed state of *apo* MBP (light gray cylinders) and *holo* MBP (dark gray cylinders) with the open form of *apo* MBP shown as a molecular surface color-coded according to electrostatic potential. Adapted from (18). See insert for color version.

the energy landscape is altered and access to the *holo* structure is energetically highly unfavorable as a result of electrostatic repulsion and lack of interdomain surface complementarity within the ligand-binding pocket that cannot be offset by bridging water molecules. These unfavorable interdomain interactions are circumvented in the partially closed *apo* state by translation of the CTD out of the sugar-binding pocket, thereby exposing several negatively charged residues, and, in addition, leaving the sugar binding surface on the CTD exposed (18).

The PRE results for *apo* MBP prove the existence of a dynamic equilibrium between a predominant (~95 percent) open state and a minor (~5 percent) partially closed state. The presence of the minor species in *apo* MBP may facilitate the transition to the *holo* conformation that is only rendered energetically accessible by intermolecular interactions between the two domains and the sugar ligand. Thus the predominant fluctuations in *apo* MBP do not involve the energetically disfavored *holo* conformation and therefore the conformational change upon ligand binding can be viewed as an example of induced fit (18).

8. CONCLUDING REMARKS

NMR is the only solution technique capable of providing high-resolution structural information on biological macromolecules and their complexes at atomic resolution. While NMR is not a high-throughput technique, recent advances have considerably enhanced the speed of NMR structure determinations and the size and complexity of the systems that can be studied. Indeed, NMR spectroscopy combined with prior knowledge on the structures of individual proteins from high-resolution X-ray crystallography promises to provide a very powerful approach for the efficient determination of three-dimensional structures of protein-protein complexes up to ~100 kDa, particularly in cases that are refractory to co-crystallization. The latter often involve relatively weak complexes ($K_D > 1$ μM) such as those involved in many signal transduction pathways. In addition to conventional structure determination, new NMR methodologies, particularly those involving the application of paramagnetic relaxation enhancement, offer unique opportunities to directly probe and explore in structural terms lowly populated regions of the free energy landscape and promise to yield fundamental new insights into biophysical processes.

ACKNOWLEDGMENTS

This work was supported by the Intramural Program of the NIDDK, National Institutes of Health.

REFERENCES

1. Wüthrich K (1986) NMR of proteins and nucleic acids. John Wiley, New York.
2. Clore GM, Gronenborn AM (1989) Determination of three-dimensional structures of proteins and nucleic acids in solution by nuclear magnetic resonance spectroscopy. CRC Crit Rev Biochem Mol Biol 24:479–564.

3. Gronenborn AM, Clore GM (1995) Structures of protein complexes by multidimensional heteronuclear NMR spectroscopy. CRC Crit Rev Biochem Mol Biol 30:351:385.
4. Cavanagh, J., Fairbrother, W. J., Palmer, A. G. and Skelton, N. J. (2007) Protein NMR Spectroscopy: Principles and Practice. 2nd edition. Elsevier Academic Press.
5. Korzhnev DM, Kay LE (2008) Probing invisible low-populated states of protein molecules by relaxation dispersion NMR spectroscopy: application to protein folding. Acc Chem Res 41:442–451.
6. Clore GM, Tang C, Iwahara J (2007) Elucidating transient macromolecular interactions using paramagnetic relaxation enhancement. Curr Op Struct Biol 17:603–616.
7. Ernst RR, Bodenhausem G, Wokaun A (1987) Principles of Nuclear Magnetic Resonance in One and Two Dimensions, Clarendon Press, Oxford.
8. Williamson MP, Havel TF, Wüthrich K (1985) Solution conformation of proteinase inhibitor IIA from bull seminal plasma by 1H nuclear magnetic resonance and distance geometry. J Mol Biol 182:295 –315.
9. Clore GM, Nilges M, Sukuraman DK, Brünger AT, Karplus M, Gronenborn AM (1986) The three-dimensional structure of α1-purothionin in solution: combined use of nuclear magnetic resonance, distance geometry and restrained molecular dynamics. EMBO J 5:2729–2735.
10. Clore GM and Gronenborn AM (1991) Structures of larger proteins in solution: three—and four-dimensional heteronuclear NMR spectroscopy. Science 252:1390–1399.
11. Clore GM, Wingfield PT, Gronenborn AM (1991) High resolution three-dimensional structure of interleukin-1β in solution by three- and four-dimensional nuclear magnetic resonance spectroscopy. Biochemistry 30:2315–2323.
12. Clore GM, Gronenborn AM (1998) Determining structures of larger proteins and protein complexes by NMR. Trends Biotechnol 16:22–34.
13. Hu J, Hu K, Williams DC, Komlosh, ME, Cai M, Clore GM (2008) Solution NMR structures of productive and non-productive complexes between the A and B domains of the cytoplasmic subunit of the mannose transporter of the *Escherichia coli* phosphotransferase system. J Biol Chem 283:11024–11037.
14. Peruvshin K, Riek R, Wider G, Wüthrich K (1997) Attenuated T2 relaxation by mutual cancellation of dipole-dipole coupling and chemical shift anisotropy indicates an avenue to NMR structures of very large biological macromolecules in solution. Proc Natl Acad Sci USA 94:12366–12371.
15. Kay LE (2005) NMR studies of protein structure and dynamics. J Magn reson 173:192–207.
16. Iwahara J, Clore GM (2006) Detecting transient intermediates in macromolecular bnding by paramagnetic NMR. Nature 440:1227–1230.
17. Tang C, Iwahara J, Clore GM (2006) Visualization of transient encounter complexes in protein-protein association. Nature 444:383–386.
18. Tang C, Schwieters CD, Clore GM (2007) Open-to-closed transition in apo maltose-bindng protein visualized by paramagnetic NMR. Nature 449:1078–1082.
19. Tang C, Ghirlando R, Clore GM (2008) Visualization of transient ultra-weak protein self-association in solution using paramagnetic relaxation enhancement. J Am Chem Soc 130:4048–4056.
20. Montelione GT (2001) Structural genomics: an approach to the protein folding problem. Proc Natl Acad Sci USA 98:13488–13489.

21. Herrmann T, Güntert P, Wüthrich K (2002) Protein NMR structure determination with automated NOE-identification in the NOESY spectra using the new software ATNOS. J Biomol NMR 24:171–189.
22. Gerstein M, Edwards A, Arrowsmith CH, Montelione GT (2003) Structural genomics: current progress. Science 299:1663.
23. Yee A, Pardee K, Christendat D, Savchenko A, Edwards AM, Arrowsmith CH (2003) Structural proteomics: toward high-throughput structural biology as a tool in functional genomics. Acc Chem Res 36:183–189.
24. Schwieters CD, Clore, GM (2001). Internal coordinates for molecular dynamics and minimization in structure determination and refinement. J Magn Reson 152:288–302.
25. Linge JP, Habeck M, Rieping W, Nilges M. (2003) ARIA: automated NOE assignment and NMR structure calculation. Bioinformatics 19:315–316.
26. Herrmann T, Güntert P, Wüthrich K (2002) Protein NMR structure determination with automated NOE assignment using the new software CANDID and the torsion angle dynamics program DYANA. J Biomol NMR 319:209–227.
27. Kuszewski J, Schwieters CD, Garrett DS, Byrd RA, Tjandra N, Clore GM (2004) Completely automated, highly error-tolerant macromolecular structure determination from multidimensional nuclear Overhauser enhancement spectra and chemical shift assignments. J. Am. Chem. Soc. 126:6258–6273.
28. Kuszewski J, Thottungal RA, Clore GM, Schwieters CD (2008) Automated error-tolerant macromlecular structure determination from multidimensional nuclear Overhauser enhancement spectra and chemical shift assignments: improved robustness and performance of the PASD algorithm. J Biomol NMR 41:221-239.
29. Garrett DS, Seok YJ, Peterkofsky A, Gronenborn AM and Clore GM (1999) Solution structure of the 40,000 M_r phosphoryl transfer complex between the N-terminal domain of enzyme I and HPr. Nature Stuct Biol 6:166–173.
30. Clore GM (2000) Accurate and rapid docking of protein-protein complexes on the basis of intermolecular nuclear Overhauser enhancement data and dipolar couplings by rigid body minimization. Proc Natl Acad Sci USA 97:9021–9025.
31. Clore GM, Bewley CA (2002) Using conjoined rigid body/torsion angle simulated annealing to determine the relative orientation of covalently linked protein domains from dipolar couplings. J Magn Reson143:329–335.
32. Bax, A, Kontaxis G, Tjandra N (2001) Dipolar couplings in macromolecular structure determination. Methods Enzymol 339: 127–174.
33. Prestegard JH, al-Hashimi HM, Tolman JR (2000) NMR structures of biomolecules using field-oriented media and residual dipolar couplings. Q. Rev. Biophys 33:371–424.
34. Clore GM, Schwieters CD (2003) Docking of protein-protein complexes on the basis of highly ambiguous distance restraints derived from $^{1}H_{N}/^{15}N$ chemical shift mapping and backbone ^{15}N-^{1}H residual dipolar couplings using conjoined rigid body/torsion angle dynamics. J. Am. Chem. Soc. 125:2902–2912.
35. Pintacuda G, John M, Su XC, Otting G (2007) NMR structure determination of protein-ligand complexes by lanthanide labeling. Acc Chem Res 40:206–212.
36. Walters KJ, Ferentz AE, Hare NJ, Hidalgo P, Jasanoff A, Matsuo H, Wagner G (2001) Characterizing protein-protein complexes and oligomers by nuclear magnetic resonance spectroscopy. Methods Enzymol 339:238–258.

37. Takahashi H, Nakanishi T, Kami K, Arata Y, Shimada Y (2000) A novel NMR method for determining the interfaces of large protein-protein complexes. Nature Struct Biol 7:220–223.
38. Bennett KL, Kussmann M, Bjork P, Godzwon M, Mikkelsen M, Sorensen, P (2000). Chemical cross-linking with thiol-cleavable reagents combined with differential mass spectrometric peptide mapping: a novel approach to assess intermolecular protein contacts. Protein Sci 9:1503–1518.
39. Sinz A, Wang K (2001). Mapping protein interfaces with a fluorogenic cross-linker and mass spectrometry: application to nebulin-calmodulin complexes. Biochemistry 40:8903–7913.
40. Schulz DM, Ihling C, Clore, G. M., Sinz A (2004) Mapping the topology of the calmodulin·melittin complex by chemical cross-linking and high resolution FTICR mass spectrometry. Biochemistry 43:4703–4715.
41. Hillisch A, Lorenz M, Diekmann S (2001) Recent advances in FRET: distance determination in protein-DNA complexes. Curr Op Struct Biol 11:201–207.
42. Altenbach C, Kusnetzow AK, Ernst OP, Hoffmann KP, Hubbell WL (2008) High-resolution distance mapping in rhodopsin reveals the pattern of helix movement due to activation. Proc Natl Acad Sci USA 105:7439–7444.
43. Wu Z, Tjandra N, Bax A (2001) ^{31}P chemical shift anisotropy as an aid in determining nucleic acid structure in liquid crystals. J. Am. Chem. Soc. 123:3617–3618.
44. Tjandra N, Garrett DS, Gronenborn, AM, Bax A, Clore, G. M. (1997) Defining long-range order in NMR structure determination from the dependence of heteronuclear relaxation times on rotational diffusion anisotropy. Nature Struct Biol 4:443–449.
45. Wang G, Louis JM, Sondej M, Seok YJ, Peterkofski A, Clore G M (2000) Solution structure of the phosphoryl transfer complex between the signal transducing proteins HPr and IIA^{Glucose} of the *Escherichia coli* phosphoenolpyruvate:sugar phosphotransferase system. EMBO J. 19:5635–5649.
46. Cornilescu G, Lee B R, Cornilescu CC, Wang G, Peterkofsky A, Clore GM (2002) Solution structure of the phosphoryl transfer complex between the cytoplasmic A domain of the mannitol transporter II^{Mannitol} and HPr of the *Escherichia coli* phosphotransferase system. J. Biol. Chem. 277:42289–42298.
47. Cai M, Williams DC, Wang G, Lee BR, Peterkofsky A, Clore GM (2003) Solution structure of the phosphoryl transfer complex between the signal-transducing protein IIA^{Glucose} and the cytoplasmic domain of the glucose transporter IICB^{Glucose} of the *Escherichia coli* glucose phosphotransferase system. J Biol Chem 278:25191–25206.
48. Williams DC, Cai M, Suh JY, Peterkofsky A, Clore GM (2005) Solution NMR structure of the 48 kDa IIA^{Mannose}-HPr complex of the *Escherichia coli* mannose phosphotransferase system. J Biol Chem 280:20775–20784.
49. Suh JY, Cai M, Williams DC, Clore GM (2006) Solution structure of a post-transition state analog of the phosphotransferase reaction between the A and B cytoplasmic domains of the mannitol transporter II^{Mannitol} of the *Escherichia coli* phosphotransferase system. J Biol Chem 281:8939–8949.
50. Williams DC, Cai M, Clore GM (2004) Molecular basis for synergistic activation by Oct1 and Sox2 revealed from the solution structure of the 42 kDa Oct1·Sox2·*Hoxb1*-DNA ternary transcription factor complex. J Biol Chem 279:1449–1457.

51. Van Montfort RL, Pijning T, Kalk KH, Hangyi I, Kouwijzer ML, Robillard GT, Dijkstra BW (1998) The structure of the *Escherichia coli* phosphotransferase $IIA^{Mannitol}$ reveals two conformations of the active site. Structure 6:377–388.
52. Deutscher J, Francke C, Postma PW (2006) How phosphotransferase system-related protein phosphorylation regulates carbohydrate metabolism in bacteria. Microbiol Mol Biol Rev 70:939–1031.
53. Miyashita O, Onuchic JN, Wolynes PG (2003) Nonlinear elasticity, proteinquakes and the energy landscapes of functional transitions in proteins. Proc. Natl Acad Sci USA 100:12570–12575.
54. Iwahara J, Schwieters CD, Clore GM (2004) Ensemble approach for NMR structure refinement against ^{1}H paramagnetic relaxation enhancement data arising from a flexible paramgnetic group attached to a macromolecule. J. Am. Chem. Soc. 126:5879–5896.
55. Schreiber G, Fersht AR (1996) Rapid electrostatically assisted association of proteins. Nature Struct Biol 3:427–431.
56. Selzer T, Albeck S, Schreiber G (2000) Rational design of faster associating and tighter binding complexes. Nature Struct Biol 7:537–541.
57. Northrup SH, Boles JO, Reynolds JCL (1988) Brownian dynamics of cytochrome c and cytochrome c peroxidase association. Science 241:67–70.
58. Spaar A, Dammer C, Gabdouilline RR, Wade RC, Helms V (2006) Diffusional encounter of barnase and barstar. Biophys J 90:1913–1924.
59. Liao DI, Silverton E, Seok YJ, Lee BR, Peterkofsky A, Davies DR (1996) The first step in sugar transport: crystal structure of the amino terminal domain of enzyme I of the *E. coli* PEP:sugar phosphotransferase system and a model of the phosphotransfer complex with HPr. Structure 4:861–872.
60. Jia Z, Quail JW, Waygood EB, Delbaere LT (1993) The 2.0 Å resolution structure of the *Escherichia coli* histidine-containing phosphocarrier protein HPr: a redetermination. J. Biol. Chem. 268:22940–22501.
61. Garrett DS, Seok YJ, Liao DI, Peterkofsky A, Gronenborn AM, Clore GM (1997) Solution structure of a 30 kDa N-terminal domain of enzyme I of the *Escherichia coli* phosphotransfer system by multidimensional NMR. Biochemistry 36:2517–2530.
62. van Nuland NA, Hangyi JW, van Schaik RD, Berendsen JH, van Gunsteren WF, Scheek RM, Robillard GT (1994) The high-resolution structure of the histidine-containing phosphocarrier protein HPr from *Escherichia coli* determined by restrained molecular dynamics from nuclear magnetic resonance nuclear Overhaser effects. J Mol Biol 237:544–559.
63. Schwieters CD, Clore GM (2002) Reweighted atomic densities to represent ensembles of NMR structures. J Biomol NMR 23:221–225.
64. Suh JY, Chang C, Clore GM (2007) Role of electrostatic interactions in transient encounter complexes in protein-protein association investigated by paramagnetic relaxation enhancement. J Am Chem Soc 129:12954–12955.
65. Levy Y, Wolynes PG, Onuchic JN (2004) Protein topology determines binding mechanism. Proc Natl Acad Sci USA 101:511–516.
66. Volkov AN, Worrall HA, Holtzmann E, Ubbink M (2006) Solution structure and dynamics of the complex between cytochrome c and cytochrome c peroxidase determined by paramagnetic NMR. Proc Natl Acad Sci USA 103:18945–18950.
67. Sharff AJ, Rodseth LE, Spurlino JC, Quiocho FA (1992) Crystallographic evidence of a large ligand-induced hinge-twist motion between the two domains

of maltodextrin binding protein involved in active transport and chemotaxis. Biochemistry 31:10657–10663.

68. Quiocho FA, Spurlino JC, Rodseth LE (1997) Extensive features of tight oligosaccharide binding revealed in high-resolution structures of maltodextrin transport/chemosensory receptor. Structure 5:997–1015.
69. Skrynnikov NR, Goto NK, Yang D, Choy WY, Tolman JR, Mueller GA, Kay LE (2000) Orienting domains in proteins using dipolar couplings measured by liquid-state NMR: differences in solution and crystal forms of maltodextrin binding protein loaded with β-cyclodextrin. J Mol Biol 295:1265–1273.
70. Evenäs J, Tugarinov V, Skrynnikov NR, Goto NK, Muhandiram R, Kay LE (2001) Ligand-induced structural changes to maltodextrin-binding protein as studied by solution NMR spectroscopy. J Mol Biol 309:961–974.
71. Millet O, Hudson RP, Kay LE (2003) The energetic cost of domain reorientation in maltose-binding protein as studied by NMR and fluorescence spectroscopy. Proc Natl Acad Sci USA 100:12700–12705.

4

Cryo-Electron Microscopy

PHOEBE L. STEWART

4.1. THE ADVENT OF CRYOELECTRON MICROSCOPY

Cryoelectron microscopy, or cryoEM, is a technique that is easy to warm up to once one sees the structures it can produce. The relatively young field of cryoEM emerged out of the more established disciplines of electron microscopy (EM) and protein X-ray crystallography. The Royal Swedish Academy of Sciences has recognized both of these older disciplines with Nobel prizes. In 1962 the Prize in Chemistry was awarded to Max Perutz and John Kendrew "for their studies of the structures of globular proteins" (1). Protein crystallography has undeniably accelerated biological and biomedical research as it provides a tool by which we can visualize the three-dimensional organization of the atoms and amino acid sidechains, which are the basic chemical constituents of a protein. From this information we can begin to fathom how proteins carry out their functions that are essential for life, and ultimately design drugs to modulate their action and treat disease. In 1986 the Prize in Physics was awarded to Ernst Ruska "for his fundamental work in electron optics, and for the design of the first electron microscope" more than fifty years earlier (2). This achievement has had a similar tremendous impact on biology and medicine as it has allowed us to image the inner workings of a cell with a resolution over a hundred times greater than the conventional light microscope. The Nobel committee proclaimed the electron microscope as one of the most important inventions of the twentieth century.

By the early 1960s EM had became a well-established research tool and a group of structural biologists began to envision that the electron microscope might be used to study the three-dimensional structure of proteins and macromolecular assemblies. One of the main attractions in using the electron microscope was that it could provide an image, as well as a diffraction pattern, and thus one could in principle avoid the "phase problem" of X-ray crystallography. The group of visionary structural biologists who dared to imagine protein structure determination by EM included Aaron Klug, who won the Nobel Prize in Chemistry in 1982 "for his development of crystallographic electron microscopy and his structural elucidation of biologically important nucleic acid-protein complexes" (3). Klug's interests included assemblies with helical symmetry, such as tobacco mosaic virus and the bacteriophage T4 tail, icosahedral viruses, and the nucleosomal structure of chromatin. The term "crystallographic electron microscopy" in the Nobel announcement referred to his adaption of Fourier techniques, developed first for protein X-ray crystallography, to the analysis of EM images. In a landmark paper that set the stage for modern day cryoEM, DeRosier and Klug laid out the foundations for determining a three-dimensional reconstruction from two-dimensional projection images obtained by EM (4).

One ever-present complication in the use of EM to study biological samples is that of radiation damage from the high-energy electrons (typically 100kV to 300kV) needed to achieve high-resolution imaging. In 1933 Ernst Ruska built a primitive two-stage electron microscope and generated experimental proof that images could be formed with higher magnifications than were possible with a light microscope. Despite this achievement Ruska asked himself "But what was the use of such images if even grids of platinum or molybdenum were burnt to cinders at the irradiation level needed for a magnification of only 17.4 X?" (5). The problem of heating and specimen damage caused by irradiation with a high-energy electron beam was a major technical challenge that had to be overcome before a commercial electron microscope could be produced.

Ruska was heartened by the observation that electron diffraction patterns of very thin specimens could be produced with less specimen damage than occurred during imaging. The advantages of diffraction over imaging are twofold. First, assuming that the sample has crystalline order, all of the signal will be concentrated into discrete diffraction spots rather than spread uniformly over the film or detector. Thus a significantly lower electron dose rate can be used for diffraction while still providing detectable signal over the baseline noise level of the detection medium. Secondly, any small translational movement of the specimen will cause blurring in a high magnification image, but it will not affect an observed diffraction pattern. This is related to the Fourier principle that a displacement in space of an object will induce a phase shift rather than a change in amplitude of waves diffracted by an object. Since the measured diffraction spot intensities are proportional only to the amplitudes and not the phases of the scattered waves, diffraction is not affected by small translational movements. This concept is true in both X-ray and electron diffraction. These physical considerations explain, in part, why the ability of cryoEM to provide near-atomic resolution information of biological macromolecules was first demonstrated with electron diffraction patterns from a crystalline specimen (6).

A major difficulty in using EM to study biological samples is that essentially all biological samples contain water and they must be dried or preserved in some manner before they can be introduced into the ultra high vacuum of an electron microscope. Traditional preparation techniques for biological tissue samples include fixation, dehydration, embedding, sectioning, and staining with heavy metals. For isolated proteins and macromolecular complexes the technique of negative staining allows one to replace the water around a biological molecule or complex with a heavy metal stain and then image the metal replica, which often retains many of the structural features of the biological object. While these preparation techniques have been of tremendous value, the resulting electron micrographs are not truly of the biological specimen itself but rather of a replica. Therefore they are limited in the resolution of structural detail that they can provide.

The central idea of cryoEM is to cryogenically freeze a biological sample so that it can retain water, albeit in a solid form, while being examined in a transmission electron microscope. CryoEM samples are often referred to as frozen hydrated and they are thought to represent biological specimens in a near native or physiologically relevant state. To best preserve the three-dimensional structure of biological molecules the freezing must be rapid and at cryogenic temperatures so that amorphous solid water, or vitreous ice, is formed. If the sample were allowed to freeze slowly the bulk water would form crystalline, hexagonal ice. Crystalline ice formation would cause mechanical stresses to be exerted on the biological macromolecules and this could result in structural distortions. Thin specimens, preferably less than 1μm thick, such as flat crystals or suspensions of isolated macromolecular complexes, can be vitrified at atmospheric pressure by plunging into a cryogen, such as liquid ethane (7). Thicker specimens, such as tissue sections, must be frozen under high pressure in order to achieve vitrification (8). Once the sample is frozen it has to be maintained below the devitrification temperature, which is approximately minus 140°C, to avoid conversion of the amorphous vitreous ice within the specimen into cubic or hexagonal ice.

Dealing with cryo-preserved samples introduces a whole new level of complexities to the process of collecting an electron micrograph (9, 10). After the cryo sample is produced it must be transferred into the electron microscope without warming it above the devitrification temperature and without collecting too much condensation from water droplets in the air. The cryo sample must also be maintained below the devitrification temperature while in the microscope and extra anti-contamination devices are necessary to avoid excessive contamination buildup on the cold specimen while it is in the microscope. Low-dose imaging techniques must be utilized in order to avoid destroying the extremely radiation sensitive cryo sample before an image can be collected. These difficulties were overcome by a group of scientific pioneers in the 1970s and 80s. Bob Glaeser and Ken Taylor, at the Lawrence Berkeley National Laboratory, and Jacques Dubochet and his coworkers at the EMBL, made tremendous contributions to the early development of cryoEM. Taylor and Glaeser have written a fascinating account of the early trials and tribulations, and finally successes in making cryoEM a practical experimental approach (11). In 1974 Taylor and Glaeser published electron diffraction patterns from frozen hydrated catalase crystals with diffraction spots extending to near-atomic (3Å) resolution, and soon thereafter published images of frozen hydrated catalase crystals that displayed optical diffraction spots extending to 11Å (6, 12).

Within the next decade Dubochet and colleagues demonstrated that it was possible to vitrify a thin layer of an unfixed and unstained specimen, such as a virus, and produce cryoEM images that revealed an impressive level of structural detail (7, 13). Their method involves placing a droplet of sample on either a bare copper mesh specimen grid, or a grid with a perforated carbon film, blotting the grid with filter paper to remove most of the liquid, and immediately freezing the specimen. This approach results in the biological particles of interest suspended in a thin layer of vitreous ice, rather than adsorbed to a solid support layer. The lack of a solid support layer increases the signal-to-noise ratio that can be obtained in electron micrographs. In 1984 Dubochet and colleagues published remarkably clear cryoEM images of several viruses, including human adenovirus (13). As the authors pointed out, the three-dimensional structures of the viruses were difficult to interpret directly from these images because of the superposition of information from both the upper and lower sides of the virus particles. Nevertheless the repeating pattern of the major

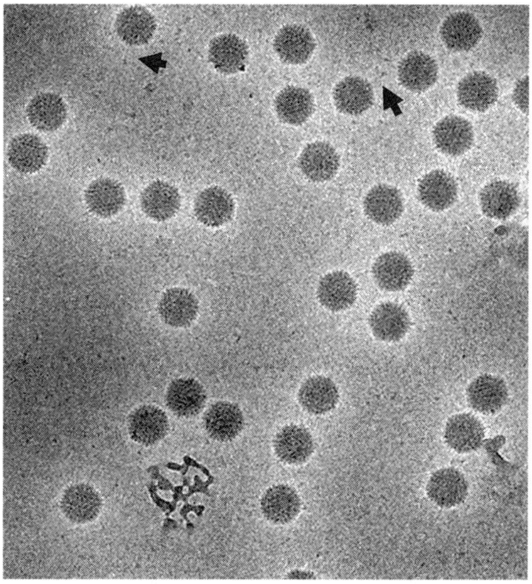

Figure 4.1 Cryoelectron micrograph of adenovirus type 2 from the same set described in 1984 by Adrian et al. (13). The adenovirus virions appear dark gray and the vitreous ice in the background appears lighter gray. For most virions a speckled pattern is visible corresponding to the projection of capsid proteins in the top and bottom surface of the virus particles. In a few favorable cases a long thin protruding fiber is visible (arrows). The fiber is characteristic of adenovirus. Each virion has twelve fibers with one fiber protruding from each vertex of the icosahedral capsid. Reprinted from (11) with permission from Elsevier.

capsid proteins of adenovirus, hexon, and penton base, was evident in the images (Figure 4.1). Remarkably, even the protruding fiber of adenovirus, which is tricky to visualize even by negative stain EM, could be seen in favorable cryoEM particle images. We now know from the crystal structure of the adenovirus fiber protein that the diameter of its long shaft is a mere 15Å (14). Dubochet predicted that as cryoEM images of vitrified samples appeared to have a reduced level of distortions compared to negative stain EM images, they would allow greater advantage to be made of image averaging. His insight proved to be correct as the field of cryoEM in combination with computer image processing has flourished in the intervening years.

At about the same time as Taylor and Glaeser were generating frozen hydrated crystals, Nigel Unwin and Richard Henderson at the MRC in Cambridge were inventing the technique of glucose embedding (15). Their idea was to exchange the bulk water in a specimen with glucose, while retaining the layer of strongly bound water at the protein surface. Although glucose embedded specimens could be examined in an electron microscope at room temperature, employing cryoEM techniques and keeping the sample at cryogenic temperatures while in the microscope did a better job of preserving high-resolution structural information. This approach led to the determination of the first atomic resolution structure of a protein, bacteriorhodopsin, based solely on EM data (16). Electron micrographs of tilted and untilted crystals were analyzed to produce phase information, which was combined with amplitude information derived from electron diffraction patterns. The synthesis of this information produced a 3.5Å resolution density map akin to an electron density map in X-ray crystallography. The three-dimensional EM map of bacteriorhodopsin showed density for large aromatic sidechains (Phe, Tyr, Trp) and was of sufficiently high resolution to allow an atomic model to be built for the protein. In this manner the phase problem of protein X-ray crystallography was entirely avoided; however, new computational challenges including correcting for distortions introduced by the electron microscope had to be overcome.

Along with the great advantage of enhanced structural preservation provided by cryo and glucose embedding sample preparation techniques came the dilemma of generating sufficient image contrast in an electron micrograph of an unstained specimen. The negative stain sample preparation technique has the advantage of a large mass variation between areas of the grid with dried regions of heavy metal stain, such as uranium based stains, and areas with little or no stain. Large mass variations create significant differences in electron scattering that lead to high image contrast. In this case the dominant mode of image formation is amplitude contrast. The situation is quite different for frozen hydrated and glucose embedded biological samples. There is little mass variation between biological macromolecules, which are formed predominantly by carbon, nitrogen, oxygen, and hydrogen atoms, and the surrounding vitreous ice (oxygen and hydrogen) or glucose (carbon, oxygen, and hydrogen). An electron micrograph collected in focus for either a frozen hydrated or glucose embedded sample shows almost no contrast. This is due in part to the similar density between the biological sample and the embedding medium (vitreous ice or glucose). In addition, cryoEM samples are more sensitive to the high-energy electron beam than negative stain samples and thus a low dose of electrons (~20 $e^-/Å^2$) has to be used for imaging cryoEM samples. The low electron dose necessary for cryoEM imaging also contributes to low signal-to-noise ratios in the micrographs.

The practical solution that emerged for the dilemma of generating contrast for unstained specimens was to collect micrographs with an intentionally applied underfocus value on the order of several microns. Klug realized in the course of studying negatively stained catalase crystals that varying the amount of underfocus would accentuate different spatial frequencies of information present in the biological structure and would create phase contrast in the image (17). Phase contrast, combined with an estimated 7 percent amplitude contrast for frozen hydrated samples (18), enables biological macromolecules to be visualized against a background of vitreous ice or glucose. Klug also suggested a way in which the distortions introduced by collecting the image out of focus could be corrected in order to computationally create a "true" image. An alternate solution for generating image contrast, which was developed more recently, is to add heavy metal stain to the cryo-preserved sample. This approach is referred to as cryo-negative staining (19).

It is now possible to buy an electron microscope that is designed and optimized specifically for cryoEM. Also available are robotic vitrification devices for preparing reproducible cryoEM grids, for example the Vitrobot from the FEI Company (Hillsboro, OR). But preparing cryoEM grids and collecting cryoelectron micrographs or electron diffraction patterns are only the first steps in the determination of a three-dimensional structure. Image processing has become an essential part of cryoEM, and it takes significant computational effort to extract high-resolution structural information from cryoelectron micrographs of unstained, cryo-preserved biological specimens.

The field of cryoEM has evolved to include many approaches for deriving three-dimensional structures of biological samples. Included under the cryoEM umbrella are the subspecialties of electron crystallography, which enabled Henderson and colleagues to build an atomic model for bacteriorhodopsin (16); helical reconstruction as demonstrated by DeRosier and Klug for T4 bacteriophage tails (4); image reconstruction of icosahedral viruses, as first developed by Tony Crowther and colleagues (20, 21); image reconstruction of asymmetric macromolecular assemblies, such as the ribosome, as pioneered by Joachim Frank and coworkers (22); image reconstruction of macromolecular complexes with intermediate levels of symmetry; and cryoelectron tomography. The approach of image reconstruction for either asymmetric or symmetric complexes is best applied to large macromolecular assemblies (~300kDa to several hundred MDa). It has the advantage that neither crystals nor helical arrays are needed. However, macromolecules or assemblies that are not part of a crystal or helical array, and that are smaller than ~300kDa, are difficult to study by cryoEM because of the low signal-to-noise ratio in cryoelectron micrographs.

All of these approaches, except for tomography, rely on averaging of data collected from a large number of identical repeating units or presumably identical biological particles. The basic principle of tomography is that many different projection views of a single object can be collected by tilting the specimen relative to the electron beam and these projection views can be synthesized into a three-dimensional structure or tomogram. As another chapter in this volume is devoted entirely to electron tomography, this chapter will focus on all of the other subspecialties of cryoEM.

4.2. CHALLENGES IN CRYOEM IMAGE PROCESSING

To convert two-dimensional projection information from electron micrographs into three-dimensional density maps one may work in real space, as is done in back-projection approaches (23, 24), or in Fourier space as put forth by DeRosier and Klug (4). According to the projection theorem in crystallography, the Fourier transform of an electron micrograph generates a central section of the full, three-dimensional Fourier transform of an object (25). Thus as DeRosier and Klug proposed "the three-dimensional transform can therefore be built up section by section using transforms of different views of the object, and the three-dimensional reconstruction then produced by Fourier inversion" (17). This approach assumes that one can determine the relative orientational angles of all of the projection views in the data set, a task which is not trivial for cryoEM images of unstained specimens.

Asymmetric or symmetric macromolecules or macromolecular complexes that are not part of a helical array or a two-dimensional crystal are referred to in the cryoEM field as "single particles." Essential steps in the reconstruction of single particles are: 1) to determine the position of the particle center in x, and y for each particle image in the data set, and 2) to determine the three orientational angles that describe the projection view of each particle image in the data set. After these five parameters are estimated for each particle image a preliminary three-dimensional reconstruction may be calculated. Once the first reconstruction is calculated, and ideally validated to be a reasonable facsimile of the true structure, it becomes possible to refine the initial estimates for the translational centers and orientational angles for each of the particle images by comparison with the reconstruction. Many iterative rounds of refinement are normally done to improve the orientational parameters, which should simultaneously improve the resolution of the reconstruction.

The following example is a case in point illustrating why this process can be challenging. Figure 4.2A shows a cryoEM particle image of human adenovirus that contributed along with 2,629 other similar particle images to a 6Å resolution structure of the virus (26). Computational binning, or averaging of adjacent pixels, during the initial phases of image processing helps to improve the image contrast. This in turn aids in the determination of the initial, relatively crude, orientational parameters. Later rounds of refinement may make use of the higher resolution information contained in lower contrast versions of the particle images with smaller binning factors and hence finer image pixel sizes. Despite applying computational binning to the adenovirus particle image in Figure 4.2A, five different algorithms for determining the initial center parameters will more than likely produce five different answers with variations in x and y of up to ~90Å. This corresponds to ~10 percent of the diameter of the adenovirus capsid, 925Å.

The conundrum this example illustrates is that although a cryoEM particle image may contain high-resolution information about a biological structure, the overall signal-to-noise ratio may be quite low. In this case the edge of the adenovirus particle is barely discernible above the background noise in the image even with computational binning. Dubochet and colleagues had optimized their imaging conditions to accentuate low-resolution (90 to 100Å) structural features of adenovirus (13). They used a conventional transmission EM with an electron beam of 80kV electrons, an early style specimen cold stage, a magnification of 12,500X, ~8μm of underfocus, and film as the detection medium. In the 6Å resolution cryoEM adenovirus study a field emission gun (FEG) transmission EM with a 300kV electron beam and a highly stable specimen cryostage (in an FEI Polara microscope) was used, with an effective magnification of 398,000X at the level of the CCD detector, and an underfocus range of 0.3 to 3.0μm (26). Imaging with an FEG microscope is generally regarded as being advantageous for cryoEM as the high coherence of the electron beam produces enhanced phase contrast in the images (9, 27, 28).

Any particle image in a data set that does not agree well with the population average represented by the data set as a whole may be excluded from the reconstruction. Poor agreement with the average might be the consequence of a structural deformity for a particular virion, a "broken" particle; shortcomings in the image quality such as blurring caused by specimen motion; or simply by poorly determined translational and angular parameters (29). In the final round of refinement for the 6Å resolution cryoEM structure of adenovirus (Figure 4.2B, C, D), a translational error of just 1.5Å for the particle image shown would be sufficient to trigger its exclusion from the final reconstruction.

In calculating the adenovirus reconstruction shown in Figure 4.2, full use was made of the icosahedral (60-fold) symmetry of the viral capsid. This effectively increases the number of asymmetric units, by a factor of 60, which are averaged to produce the final density map. In doing so, any features of the virus that follow icosahedral symmetry will be strengthened and any features that do not follow this symmetry will be weakened or averaged away. For example, the thin protruding fiber is flexible for many adenovirus serotypes. In the reconstruction shown in Figure 4.2, only a short portion of the fiber close to the icosahedral capsid is reconstructed due

Figure 4.2 CryoEM particle image and 6Å resolution reconstruction of adenovirus. (A) A particle image binned to have a pixel size of 1.5Å and a box edge length of 750 pixels. (B) A 6Å resolution reconstruction of adenovirus with imposed icosahedral symmetry (26). The view is along an icosahedral three-fold symmetry axis. Only a short portion of the protruding fiber is reconstructed due to flexibility. The surface of the virion is shown radially shaded (radius 480Å in dark gray; radius 380Å in light gray). The scale bar represents 100Å. (C) Enlarged view of one penton base with a short portion of the fiber protruding from the center. The penton base is a pentameric protein composed of five identical subunits and the fiber is a trimer composed of three identical subunits. (D) Enlarged view of one hexon. Hexon is a trimeric protein composed of three identical subunits.

to flexibility of the fiber beyond this point. The DNA and protein packaged within the adenovirus core also do not follow icosahedral symmetry and thus their density is not well represented in the reconstruction. Sometimes it is useful to impose different symmetries on the same cryoEM data set, as has been done for GroEL (30). Use of the higher fold symmetry will tend to produce a higher resolution density map at the expense of eliminating variations between presumably equivalent repeating units. Application of lower fold symmetry may bring out true structural variations within the biological complex.

There is a common misperception that cryoEM single-particle reconstruction can only determine the external structural features of a biological macromolecule or complex. This notion has probably arisen in part because icosahedral virus reconstructions often reveal disordered cores, and because moderate resolution cryoEM density maps are often represented as solid surfaces giving the false impression that only the three-dimensional surface structure has been determined. However, cryoEM images are projection images of all of the mass within a biological particle, and therefore they contain information about the external and internal structural features of a biological particle. As long as the internal structural features are the same in the majority of the particles represented by a cryoEM dataset, a reconstruction will provide the three-dimensional structure of the interior as well as the exterior of the particle.

Another computational task that generally must be done in order to achieve the highest resolution possible from a given cryoEM data set is to correct for the contrast transfer function (CTF) of the microscope (23, 31, 32). CTF is defined as a "mathematical function that expresses the ability of an optical or electronic device to transfer signals faithfully as a function of the spatial or temporal frequency of the signal" (www.biology-online.org). In the case of cryoEM the device is of course the electron microscope, and the relevant frequency is spatial frequency, or the resolution content, of the signal. Clearly every type of microscope, light or electron, will produce some level of information degradation or distortion in the images it generates. The situation is exacerbated in cryoEM by the fact that the widely adopted solution for obtaining sufficient image contrast to allow image processing is to collect electron micrographs that are underfocused. This procedure results in the accentuation of certain resolution bands and the concurrent de-accentuation of other resolution bands in a cryoelectron micrograph. The amount of underfocus can be chosen so that certain repeating features in the biological structure, for example the capsid proteins of an icosahedral virus, will be accentuated (9). This can be helpful in determining orientational angles for an initial cryoEM data set. However, accentuating certain resolution bands does not give a true, undistorted projection image of a biological particle. Computational correction for the CTF can be accomplished with a mathematical deconvolution procedure performed on either the set of two-dimensional projection images or the three-dimensional reconstruction. This requires determination or estimation of the amount of underfocus in each micrograph. The information in cryoelectron micrographs is also distorted by astigmatism in the objective lens of the microscope. Some CTF correction algorithms include estimation and correction for astigmatism (33, 34).

A third difficulty in the field of single-particle reconstruction is to determine the absolute hand of the structure of a protein or biological assembly. Biological macromolecules have an intrinsic handedness, for example amino acids can exist in one of two possible isomeric forms, D or L. The vast majority of amino acids in proteins are L-amino acids. The problem in determining the hand of a structure arises because cryoEM images are projections images, and projections of the true structure and its mirror image will appear the same. It is possible to resolve this ambiguity at atomic resolution, but if the cryoEM reconstruction is only at moderate resolution, other approaches are necessary. These include comparing the cryoEM density to an atomic resolution structure of a component molecule or comparing the orientations of tilted and untilted cryoEM particle images of the same particle (29, 35).

The three computational challenges outlined above for single-particle reconstructions are by no means an exhaustive list. They are only meant to give a sense of the amount of work that is necessary in order to turn two-dimensional cryoEM data into three-dimensional structures. The subspecialty of electron crystallography has its own set of image processing challenges, including correcting for stretching and bending within the crystal and a variety of other crystalline defects (16). Helical reconstruction has its own unique complexities, including dealing with heterogeneity within and between filaments (36). The success that the field of cryoEM has achieved in generating structural information on a wide variety of biological systems is due in large part to the creative software development by many talented people (reviewed in (37)). It is worth noting that cryoEM software is still undergoing continual development, and is not yet as mature as software for protein crystallography.

4.3. DRAMATIC IMPROVEMENT IN CRYOEM RESOLUTION

The vision that EM might be used to determine protein structure has now come to fruition with the determination of structures at atomic or near-atomic resolution by electron crystallography, helical image processing methods, and single-particle reconstruction. Atomic resolution in electron crystallography was achieved in 1990 with the 3.5Å resolution bacteriorhodopsin structure (16). Since then several protein structures have been determined by electron crystallography at atomic resolution, including αβ-tubulin dimer (38) and the plant light harvesting complex (39).

Within the last several years a structure of lens-specific aquaporin-0 (AQP0) in a closed pore state was determined by electron crystallography to an impressive 1.9Å resolution, setting a new precedent in resolution for the technique (40). This structure facilitated an analysis of both the membrane protein and the protein-lipid interactions at atomic resolution. Water molecules were clearly identifiable and three were observed in the center water pathway of the molecule (Figure 4.3). The two-dimensional crystals used for the structural study were actually double layered crystals, with two lipid bilayers, each of which contain embedded AQP0 tetramers. The tetramers in the crystal interact with each other through their extracellular surfaces as they would in thin junctions between lens fiber cells. The double-layered nature of these crystals may have contributed to the high degree of crystalline order necessary to observe diffraction to 1.9Å resolution. Quite strikingly an annular ring of lipid molecules around the AQP0 tetramers was found to be ordered. Atomic models were built for nine lipid molecules in the asymmetric unit. The acyl chains of the lipid molecules next to the protein surface adopt conformations that effectively fill in depressions on the protein surface and create a smoother interface for the surrounding bulk lipids.

A landmark resolution point was reached for helical structure determination in 2003 with the publication of two papers at ~4Å resolution reporting structures of the acetylcholine receptor and the bacterial flagellar filament (41, 42). The acetylcholine receptor from *Torpedo marmorata*, the electric ray, has a membrane-spanning gated pore which serves to control electrical signaling between nerve and muscle cells. Tubular crystals composed of protein and lipid can be grown from postsynaptic membranes of the *Torpedo* electric organ. These tubes have helical symmetry, although different tubes are of different helical families and have varying symmetry elements and unit cell dimensions. A few hundred images of helical tubes, representing ~10^6 receptors, were processed to yield a 4Å resolution density map from which an atomic model for the receptor was built. A refined structure was published two

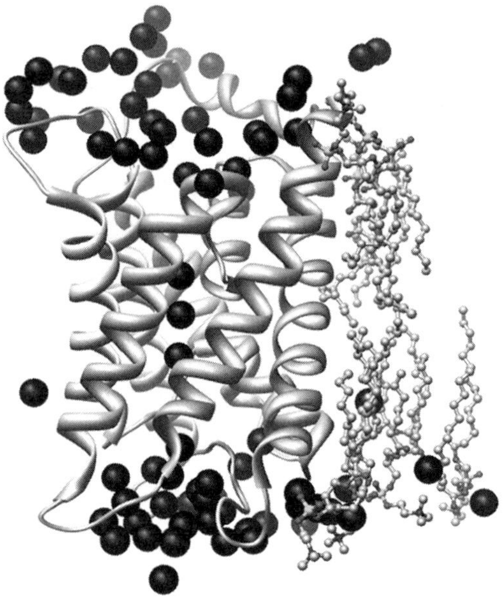

Figure 4.3 Structure of the lens-specific aquaporin-0 (AQP0) in a closed pore state determined by electron crystallography at 1.9Å resolution (PDB ID 2B6O) (40). Although only a ribbon diagram is displayed for the protein, the atomic model does include sidechain atoms. A monomer of AQP0 is shown (ribbon representation) with the water molecules (dark spheres) and lipid molecules (ball and stick representation) in the asymmetric unit.

years later (43) and is shown in (Figure 4.4). The protein is quite elongated with an extracellular acetylcholine binding domain, a membrane spanning domain formed by α-helices, and an intracellular domain.

The bacterial flagellar filament is an equally impressive structure (42). At the time of this cryoEM work an X-ray crystal structure had already been determined for a large fragment of the protein flagellin, which is the basic subunit in the filament. This information helped in analyzing the cryoEM density of the intact flagellar filament. The filament is a helical structure formed by eleven protofilaments and it can adopt two distinct supercoiled states, the L- and R- states, necessary for the running and tumbling movements of bacteria. The cryoEM image processing involved averaging information from 102 filaments, corresponding to ~41,000 molecular images. Analysis of the resulting density map led to a complete atomic model for the R-type filament, including the flagellin termini that had been missing from the X-ray structure. In particular, the cryoEM structure revealed the interactions between the flagellin termini in the inner core of the filament. This is a good example of the power of hybrid methods, in this case X-ray crystallography and cryoEM, to study the structure of a large macromolecular complex.

In the single-particle reconstruction field, 1997 was the year that the "10 Å resolution barrier" was broken. This was significant because it enabled α-helices to be resolved (44, 45). In 2008 the single-particle reconstruction approach arrived at near-atomic resolution with four structures determined at 3.8 to 4.5Å resolution (30, 46–48). These include GroEL (30), and three icosahedral viruses: cytoplasmic

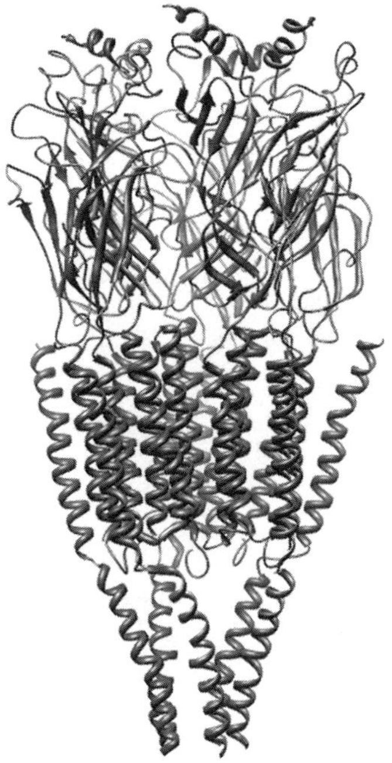

Figure 4.4 Structure of the nicotinic acetylcholine receptor determined from cryoEM images of helical tubes at 4Å resolution (PDB ID 2BG9) (43). Although only a ribbon diagram is displayed, the atomic model does include sidechain atoms. The receptor has a large extracellular acetylcholine binding domain (top), a membrane spanning domain formed by 20 α-helices (middle), and an intracellular domain (bottom).

polyhedrosis virus (CPV) (47), epsilon 15 bacteriophage (46), and rotavirus, double-layered particles (48). The number of subunits averaged in these structures ranged from $\sim 3\times10^5$ to $\sim 7\times10^6$. The cryoEM data acquisition methods and image processing techniques used to determine these structures varied, but all four are based on cryoEM data collected with 300kV electrons from a FEG electron source (49).

In the case of the rotavirus particle an additional thirteen-fold symmetry averaging was applied to the icosahedral density map to average the thirteen viral protein 6 (VP6) monomers within the icosahedral asymmetric unit. This led to an improvement in the resolution and interpretable density for most of the large VP6 sidechains. The authors noted that the quality of the density map with the additional nonicosahedral averaging was comparable to an X-ray crystallographic map at 3.8Å resolution. For GroEL the cryoEM density map allowed de novo building of an atomic model that compares favorably with existing X-ray crystal structures. In the remaining two near-atomic resolution single particle structures, CPV and epsilon 15 phage, no prior atomic structural information was available and atomic models were built from the cryoEM density.

Once an atomic model can be built for a macromolecular complex, this enables interpretation of the structure at the level of amino acid sidechain chemistry. This in turn often provides answers for questions such as how the macromolecular complex is formed and how it carries out its various biological functions. Atomic and

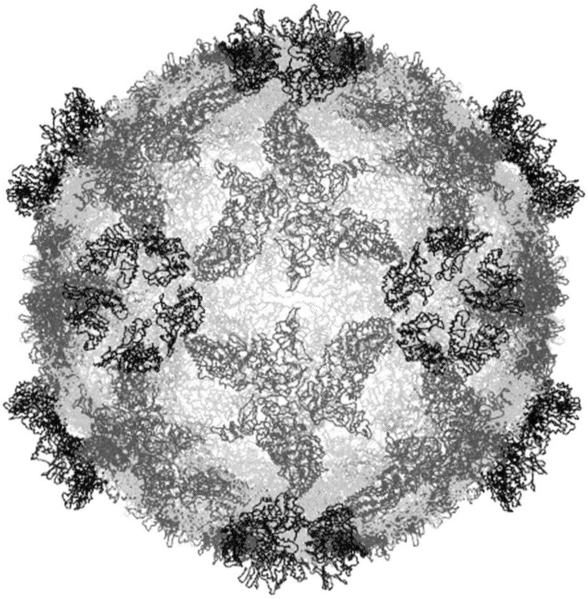

Figure 4.5 Structure of cytoplasmic polyhedrosis virus (CPV) determined by cryoEM single-particle reconstruction at 3.88Å resolution (PDB ID 3CNF) (47). The atomic model contains only Cα backbone atoms. The reconstruction was calculated from 12,814 particle images with imposed icosahedral symmetry. The two capsid shell proteins, CSP-A and CSP-B, are shown in light gray and dark gray and the turret protein is shown in black.

near-atomic resolution structural information can also be enormously valuable in the formulation of new testable hypotheses designed to address additional biological questions not immediately answered by the structure. The cryoEM derived atomic model for CPV (Figure 4.5) helped to answer several biological questions about the virus, including issues related to RNA packaging, replication, and release (47, 49).

4.4. CRYOEM AND HYBRID STRUCTURAL METHODS

Our present understanding of ribosome dynamics during protein translation is due to both X-ray crystal structures and cryoEM structures of the ribosome trapped in different functional states (50). This body of work is an excellent example of the power of hybrid structural methods for studying large macromolecular complexes. By far the most common hybrid combination has been cryoEM and X-ray crystallography (51). However, there are growing numbers of other combinations that are proving to be advantageous. Information from NMR, X-ray crystallography, and EM has led to insight into the interaction of the cyanobacterial circadian rhythm proteins KaiA and KaiC from the model organism *Synechococcus elongatus* (52). Electron paramagnetic resonance (EPR) and cryoEM have provided complementary structural information for small heat-shock protein complexes (53, 54). The combination of EPR and cryoEM can be useful for gaining insight into regions of protein that are disordered in X-ray crystal structures. Small angle X-ray scattering (SAXS) has been used to distinguish between two different electron microscopy models for the CaMKII holoenzyme (55).

The combination of cryoEM structures of large complexes with X-ray crystal structures of component proteins was first utilized in the 1990s. A few early examples include analysis of the complex of cowpea mosaic virus with Fab fragments from monoclonal antibodies (56), the actin-myosin complex (57), and adenovirus (58). Even at a moderate resolution of ~25Å, the characteristic molecular shape of the adenovirus hexon protein, known by X-ray crystallography, was clearly discernible in a cryoEM reconstruction (Figure 4.6A) (58). Recognition of the hexon molecular features allowed unique placement and refinement of the hexon crystal structure within the cryoEM density to produce a pseudo-atomic model for all of the hexons in the viral capsid. This in turn facilitated a difference analysis between the cryoEM density of the intact virus and the pseudo-atomic model with 240 hexons. Combined with information on the mass and copy numbers for the additional capsid proteins it was possible to make tentative assignments for the locations of several minor capsid proteins.

With the determination of an adenovirus cryoEM structure at 6Å resolution in 2006, a region of density that had previously appeared solid resolved into the shape of a coiled coil with four α-helices (Figure 4.6) (26). This enabled a more conclusive assignment to be made for this region, as only one protein within adenovirus has a predicted coiled coil. The C-terminal region of adenovirus protein IX has a strongly indicated propensity for coiled coil formation. In addition the length of the predicted coiled coil, ~40 amino acids, matches the observed length of the coiled coil cryoEM density. As this adenovirus example illustrates, cryoEM structures of large complexes at even moderate (10–25Å) resolution can reveal the molecular architecture of the assembly if atomic resolution structures are available for component proteins. At higher resolutions of 6 to 10Å α-helices can be resolved, and when combined with secondary structure prediction, this can facilitate identification of specific protein domains.

4.5. CURRENT FRONTIERS IN CRYOEM

Until recently the resolution of the transmission electron microscope has been limited by both spherical and chromatic lens aberrations. We have now entered an era of aberration corrected electron microscopes (59). The benefit of aberration correction for materials characterization has already been demonstrated and has allowed image interpretation to sub-Angstrom levels (60). In addition, active work is ongoing in the area of phase plate development for the electron microscope (61–63). The idea of a phase plate is that it would create phase contrast for specimens, such as unstained vitrified biological samples which fall into the category of "weak-phase objects." It will be interesting to see in the coming years how these new technical advances will impact high-resolution cryoEM structure determination.

On the software front, the development of automated and semi-automated methods for data acquisition (64–66) to facilitate collection of large EM datasets has already made an impact on cryoEM structure determination (67, 68). Ongoing software development includes methods for dealing with structural heterogeneity in single particle datasets (69, 70), and the flexible fitting of atomic resolution structures

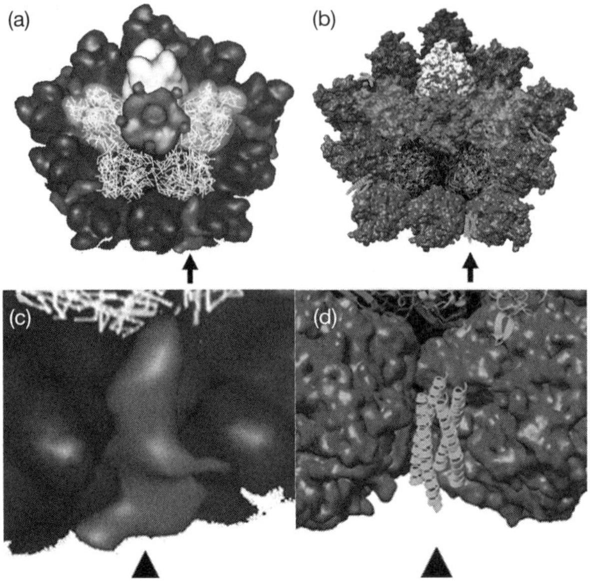

Figure 4.6 Comparison of pseudo-atomic models of adenovirus based on cryoEM structures at 25Å and at 6Å resolution. (A) The vertex region from the pseudo-atomic model at 25Å resolution (58). The penton base and fiber are at the center, surrounded by five hexons displayed in both a molecular surface and a stick representation with every tenth backbone Cα atom. Within the outer ring of ten hexons are five copies of an elongated density region (medium gray), one of which is marked by an arrow. (B) The vertex region from the pseudo-atomic model at 6Å resolution (26). The elongated density region is now resolved as a coiled coil with four α-helices, shown in a ribbon representation. One coiled coil is marked by an arrow. (C) Enlarged view of one elongated density region from panel A. (D) Enlarged view of one coiled coil from panel B. (Panel A) Reprinted with permission from *EMBO Journal*, volume 12, number 7, cover image, 1993 (58). (Panel B) Reprinted with permission from Journal of Virology. Copyright American Society for Microbiology, *Journal of Virology*, volume 80, number 20, cover image, 2006 (26).

into EM density maps (71, 72). Structural heterogeneity between particles in a cryoEM single particle dataset is probably a major limiting factor in reaching high resolution for many biological complexes. Collection of large datasets and sorting into more homogeneous subsets should help to improve the resolution of resulting reconstructions. Rigid body docking of atomic resolution structures into cryoEM density maps of large complexes has played an important role in the field (73). New tools for flexible fitting of macromolecules that undergo conformational changes between various stages of assembly will help in the analysis of cryoEM structures. One additional frontier in cryoEM that remains to be fully addressed is dealing with the complications that arise when the sample thickness exceeds the depth of field of the microscope (74). This problem is also referred to as correction for the curvature of the Ewald sphere. The problem is particularly important for large particles, on the order of ~1,000Å in diameter, and it leads to attenuation of high-resolution signal.

Although several procedures have been proposed for Ewald sphere correction (74, 75), none have been successfully applied to experimental datasets as of yet.

Achievement of atomic and near-atomic resolution by three cryoEM subspecialties of electron crystallography, helical image processing, and single-particle reconstruction, is a tremendous milestone for the field. The single-particle reconstruction approach offers a way to derive an atomic model for a biological macromolecular complex that does not form a diffraction quality crystal suitable for X-ray crystallography, and that is way beyond the current mass limitations in NMR structure determination. Virus particles with molecular masses well over 100 million Daltons and protein capsids as large as ~5,000Å in diameter such as the gigantic mimivirus (76) have been studied by cryoEM reconstruction. Clearly achieving atomic resolution by cryoEM methods provides the most information about a biological system. However, more moderate resolution cryoEM structures can also provide valuable information, particularly when combined with hybrid structural approaches (77). With the most recent successes in single-particle reconstruction the future for cryoEM structure determination looks especially bright (11).

ACKNOWLEDGEMENTS

The author sincerely thanks her collaborator Glen Nemerow, and members of her laboratory, especially Susan Saban, Mariena Silvestry, Steffen Lindert, and Dewight Williams, who contributed to the adenovirus cryoEM studies highlighted here. Support from the National Institutes of Health is acknowledged (R01-AI42929 to P.L.S).

REFERENCES

1. Grandin K ed (1963) *Les Prix Nobel* (The Nobel Foundation, Stockholm).
2. Ekspong G ed (1993) *Nobel Lectures in Physics 1981-1990* (World Scientific Publishing Co., Singapore).
3. Malmström BG ed (1993) *Nobel Lectures in Chemistry 1981-1990* (World Scientific Publishing Co., Singapore).
4. DeRosier DJ, Klug A (1968) Reconstruction of three dimensional structures from electron micrographs. *Nature 217*:130–134.
5. Ruska E (1986) The development of the electron microscope and of electron microscopy in *Nobel Lectures in Physics 1981-1990*, ed Ekspong G (World Scientific Publishing Co., Singapore).
6. Taylor KA, Glaeser RM (1974) Electron diffraction of frozen, hydrated protein crystals. *Science* 186:1036–1037.
7. Dubochet J, Booy FP, Freeman R, Jones AV, Walter CA (1981) Low temperature electron microscopy. *Annu Rev Biophys Bioeng* 10:133–149.
8. Robards AW, Sleytr UB (1991) *Low Temperature Methods in Biological Electron Microscopy* (Elsevier, Amsterdam).
9. Baker TS, Olson NH, Fuller SD (1999) Adding the third dimension to virus life cycles: three-dimensional reconstruction of icosahedral viruses from cryo-electron micrographs. *Microbiol Mol Biol Rev* 63:862–922.

10. Dubochet J, et al. (1988) Cryo-electron microscopy of vitrified specimens. *Q Rev Biophys* 21:129–228.
11. Taylor KA, Glaeser RM (2008) Retrospective on the early development of cryo-electron microscopy of macromolecules and a prospective on opportunities for the future. *J Struct Biol* 163:214–223.
12. Taylor KA, Glaeser RM (1976) Electron microscopy of frozen hydrated biological specimens. *J Ultrastruct Res* 55:448–456.
13. Adrian M, Dubochet J, Lepault J, McDowall AW (1984) Cryo-electron microscopy of viruses. *Nature* 308: 32–36.
14. van Raaij MJ, Mitraki A, Lavigne G, Cusack S (1999) A triple beta-spiral in the adenovirus fibre shaft reveals a new structural motif for a fibrous protein. *Nature* 401:935–938.
15. Unwin PN, Henderson R (1975) Molecular structure determination by electron microscopy of unstained crystalline specimens. *J Mol Biol* 94:425–440.
16. Henderson R, et al. (1990) Model for the structure of bacteriorhodopsin based on high-resolution electron cryo-microscopy. *J Mol Biol* 213:899–929.
17. Klug A (1993) From macromolecules to biological assemblies in *Nobel Lectures in Chemistry 1981-1990*, ed Malmström BG (World Scientific Publishing Co., Singapore).
18. Toyoshima C, Unwin N (1988) Contrast transfer for frozen-hydrated specimens: determination from pairs of defocused images. *Ultramicroscopy* 25:279–291.
19. Adrian M, Dubochet J, Fuller SD, Harris JR (1998) Cryo-negative staining. *Micron* 29:145–160.
20. Crowther RA (1971) Procedures for three-dimensional reconstruction of spherical viruses by Fourier synthesis from electron micrographs. *Philos Trans R Soc Lond B Biol Sci* 261:221–230.
21. Crowther RA, Amos LA, Finch JT, De Rosier DJ, Klug A (1970) Three dimensional reconstructions of spherical viruses by Fourier synthesis from electron micrographs. *Nature* 226:421–425.
22. Verschoor A, Frank J, Radermacher M, Wagenknecht T, Boublik M (1984) Three-dimensional reconstruction of the 30 S ribosomal subunit from randomly oriented particles. *J Mol Biol* 178:677–698.
23. Frank J, et al. (1996) SPIDER and WEB: processing and visualization of images in 3D electron microscopy and related fields. *J Struct Biol* 116:190–199.
24. van Heel M, Harauz G, Orlova EV, Schmidt R, Schatz M (1996) A new generation of the IMAGIC image processing system. *J Struct Biol* 116:17–24.
25. Glaeser RM (1985) Electron crystallography of biological macromolecules. *Ann Rev Phys Chem* 36:243–275.
26. Saban SD, Silvestry M, Nemerow GR, Stewart PL (2006) Visualization of alpha-helices in a 6-Å resolution cryoelectron microscopy structure of adenovirus allows refinement of capsid protein assignments. *J Virol* 80:12049–12059.
27. Chiu W, Schmid MF (1997) Pushing back the limits of electron cryomicroscopy. *Nat Struct Biol* 4:331–333.
28. Mancini EJ, de Haas F, Fuller SD (1997) High-resolution icosahedral reconstruction: fulfilling the promise of cryo-electron microscopy. *Structure* 5:741–750.
29. Rosenthal PB, Henderson R (2003) Optimal determination of particle orientation, absolute hand, and contrast loss in single-particle electron cryomicroscopy. *J Mol Biol* 333:721–745.

30. Ludtke SJ, et al. (2008) De novo backbone trace of GroEL from single particle electron cryomicroscopy. *Structure* 16:441–448.
31. Ludtke SJ, Baldwin PR, Chiu W (1999) EMAN: semiautomated software for high-resolution single-particle reconstructions. *J Struct Biol* 128:82–97.
32. Mindell JA, Grigorieff N (2003) Accurate determination of local defocus and specimen tilt in electron microscopy. *J Struct Biol* 142:334–347.
33. Grigorieff N (2007) FREALIGN: high-resolution refinement of single particle structures. *J Struct Biol* 157:117–125.
34. Mallick SP, Carragher B, Potter CS, Kriegman DJ (2005) ACE: automated CTF estimation. *Ultramicroscopy* 104:8–29.
35. Belnap DM, Olson NH, Baker TS (1997) A method for establishing the handedness of biological macromolecules. *J Struct Biol* 120:44–51.
36. Egelman EH (2007) The iterative helical real space reconstruction method: surmounting the problems posed by real polymers. *J Struct Biol* 157:83–94.
37. Smith R, Carragher B (2008) Software tools for molecular microscopy. *J Struct Biol* 163:224–228.
38. Nogales E, Wolf SG, Downing KH (1998) Structure of the alpha beta tubulin dimer by electron crystallography. *Nature* 391:199–203.
39. Kuhlbrandt W, Wang DN (1991) Three-dimensional structure of plant light-harvesting complex determined by electron crystallography. *Nature* 350:130–134.
40. Gonen T, et al. (2005) Lipid-protein interactions in double-layered two-dimensional AQP0 crystals. *Nature* 438:633–638.
41. Miyazawa A, Fujiyoshi Y, Unwin N (2003) Structure and gating mechanism of the acetylcholine receptor pore. *Nature* 423:949–955.
42. Yonekura K, Maki-Yonekura S, Namba K (2003) Complete atomic model of the bacterial flagellar filament by electron cryomicroscopy. *Nature* 424:643–650.
43. Unwin N (2005) Refined structure of the nicotinic acetylcholine receptor at 4Å resolution. *J Mol Biol* 346:967–989.
44. Bottcher B, Wynne SA, Crowther RA (1997) Determination of the fold of the core protein of hepatitis B virus by electron cryomicroscopy. *Nature* 386:88–91.
45. Conway JF, et al. (1997) Visualization of a 4-helix bundle in the hepatitis B virus capsid by cryo-electron microscopy. *Nature* 386:91–94.
46. Jiang W, et al. (2008) Backbone structure of the infectious epsilon15 virus capsid revealed by electron cryomicroscopy. *Nature* 451:1130–1134.
47. Yu X, Jin L, Zhou ZH (2008) 3.88 Å structure of cytoplasmic polyhedrosis virus by cryo-electron microscopy. *Nature* 453:415–419.
48. Zhang X, et al. (2008) Near-atomic resolution using electron cryomicroscopy and single-particle reconstruction. *Proc Natl Acad Sci USA* 105:1867–1872.
49. Zhou ZH (2008) Towards atomic resolution structural determination by single-particle cryo-electron microscopy. *Curr Opin Struct Biol* 18:218–228.
50. Steitz TA (2008) A structural understanding of the dynamic ribosome machine. *Nat Rev Mol Cell Biol* 9:242–253.
51. Rossmann MG, Morais MC, Leiman PG, Zhang W (2005) Combining X-ray crystallography and electron microscopy. *Structure* 13:355–362.
52. Pattanayek R, et al. (2006) Analysis of KaiA-KaiC protein interactions in the cyano-bacterial circadian clock using hybrid structural methods. *EMBO J* 25:2017–2028.
53. Koteiche HA, Chiu S, Majdoch RL, Stewart PL, Mchaourab HS (2005) Atomic models by cryo-EM and site-directed spin labeling: application to the N-terminal region of Hsp16.5. *Structure* 13:1165–1171.

54. Shi J, Koteiche HA, Mchaourab HS, Stewart PL (2006) Cryoelectron microscopy and EPR analysis of engineered symmetric and polydisperse Hsp16.5 assemblies reveals determinants of polydispersity and substrate binding. *J Biol Chem* 281:40420–40428.
55. Rosenberg OS, Deindl S, Sung RJ, Nairn AC, Kuriyan J (2005) Structure of the autoinhibited kinase domain of CaMKII and SAXS analysis of the holoenzyme. *Cell* 123:849–860.
56. Wang GJ, Porta C, Chen ZG, Baker TS, Johnson JE (1992) Identification of a Fab interaction footprint site on an icosahedral virus by cryoelectron microscopy and X-ray crystallography. *Nature* 355:275–278.
57. Rayment I, et al. (1993) Structure of the actin-myosin complex and its implications for muscle contraction. *Science* 261:58–65.
58. Stewart PL, Fuller SD, Burnett RM (1993) Difference imaging of adenovirus: bridging the resolution gap between X-ray crystallography and electron microscopy. *EMBO J* 12:2589–2599.
59. Haider M, et al. (1998) Electron microscopy image enhanced. *Nature* 392:768–769.
60. Smith DJ (2008) Development of aberration-corrected electron microscopy. *Microsc Microanal* 14:2–15.
61. Gamm B, Schultheiss K, Gerthsen D, Schroder RR (2008) Effect of a physical phase plate on contrast transfer in an aberration-corrected transmission electron microscope. *Ultramicroscopy* 108:878–884.
62. Hosokawa F, Danev R, Arai Y, Nagayama K (2005) Transfer doublet and an elaborated phase plate holder for 120 kV electron-phase microscope. *J Electron Microsc (Tokyo)* 54:317–324.
63. Majorovits E, et al. (2007) Optimizing phase contrast in transmission electron microscopy with an electrostatic (Boersch) phase plate. *Ultramicroscopy* 107:213–226.
64. Marsh MP, et al. (2007) Modular software platform for low-dose electron microscopy and tomography. *J Microsc* 228:384–389.
65. Shi J, Williams DR, Stewart PL (2008) A script-assisted microscopy (SAM) package to improve data acquisition rates on FEI Tecnai electron microscopes equipped with Gatan CCD cameras. *J Struct Biol* 164:166–169.
66. Suloway C, et al. (2005) Automated molecular microscopy: the new Leginon system. *J Struct Biol* 151:41–60.
67. Stagg SM, et al. (2006) Automated cryoEM data acquisition and analysis of 284742 particles of GroEL. *J Struct Biol* 155:470–481.
68. Williams DR, Lee KJ, Shi J, Chen DJ, Stewart PL (2008) Cryo-EM structure of the DNA-dependent protein kinase catalytic subunit at subnanometer resolution reveals alpha helices and insight into DNA binding. *Structure* 16:468–477.
69. Penczek PA, Frank J, Spahn CM (2006) A method of focused classification, based on the bootstrap 3D variance analysis, and its application to EF-G-dependent translocation. *J Struct Biol* 154:184–194.
70. Scheres SH, et al. (2007) Modeling experimental image formation for likelihood-based classification of electron microscopy data. *Structure* 15:1167–1177.
71. Trabuco LG, Villa E, Mitra K, Frank J, Schulten K (2008) Flexible fitting of atomic structures into electron microscopy maps using molecular dynamics. *Structure* 16:673–683.
72. Velazquez-Muriel JA, Valle M, Santamaria-Pang A, Kakadiaris IA, Carazo JM (2006) Flexible fitting in 3D-EM guided by the structural variability of protein superfamilies. *Structure* 14:1115–1126.

73. Wriggers W, Chacon P (2001) Modeling tricks and fitting techniques for multiresolution structures. *Structure* 9:779–788.
74. DeRosier DJ (2000) Correction of high-resolution data for curvature of the Ewald sphere. *Ultramicroscopy* 81:83–98.
75. Wolf M, DeRosier DJ, Grigorieff N (2006) Ewald sphere correction for single-particle electron microscopy. *Ultramicroscopy* 106:376–382.
76. Xiao C, et al. (2005) Cryo-electron microscopy of the giant Mimivirus. *J Mol Biol* 353:493–496.
77. Steven AC, Baumeister W (2008) The future is hybrid. *J Struct Biol* 163:186–195.

5

Single-Molecule Imaging and Force Spectroscopy by Atomic Force Microscopy

K. TANUJ SAPRA AND DANIEL J. MULLER

INTRODUCTION

In the early 1980s, Binnig and Rohrer revolutionized microscopy by inventing the scanning tunneling microscope (STM) (Binnig et al., 1982). The STM arose from an interest in the study of electrical properties of thin insulating layers, and it showed for the first time the atomic structure of silicon at the crystalline surface in real space. The dawning of scanning probe microscopy (SPM), thus, began a whole new exciting era. For the curious, opportunities abound. Soon after this, it was discovered that the STM can be used to image and manipulate single atoms. The importance of this discovery was recognized through the award of the Nobel Prize in Physics in 1986. In the same year, Binnig, together with Quate and Gerber, showed that short-range van der Waals interactions between a scanning probe and an object could also be used to contour non-conducting and conducting surfaces at atomic or close to atomic resolution (Binnig et al., 1986). Thus was the birth of the scanning force microscope (SFM) or the atomic force microscope (AFM), the second member of the SPM family (Gerber and Lang, 2006).

The evolution of the AFM has seen its growth from a specific tool in the semiconductor industry to a general nanotechnological tool for imaging biological specimens and, at the same time, to specifically probe biological, chemical and physical information (Gerber and Lang, 2006; Muller and Dufrene, 2008). The AFM provides novel ways to characterize structure-function relationships of live cells and cellular compartments down to the resolution of single molecules (Drake et al., 1989; Engel and Müller, 2000; Muller et al., 2006). The current AFM methodologies readily encompass its use to sense and locate inter- and intramolecular interactions of tissues, living cells and single molecules.

An important advantage of the AFM for biological studies is that it can be operated in a buffered solution at ambient temperatures. Also, the AFM does not require labeling of the biological specimen. The initial notion that the AFM would only complement conventional structural biology methods, such as electron microscopy, X-ray crystallography and NMR, has been proven naïve. Membrane proteins make excellent examples demonstrating that the AFM provides structural and functional insights (e.g., into the different oligomeric and conformational states of membrane proteins in their native environment) that are difficult to obtain with conventional approaches. Although the AFM images only surfaces, a lateral resolution of ~0.5 nm and a vertical resolution of ~0.1 nm with an outstanding signal-to-noise ratio enable the observation of single membrane proteins in great detail. The conformational flexibility of individual membrane proteins can be observed, and their functional states directly imaged. Moreover, time-lapse imaging of bacterial surface-layer proteins (Müller et al., 1996), nuclear pore complexes (Stöffler et al., 1999), gap junction hemichannels (Müller et al., 2002a; Yu et al., 2007), and cytolysin pores (Czajkowsky et al., 2004) with the AFM was an important technological step to reveal the functional conformations of these proteins.

In conjunction with high-resolution imaging, the AFM as a single-molecule force spectroscope has gained tremendous importance in recent years (Engel and Gaub, 2008; Kedrov et al., 2007). The associated technique, single-molecule force spectroscopy (SMFS), allows the characterization of stabilizing interactions in functional proteins, as well as destabilizing interactions that lead to misfolding and malfunction (Janovjak et al., 2006). Because of its practical use to characterize various parameters of cells and proteins in their native environment, the AFM has been aptly described as a "lab on a tip" device (Muller and Dufrene, 2008). In the forthcoming sections, we will first describe the working principle of the AFM. This will be followed by sections on the application of the AFM in the fields of single-molecule high-resolution imaging of membrane proteins, and protein folding and unfolding. Finally, we conclude with our opinion on the future of the AFM in biology.

ATOMIC FORCE MICROSCOPE

Commercial AFMs began to appear in the early 1990s and have evolved through several generations. A milestone for biological applications was the invention of a liquid cell that allows specimens to be investigated in buffered solutions (Drake et al., 1989). Further inventions introduced an AFM combined with light microscopy, intermittent contact AFM in air (Zhong et al., 1993), and later in liquid (Hansma et al., 1994; Putman et al., 1994), cryo-AFM (Han et al., 1995; Shao and Zhang, 1996), AFM to detect ionic currents—scanning ion conductance microscopy (SICM) (Hansma et al., 1989; Marti et al., 1988; Proksch et al., 1996), scanning electrochemical microscopy (Kueng et al., 2005), molecular recognition microscopy (Hinterdorfer and Dufrene, 2006), AFM-surface plasmon resonance (AFM-SPR) (Chen et al., 1996), "submarine AFM" or the combined AFM-Langmuir trough (Eng et al., 1996), with patch-clamp and scanning near-field optical microscope (SNOM). Many types of SPMs have been developed and can be used to measure surface topologies and also to measure various material properties at or close to surfaces. This can be done over a broad range of pHs and temperatures with a resolution down to the nanometer (nm) or even the atomic scale.

Principle and Setup

The AFM is designed on a very simple principle, and is probably one of the easiest forms of microscopy to understand. The main parts of the AFM setup (illustrated in Figure 5.1) are easily described. A micromachined stylus or the tip is mounted at the end of a soft cantilever spring. The piezoelectric transducer/actuator has a sample stage on the top. A so-called laser-head houses the optical deflection system and consists of a laser diode and a quadrant photodetector. All of this is connected to a computer. The fluid cell with the cantilever is mounted inside (or attached to) the head of the AFM. The piezoelectric transducer is voltage driven and can be moved with sub-nanometer accuracy in the *x, y, z* directions, thus ensuring three-dimensional positioning of the sample at a high resolution. In the most common scenario, AFM images are created by scanning a sharp stylus over the sample surface in its *x-y* plane. The interaction force between the stylus and the sample is used to map the topography of the surface. The force between the stylus and the sample is monitored by measuring the deflection (vertical bending) of the cantilever, which is usually detected by a laser beam focused on the free end of the cantilever and reflected onto the quadrant photodiode (Figure 5.1). Because the cantilever is a Hookean spring, the deflection is linearly proportional to the force.

During AFM imaging of surfaces, the piezoelectric transducer or the cantilever is raster scanned or oscillated along the surface (Figure 5.2, please see later sections), and for SMFS measurements, the transducer is moved only in the *z* (up-down) direction (Figure 5.9), thus enabling the unfolding of a molecule that is picked.

Cantilever

Inarguably, the most essential component of an AFM is the cantilever with a stylus mounted at its end that does the "sensing." Modern AFM styli and cantilevers are made from silicon, silicon-nitride or diamond by microfabrication. The fabrication of cantilevers employs techniques developed for integrated circuit manufacture, such as lithographic photo-masking, etching and vapor deposition. By the micro-machining approach, a sharp stylus can be integrated onto cantilevers, a prerequisite for high-resolution imaging of biological samples. The structure of the stylus determines the type of interaction with the surface, and its geometry the area of interaction.

A silicon-nitride stylus, used for most routine experiments, is covered by a non-conducting layer of native oxide and is often conical or pyramidal in geometry with a high aspect ratio. The pyramidal stylus has a macroscopic cone angle of 50°, which reduces at the apex. To increase the aspect ratio, apart from the sophisticated etching techniques, selective deposition of contaminants on the stylus has been done. Carbon nanotubes have been used as well (Dai et al., 1996). The stylus may range in size from 2 μm to tens of microns in height, with a radius of 2–60 nm, and can be conducting or non-conducting.

Cantilevers are usually triangular or "V"-shaped, and rectangular or beam shaped with a length of ≈ 20–450 μm and a thickness of ≈ 0.1–5 μm.[1] Cantilevers

1. Information on cantilevers and styli can be obtained at http://www.brukerafmprobes.com, http://www.asylumreasearch.com, http://probe.olympus-global.com.

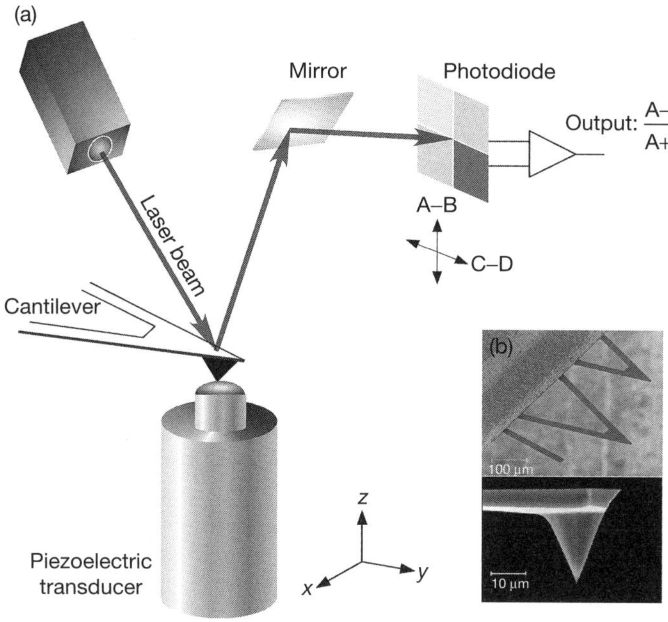

Figure 5.1 AFM setup. (A) The scheme shows the different parts and working of an AFM. Except the piezoelectric transducer, all other components are housed in a so-called "head" of the microscope (not drawn). The cantilever (B) is fixed in a fluid cell (not shown) and placed in the head, which is then mounted on the piezoelectric actuator. A laser beam is adjusted on the edge of the cantilever to obtain a maximum sum signal on the photodiode. During sample scanning with the AFM stylus, the beam is deflected with the up-down movement of the cantilever and the voltage signal is converted to force (pN) or deflection (nm) with the help of software. The output of the deflection signal varies with the AFM mode. (B) V-shaped cantilevers (top) with pyramidal styli (bottom) are most commonly used for high-resolution AFM imaging and single-molecule unfolding measurements (cantilever images from www.veecoprobes.com).

are often coated with a thin layer of gold or aluminum to provide high reflectivity for the laser beam deflection. The V-shaped geometry minimizes the torsional motion or twisting of the cantilever while scanning a sample, making it the lever of choice for topographical imaging. A simple beam or rectangular geometry cantilever can be used for frictional measurements due to a greater degree of rotational freedom making it sensitive to lateral forces. Irrespective of the geometry, the force contribution, F, on the sample from the bending of the cantilever, x, is determined by Hooke's law,

$$F = -\kappa x \tag{1}$$

where κ is the spring constant of the cantilever, and x the displacement experienced. The typical values of stiffness for AFM cantilevers used in contact mode are 6–100 pN/nm, in non-contact mode between 0.5–5 pN/nm, and in tapping mode 30–60 pN/nm. Low values of cantilever stiffness represent high sensitivity to force for each nanometer deflection of the piezoelectric actuator,

$$\partial x^2 \sim \frac{k_B T}{\kappa} \tag{2}$$

where k_B is the Boltzmann constant and T is the temperature. On the other hand, high cantilever stiffness results in large thermal fluctuations of the applied force (Evans, 2001),

$$\partial F^2 \sim k_B T \cdot \kappa \tag{3}$$

Low stiffness cantilevers have the minor drawback that they are susceptible to thermal fluctuations, and have a slow response time in an over-damped viscous-water environment. Soft cantilevers are best suited for contact mode AFM imaging (see AFM imaging modes) because they deflect while minimally deforming the soft sample surface. Stiff cantilevers, on the other hand, are suitable for non-contact mode imaging because of their high-resonance frequencies. In addition to the stiffness of a cantilever, the resonance frequency, f_r, the quality factor, *Q*, and the measurement bandwidth, *B*, also contribute to the choice of a cantilever for a specific application (Viani et al., 1999a). The force resolution of the AFM is, in first approximation, limited by the thermal noise of the cantilever, which can be estimated by its spring constant and dimensions (such as shape, length and thickness). However, the minimal force, F_{min}, that a cantilever can measure in liquid can be expressed as a function of κ, f_r, *Q*, and the measured bandwidth *B*,

$$F_{min} = \sqrt{\frac{4 \cdot k_B \cdot T \cdot B}{2\pi} \cdot \frac{\kappa}{f_r \cdot Q}} \tag{4}$$

Calibrating the Cantilever

Because it is such an important parameter of the cantilever, the spring constant needs to be accurately determined for quantitative studies. As stated above, the force resolution, the thermal noise of the cantilever and its spring constant are all correlated. Moreover, to translate the deflection of the cantilever, *x*, to the units of force, *F*, in SMFS ($F = -\kappa x$), it is necessary to determine the spring constant, κ, of the cantilever.

κ for normal bending of a rectangular cantilever, as calculated from its geometry, is given by (Meyer et al., 2003),

$$\kappa = \frac{Ewt^3}{4l^3} \tag{5}$$

where *w* is the width, *l* the length, *t* the thickness of the cantilever, and *E* the Young's modulus of the material.

To account for the effect of coating the cantilever, chemical modification of the stylus, and/or manufacturer variability from cantilever to cantilever, it is necessary to determine κ with empirical methods. Although a number of reliable spring constant calibration methods have been devised (Cleveland et al., 1993; Gibson et al., 1996; Sader et al., 1995; Torii et al., 1996), the analysis of thermal fluctuations of the vibrating lever,based on the equipartition theorem, is routinely used to determine the cantilever stiffness (Florin et al., 1995). The cantilever stylus is treated as a simple harmonic oscillator whose power spectrum of the thermal fluctuations is used to derive the spring constant. To avoid overestimation of the displacement of the lever and hence an underestimation of the measured stiffness, this approach requires that no additional noise is added to the thermal noise. Therefore, the cantilever is raised several microns from the surface, its resonance frequency monitored for 5–10 s, and the measured variance of deflection, $\langle x^2 \rangle$, used to calculate the spring constant,

$$\frac{1}{2}k_B T = \frac{1}{2}\kappa\langle x^2 \rangle \tag{6}$$

By this approach, the spring constant of a cantilever can be calibrated in either air or in solution.

AFM Imaging Modes

In the following sections, we will discuss the most commonly used AFM imaging modes to image biological specimens. These are the constant force mode and the oscillation mode. The non-contact mode will be introduced as well.

Constant Force Mode

In the constant force mode, frequently referred to as the contact mode, topographic images are recorded as the stylus makes physical contact with the sample and scans the surface to monitor changes in the cantilever deflection (Figure 5.2). A feedback loop causes the piezoelement to move whenever the predefined deflection of the cantilever changes during scanning. Within certain limits, the feedback loop holds the cantilever deflection, and therefore, the force applied between the stylus and the object remains constant. In principle, this method is sufficiently sensitive to detect height differences at the sub-atomic scale. However, biological samples show thermal fluctuations and mechanical properties (elasticity and friction) that in most cases blur the lateral resolution to $\approx$ 1 nm. In addition, sample topographies showing a roughness on the scale of a few nm and higher can be imaged only with artifacts, since then the contouring stylus nonlinearly convolutes with the sample.

The interaction range of the different types of forces is of great importance for force microscopy because different parts of the stylus and cantilever contribute differently to the total measured force. Long-range van der Waals and electrostatic forces are always to be taken into account in force microscopy even though they may not be of any interest for the actual measurement. Thus, the typical situation in contact mode is determined by an equilibrium between the attractive long-range forces and the repulsive short-range forces, where only the latter provide good resolution (Figure 5.2) (Meyer et al., 2003).

In the constant force mode and in other imaging modes the force between the AFM stylus and the object is a crucial component of the information used by the

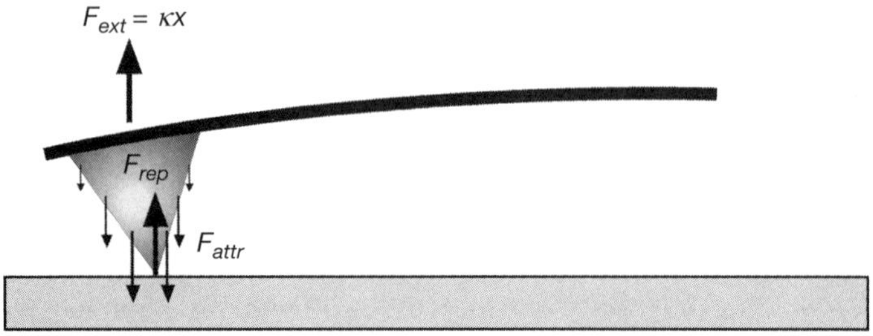

Figure 5.2 Equilibrium of forces in constant force mode AFM. The position of the AFM stylus is given by an equilibrium of forces: the attractive force between the stylus and the sample has to be balanced by the repulsive force between the stylus apex and the sample, and the external force exerted by the cantilever spring has to be added. The attractive long-range force, F_{attr}, between the cantilever stylus and the sample is balanced by the short-range repulsive force, F_{rep}, at the contact and the external force, F_{ext}, exerted by the cantilever bending.

feedback loop. Thus, the topography in principle reflects a constant force map of the sample surface. Assuming a sample surface that interacts homogeneously with the stylus, the AFM topograph depicts the real structural topography of the sample. However, in most cases the surfaces of biological samples have heterogeneous properties; different regions may show different flexibilities, charges, hydrophilic or—phobic properties. Thus, when scanning such surfaces the force detected by the deflecting cantilever may be composed of different interactions acting simultaneously between the AFM stylus and the sample. For example, the apparent height difference measured by an AFM stylus on a flat surface that is negatively charged in one part but positively charged in another would be based on different electrostatic interactions and have no structural origin. The AFM user must be aware of such artifacts and how to avoid them, for example, by screening the electrostatic interactions with ions or pH (Müller and Engel, 1997; Müller et al., 1999). However, the knowledge of such effects brings along the possibility to structurally map the electrostatic interactions of the protein. Besides the electrostatic interactions, the interaction forces between the AFM stylus and the biological sample can be of manifold nature (Israelachvili, 1991).

A biological sample is much softer than an AFM cantilever. Applying a high force while contouring the biological object can easily lead to reversible or irreversible sample deformation. Imaging living cells requires an applied force ≤ 100 pN. Otherwise, the membrane will be pressed onto the cytoskeleton, and the resulting topography will reflect the membrane tightly wrapping the cell interior. Similarly, when imaging single native membrane proteins at high resolution, individual polypeptide loops can be easily deformed at forces > 100 pN. Structural comparison of AFM topographs with atomic structures obtained by X-ray and electron crystallography demonstrates that the imaging process minimally disturbs the protein structures if the contact forces are kept < 100 pN. Therefore, the forces applied for high-resolution imaging of proteins should be as low as 50–100 pN to avoid sample deformation (Müller et al., 1995; Müller and Engel, 2002).

Oscillation Mode AFM

In the oscillation mode, the AFM cantilever is oscillated while scanning the surface of an object (Figure 5.2). The oscillating stylus touches the sample only at its very end of the downward movement. Forces interacting between the stylus and the sample change the oscillation behavior of the cantilever. A small sinusoidal electrical signal applied to the main piezoelectric transducer causes the transducer to vibrate in the vertical z direction, which in turn leads to the vibration of the cantilever owing to viscous coupling. Alternatively, a small piezoelectric transducer mounted at the cantilever end, or an oscillating magnetic field where the cantilever must be coated with a magnetic material, can be used to drive the cantilever oscillation too. As in contact force AFM, a feedback loop moves the piezo actuator to keep the changes of the oscillation amplitude constant. The feedback loop maintains constant oscillation amplitude by maintaining a constant RMS (root mean square) of the oscillation signal and, thus, a constant stylus-sample interaction during imaging. For optimal sensitivity, the cantilever is oscillated close to its resonance frequency (10–400 kHz) with an amplitude range of 1–100 nm. Typically, small amplitudes are used to contour samples with almost no corrugation, and high amplitudes are used to image rough samples.

Contact mode AFM is not suitable for imaging weakly immobilized or extremely soft structures, such as single macromolecules, because these are often swept away by the AFM stylus while raster scanning the surface (Karrasch et al., 1993). Oscillation-mode AFM overcomes this disadvantage. Because the stylus touches the sample only at the end of its downward movement, the lateral forces are minimized to values of ≈ 1 pN. This makes the oscillation mode ideal for studying soft or elastic samples. A variety of macromolecules which could not be imaged before have been observed using oscillation mode AFM (Bezanilla et al., 1994; Dunlap et al., 1997; Fritz et al., 1995; Guthold et al., 1999; Möller et al., 1999).

Non-Contact AC Mode

In contrast to the oscillation mode, the non-contact AC mode oscillates the cantilever at a frequency slightly above the cantilever's resonance frequency with an amplitude of < 10 nm to obtain an AC signal from the cantilever. As the name implies, in this mode, the oscillating stylus never actually touches the surface of the sample but hovers a few nm above it. A clear advantage of this technique is that the force exerted between the stylus and the sample remains at its very minimum. The initial use of this mode was limited to samples under vacuum, but recent developments have made it possible to image surfaces in aqueous environment (Frederix et al., 2003; Fukuma et al., 2005).

Prerequisites to Sample Preparation for High-Resolution Single-Molecule Measurements

Irrespective of the imaging mode used, high-resolution AFM imaging offers some practical advantages over the conventional optical and electron microscopy techniques. A major advantage is the exceptionally high signal-to-noise ratio of AFM topographs. However, this feature is not without some disadvantages. A buffer solution contaminated with macromolecules would introduce image artifacts. This

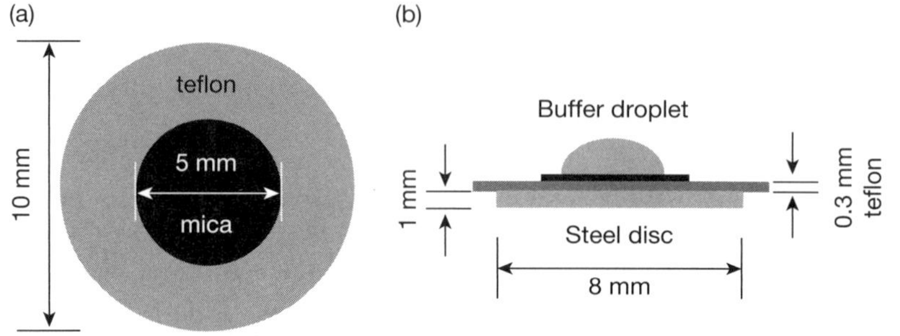

Figure 5.3 Sample support for high-resolution AFM imaging and SMFS of biological macromolecules. (A) Top-view and (B) side-view showing the dimensions and arrangement of mica and teflon discs on a magnetic steel disc. Teflon being highly hydrophobic prevents the buffer droplet from spilling onto the piezoelectric transducer.

necessitates meticulous preparation of the biological samples for high-resolution imaging and SMFS by carefully cleaning all parts in contact with the buffer solution, and filtering the buffer solution through a 0.22 μm pore-size filter. Another prerequisite for obtaining reliable information from imaging and SMFS experiments is the use of a specially designed support to immobilize the sample (Figure 5.3).

Importantly, the adsorption of a biological sample to the support should not impair its function. For example, hydrophobic surfaces denature hydrophilic proteins, whereas hydrophilic surfaces show much less functional interference. To unambiguously resolve the structural features of the biological sample, the roughness of the supporting surface should be much smaller than the size of the biological object (e.g., proteins, DNA, bilayers). For single-molecule AFM imaging and SMFS, a good support surface should be hydrophilic and ideally atomically flat. Commonly used materials that fulfill these criteria are mica, graphite, gold and ultraflat glass.

For high-resolution AFM imaging and SMFS measurements of membrane proteins, mica is routinely used as the supporting surface (Müller et al., 1997). Just before the experiments, the mica surface is cleaved with Scotch® tape to obtain an atomically flat surface. To check the integrity of the surface and the cantilever stylus, the freshly cleaved surface is imaged. This helps control the flatness of the surface, to ensure if the cantilever has been mounted properly and the stylus has the desirable properties. Only those supports that show a height variation of ≤ 1 nm are useful as supporting surfaces to image single molecules at a resolution better than 2 nm. An excellent source that discusses the methodologies developed in high-resolution AFM imaging and SMFS of membrane proteins is a recent review by Müller and Engel (Muller and Engel, 2007).

High-Resolution AFM Imaging

AFM Imaging of Cells

Owing to its superb signal-to-noise ratio, AFM imaging requires no staining or labeling of the biological sample. Besides simplifying the sample preparation significantly, bypassing tedious labeling steps enable live cell imaging in its native state without altering the structural and functional dynamics of the proteins to be studied. However, it also brings some disadvantages. The AFM provides topographs in which all features of a cell surface are imaged at the same time. Thus, AFM topographs of a complex biological system are not as easy to interpret as, for example, fluorescence images, which basically image only the fluorescently labeled structures. Consequently, structures recorded in an AFM topograph need to be carefully identified. In straightforward cases, the AFM can be combined preferably with fluorescence microscopy, which helps in the identification of specifically labeled structures of interest. In such a combination, light microscopy enables imaging of the biological object down to a resolution of ≤ 200 nm, and AFM topographs reveal structural features at a resolution of a few tens of nm.

Below, we discuss an example of combined AFM and fluorescence microscopy, imaging of the cell surface of zebrafish stratified epithelium (Sharma et al., 2005). Figure 5.4 shows the cell surface of living epithelial cells imaged under conditions that simulate wound healing. The AFM topograph clearly reveals ridge-like structures rising on average 100 nm above the surface. Time-lapse microscopy revealed these ridges to form a highly dynamic network (not shown). Laser scanning confocal fluorescence microscopy of phalloidin stained cells unambiguously correlated the ridges to bundles of actin filaments. It would not have been possible to characterize the ridges solely from the AFM topographs.

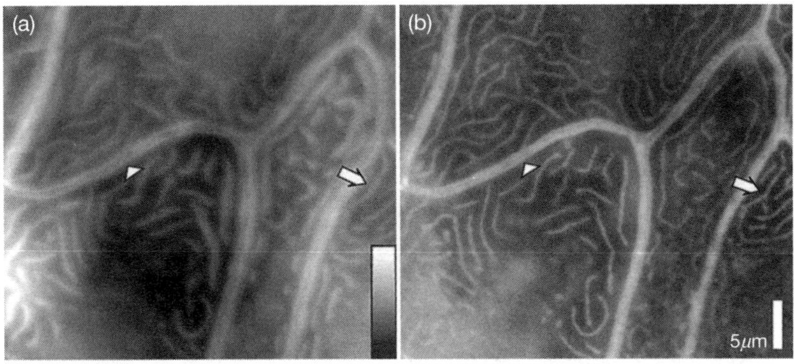

Figure 5.4 AFM topograph and fluorescence image of surface ridges of zebrafish stratified epithelium. (A) The AFM topograph reveals ridges occurring at the cell surface. (B) Fluorescent image of the same region stained with phalloidin demonstrates that the ridges on the cell surface are composed of F-actin. Cell–cell junctions are pointed by arrowheads. In the AFM topograph, cell junctions appear as two ridges each belonging to a different cell. Arrows point to the same ridge in both images. Individual ridges appear narrower in the fluorescence image and broader in the AFM topograph. This is because the cell membrane wraps the ridges. This wrapping is not observed by light microscopy but by AFM. Nevertheless, ridges are better resolved at cell junctions in the constant force mode AFM image (Sharma et al., 2005). The grey level bar in (A) reflects the vertical height scale of the AFM topograph. It ranges from 0 nm (black) to 500 nm (white).

HIGH-RESOLUTION IMAGING OF SINGLE NATIVE PROTEINS AND PROTEIN ASSEMBLIES

Although the AFM images only surfaces, a lateral resolution of ≈ 0.5 nm and a vertical resolution of ≈ 0.1 nm combined with an outstanding signal-to-noise ratio enables the observation of single water-soluble and membrane proteins in great details. However, to ensure that the vertical information reflects the true topographic information, controls may have to be performed as described in detail elsewhere (Heymann et al., 2002; Müller et al., 1995; Müller et al., 1999). Whereas oscillation mode imaging is in most cases used to reveal high-resolution topographs of water-soluble proteins (Cisneros et al., 2006; Elie-Caille et al., 2007; Thomson, 2005; Vinckier et al., 1998; Wagner et al., 2004), contact mode imaging is used to reveal topographs of membrane proteins (Fotiadis et al., 2003; Müller et al., 2002b; Yu et al., 2007). In general, contact mode topographs of membrane proteins approach a much higher resolution than oscillation mode images of water-soluble proteins. This difference in resolution might be due to the fact that membrane proteins are embedded in a lipid bilayer that orients and holds them at specific positions, but water-soluble proteins lack such a stable constrained environment. However, the resolution achieved with oscillation mode AFM on membrane proteins does not match that of contact mode AFM. Figures 5.5 and 5.6 show examples of high-resolution topographs of a water-soluble protein and a membrane protein, respectively, in buffer solutions. The topographs allow easy resolution of the characteristic substructures of the proteins. Figures 5.5 and 5.6 show only a selection of a plethora of proteins and nucleic acids that have been imaged by AFM. For an overview we recommend the following reviews (Czajkowsky et al., 2000; Hansma and Hoh, 1994; Muller and Dufrene, 2008; Müller et al., 2002c; Muller et al., 2006; Scheuring et al., 2005).

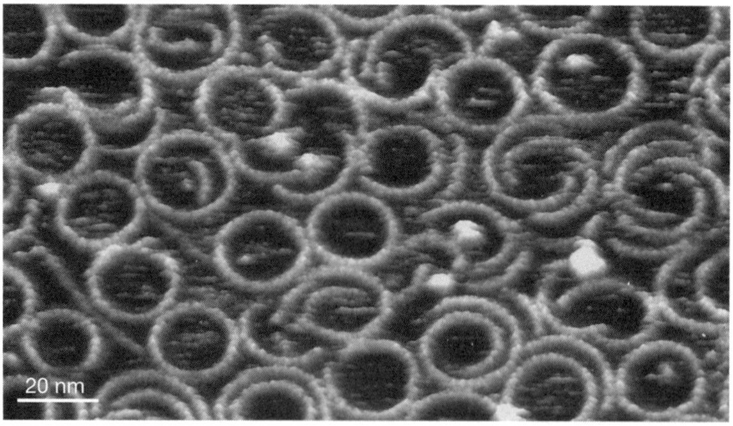

Figure 5.5 High-resolution AFM topographs of tubulin protofilaments imaged in a buffer solution. Shortly after preparation (< 0.5 h), microtubules depolymerize into protofilaments, linear chains of alpha- and beta-tubulin dimers (Elie-Caille et al., 2007).

Figure 5.6 High-resolution AFM topographs of sodium-driven rotors from the F_oF_1-ATP synthase of *Ilyobacter tartaricus*. The topograph clearly reveals the densely packed assembly of membrane embedded rotors. The exceptional signal-to-noise ratio allows structural analysis of individual rotors. It appears that the intact rotors are composed of eleven subunits. The constant force mode AFM topograph was recorded in a buffer solution (Stahlberg et al., 2001).

HIGH-RESOLUTION TIME-LAPSE IMAGING OF PROTEINS AT WORK

AFM-based high-resolution single-molecule imaging can be used to obtain information on membrane protein dynamics, such as conformational changes, assembly and complex formation. Most importantly, the dynamics of membrane proteins can be characterized in the context of the functionally important lipid membranes. Combined with high-resolution, time-lapse imaging allows to take sequential snap-shots of a biological system. Early examples applied time-lapse AFM to image nucleic acids, water-soluble and membrane proteins at work (Engel et al., 1999; Engel and Müller, 2000). At a high spatial resolution of ≈ 1 nm, time-lapse AFM could directly monitor how the individual subunits of gap junction hemichannels move to facilitate the opening and closing of their communication channels (Figure 5.7). Recently, high-resolution time-lapse AFM could even observe that different ligands binding to human gap junctions may result in the same functional change, but apparently use different mechanisms and conformational changes to switch the functional states of the channels (Müller et al., 2002a; Yu et al., 2007). Besides the gap junction hemichannels (Müller et al., 2002a; Yu et al., 2007), other examples of time-lapse AFM include characterization of bacterial surface-layer proteins (Müller et al., 1996), nuclear pore complexes (Stöffler et al., 1999), outer membrane porins (Müller and Engel, 1999), cytolysin pores (Czajkowsky et al., 2004), and potassium channel KirBac3.1 (Jaroslawski et al., 2007).

HIGH-RESOLUTION IMAGING OF PROTEIN ASSEMBLIES

High-resolution AFM can be conveniently applied to image native membrane patches extracted from cells to look at how different membrane proteins form complexes and assemble into functional units (Fotiadis et al., 2003; Scheuring et al., 2004). In pioneering work, Scheuring and co-workers imaged the assembly of membrane proteins in native-like preparations of the photosynthetic membrane patches (Scheuring and Sturgis, 2005). The assembly of supramolecular light-harvesting

Figure 5.7 Imaging communication channels at work. High-resolution AFM topograph (raw data) of the extracellular surface of Cx26 gap junction hemichannels. Gap junction membranes extracted from HeLa cells were dissected with the AFM probe to expose their extracellular surface (Yu et al., 2007). The right row shows correlation averages of hemichannels. The averages reveal the conformational changes of the hemichannels in response to Ca^{2+}. The closed channels (top) switch, via an intermediate conformation (middle), to the open state (bottom) in the presence of 0.5 mM Ca^{2+} (Müller et al., 2002b).

complexes I and II from the photosynthetic bacterium *Rhodospirillum photometricum* and its dependence to high light and low light conditions could be studied. Until recently, most native membrane preparations characterized by high-resolution AFM were extracted from bacteria (Muller et al., 1995; Scheuring et al., 1999; Stahlberg et al., 2001). However, it has now become possible to image native membrane patches extracted from mitochondrial outer membranes. Two examples of AFM imaging of the outer membranes extracted from yeast (Goncalves et al., 2007) and potato (Hoogenboom et al., 2007) revealed regions of membrane proteins that represented the voltage-dependent anion channels (VDAC). These VDAC channels were observed to occur as monomers that could assemble into dimers, trimers, tetramers and hexamers.

TIME-LAPSE HIGH-RESOLUTION IMAGING OF PROTEIN ASSEMBLY

Beyond the evident advantages of high spatial resolution, high temporal resolution can also be achieved during single-molecule AFM imaging. In the realm of soluble proteins, time-lapse AFM was employed to directly observe the self-assembly kinetics of collagen type I molecules into collagen fibrils (Cisneros et al., 2006). Such studies may find promising biotechnological applications (Cisneros et al., 2007). Applied to membrane proteins, diffusion of individual membrane proteins in the plane of the lipid bilayer could be followed (Müller et al., 2003). The analysis of single trajectories of diffusing sodium-driven ATP synthase rotors from *Ilyobacter tartaricus* allowed to determine their diffusion behavior in a supported lipid bilayer (Figure 5.8). We

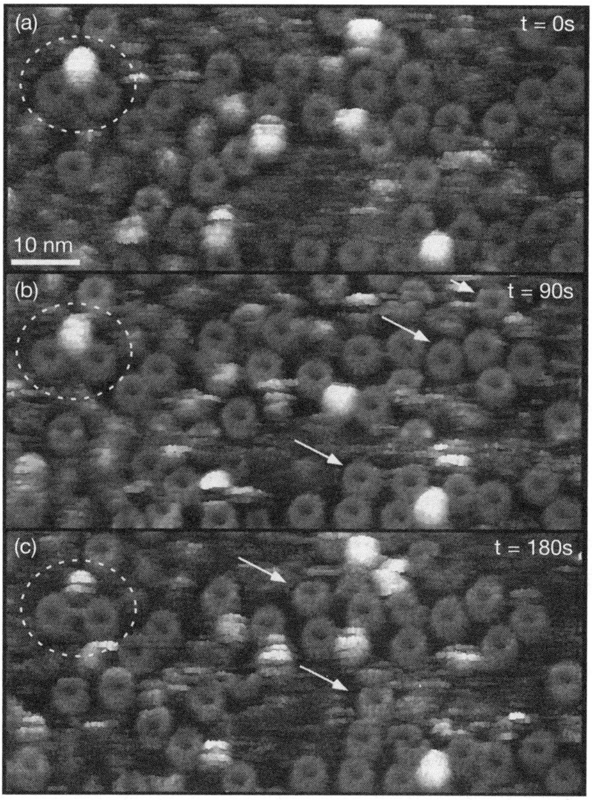

Figure 5.8 High-resolution time-lapse AFM topographs reveal the diffusion of single sodium-driven rotors in the lipid bilayer. Arrows point to membrane proteins that changed their positions during the recording time. The dotted ellipses frame a small cluster of membrane proteins that did not change its position within the time frame shown. The constant force mode AFM topographs were recorded at a minimally applied force of < 100 pN in a buffer solution (Müller et al., 2003).

envisage that in the near future time-lapse AFM will be used to study the assembly and disassembly kinetics of pharmacological targets such as G-protein-coupled receptors.

The power of high-resolution time-lapse AFM imaging of biological macromolecules in their native environment is exemplified by studies directly observing cell machineries at work. Relevant examples include binding-dissociation of single chaperonin GroES molecules from individual GroEL proteins (Viani et al., 2000), RNA polymerases (Guthold et al., 1999; Hughes et al., 2004; Kasas et al., 1997), motor proteins (Ando et al., 2003), bacterial surface-layers (Müller et al., 1996), pore-forming membrane proteins (Müller and Engel, 1999), and gap-junction communication channels (Müller et al., 2002b). Currently, the commercially available AFMs require ≥ 60 s to record a single high-resolution topograph. In contrast, most biological processes take place on a timescale of micro to milliseconds. To circumvent this limitation, high-speed AFMs have been developed and are commercially available (Ando et al., 2001; Humphris et al., 2005; Viani et al., 1999b). Depending on the approach, these high-speed AFMs can record up to 200 topographs per second. Clearly, such temporal resolution, combined with the high-resolution imaging capabilities of the AFM, will open new avenues to characterize dynamic processes of biological systems in the future (Junge & Muller, 2011; Uchihashi et al., 2011).

SINGLE-MOLECULE FORCE SPECTROSCOPY

The mechanistic basis of biological processes can only be understood with a complete knowledge of the underlying interactions. Single-molecule measurements with the AFM enable measuring forces at picoNewton (10^{-12} N = pN) resolution, which is sufficiently sensitive to detect interactions of individual proteins and nucleic acids (Borgia et al., 2008; Clausen-Schaumann et al., 2000; Engel and Gaub, 2008). This approach provides fundamental insights into the molecular basis of biological phenomena and properties as diverse as molecular recognition (Florin et al., 1994; Hinterdorfer et al., 1996; Lee et al., 1994), protein folding and unfolding (Kedrov et al., 2007; Oberhauser and Carrion-Vazquez, 2008; Rief et al., 1997a), DNA mechanics (Rief et al., 1999a), and cell-adhesion processes (Benoit et al., 2000; Franz et al., 2007; Helenius et al., 2008; Krieg et al., 2008).

To detect interaction forces, the stylus of the AFM cantilever anchoring a molecule or a molecular assembly, via molecular linkers, is brought into contact with its interacting partner substrate. Withdrawal of the stylus from the substrate exerts a force on the bond formed between the molecules resulting in cantilever deflection (Figures 5.9 and 5.10). As soon as the force on the cantilever supersedes the strength of the bond, the bond is ruptured and the cantilever springs back. Thus, the maximal deflection provides a measure of the specific interaction formed between the molecule and its substrate. In recent times, this mechanical approach, named single-molecule spectroscopy (SMFS), has helped us to learn about the interactions and mechanical properties of various biological systems (Cui and Bustamante, 2000; Florin et al., 1994; Merkel et al., 1999; Perez-Jimenez et al., 2011; Oesterhelt et al., 2000; Rief et al., 1997a).

In the case of protein unfolding, during certain approach-retract cycles, a protein molecule adsorbed on the supporting surface may attach to the stylus, and the force applied by the cantilever may unfold and stretch the unfolded polypeptide chain. Pulling induces a mechanical stress in the molecular system, forcing a structural transition, and consequently unfolding of the protein. The structural transitions during protein unfolding or the detachment of a molecular bond under mechanical stress are a sequence of physical events—elongation, unfolding, and subsequent relaxation of the polypeptide chain—recorded in the so-called force-distance (F-D) curve (Figure 5.9).

F-D curves are obtained by monitoring the cantilever deflection, x, as a function of the vertical displacement of the piezoelectric transducer, z (Figure 5.10). This yields a raw "voltage-displacement" curve, which can be converted into a "force-distance" curve by two simple conversions. First, the deflection sensitivity of the AFM cantilever (i.e., the slope of the retraction curve in the region where the stylus and a hard surface are in contact) is used to convert the voltage into cantilever deflection. Second, the cantilever deflection is converted into a force, F, using Hooke's law as explained earlier.

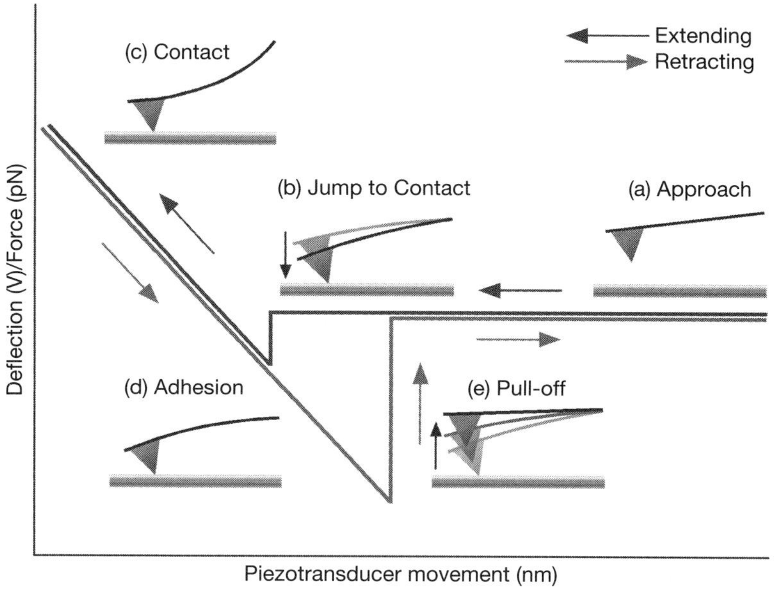

Figure 5.9 A typical force-distance (F-D) curve. Schematic representation of a F-D curve with the corresponding stages in stylus-sample interaction during the up-down movement of the piezoelectric transducer. The slope of the region between (C) and (D) is used to determine the deflection of the cantilever which is required to convert a "voltage-displacement" curve into a "force-displacement" curve if the spring constant of the cantilever is known. It should be noted that the F-D curve (light grey) shown here does not represent the unfolding of a polypeptide chain, but rather denotes an adhesion peak between the cantilever stylus and the sample support.

What Do Mechanical Forces Tell Us about Molecular Interactions?

What exactly does the rupture force signify? Crucially, it is not an absolute value that can be related to the strength of a bond. The conceptual basis of forced unfolding lies in the lifetime of a bond. Weak bonds and structures have limited lifetimes, t_{off}, and will dissociate under almost any force if pulled for sufficient time. If pulled apart faster than t_{off}, a bond resists detachment leading to a higher unfolding force. The magnitude of the detachment or unfolding force is thus dictated by how fast a bond is loaded (i.e., the loading rate). At higher loading rates, the bond ruptures with higher unfolding or detachment forces, and when pulled slowly, the bond exhibits lower unfolding forces (Evans, 1998, 2001).

To understand a reaction under force, let us summon the classical two-state model. An energy profile diagram of a two-state reaction involves a low-energy folded state and a high-energy unfolded state separated by a barrier that should be overcome for interconversion between the two states (Figure 5.11). Folding and unfolding reactions of many small water-soluble proteins under no force can be described using this model (Baldwin and Rose, 1999a, b). The mechanical unfolding of small water-soluble proteins has also been often described by a two-state model where a folded protein reaches an unfolded state on crossing an intervening transition state (Evans and Ritchie, 1997; Rief et al., 1998a). In the context of a two-state model, a

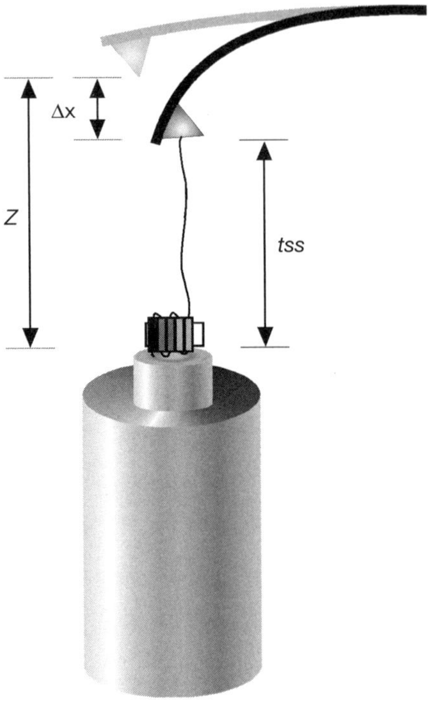

Figure 5.10 Schematic drawing showing the cantilever deflection. The actual length of the unfolded polypeptide is given by the tip-sample separation (*tss*), and not the movement of the piezo transducer (*z*), due to the deflection of the cantilever by Δx.

very useful way of thinking about a reaction under force is that an external load lowers the energy barrier of the underlying energy landscape (Figure 5.11), i.e., the external pulling force, F, lowers the energy barrier, G^*, at the transition state ($x = x_{ts}$),

$$\Delta G^*_{u(f)} = \Delta G^*_u - Fx_u \tag{7}$$

where $\Delta G^*_{u(f)}$ and ΔG^*_u represent the free energy difference under applied force and zero force, respectively, and $x_u = \langle x_{ts} \cos\theta \rangle$ is the thermally averaged projection of the width of the potential barrier (Evans, 1998).

If the transition state has a free energy, G^*, then the unfolding rate constant, k_u, of a reaction under zero load is,

$$k_u = \left(\frac{1}{t_D}\right) \exp\left(-\frac{\Delta G^*_u}{k_B T}\right) \Delta G^*_u = G^* - G_n \tag{8}$$

where t_D is the relaxation time. As first postulated by Bell thirty years ago (Bell, 1978), the rate of escape, k_{off}, increases exponentially with force,

$$k_{off} \approx \left(\frac{1}{t_{off}}\right) \exp\left(\frac{F}{F_u}\right) \tag{9}$$

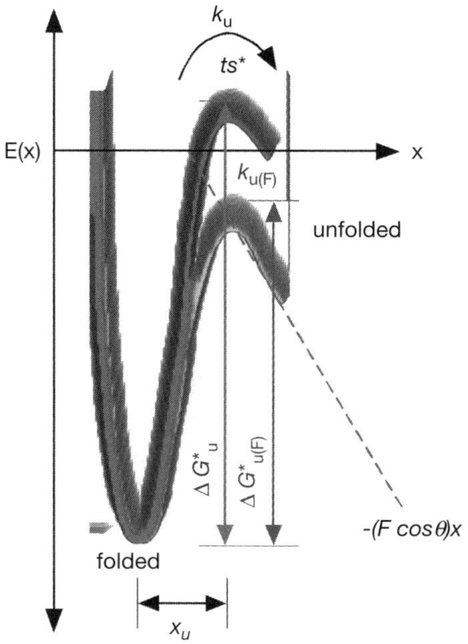

Figure 5.11 Energy barrier under no force and an external force. A folded protein confined in a well by an energetic barrier, ΔG^*_u, can be unfolded with an unfolding rate of k_u via a transition state *ts** at a distance x_u from the folded state. An external force, *F*, at an angle θ to the molecular coordinate *x* adds a mechanical potential—$(F\cos\theta)x$ that tilts the landscape and lowers the energy barrier to $\Delta G^*_{u(F)}$. The new unfolding rate under force is $k_{u(F)}$. However, the distance to the transition state, x_u, and the shape of the barrier remains unchanged.

where $F_u = \dfrac{k_B T}{x_u}$. The rate of a reaction under force can be similarly determined by combining equations (8) and (9),

$$k_{u(F)} \cong \left(\frac{1}{t_D}\right)\exp\left(-\frac{\Delta G^*_u - Fx_u}{k_B T}\right) = k_u \exp\left(\frac{Fx_u}{k_B T}\right) \quad (10)$$

A Variety of Proteins Ripped Apart by SMFS

Nature has ingeniously designed load-bearing proteins. The modular structure of proteins seems to be a general strategy for resistance against mechanical stress. The mechanical properties of several modular proteins, like titin, have already been investigated by SMFS (Rief et al., 1997a). The initial experiments demonstrated that the elongation events observed during the stretching of single multi-modular proteins can be attributed to the unfolding of individual domains. Experiments with optical tweezers corroborated these results (Kellermayer et al., 1997; Tskhovrebova et al., 1997). Since these pioneering experiments (Rief et al., 1997a), SMFS has been used extensively to characterize the mechanical properties of a range of biomolecules, including titin and its intermediates (Carrion-Vazquez et al., 1999; Marszalek et al., 1999; Rief et al., 1998b), the triple helical coiled-coils of spectrin (Rief et al., 1999b), FN-III domains of tenascin (Oberhauser et al., 2002; Oberhauser et al., 1998), polysaccharides (Marszalek et al., 2002; Rief et al., 1997b), integration strength of transmembrane α-helical peptides in lipid bilayers (Ganchev et al., 2004), diatom single adhesive nanofibers (Dugdale et al., 2005), myomesin (the molecular spring made of immunoglobulin (Ig) and fibronectin type III (FN-III) domains) (Schoenauer et al., 2005), the nanospring behavior of multidomain protein ankyrin (Lee et al., 2006), the ubiquitous collagen (Gutsmann et al., 2004), spider silk (Oroudjev et al., 2002), and even a non-mechanical protein barnase (Best et al., 2001).

Experiments at the nanometer scale provide complete new insights into processes, which before the development of the AFM and related instruments were not accessible by methods based on ensemble averaging. The single molecule techniques have been applied to novel uses, including the unbinding properties of ligand-receptor systems (e.g., avidin/biotin, antibody/antigen and *p*-selectin/carbohydrate pairs) (Fritz et al., 1998; Hinterdorfer et al., 1996; Merkel et al., 1999), determining misfolding events of single proteins (Oberhauser et al., 1999), the strength of a covalent bond (Grandbois et al., 1999), studying how point mutations determine the mechanical energy landscape of Ig modules (Li et al., 2000), unfolding kinetics of ubiquitin (Schlierf et al., 2004), mapping the energy landscape of GFP (Dietz and Rief, 2004), elucidating the material properties of a protein due to temperature changes (Schlierf and Rief, 2005) and pulling direction (Dietz et al., 2006a), measurement of protein energy landscape roughness (Nevo et al., 2005), viscoelastic properties of single polysaccharide (Kawakami et al., 2004) and protein molecules (Janovjak et al., 2005), and determining amino acid positions using a method termed "mechanical triangulation" (Dietz and Rief, 2006). These studies have led us to territories that will keep us busy exploring for many years.

The advancement of these studies, however, would have not been possible without technical tinkering in both instrumentation and protein engineering. Different modes of force application have been developed for single-molecule unfolding experiments. The constant velocity mode is most commonly used for unfolding single protein molecules. In 2001, Fernandez and co-workers reported the construction of a force-clamp AFM instrument that allowed them to mechanically unfold single titin molecules under a constant applied force or alternatively increasing the force linearly with time[2] (Oberhauser et al., 2001). Using these modes, the natural lifetimes of the folded states and the refolding kinetics of proteins (e.g., ubiquitin (Schlierf et al., 2004), the insect flight muscle proteins projectin and kettin (Bullard et al., 2006)) have been estimated. The most interesting breakthrough came with the refolding of ubiquitin using the force-clamp mode. Though no details or discrete folding steps were resolved in the rate-determining stage, a more or less continuous process akin to polymer collapse was observed (Fernandez and Li, 2004). The use of force-clamp experiments has been extended to more novel applications like determining the kinetics of disulfide exchange on the stability of folded domains as a

2. Later the two modes were correctly named as "force-clamp" and "force-ramp," respectively.

function of constant force (Wiita et al., 2006), and the role of thioredoxin catalysis in the reduction of a disulfide bond under an external load (Wiita et al., 2007).

Single-molecule studies are statistically demanding. Most mechanical proteins that occur in nature have a modular architecture. Fortunately, the modular structure helps to increase the throughput of the experiments by increasing the number of measured events by a factor that depends on the modules in the protein. The modular design of naturally occurring proteins has inspired the re-designing of non-modular proteins in a repeat pattern for high-throughput mechanical unfolding. Another useful strategy has been to insert a non-modular protein in the middle of a well-characterized modular protein (Dietz and Rief, 2004). The Rief laboratory has devised an ingenious way of constructing disulfide linked polyprotein chains of non-modular proteins by engineering cysteine residues at specific positions in a protein (Dietz et al., 2006b). This method has been used for site-directed force application to a protein (Dietz et al., 2006a).

Relevance of Studying Protein (Un)Folding under Constrained Load

Protein folding and unfolding are conventionally studied with bulk assays employing pH, temperature or chemical denaturants. The final results from bulk measurements include the ensemble behavior of molecules masking the behavior of individual molecules. The advent of single-molecule techniques has revealed interesting, hitherto hidden, aspects of biochemical processes (Bustamante, 2008; Herbert et al., 2008). Crucially, studying the stability of proteins designed to fulfill a mechanical purpose under extreme conditions of pH, temperature or chemicals may not be relevant (Brockwell, 2007; Oberhauser and Carrion-Vazquez, 2008). The nature of spontaneous protein unfolding is very different under those parameters compared to under force or even *in vivo* (Prakash and Matouschek, 2004).

In the realm of membrane biology, membrane protein unfolding and folding has been studied in SDS and lipid vesicles (Bowie, 2001; Seddon et al., 2004). While lipid vesicles offer a close to native-like environment, nevertheless, both, detergents as well as lipid vesicles, do not mimic the *in vivo* folding environment (Booth, 2003; McKibbin et al., 2007). Single-molecule AFM assays partially circumvent these problems. The experiments are always done with the membrane protein inserted in the lipid bilayer. Although the final stages of helix insertion in a lipid bilayer may involve lateral release of a helix (Sadlish et al., 2005; Xie et al., 2007), insertion of membrane proteins is a vectorial process facilitated by translocons (Rapoport, 2007), Sec61 in the rough endoplasmic reticulum (ER) (Van den Berg et al., 2004), or the bacterial SecYEG complex machinery (Breyton et al., 2002). Of crucial importance is the vectorial nature of mechanical unfolding, which is a close approximation to the extraction of membrane proteins from mitochondrial and ER membranes *in vivo*, a mechanical process mediated by AAA proteases (Leonhard et al., 2000; Ye et al., 2001). Complementing the finding, on a different note, *Escherichia coli* ClpXP, a hexameric AAA$^+$ ATPase that catalyzes the ATP-dependent denaturation of proteins and their energy-dependent translocation into ClpP, was proposed to mediate degradation by repetitive applying a uniform, mechanical force (Kenniston et al., 2003). This and other numerous examples (Bustamante et al., 2004; Neupert and

Herrmann, 2007) illustrate the importance of mechanics in biology justifying the necessity of using mechanical load to understand biological processes.

What Can Be Learned from SMFS on Membrane Proteins?

The molecular-level understanding of membrane proteins lags far behind that of the water-soluble proteins. Their strict confinement in the lipid bilayer makes it difficult to obtain high-resolution structural information. In addition, membrane proteins often have a metastable nature once outside the membrane, and tend to aggregate in solution losing their activity (Booth, 2003; Bowie, 2001). It is, therefore, difficult to sustain hypothesis-driven experimentation aimed at understanding structure-function relationship of integral membrane proteins (Fleishman et al., 2006). The sensitivity and ease of employing AFM imaging and SMFS for studying membrane proteins in lipid bilayers make these techniques relevant for probing interactions of membrane proteins inside the membrane bilayer, and for providing a framework for the mechanistic understanding of membrane protein structure and function.

Bacteriorhodopsin has been an excellent model system for studying membrane protein folding (Curnow and Booth, 2007; Joh et al., 2008). This made it the protein of choice for the first proof of principle experiment to demonstrate the mechanical unfolding of single membrane proteins (Oesterhelt et al., 2000). Further detailed experiments revealed a number of interesting facts about the mechanical unfolding pathways of bacteriorhodopsin (Müller et al., 2002d). For example, when unfolded by mechanical force, bacteriorhodopsin took different unfolding routes on the energy landscape (Figure 5.12). The distance between distinct peaks in a F-D curve reflects the gain of distance after unfolding and stretching of a folded structure in the protein, and the amplitude is a measure of the force required to unfold the polypeptide chain. An unfolding event occurring at a given extension is specified by the amplitude of the force peak at this extension and the distance from the same peak to the next one. These "fingerprints" of the protein reveal different populations of unfolding events.

The varied pathways observed during unfolding of single membrane proteins supports the "new view" of protein folding in which proteins slide along the barriers and ridges of the folding funnel to reach their final native folded states (Frauenfelder et al., 1991; Wolynes et al., 1995). Albeit these or similar unfolding pathways may occur during chemical unfolding in bulk measurements, it may not be possible to determine their individual populations as in SMFS experiments. Single-molecule mechanical unfolding of membrane proteins offers such and many other vantage points for studying and correlating the unfolding, folding and functional dynamics of membrane proteins. In this and the forthcoming sections we will give a comprehensive overview of what has been learnt thus far by unfolding membrane proteins using AFM, where we stand now, and what the open vistas are for the future.

Under an applied mechanical load, single membrane proteins unfold via individual unfolding of their helices (Kedrov et al., 2007), in contrast to water-soluble proteins, which unfold in an all-or-none fashion. The molecular interactions established by a variable number of amino acids in the helices act collectively to form stable structural segments or the so-called unfolding barriers. These amino acids unfold in a cooperative manner over a certain threshold of applied force. Stable structural segments do not

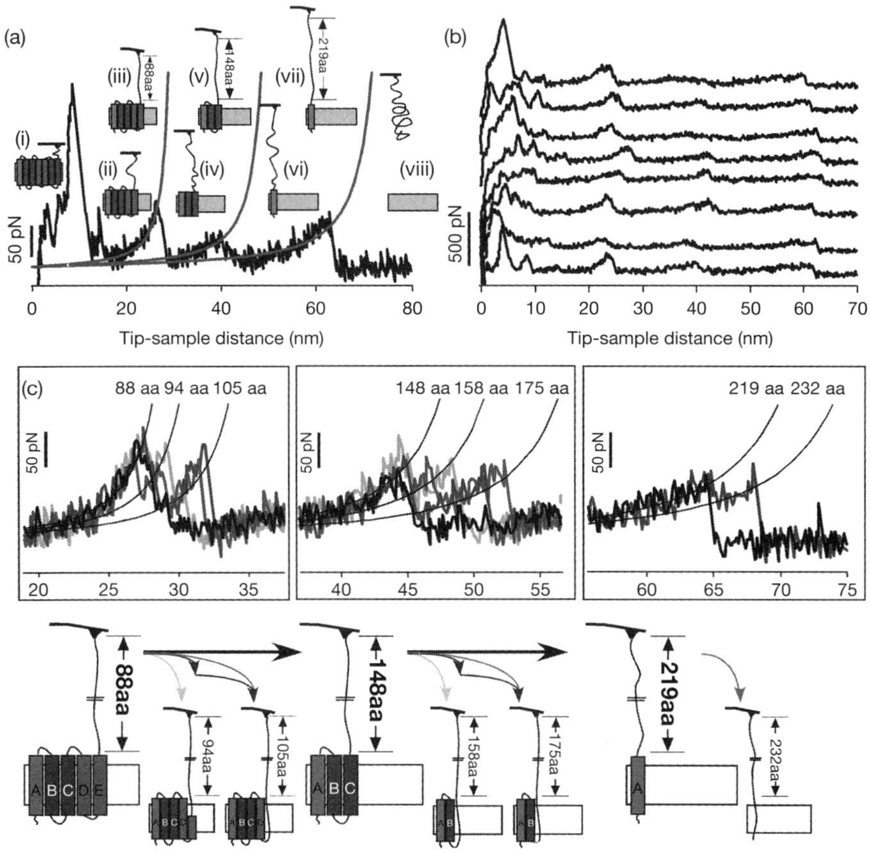

Figure 5.12 Unfolding pathways of bacteriorhodopsin. (A) The spectrum shows a representative F-D trace depicting the pairwise unfolding pathway of a single bacteriorhodopsin molecule. The schematic drawings show the unfolding pathways at different positions as unfolding proceeds. Individual force peaks were fitted to the WLC model.. (B) Individual force spectrum of single bacteriorhodopsin molecules. (C) Unfolding pathways of individual secondary structure elements of bacteriorhodopsin. Black smooth curves represent average WLC fits of force peaks. Single peaks deviate from the average values within the standard deviation. Occasionally, the first unfolding peak at 88 amino acids (aa) shows two shoulder peaks (first column), indicating the stepwise unfolding of the helical pair. If both shoulders occur, the peak at 88 aa indicates the unfolding of helix E, that at 94 of loop DE, and the peak at 105 corresponds to the unfolding of helix D. The shoulder peaks of the second major peak at 148 aa indicate the stepwise unfolding of helices C and B and loop BC. The peak at 148 aa indicates the unfolding of helix C, that at 158 of loop BC, and the peak at 175 aa represents unfolding of helix B. The unfolding scheme via the different pathways is shown at the bottom. The arrows indicate the observed unfolding pathways. In certain pathways (black arrows), a pair of transmembrane helices and their connecting loop unfolded in a single step. In other unfolding pathways (grey arrows), these structural elements unfolded in several intermediate steps.

necessarily correlate to the secondary structure elements of the protein. The segments can be constituted of a single transmembrane helix, a pair of transmembrane helices, a transmembrane helix together with a polypeptide loop, or just a small extramembranous loop. The results from unfolding membrane proteins indicate that the smallest segment contains ≈ 6 amino acids and the largest one ≈ 40 amino acids (Müller et al., 2002d).

By mapping the structural segments on the secondary structure of a protein, interactions that stabilize secondary structures such as whole or parts of transmembrane α-helices, polypeptide loops can be localized (Kedrov et al., 2007). Measuring the force required to unfold the structural segments gives a direct measure of the strength of molecular interactions stabilizing these regions. Besides the unfolding force of a structural segment, the probability to unfold a structural segment is an indicator of changed molecular interactions. Because protein activity, function and structure are directly related to the changes in their molecular interactions, it is possible to detect regions of altered molecular interactions under varied environmental factors (e.g., pH, temperature, ligands, metal ions, lipids, point mutations).

The combination of AFM imaging and SMFS has yielded detailed insights into the inter and intra-molecular interactions stabilizing membrane proteins like bacteriorhodopsin (Janovjak et al., 2003; Müller et al., 2002d; Sapra et al., 2006a) and halorhodopsin (Cisneros et al., 2005) from *Halobacter salinarum*, human aquaporin-1 from red blood cells (Möller et al., 2003), the Na^+/H^+ antiporter NhaA from *E. coli* (Kedrov et al., 2004), native bovine rhodopsin (Fotiadis et al., 2003; Sapra et al., 2006b), β_2-adrenergic receptor (Zocher et al., 2012), the amino acid anti-porter SteT (Bippes et al., 2009), and the β-barrel membrane proteins OmpG (Sapra et al., 2009), OmpA (Bosshardt et al., 2012) and FhuA (Thoma et al., 2012) to mention a few.

Insights into Folding of Proteins from Single-Molecule Mechanical Unfolding

The wide scope of opportunities offered by SMFS has helped to answer new relevant questions and also to look at old ones in a new light. A relevant example is that of the topologically similar bacteriorhodopsin and halorhodopsin (Cisneros et al., 2005). In spite of ≈ 30% sequence identity between the two structures, their mechanical unfolding patterns are remarkably conserved. The similar structural segments and unfolding pathways detected for bacteriorhodopsin (Müller et al., 2002d), halorhodopsin (Cisneros et al., 2005), sensory rhodopsin II (Cisneros et al., 2008), and proteorhodopsin (Klyszejko et al., 2008) in SMFS experiments suggest that, despite the low sequence identity between these proteins, very similar mechanisms are utilized to reach the final functional folded state. Thus, nature seems to devise elegant ways to maximize the functional pool of proteins from a minimum set of protein folds by utilizing the same architectural scaffold for different functions (Vendruscolo and Dobson, 2005).

Interactions of the G-Protein-Coupled Receptor (GPCR) Rhodopsin Revealed by SMFS

Understanding the molecular interactions that stabilize or destabilize proteins is fundamental to understanding their function. Mechanical unfolding of the pharmacologically important rhodopsin was performed to characterize its stabilizing interactions. An SMFS assay permitted the unfolding of rhodopsin from its native rod-outer segment disc membranes obtained from bovine retina. When unfolded by applying a force at its N-terminal, the resulting F-D curves of rhodopsin showed two very distinct patterns distinguished by a difference in their lengths (Figure 5.13). The shorter (≈ 65 nm) F-D curves represented a molecular state in which the highly conserved disulfide bridge between Cys110 and Cys187 was intact, and the longer (≈ 95 nm) curves represented a state in which the disulfide bond between Cys110 and Cys187 was broken. Control experiments suggested that the native Cys110-Cys187 bond breaks and Cys187 forms a rearranged bridge with an

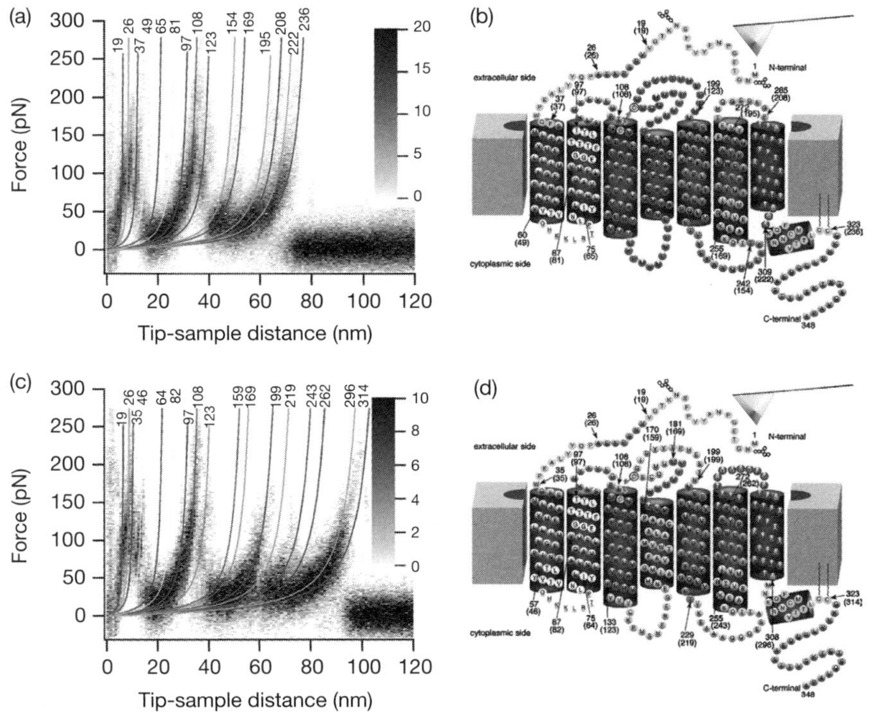

Figure 5.13 Single-molecule force spectroscopy curves recorded on native ROS disc membranes. (A) The superimposition of several force curves (n = 42) enhanced common features among the curves. Major (black) and minor peaks (grey shaded) were fitted using the WLC model as above. (B) Secondary structure of rhodopsin mapped with structural segments that constitute the unfolding barriers in SMFS. (C) The superimposition of several F-D curves (n = 25) enhanced the common features among the curves. Major (black) and minor peaks (grey shaded) were fitted using the WLC model as above. The analysis of force curves revealed the strengths and locations of molecular interactions established within rhodopsin (D). (D) Secondary structure of rhodopsin mapped with structural segments observed by SMFS. Arrows locate the starting and the end of each structural segment. Numbers in brackets are the amino acid values obtained from WLC fitting and the numbers without brackets are the corresponding residue number in the rhodopsin sequence. See insert for color version.

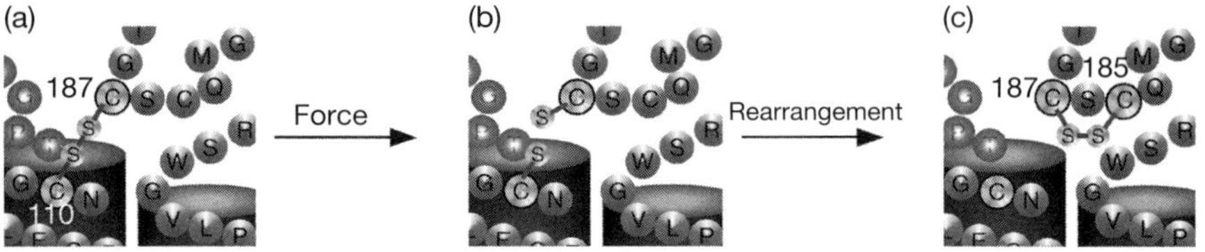

Figure 5.14 Effect of force on the native disulfide bridge during rhodopsin unfolding. (A) The intact disulfide bridge between Cys110 and Cys187 in native rhodopsin. It is assumed that, under an external force during unfolding, this bond might break (B), possibly rearranging to form a non-native disulfide bond between Cys185 and Cys187 (C). The ratio of the two populations of F-D curves shown in (A) & (C) in Figure 5.13, in the presence of DTT and N-ethylmaleimide suggested that the disulfide bridge between Cys110-Cys187 rearranged to Cys185-Cys187 (Sapra et al., 2006b).

adjacent Cys185 (Figure 5.14) (Sapra et al., 2006b). The structural segments when mapped onto the secondary structure of rhodopsin (Figure 5.13) helped characterize the molecular interactions of rhodopsin in the presence and absence of the stabilizing Cys110-Cys187 bond. The structural segments changed their strengths and locations in the absence of the native Cys110-Cys187 bond. Such changes may build a molecular mechanism of protein destabilization, misfolding and malfunction.

The effects observed in rhodopsin lacking the native disulfide bond may underlie the disruption of molecular interactions that occur in point mutations resulting in the replacement of the native Cys110-Cys187 bond with the non-native Cys185-Cys187 bond and causing complete or partial misfolding of the protein (Garriga et al., 1996; Hwa et al., 1997; Liu et al., 1996). In addition, these results suggest that the native disulfide bond (Cys110-Cys187), extracellular and cytoplasmic loops, and the transmembrane domains are structurally coupled (Hwa et al., 1997) since the absence of the native disulfide bond leads to altered interactions in most regions of the molecule.

Screening Factors That Stabilize GPCRs

Recent studies have identified zinc (Zn^{2+}) as an allosteric modulator of the structure and function of a number of GPCRs (Holst et al., 2002; Schetz and Sibley, 1997; Swaminath et al., 2003). Zn^{2+} appears to play an important functional and structural role in rhodopsin too (del Valle et al., 2003; Shuster et al., 1996; Stojanovic et al., 2004). Despite the numerous studies, the exact role of Zn^{2+} in the visual cycle and specifically in its interaction with rhodopsin remains unclear.

SMFS was employed to monitor the effect of Zn^{2+} binding on the molecular interactions stabilizing dark state rhodopsin (Park et al., 2007). Unfolding of rhodopsin at increasing Zn^{2+} concentrations (0–400 μM) showed a clear increasing trend in the mechanical stability of most structural segments with increasing concentration, reaching a saturation around 200 μM. The most dramatic increase in the stability was observed for structural segments hosting the Zn^{2+} binding sites (Figure 5.15). From the titration curves the EC_{50} value for Zn^{2+} binding to rhodopsin was determined to be ≈ 40 μM for the two structural segments hosting the Zn^{2+} binding sites. This is in close corroboration with the dissociation constant value of 2–10 μM obtained from

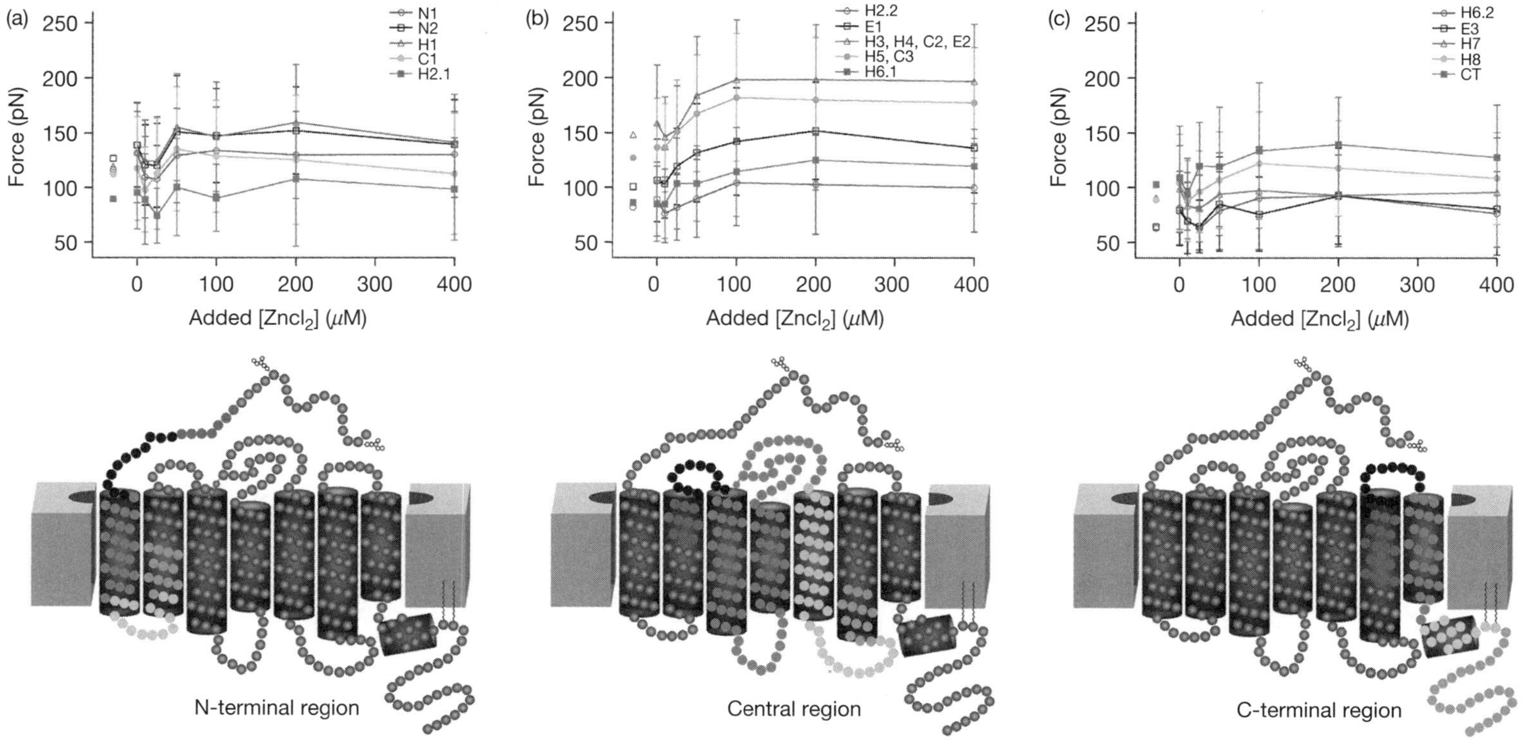

Figure 5.15 Rhodopsin unfolding forces depend on $ZnCl_2$ concentrations. SMFS was performed at $ZnCl_2$ concentrations of 0–400 μM, and the average unfolding force was determined for each structural segment (A–C). Data points below 0 μM added $ZnCl_2$ on the *x*-axis denote average unfolding forces from experiments done in the presence of 1 mM EDTA. Error bars are the standard deviations of the unfolding forces of each structural segment at different Zn^{2+} concentrations. See insert for color version.

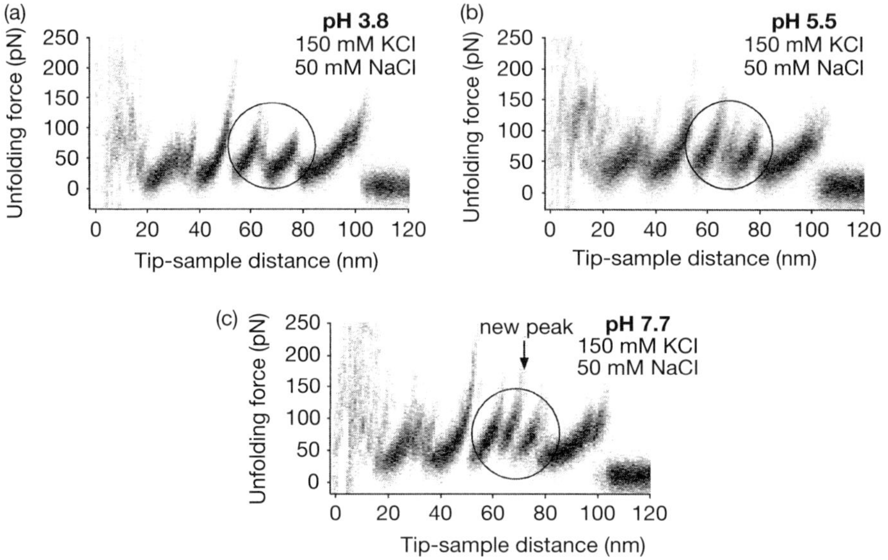

Figure 5.16 F-D curves of NhaA obtained from SMFS in its inactive, intermediate and active states. NhaA was unfolded at three different pH values of 3.8, 5.5 and 7.7 where it exists in the inactive, intermediate and active states, respectively. (A), (B) and (C) show the superimpositions of the F-D traces obtained at different pHs. As the pH was increased from 3.8 (inactive) to 7.7 (active), a new peak (circled region), corresponding to the ligand binding region in helix V, appeared. All the experiments were done in the presence of NhaA's ligand Na^+ and 150 mM KCl.

biochemical measurements (Shuster et al., 1992). Zn^{2+} binding was hypothesized to occur along the interface of two rhodopsin monomers. Molecular dynamic simulations provided the first crucial support for this hypothesis suggesting that Zn^{2+} binding may play a role in the dimerization process of rhodopsin (Park et al., 2007). Thus, the SMFS experiments suggest that Zn^{2+} not only improves the stability of rhodopsin but may also help it to attain its optimum functional state.

These initial experiments demonstrate the untapped potential of SMFS as a general tool for understanding the factors that affect molecular interactions involved in the folding and function of membrane proteins. It is hoped that such experiments will provide deeper insights into the function and dysfunction of membrane proteins.

Potential Use of SMFS in Analyte Screening

Regulation of proton (H^+) and sodium (Na^+) gradients is involved in virtually every physiological process. Na^+/H^+ antiporters regulate intracellular pH, cellular Na^+ concentration, and cell volume of eukaryotic and prokaryotic organisms. In *E. coli*, two antiporters, NhaA and NhaB, specifically exchange Na^+ and Li^+ ions for H^+, helping the cell to adapt to high environmental salinity and to grow at an alkaline pH (Padan et al., 2001).

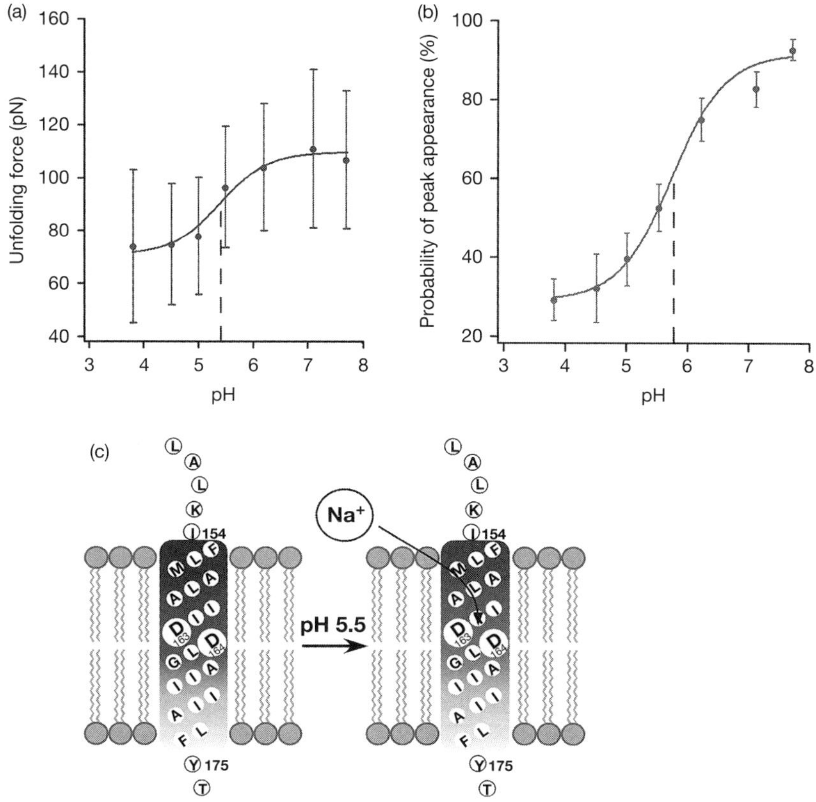

Figure 5.17 Detecting the activation of NhaA upon binding of its natural ligand. (A) The unfolding force of helix V, which hosts the Na^+ binding pocket of NhaA, showed a sigmoidal increase as the pH was increased from 3.8 to 7.7. A similar trend was observed in the frequency of peak appearance assigned to helix V (B). In both cases, the transition pH ≈ 5.5 indicated the interactions maturing towards a complete activation of NhaA indicated by the plateau of the sigmoidal trace between pH 7 and 8. (C) The pH-dependent ligand binding between Asp163 and Asp164 in helix V.

The recent application of SMFS to detect the binding of Na^+ ions to the Na^+/H^+ antiporter NhaA shows another promising aspect of the technique (Kedrov et al., 2005). Na^+, which is a natural ligand of NhaA, binds to helix V bridging amino acids Asp163 and Asp164 at pH 7 (Figure 5.17). This binding manifested itself as an extra force peak in the F-D spectra of NhaA (Figure 5.16). The appearance of a new force peak suggests new interactions upon ligand binding to the protein. The conformity of these results was shown by the reversible nature of the force peak occurrence. When the pH was reduced from 7.7 to 3.8 in the same experiment, the new force peak was not detected.

The force required to unfold the Na^+ binding region showed a sigmoidal dependence on pH with a transition at pH 5.4 and saturating at pH 7 (Figure 5.17). The frequency of the new peak appearance exhibited a similar sigmoidal effect as a function of pH with a transition at pH 5.7 and a plateau at pH 7. NhaA has an optimum activity at pH 7.5 (Taglicht et al., 1991). To explain the difference of pH ≈ 2 between the optimal pH from biochemical experiments and the transition in SMFS experiments, the transition was suggested to mark the formation of interactions crucial to the first steps in the activation process of NhaA. The final activation is, however, achieved at pH 7–8 denoted by the plateau in the sigmoidal rise (Kedrov et al., 2005). These results have been further substantiated by SMFS experiments detecting inhibitor binding, differentiating ligand and inhibitor binding (Kedrov et al., 2006b), estimating binding constants, and even elucidating the inhibition mechanism (Kedrov et al., 2006b). These and similar studies on the water-soluble protein mouse dihydrofolate reductase (Ainavarapu et al., 2005; Junker et al., 2005) conclusively demonstrate the use of SMFS as a possible method to screen binding of small molecules.

The pharmaceutical importance of membrane proteins stems from the fact that they include structural proteins, channels and receptors that are accessible through the exterior of cells and thus are formidable drug targets. This has necessitated development of high-throughput assays to screen drugs, ligands or other small-molecule libraries. Screening potential lead compounds from a library of molecules is indeed a challenging task. The hurdle is not only to get the right compound but also to expedite the process. To overcome this difficulty, an ingenious AFM was designed recently (Struckmeier et al., 2008). We foresee the fully automated SMFS as a potential tool for screening molecular interactions of biological molecules and complexes. In particular, this approach may be suitable to screen small libraries of compounds to identify potential drug targets for GPCRs and ion channels.

Probing the Dynamics of Energy Landscape with SMFS

Ingenious strategies employed by single-molecule based approaches present opportunities to test various aspects of the energy landscape model as well as to discover new ones (Hyeon and Thirumalai, 2003). The available experimental techniques, however, only permit to chart out the 1-D energy landscape of a biological system. An extension of the SMFS technique, dynamic force spectroscopy (DFS), is a very relevant experimental approach to map the 1-D energy landscape. DFS involves measuring the mechanical strength of a molecular system at increasing pulling speeds (Evans, 1999). When plotted as the most probable unfolding or detachment force, F_p, *vs.* $\log_e$ (loading rate), the force spectrum maps the most prominent energy barriers in the energy landscape along the force-driven pathway (Evans, 1998; Evans and Ritchie, 1997). F_p can be described as,

$$F_p = \frac{k_B T}{x_u} \ln\left(\frac{x_u \cdot r_f}{k_B T \cdot k_u}\right) \tag{11}$$

where k_B is the Boltzmann constant, T is the temperature, r_f is the loading rate, x_u denotes the distance between the intermediate folded state (or the initial native state) and the transition state, and k_u is the unfolding rate (or the detachment rate, k_{off}) under zero load. k_u (or k_{off}) and x_u serve as useful parameters for reconstructing the 1-D energy landscape of a protein, DNA, RNA, ligand-receptor pair or even a single bond (Evans, 1998, 2001; Greenleaf et al., 2008; Merkel et al., 1999; Woodside et al., 2006).

The mechanical unfolding energy landscape of bacteriorhodopsin revealed some detailed insights into the stability and folding of the protein (Janovjak et al., 2004). Unfolding of all the structural segments—single helices, paired helices with connecting loops, and individual loops—depicted a two-state process. The unfolding rates of individual or pairs of helices under no external load ranged between 10^{-6}–10^{-2} s^{-1}. Assuming the principle of microscopic reversibility (McCully et al., 2008), these values suggest that individual helices form stable structural intermediates during the folding of bacteriorhodopsin. Moreover, the unfolding of the helices was observed to be highly dependent on the loading rate. At higher loading rates the helices unfolded individually, whereas at lower unfolding rates the unfolding occurred in a pairwise fashion in confirmation with the "helical hairpin hypothesis" (Engelman and Steitz, 1981).

An extension of the DFS methodology is its application to determine the energy landscape roughness. Based on the theory of diffusion in rough potentials (Zwanzig, 1988), Hyeon and Thirumalai recently proposed that the energy landscape roughness can be estimated from single-molecule DFS experiments at different temperatures (Hyeon and Thirumalai, 2003). This was first demonstrated successfully by Nevo et al. (Nevo et al., 2005) for the protein complex constituted of the small GTPase Ran and the nuclear transport receptor importin-β, followed by Rief and co-workers (Schlierf and Rief, 2005) for the F-actin crosslinker filamin of *Dictyostelium discoideum*, ddFLN. More recently, the degree of roughness of bacteriorhodopsin, as an estimate of its conformational flexibility, was estimated using DFS (Janovjak et al., 2007). The energy surface ruggedness of bacteriorhodopsin helices was estimated to be ≈ 5 k_BT. The power of DFS lies in extending the scope of mechanical unfolding to divulge the physical aspects of interactions.

Estimating Protein Rigidity from Single-Molecule Mechanical Unfolding

Proteins exhibit both flexible and rigid regions that contribute to their functional purposes. The DFS methodology is a unique way to directly probe the mechanical properties of different regions within a protein (Figure 5.18). The parameters x_u and k_u can be used to estimate the stiffness of individual structural segments of a membrane protein (Sapra et al., 2008) using the formula,

$$\Delta G_u^* = -k_B T \cdot \ln\left(\frac{\tau_A}{\tau_0}\right) \tag{12}$$

where τ_0 denotes the unloaded lifetime of the state ($\tau_0 = 1/k_u$) and $1/\tau_A$ the Arrhenius frequency factor. For protein dynamics τ_A has a value of 10^{-9} s (Bieri et al., 1999).

The stiffness measurements of rhodopsin clearly showed that the protein is interspersed with rigid and flexible structural segments. This led to the speculation that the presence of both rigid and flexible regions in rhodopsin may help the protein to work as an efficient signaling molecule. The signaling mechanism of a GPCR involves the transfer of stimulus from its extracellular to the cytoplasmic domain through an

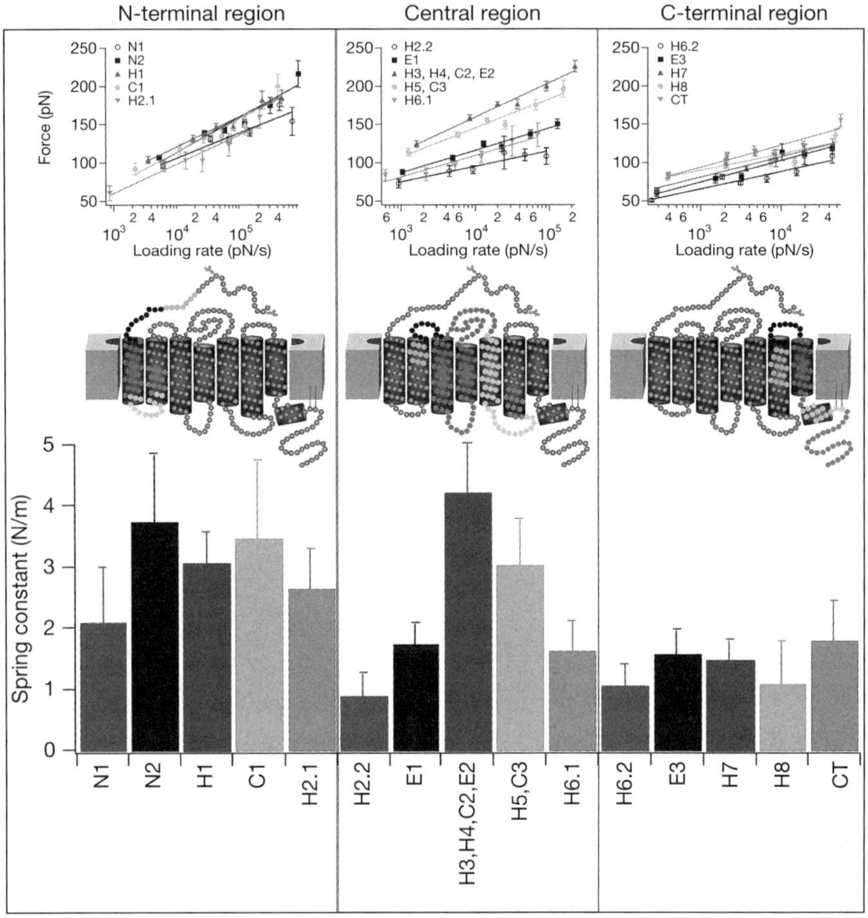

Figure 5.18 Rigidity of structural segments of rhodopsin as determined by DFS. Rigidity of structural segments (lower panel) was estimated by calculating the spring constant, κ, from x_u and $\Delta G_u^{\ddagger}$ values derived from DFS experiments (upper panel). The structural segment, H2.2, in helix II is the most flexible region ($\kappa = 0.9$ N/m), and the structural segment, H3,H4,C2,E2, constituted of helices III and IV with their connecting cytoplasmic loop CII and the extracellular loop EII is the most rigid region ($\kappa = 4.2$ N/m). See insert for color version.

event of well-coordinated conformational changes or changes in dynamic interactions. The conformational changes demand a precise orchestration of helix movements. A molecular design consisting of helices of different stiffness, coupled to a retinal chromophore which photoisomerizes from 11-*cis* to all-*trans* on absorption of a photon, seems optimal for maintaining the functional framework of rhodopsin. The evolution of different regions of a protein to different rigidities may help the protein towards optimization of a synergistic working mechanism.

In the case of water-soluble proteins, Rief and co-workers have shown, using GFP as a model protein, that the pulling coordinates determine the rigidity of a protein

(Dietz et al., 2006a). It remains to be shown how the photobleaching or the fluorescent properties of GFP are related to its mechanical properties; this may help to establish GFP as an *in situ* force sensor calibrated to its fluorescence intensity.

AFM Cantilever as a Handle to Refold Membrane Proteins

The folding of membrane proteins in conventional bulk assays is studied in detergents or lipid micelles (Seddon et al., 2004). In these systems, the denatured state is undefined; in many cases the residual helical content is estimated to be up to 50% (Hunt et al., 1997; Riley et al., 1997). In the absence of a well-defined denatured state, determining the thermodynamic and kinetic stability may be ambiguous. Also, in many cases the assumed final folded state of a membrane protein is determined by a change in one of its physical properties, for instance, by monitoring a change in fluorescence intensity. A change in the physical property of a membrane protein may not be a realistic indicator of its final natively folded functional state. It is therefore crucial to have the necessary boundary conditions *in vitro* that mimic the constraints experienced by a membrane protein *in vivo*.

Mechanical unfolding using AFM serves a dual advantage, *viz.*, (i) the stretched polypeptide state is a well-defined unfolded state thus defining similar energy states for comparative studies, and (ii) using the AFM stylus as an anchor it is possible to bring the polypeptide chain close to the lipid bilayer to refold a membrane protein inside the membrane. Refolding with the AFM involves simple steps of partial mechanical unfolding of a membrane protein followed by relaxation of the unfolded polypeptide, waiting for different time intervals at a specific external load, and finally unfolding the protein completely to monitor the extent of folding of the structural segments during the wait period (Figure 5.19).

The refolding kinetics of NhaA (Kedrov et al., 2006a) and later bacteriorhodopsin (Kessler et al., 2006) were determined using this strategy. Characteristic folding rates of NhaA structural segments ranged from 0.3 to 2.2 s^{-1} and were comparable to the folding rates of bacteriorhodopsin obtained from ensemble folding studies (0.002–13 s^{-1}) (Booth, 2000). However, helix V of NhaA had the fastest folding rate of $\approx 50\ s^{-1}$, and could refold even under an external load of ≈ 30 pN. This structural segment formed the folding core, and its refolding was followed by the folding of other structural segments (Kedrov et al., 2006a).

A similar approach was employed to study the refolding of bacteriorhodopsin under an external load of several picoNewtons (Kessler et al., 2006). Although these were not kinetic measurements, insights into the refolding process of individual helices could be derived. For example, it was found that helices E and D exerted a force of ≈ 50 pN on the AFM cantilever during insertion into the lipid bilayer. By analyzing the area between the snap-in curve and the WLC fit it was shown that the work done during the folding of the helical hairpin constituted of helices E and D ranged from 8 to 43 k_BT. A free energy gain of 0.84 k_BT/ residue was approximated for this helical hairpin of 51 residues (Kessler et al., 2006). These results show good agreement with those estimated at a membrane/water interface (Ladokhin and White, 1999; Wieprecht et al., 2002). Free energy values of 0.2–0.69 k_BT/ residue were obtained from experiments where no insertion of the polypeptide was monitored inside the lipid bilayer but only association with the vesicle surface. The free energy values

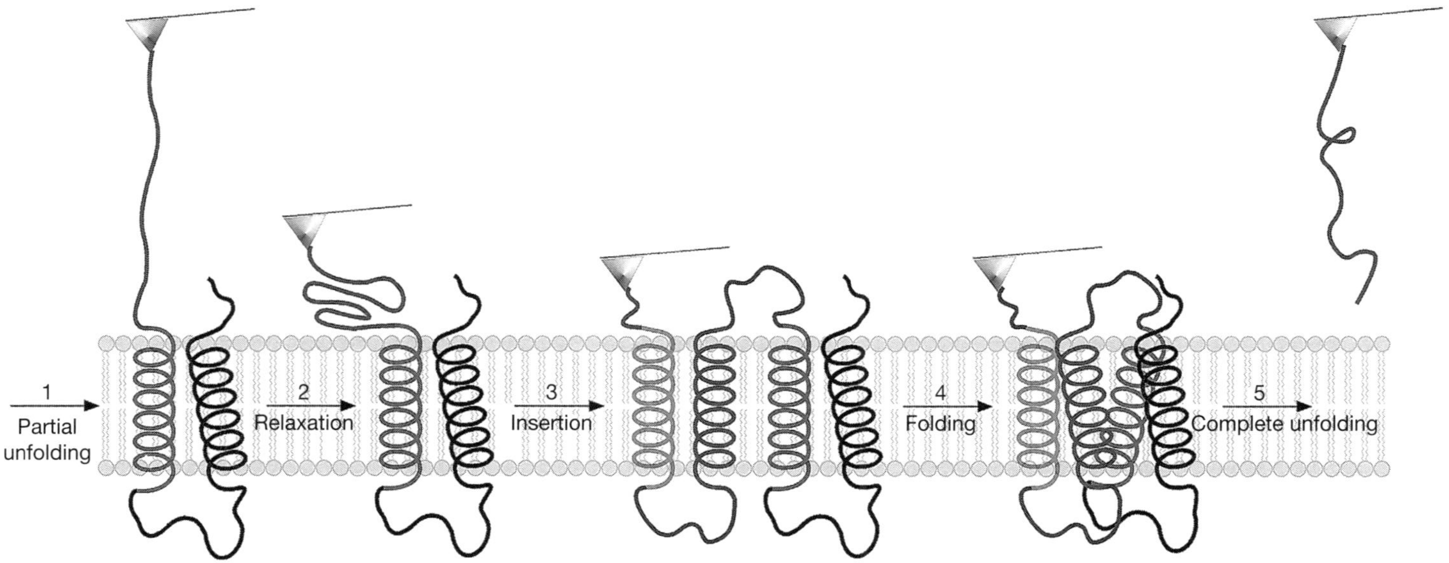

Figure 5.19 Refolding of membrane proteins using the AFM cantilever. Besides unfolding, the AFM cantilever can be used as a handle to refold membrane proteins inside the lipid bilayer. This approach involves (1) unfolding the membrane protein partially, (2) relaxing the unfolded polypeptide chain close to the bilayer-water interface, waiting for specific time intervals allowing the protein to (3) insert inside the membrane and (4) subsequently refold, and finally (5) completely unfolding the protein to determine the folding kinetics.

reported for helix insertion, 0.17–0.58 k_BT/ residue (Ladokhin and White, 2004; Soekarjo et al., 1996), are also consistent with the AFM refolding measurements.

Future Endeavors

The work reviewed here goes beyond satiating our basic curiosity on protein folding to an essential role in pharmacology and drug discovery (Muller et al., 2008). It has been shown that a number of proteins not implicated in diseases could form amyloid like plaques if given the optimum environmental factors like pH, temperature and solvent (Dobson, 2003; Fandrich et al., 2003). Because the (mis)folding and functional states of membrane proteins can be determined with SMFS (Kedrov et al., 2007), the presented approach may be applied in the future to determine which energetic pathway favors membrane protein malfunction, destabilization and misfolding. At present it is difficult to ascertain which pathways lead to the native folded state and which to the non-native or misfolded states. More elaborate experimental work in conjunction with molecular dynamic simulations is needed to shed light on the contribution of each pathway in the folding, misfolding and misassembly of a protein.

Probing the Relationship between Membrane Protein Stability and Folding Efficiency

With the advent of force-clamp AFM (Fernandez and Li, 2004; Schlierf et al., 2004) it is now possible to measure the lifetimes of different conformational states of a protein, and the data corroborated with *in vivo* studies. One of the many exciting opportunities that force-clamp presents is measuring the lifetimes of (un)folding intermediates in a stable, functional protein and in an unstable, nonfunctional protein. In particular, such measurements are relevant for having a comprehensive understanding of the behavioral dynamics of proteins *in vivo*. Such measurements would help identify the kinetically trapped misfolded states, which may be difficult for the cell to degrade.

The emerging success of chemical and pharmacological chaperones in "rescuing" membrane proteins from inefficient assembly (Bernier et al., 2004; Morello et al., 2000) is an opportunity in waiting with tremendous medical applications to exploit the potential of high-resolution imaging in combination with SMFS. Using these techniques in tandem, it may be easy to identify trapped conformations or misassembled proteins, and help design and study the effect of small molecules and mutations to test new conditions where abnormal conformational states can be reversed or avoided.

Another possible use of the high-resolution imaging and SMFS technology that we foresee is its potential use as a tool to predict membrane protein topology (Elofsson and von Heijne, 2007). Current topology prediction routines are based on sequence statistics and rely heavily on computations. It was only very recently that the first membrane protein topology prediction methods based on the physicochemical principles were presented (Bernsel et al., 2008). The similarity in the unfolding patterns of the topological siblings BR, HR, PR and SR stresses the importance of the possibility of using SMFS as a topology predictor.

Improving the Experimental Setup

An obvious shortcoming of the SMFS setup for membrane proteins is the use of supported lipid bilayers. Membrane proteins in lipid bilayers adsorbed directly on a mica or a gold support may interact with the surface. This is, however, not a common scenario as suggested by comparative experiments done on lipid bilayers adsorbed on mica and those done on lipid bilayers adsorbed on "empty" lipid membranes on mica (Kedrov et al., 2004; Müller et al., 2002d). Nevertheless, overcoming this drawback is an important issue in SMFS. Employing fabricated surfaces with an array of nano holes (200–500 nm in diameter) offers a convenient solution to this problem (Goncalves et al., 2006). This methodology would ease establishing ion gradients across lipid bilayers for studying ion channels or pH gradients for membrane proteins that require a proton gradient for their function. Another possible advantage of fabricated surfaces would be in the study of membrane protein unfolding and refolding in the presence of their soluble partners.

To conclude, the AFM-related techniques presented here, in our opinion, will not only help us to understand the basic principles of water-soluble and membrane protein (un)folding, but also to manipulate and tune energy landscapes with the help of small molecules, proteins, or mutations, thus opening up new vistas in medicine and pharmacology.

REFERENCES

Ainavarapu, S.R., Li, L., Badilla, C.L., and Fernandez, J.M. (2005). Ligand binding modulates the mechanical stability of dihydrofolate reductase. BiophysJ *89*, 3337–3344.

Ando, T., Kodera, N., Takai, E., Maruyama, D., Saito, K., and Toda, A. (2001). A high-speed atomic force microscope for studying biological macromolecules. Proc Natl Acad Sci USA *98*, 12468–12472.

Ando, T., Kodera, N., Naito, Y., Kinoshita, T., Furuta, K., and Toyoshima, Y.Y. (2003). A high-speed atomic force microscope for studying biological macromolecules in action. Chemphyschem *4*, 1196–1202.

Baldwin, R.L., and Rose, G.D. (1999a). Is protein folding hierarchic? I. Local structure and peptide folding. Trends Biochem Sci *24*, 26–33.

Baldwin, R.L., and Rose, G.D. (1999b). Is protein folding hierarchic? II. Folding intermediates and transition states. Trends Biochem Sci *24*, 77–83.

Bell, G.I. (1978). Models for the specific adhesion of cells to cells. Science *200*, 618–627.

Benoit, M., Gabriel, D., Gerisch, G., and Gaub, H.E. (2000). Discrete interactions in cell adhesion measured by single-molecule force spectroscopy. Nat Cell Biol *2*, 313–317.

Bernier, V., Bichet, D.G., and Bouvier, M. (2004). Pharmacological chaperone action on G-protein-coupled receptors. Curr Opin Pharmacol *4*, 528–533.

Bernsel, A., Viklund, H., Falk, J., Lindahl, E., von Heijne, G., and Elofsson, A. (2008). Prediction of membrane-protein topology from first principles. Proc Natl Acad Sci U S A *105*, 7177–7181.

Best, R.B., Li, B., Steward, A., Daggett, V., and Clarke, J. (2001). Can non-mechanical proteins withstand force? Stretching barnase by atomic force microscopy and molecular dynamics simulation. Biophys J *81*, 2344–2356.

Bezanilla, M., Drake, B., Nudler, E., Kashlev, M., Hansma, P.K., and Hansma, H.G. (1994). Motion and enzymatic degradation of DNA in the atomic force microscope. Biophys J *67*, 2454–2459.

Bieri, O., Wirz, J., Hellrung, B., Schutkowski, M., Drewello, M., and Kiefhaber, T. (1999). The speed limit for protein folding measured by triplet-triplet energy transfer. Proc Natl Acad Sci U S A *96*, 9597–9601.

Binnig, G., Quate, C.F., and Gerber, C. (1986). Atomic force microscope. Phys Rev Lett *56*, 930–933.

Binnig, G., Rohrer, H., Gerber, C., and Weibel, E. (1982). Tunneling through a controllable vacuum gap. Appl Phys Lett *40*, 178.

Bippes, C.A., Zeltina, A., Casagrande, F., Ratera, M., Palacin, M., Muller, D.J., and Fotiadis, D. (2009). Substrate binding tunes conformational flexibility and kinetic stability of an amino acid antiporter. J Biol Chem *284*, 18651-18663.

Booth, P.J. (2000). Unravelling the folding of bacteriorhodopsin. Biochim Biophys Acta *1460*, 4–14.

Booth, P.J. (2003). The trials and tribulations of membrane protein folding in vitro. Biochim Biophys Acta *1610*, 51–56.

Borgia, A., Williams, P.M., and Clarke, J. (2008). Single-Molecule Studies of Protein Folding. Annu Rev Biochem *77*, 101–125.

Bosshart, P.D., Iordanov, I., Garzon-Coral, C., Demange, P., Engel, A., Milon, A., and Müller, D.J. (2012). The transmembrane protein KpOmpA anchoring the outer membrane of *Klebsiella pneumoniae* unfolds and refolds in response to tensile load. Structure *20*, 121–127.

Bowie, J.U. (2001). Stabilizing membrane proteins. Curr Opin Struct Biol *11*, 397–402.

Breyton, C., Haase, W., Rapoport, T.A., Kuhlbrandt, W., and Collinson, I. (2002). Three-dimensional structure of the bacterial protein-translocation complex SecYEG. Nature *418*, 662–665.

Brockwell, D.J. (2007). Probing the mechanical stability of proteins using the atomic force microscope. Biochem Soc Trans *35*, 1564–1568.

Bullard, B., Garcia, T., Benes, V., Leake, M.C., Linke, W.A., and Oberhauser, A.F. (2006). The molecular elasticity of the insect flight muscle proteins projectin and kettin. Proc Natl Acad Sci U S A *103*, 4451–4456.

Bustamante, C. (2008). In singulo Biochemistry: When Less Is More. Annu Rev Biochem *77*, 45–50.

Bustamante, C., Chemla, Y.R., Forde, N.R., and Izhaky, D. (2004). Mechanical processes in biochemistry. Annu Rev Biochem *73*, 705–748.

Carrion-Vazquez, M., Oberhauser, A.F., Fowler, S.B., Marszalek, P.E., Broedel, S.E., Clarke, J., and Fernandez, J.M. (1999). Mechanical and chemical unfolding of a single protein: a comparison. Proc Natl Acad Sci U S A *96*, 3694–3699.

Chen, X., Davies, M.C., Roberts, C.J., Shakesheff, K.M., Tendler, S.J.B., and Williams, P.M. (1996). Dynamic surface events measured by simultaneous probe microscopy and surface plasmon detection. Anal Chem *68*, 1451–1455.

Cisneros, D.A., Friedrichs, J., Taubenberger, A., Franz, C.M., and Muller, D.J. (2007). Creating ultrathin nanoscopic collagen matrices for biological and biotechnological applications. Small *3*, 956–963.

Cisneros, D.A., Hung, C., Franz, C.M., and Muller, D.J. (2006). Observing growth steps of collagen self-assembly by time-lapse high-resolution atomic force microscopy. J Struct Biol *154*, 232–245.

Cisneros, D.A., Oberbarnscheidt, L., Pannier, A., Klare, J.P., Helenius, J., Engelhard, M., Oesterhelt, F., and Muller, D.J. (2008). Transducer binding establishes localized interactions to tune sensory rhodopsin II. Structure *16*, 1206-1213..

Cisneros, D.A., Oesterhelt, D., and Müller, D.J. (2005). Probing origins of molecular interactions stabilizing the membrane proteins halorhodopsin and bacteriorhodopsin. Structure *13*, 235–242.

Clausen-Schaumann, H., Seitz, M., Krautbauer, R., and Gaub, H.E. (2000). Force spectroscopy with single bio-molecules. Curr Opin Chem Biol *4*, 524–530.

Cleveland, J.P., Manne, S., Bocek, D., and Hansma, P.K. (1993). A nondestructive method for determining the spring constant of cantilevers for scanning force microscopy. Rev Sci Instrum *64*, 403–405.

Cui, Y., and Bustamante, C. (2000). Pulling a single chromatin fiber reveals the forces that maintain its higher-order structure. Proc Natl Acad Sci U S A *97*, 127–132.

Curnow, P., and Booth, P.J. (2007). Combined kinetic and thermodynamic analysis of alpha-helical membrane protein unfolding. Proc Natl Acad Sci U S A *104*, 18970–18975.

Czajkowsky, D.M., Hotze, E.M., Shao, Z., and Tweten, R.K. (2004). Vertical collapse of a cytolysin prepore moves its transmembrane beta-hairpins to the membrane. EMBO J *23*, 3206–3215.

Czajkowsky, D.M., Iwamoto, H., and Shao, Z. (2000). Atomic force microscopy in structural biology: from the subcellular to the submolecular. J Electron Microsc (Tokyo) *49*, 395–406.

Dai, H., Hafner, J.H., Rinzler, A.G., Colbert, D.T., and Smalley, R.E. (1996). Nanotubes as nonoprobes in scanning probe microscopy. Nature *384*, 147–150.

del Valle, L.J., Ramon, E., Canavate, X., Dias, P., and Garriga, P. (2003). Zinc-induced decrease of the thermal stability and regeneration of rhodopsin. J Biol Chem *278*, 4719–4724.

Dietz, H., Berkemeier, F., Bertz, M., and Rief, M. (2006a). Anisotropic deformation response of single protein molecules. Proc Natl Acad Sci U S A *103*, 12724–12728.

Dietz, H., Bertz, M., Schlierf, M., Berkemeier, F., Bornschlogl, T., Junker, J.P., and Rief, M. (2006b). Cysteine engineering of polyproteins for single-molecule force spectroscopy. Nat Protoc *1*, 80–84.

Dietz, H., and Rief, M. (2004). Exploring the energy landscape of GFP by single-molecule mechanical experiments. Proc Natl Acad Sci U S A *101*, 16192–16197.

Dietz, H., and Rief, M. (2006). Protein structure by mechanical triangulation. Proc Natl Acad Sci U S A *103*, 1244–1247.

Dobson, C.M. (2003). Protein folding and misfolding. Nature *426*, 884–890.

Drake, B., Prater, C.B., Weisenhorn, A.L., Gould, S.A., Albrecht, T.R., Quate, C.F., Cannell, D.S., Hansma, H.G., and Hansma, P.K. (1989). Imaging crystals, polymers, and processes in water with the atomic force microscope. Science *243*, 1586–1589.

Dugdale, T.M., Dagastine, R., Chiovitti, P., Mulvaney, P., and Wetherbee, R. (2005). Single adhesive nanofibers from a live diatom have the signature fingerprint of modular proteins. Biophys J *89*, 4252–4260.

Dunlap, D.D., Maggi, A., Soria, M.R., and Monaco, L. (1997). Nanoscopic structure of DNA condensed for gene delivery. Nucl Acids Res *25*, 3095.

Elie-Caille, C., Severin, F., Helenius, J., Howard, J., Muller, D.J., and Hyman, A.A. (2007). Straight GDP-tubulin protofilaments form in the presence of taxol. Curr Biol *17*, 1765–1770.

Elofsson, A., and von Heijne, G. (2007). Membrane protein structure: prediction versus reality. Annu Rev Biochem *76*, 125–140.

Eng, L.M., Seuret, C., Looser, H., and Günter, P. (1996). Approaching the liquid/air interface with scanning force microscopy. J Vac Sci & Technol B *14*, 1386–1389.

Engel, A., and Gaub, H.E. (2008). Structure and mechanics of membrane proteins. Annu Rev Biochem *77*, 127–148.

Engel, A., Lyubchenko, Y., and Müller, D.J. (1999). Atomic force microscopy: a powerful tool to observe biomolecules at work. Trends in Cell Biol *9*, 77–80.

Engel, A., and Müller, D.J. (2000). Observing single biomolecules at work with the atomic force microscope. Nat Struct Biol *7*, 715–718.

Engelman, D.M., and Steitz, T.A. (1981). The spontaneous insertion of proteins into and across membranes: the helical hairpin hypothesis. Cell *23*, 411–422.

Evans, E. (1998). Energy landscapes of biomolecular adhesion and receptor anchoring at interfaces explored with dynamic force spectroscopy. Faraday Discuss *111*, 1–16.

Evans, E. (1999). Looking inside molecular bonds at biological interfaces with dynamic force spectroscopy. Biophys Chem *82*, 83–97.

Evans, E. (2001). Probing the relation between force-lifetime-and chemistry in single molecular bonds. Annu Rev Biophys Biomol Struct *30*, 105–128.

Evans, E., and Ritchie, K. (1997). Dynamic strength of molecular adhesion bonds. Biophys J *72*, 1541–1555.

Fandrich, M., Forge, V., Buder, K., Kittler, M., Dobson, C.M., and Diekmann, S. (2003). Myoglobin forms amyloid fibrils by association of unfolded polypeptide segments. Proc Natl Acad Sci U S A *100*, 15463–15468.

Fernandez, J.M., and Li, H. (2004). Force-clamp spectroscopy monitors the folding trajectory of a single protein. Science *303*, 1674–1678.

Fleishman, S.J., Unger, V.M., and N., B.-T. (2006). Transmembrane protein structures without X-rays. Trends Biochem Sci *31*, 106–113.

Florin, E.-L., Moy, V.T., and Gaub, H.E. (1994). Adhesion forces between individual ligand-receptor pairs. Science *264*, 415–417.

Florin, E.-L., Rief, M., Lehmann, H., Ludwig, M., Dornmair, C., Moy, V.T., and Gaub, H.E. (1995). Sensing specific molecular interactions with the atomic force microscope. Biosens Bioelectron *10*, 895–901.

Fotiadis, D., Liang, Y., Filipek, S., Saperstein, D.A., Engel, A., and Palczewski, K. (2003). Atomic-force microscopy: Rhodopsin dimers in native disc membranes. Nature *421*, 127–128.

Franz, C.M., Taubenberger, A., Puech, P.H., and Muller, D.J. (2007). Studying integrin-mediated cell adhesion at the single-molecule level using AFM force spectroscopy. Sci STKE *2007*, pl5.

Frauenfelder, H., Sligar, S.G., and Wolynes, P.G. (1991). The energy landscapes and motions of proteins. Science *254*, 1598–1603.

Frederix, P.L., Akiyama, T., Staufer, U., Gerber, C., Fotiadis, D., Muller, D.J., and Engel, A. (2003). Atomic force bio-analytics. Curr Opin Chem Biol *7*, 641–647.

Fritz, J., Katopodis, A.G., Kolbinger, F., and Anselmetti, D. (1998). Force-mediated kinetics of single P-selectin/ligand complexes observed by atomic force microscopy. Proc Natl Acad Sci U S A *95*, 12283–12288.

Fritz, M., Radmacher, M., Allersma, M.W., Cleveland, J.P., Stewart, R.J., Hansma, P.K., and Schmidt, C.F. (1995). Imaging microtubles in buffer solution using tapping mode atomic force microscopy. SPIE *2384*, 150–157.

Fukuma, T., Kobayashi, K., Matsushige, K., and Yamada, H. (2005). True atomic resolution in liquid by frequency-modulation atomic force microscopy. Appl Phys Lett *87*, 034101.

Ganchev, D.N., Rijkers, D.T., Snel, M.M., Killian, J.A., and de Kruijff, B. (2004). Strength of integration of transmembrane alpha-helical peptides in lipid bilayers as determined by atomic force spectroscopy. Biochemistry *43*, 14987–14993.

Garriga, P., Liu, X., and Khorana, H.G. (1996). Structure and function in rhodopsin: correct folding and misfolding in point mutants at and in proximity to the site of the retinitis pigmentosa mutation Leu-125-->Arg in the transmembrane helix C. Proc Natl Acad Sci U S A *93*, 4560–4564.

Gerber, C., and Lang, H.P. (2006). How the doors to the nanoworld were opened. Nat Nanotechnol *1*, 3–5.

Gibson, C.T., Watson, G.S., and Myhra, S. (1996). Determination of the spring constants of probes for force microscopy/spectroscopy. Nanotechnology *7*, 259–262.

Goncalves, R.P., Agnus, G., Sens, P., Houssin, C., Bartenlian, B., and Scheuring, S. (2006). Two-chamber AFM: probing membrane proteins separating two aqueous compartments. Nat Methods *3*, 1007–1012.

Goncalves, R.P., Buzhynskyy, N., Prima, V., Sturgis, J.N., and Scheuring, S. (2007). Supramolecular assembly of VDAC in native mitochondrial outer membranes. J Mol Biol *369*, 413–418.

Grandbois, M., Beyer, M., Rief, M., Clausen-Schaumann, H., and Gaub, H.E. (1999). How strong is a covalent bond? Science *283*, 1727–1730.

Greenleaf, W.J., Frieda, K.L., Foster, D.A., Woodside, M.T., and Block, S.M. (2008). Direct observation of hierarchical folding in single riboswitch aptamers. Science *319*, 630–633.

Guthold, M., Zhu, X., Rivetti, C., Yang, G., Thomson, N.H., Kasas, S., Hansma, H.G., Smith, B., Hansma, P.K., and Bustamante, C. (1999). Direct observation of one-dimensional diffusion and transcription by *Escherichia coli* RNA polymerase. Biophys J *77*, 2284–2294.

Gutsmann, T., Fantner, G.E., Kindt, J.H., Venturoni, M., Danielsen, S., and Hansma, P.K. (2004). Force spectroscopy of collagen fibers to investigate their mechanical properties and structural organization. Biophys J *86*, 3186–3193.

Han, W., Mou, J., Sheng, J., Yang, J., and Shao, Z. (1995). Cryo-Atomic Force Microscopy: a new approach for biological imaging at high resolution. Biochemistry *34*, 8215–8220.

Hansma, H.G., and Hoh, J.H. (1994). Biomolecular imaging with the atomic force microscope. Annu Rev Biophys Biomol Struct *23*, 115–139.

Hansma, P.K., Cleveland, J.P., Radmacher, M., Walters, D.A., Hillner, P.E., Bezanilla, M., Fritz, M., Vie, D., Hansma, H.G., Prater, C.B., et al. (1994). Tapping mode atomic force microscopy in liquids. Appl Phys Lett *64*, 1738–1740.

Hansma, P.K., Drake, B., Marti, O., Gould, S.A., and Prater, C.B. (1989). The scanning ion-conductance microscope. Science *243*, 641–643.

Helenius, J., Heisenberg, C.P., Gaub, H.E., and Muller, D.J. (2008). Single-cell force spectroscopy. J Cell Sci *121*, 1785–1791.

Herbert, K.M., Greenleaf, W.J., and Block, S.M. (2008). Single-Molecule Studies of RNA Polymerase: Motoring Along. Annu Rev Biochem *77*, 149–176.

Heymann, J.B., Moller, C., and Muller, D.J. (2002). Sampling effects influence heights measured with atomic force microscopy. J Microsc *207*, 43–51.

Hinterdorfer, P., Baumgartner, W., Gruber, H.J., Schilcher, K., and Schindler, H. (1996). Detection and localization of individual antibody-antigen recognition events by atomic force microscopy. Proc Natl Acad Sci U S A *93*, 3477–3481.

Hinterdorfer, P., and Dufrene, Y.F. (2006). Detection and localization of single molecular recognition events using atomic force microscopy. Nat Methods *3*, 347–355.

Holst, B., Elling, C.E., and Schwartz, T.W. (2002). Metal ion-mediated agonism and agonist enhancement in melanocortin MC1 and MC4 receptors. J Biol Chem *277*, 47662–47670.

Hoogenboom, B.W., Suda, K., Engel, A., and Fotiadis, D. (2007). The supramolecular assemblies of voltage-dependent anion channels in the native membrane. J Mol Biol *370*, 246–255.

Howard, J. (2001). Mechanics of motor proteins and the cytoskeleton (Sunderland, Massachusetts, Sinauer Associates Inc.).

Hughes, T., Strongin, B., Gao, F.P., Vijayvergiya, V., Busath, D.D., and Davis, R.C. (2004). AFM visualization of mobile influenza A M2 molecules in planar bilayers. Biophys J *87*, 311–322.

Humphris, A.D., Miles, M., and Hobbs, J.K. (2005). A mechanical microscope: High-speed atomic force microscopy. Appl Phys Lett *86*, 034106.

Hunt, J.F., Earnest, T.N., Bousche, O., Kalghatgi, K., Reilly, K., Horvath, C., Rothschild, K.J., and Engelman, D.M. (1997). A biophysical study of integral membrane protein folding. Biochemistry *36*, 15156–15176.

Hwa, J., Garriga, P., Liu, X., and Khorana, H.G. (1997). Structure and function in rhodopsin: packing of the helices in the transmembrane domain and folding to a tertiary structure in the intradiscal domain are coupled. Proc Natl Acad Sci U S A *94*, 10571–10576.

Hyeon, C., and Thirumalai, D. (2003). Can energy landscape roughness of proteins and RNA be measured by using mechanical unfolding experiments? Proc Natl Acad Sci U S A *100*, 10249–10253.

Israelachvili, J. (1991). Intermolecular & surface forces, Second edn (London, Academic Press Limited).

Janovjak, H., Kedrov, A., Cisneros, D.A., Sapra, K.T., and Müller, D.J. (2006). Imaging and detecting molecular interactions of single transmembrane proteins. Neurobiol of Aging *27*, 546–561.

Janovjak, H., Kessler, M., Oesterhelt, D., Gaub, H.E., and Müller, D.J. (2003). Unfolding pathways of native bacteriorhodopsin depend on temperature. EMBO J *22*, 5220–5229.

Janovjak, H., Knaus, H., and Muller, D.J. (2007). Transmembrane helices have rough energy surfaces. J Am Chem Soc *129*, 246–247.

Janovjak, H., Müller, D.J., and Humphris, A.D. (2005). Molecular force modulation spectroscopy revealing the dynamic response of single bacteriorhodopsins. Biophys J *88*, 1423–1431.

Janovjak, H., Struckmeier, J., Hubain, M., Kedrov, A., Kessler, M., and Müller, D.J. (2004). Probing the energy landscape of the membrane protein bacteriorhodopsin. Structure *12*, 871–879.

Jaroslawski, S., Zadek, B., Ashcroft, F., Venien-Bryan, C., and Scheuring, S. (2007). Direct visualization of KirBac3.1 potassium channel gating by atomic force microscopy. J Mol Biol *374*, 500–505.

Joh, N.H., Min, A., Faham, S., Whitelegge, J.P., Yang, D., Woods, V.L., and Bowie, J.U. (2008). Modest stabilization by most hydrogen-bonded side-chain interactions in membrane proteins. Nature *453*, 1266-1270.

Junge, W., and Muller, D.J. (2011). Seeing a molecular motor at work. Science *333*, 704-705.

Junker, J.P., Hell, K., Schlierf, M., Neupert, W., and Rief, M. (2005). Influence of substrate binding on the mechanical stability of mouse dihydrofolate reductase. Biophys J *89*, L46–L48.

Karrasch, S., Dolder, M., Hoh, J., Schabert, F., Ramsden, J., and Engel, A. (1993). Covalent binding of biological samples to solid supports for scanning probe microscopy in buffer solution. Biophys J *65*, 2437–2446.

Kasas, S., Thomson, N.H., Smith, B.L., Hansma, H.G., Zhu, X., Guthold, M., Bustamante, C., Kool, E.T., Kashlev, M., and Hansma, P.K. (1997). Escherichia coli RNA ploymerase activity observed using atomic force microscopy. Biochem *36*, 461–468.

Kawakami, M., Byrne, K., Khatri, B., Mcleish, T.C., Radford, S.E., and Smith, D.A. (2004). Viscoelastic properties of single polysaccharide molecules determined by analysis of thermally driven oscillations of an atomic force microscope cantilever. Langmuir *20*, 9299–9303.

Kedrov, A., Appel, M., Baumann, H., Ziegler, C., and Muller, D.J. (2008). Examining the dynamic energy landscape of an antiporter upon inhibitor binding. J Mol Biol *375*, 1258–1266.

Kedrov, A., Janovjak, H., Sapra, K.T., and Muller, D.J. (2007). Deciphering molecular interactions of native membrane proteins by single-molecule force spectroscopy. Annu Rev Biophys Biomol Struct *36*, 233–260.

Kedrov, A., Janovjak, H., Ziegler, C., Kühlbrandt, W., and Müller, D.J. (2006a). Observing folding kinetics and pathways of single antiporters. J Mol Biol *355*, 2–8.

Kedrov, A., Krieg, M., Ziegler, C., Kuhlbrandt, W., and Müller, D.J. (2005). Locating ligand binding and activation of a single antiporter. EMBO Rep *6*, 668–674.

Kedrov, A., Ziegler, C., Janovjak, H., Kuhlbrandt, W., and Müller, D.J. (2004). Controlled unfolding and refolding of a single sodium-proton antiporter using atomic force microscopy. J Mol Biol *340*, 1143–1152.

Kedrov, A., Ziegler, C., and Müller, D.J. (2006b). Differentiating ligand and inhibitor interactions of a single antiporter. J Mol Biol *362*, 925–932.

Kellermayer, M.S., Smith, S.B., Granzier, H.L., and Bustamante, C. (1997). Folding-unfolding transitions in single titin molecules characterized with laser tweezers. Science *276*, 1112–1116.

Kenniston, J.A., Baker, T.A., Fernandez, J.M., and Sauer, R.T. (2003). Linkage between ATP consumption and mechanical unfolding during the protein processing reactions of an AAA+ degradation machine. Cell *114*, 511–520.

Kessler, M., Gottschalk, K.E., Janovjak, H., Muller, D.J., and Gaub, H.E. (2006). Bacteriorhodopsin folds into the membrane against an external force. J Mol Biol *357*, 644–654.

Klyszejko, A.L., Shastri, S., Mari, S.A., Grubmuller, H., Muller, D.J., and Glaubitz, C. (2008). Folding and assembly of proteorhodopsin. J Mol Biol *376*, 35–41.

Krieg, M., Arboleda-Estudillo, Y., Puech, P.H., Kafer, J., Graner, F., Muller, D.J., and Heisenberg, C.P. (2008). Tensile forces govern germ-layer organization in zebrafish. Nat Cell Biol *10*, 429–436.

Kueng, A., Kranz, C., Lugstein, A., Bertagnolli, E., and Mizaikoff, B. (2005). Nanoelectrodes integrated in atomic force microscopy cantilevers for imaging of in situ enzyme activity. Methods Mol Biol *300*, 403–415.

Ladokhin, A.S., and White, S.H. (1999). Folding of amphipathic alpha-helices on membranes: energetics of helix formation by melittin. JMolBiol *285*, 1363–1369.

Ladokhin, A.S., and White, S.H. (2004). Interfacial folding and membrane insertion of a designed helical peptide. Biochemistry *43*, 5782–5791.

Lee, G., Abdi, K., Jiang, Y., Michaely, P., Bennett, V., and Marszalek, P.E. (2006). Nanospring behaviour of ankyrin repeats. Nature *440*, 246–249.

Lee, G.U., Chrisey, L.A., and Colton, R.J. (1994). Direct measurement of the forces between complementary strands of DNA. Science *266*, 771–773.

Leonhard, K., Guiard, B., Pellecchia, G., Tzagoloff, A., Neupert, W., and Langer, T. (2000). Membrane protein degradation by AAA proteases in mitochondria: extraction of substrates from either membrane surface. Mol Cell *5*, 629–638.

Li, H., Carrion-Vazquez, M., Oberhauser, A.F., Marszalek, P.E., and Fernandez, J.M. (2000). Point mutations alter the mechanical stability of immunoglobulin modules. Nat Struct Biol *7*, 1117–1120.

Liu, X., Garriga, P., and Khorana, H.G. (1996). Structure and function in rhodopsin: correct folding and misfolding in two point mutants in the intradiscal domain of rhodopsin identified in retinitis pigmentosa. Proc Natl Acad Sci U S A *93*, 4554–4559.

Marszalek, P.E., Li, H., Oberhauser, A.F., and Fernandez, J.M. (2002). Chair-boat transitions in single polysaccharide molecules observed with force-ramp AFM. Proc Natl Acad Sci U S A *99*, 4278–4283.

Marszalek, P.E., Lu, H., Li, H., Carrion-Vazquez, M., Oberhauser, A.F., Schulten, K., and Fernandez, J.M. (1999). Mechanical unfolding intermediates in titin modules. Nature *402*, 100–103.

Marti, O., Elings, V., Haugan, M., Bracker, C.E., Schneir, J., Drake, B., Gould, S.A., Gurley, J., Hellemans, L., Shaw, K., et al. (1988). Scanning probe microscopy of biological samples and other surfaces. J Microsc *152*, 803–809.

McCully, M.E., Beck, D.A., and Daggett, V. (2008). Microscopic reversibility of protein folding in molecular dynamics simulations of the engrailed homeodomain. Biochemistry *47*, 7079–7089.

McKibbin, C., Farmer, N.A., Jeans, C., Reeves, P.J., Khorana, H.G., Wallace, B.A., Edwards, P.C., Villa, C., and Booth, P.J. (2007). Opsin stability and folding: modulation by phospholipid bicelles. J Mol Biol *374*, 1319–1332.

Merkel, R., Nassoy, P., Leung, A., Ritchie, K., and Evans, E. (1999). Energy landscapes of receptor-ligand bonds explored with dynamic force spectroscopy. Nature *397*, 50–53.

Meyer, E., Hug, H.J., and Bennewitz, R. (2003). Scanning Probe Microscopy—The Lab on a Tip (Berlin, Springer).

Möller, C., Allen, M., Elings, V., Engel, A., and Müller, D.J. (1999). Tapping-mode atomic force microscopy produces faithful high-resolution images of protein surfaces. Biophys J *77*, 1150–1158.

Möller, C., Fotiadis, D., Suda, K., Engel, A., Kessler, M., and Müller, D.J. (2003). Determining molecular forces that stabilize human aquaporin-1. J Struct Biol *142*, 369–378.

Morello, J.P., Salahpour, A., Laperriere, A., Bernier, V., Arthus, M.F., Lonergan, M., Petaja-Repo, U., Angers, S., Morin, D., Bichet, D.G., et al. (2000). Pharmacological chaperones rescue cell-surface expression and function of misfolded V2 vasopressin receptor mutants. J Clin Invest *105*, 887–895.

Müller, D.J., Amrein, M., and Engel, A. (1997). Adsorption of biological molecules to a solid support for scanning probe microscopy. J Struct Biol *119*, 172–188.

Müller, D.J., Baumeister, W., and Engel, A. (1996). Conformational change of the hexagonally packed intermediate layer of Deinococcus radiodurans monitored by atomic force microscopy. J Bacteriol *178*, 3025–3030.

Müller, D.J., Büldt, G., and Engel, A. (1995). Force-induced conformational change of bacteriorhodopsin. J Mol Biol *249*, 239–243.

Muller, D.J., and Dufrene, Y.F. (2008). Atomic force microscopy as a multifunctional molecular toolbox in nanobiotechnology. Nat Nanotechnol *3*, 261–269.

Muller, D.J., and Engel, A. (2007). Atomic force microscopy and spectroscopy of native membrane proteins. Nat Protoc *2*, 2191–2197.

Müller, D.J., and Engel, A. (1997). The height of biomolecules measured with the atomic force microscope depends on electrostatic interactions. Biophys J *73*, 1633–1644.

Müller, D.J., and Engel, A. (1999). Voltage and pH-induced channel closure of porin OmpF visualized by atomic force microscopy. J Mol Biol *285*, 1347–1351.

Müller, D.J., and Engel, A. (2002). Conformations, flexibility, and interactions observed on individual membrane proteins by atomic force microscopy. Methods Cell Biol *68*, 257–299.

Müller, D.J., Engel, A., Matthey, U., Meier, T., Dimroth, P., and Suda, K. (2003). Observing membrane protein diffusion at subnanometer resolution. J Mol Biol *327*, 925–930.

Müller, D.J., Fotiadis, D., Scheuring, S., Müller, S.A., and Engel, A. (1999). Electrostatically balanced subnanometer imaging of biological specimens by atomic force microscopy. Biophys J *76*, 1101–1111.

Müller, D.J., Hand, G.M., Engel, A., and Sosinsky, G. (2002a). Conformational changes in surface structures of isolated connexin26 gap junctions. EMBO J *21*, 3598–3607.

Müller, D.J., Hand, G.M., Engel, A., and Sosinsky, G. (2002b). Conformational changes in surface structures of isolated connexin26 gap junctions. EMBO J *21*, 3598–3607.

Müller, D.J., Janovjak, H., Lehto, T., Kuerschner, L., and Anderson, K. (2002c). Observing structure, function and assembly of single proteins by AFM. Prog Biophys Mol Biol *79*, 1–43.

Müller, D.J., Kessler, M., Oesterhelt, F., Möller, C., Oesterhelt, D., and Gaub, H. (2002d). Stability of bacteriorhodopsin alpha-helices and loops analyzed by single-molecule force spectroscopy. Biophys J *83*, 3578–3588.

Muller, D.J., Sapra, K.T., Scheuring, S., Kedrov, A., Frederix, P.L., Fotiadis, D., and Engel, A. (2006). Single-molecule studies of membrane proteins. Curr Opin Struct Biol *16*, 489–495.

Muller, D.J., Schabert, F.A., Buldt, G., and Engel, A. (1995). Imaging purple membranes in aqueous solutions at sub-nanometer resolution by atomic force microscopy. Biophys J *68*, 1681–1686.

Muller, D.J., Wu, N., and Palczewski, K. (2008). Vertebrate membrane proteins: structure, function, and insights from biophysical approaches. Pharmacol Rev *60*, 43–78.

Neupert, W., and Herrmann, J.M. (2007). Translocation of proteins into mitochondria. Annu Rev Biochem *76*, 723–749.

Nevo, R., Brumfeld, V., Kapon, R., Hinterdorfer, P., and Reich, Z. (2005). Direct measurement of protein energy landscape roughness. EMBO Rep *6*, 482–486.

Oberhauser, A.F., Badilla-Fernandez, C., Carrion-Vazquez, M., and Fernandez, J.M. (2002). The mechanical hierarchies of fibronectin observed with single-molecule AFM. J Mol Biol *319*, 433–447.

Oberhauser, A.F., and Carrion-Vazquez, M. (2008). Mechanical biochemistry of proteins one molecule at a time. J Biol Chem *283*, 6617–6621.

Oberhauser, A.F., Hansma, P.K., Carrion-Vazquez, M., and Fernandez, J.M. (2001). Stepwise unfolding of titin under force-clamp atomic force microscopy. Proc Natl Acad Sci U S A *98*, 468–472.

Oberhauser, A.F., Marszalek, P.E., Carrion-Vazquez, M., and Fernandez, J.M. (1999). Single protein misfolding events captured by atomic force microscopy. Nat Struct Biol *6*, 1025–1028.

Oberhauser, A.F., Marszalek, P.E., Erickson, H.P., and Fernandez, J.M. (1998). The molecular elasticity of the extracellular matrix protein tenascin. Nature *393*, 181–185.

Oesterhelt, F., Oesterhelt, D., Pfeiffer, M., Engel, A., Gaub, H.E., and Müller, D.J. (2000). Unfolding pathways of individual bacteriorhodopsins. Science *288*, 143–146.

Oroudjev, E., Soares, J., Arcdiacono, S., Thompson, J.B., Fossey, S.A., and Hansma, H.G. (2002). Segmented nanofibers of spider dragline silk: atomic force microscopy and single-molecule force spectroscopy. Proc Natl Acad Sci U S A *99*, 6460–6465.

Padan, E., Venturi, M., Gerchman, Y., and Dover, N. (2001). Na(+)/H(+) antiporters. Biochim Biophys Acta *1505*, 144-157.

Park, P.S., Sapra, K.T., Kolinski, M., Filipek, S., Palczewski, K., and Muller, D.J. (2007). Stabilizing effect of Zn2+ in native bovine rhodopsin. J Biol Chem *282*, 11377–11385.

Perez-Jimenez, R., Ingles-Prieto, A., Zhao, Z.M., Sanchez-Romero, I., Alegre-Cebollada, J., Kosuri, P., Garcia-Manyes, S., Kappock, T.J., Tanokura, M., Holmgren, A., *et al.* (2011). Single-molecule paleoenzymology probes the chemistry of resurrected enzymes. Nat Struct Mol Biol *18*, 592-596.

Prakash, S., and Matouschek, A. (2004). Protein unfolding in the cell. Trends Biochem Sci *29*, 593–600.

Proksch, R., Lal, R., Hansma, P.K., Morse, D., and Stucky, G. (1996). Imaging the internal and external pore structure of membranes in fluid: TappingMode scanning ion conductance microscopy. Biophys J *71*, 2155–2157.

Putman, C.A.J., Vanderwerf, K.O., Degrooth, B.G., Vanhulst, N.F., and Greve, J. (1994). Tapping mode atomic force microscopy in liquid. Appl Phys Lett *64*, 2454–2456.

Rapoport, T.A. (2007). Protein translocation across the eukaryotic endoplasmic reticulum and bacterial plasma membranes. Nature *450*, 663–669.

Rief, M., Clausen-Schaumann, H., and Gaub, H.E. (1999a). Sequence-dependent mechanics of single DNA molecules. Nat Struct Biol *6*, 346–349.

Rief, M., Fernandez, J.M., and Gaub, H.E. (1998a). Elastically coupled two-level-systems as a model for biopolymer extensibility. Phys Rev Lett *81*, 4764–4767.

Rief, M., Gautel, M., Oesterhelt, F., Fernandez, J.M., and Gaub, H.E. (1997a). Reversible unfolding of individual titin immunoglobulin domains by AFM. Science *276*, 1109–1112.

Rief, M., Gautel, M., Schemmel, A., and Gaub, H.E. (1998b). The mechanical stability of immunoglobulin and fibronectin III domains in the muscle protein titin measured by atomic force microscopy. Biophys J *75*, 3008–3014.

Rief, M., Oesterhelt, F., Heymann, B., and Gaub, H. (1997b). Single molecule force spectroscopy on polysaccharides by atomic force microscopy. Science *275*, 1295–1297.

Rief, M., Pascual, J., Saraste, M., and Gaub, H.E. (1999b). Single molecule force spectroscopy of spectrin repeats: low unfolding forces in helix bundles. J Mol Biol *286*, 553–561.

Riley, M.L., Wallace, B.A., Flitsch, S.L., and Booth, P.J. (1997). Slow alpha helix formation during folding of a membrane protein. Biochemistry *36*, 192–196.

Sader, J.E., Larson, I., Mulvaney, P., and White, L.R. (1995). Method for calibration of atomic force cantilevers. Rev Sci Instrum *60*, 3131–3134.

Sadlish, H., Pitonzo, D., Johnson, A.E., and Skach, W.R. (2005). Sequential triage of transmembrane segments by Sec61alpha during biogenesis of a native multispanning membrane protein. Nat Struct Mol Biol *12*, 870–878.

Sapra, K.T., Besir, H., Oesterhelt, D., and Muller, D.J. (2006a). Characterizing molecular interactions in different bacteriorhodopsin assemblies by single-molecule force spectroscopy. J Mol Biol *355*, 640–650.

Sapra, K.T., Park, P.S., Filipek, S., Engel, A., Müller, D.J., and Palczewski, K. (2006b). Detecting molecular interactions that stabilize native bovine rhodopsin. J Mol Biol *358*, 255–269.

Sapra, K.T., Park, P.S., Palczewski, K., and Muller, D.J. (2008). Mechanical properties of bovine rhodopsin and bacteriorhodopsin: possible roles in folding and function. Langmuir *24*, 1330–1337.

Sapra, K.T., Damaghi, M., Koster, S., Yildiz, O., Kuhlbrandt, W., and Muller, D.J. (2009). One beta hairpin after the other: exploring mechanical unfolding pathways of the transmembrane beta-barrel protein OmpG. Angew Chem Int Ed Engl *48*, 8306-8308.

Schetz, J.A., and Sibley, D.R. (1997). Zinc allosterically modulates antagonist binding to cloned D1 and D2 dopamine receptors. J Neurochem *68*, 1990–1997.

Scheuring, S., Levy, D., and Rigaud, J.L. (2005). Watching the components of photosynthetic bacterial membranes and their in situ organisation by atomic force microscopy. Biochim Biophys Acta *1712*, 109–127.

Scheuring, S., Ringler, P., Borgina, M., Stahlberg, H., Müller, D.J., Agre, P., and Engel, A. (1999). High resolution topographs of the Escherichia coli waterchannel aquaporin Z. EMBO J *18*, 4981–4987.

Scheuring, S., and Sturgis, J.N. (2005). Chromatic adaptation of photosynthetic membranes. Science *309*, 484–487.

Scheuring, S., Sturgis, J.N., Prima, V., Bernadac, A., Lévy, D., and Rigaud, J.-L. (2004). Watching the photosynthetic apparatus in native membranes. Proc Natl Acad Sci USA *101*, 11293–11297.

Schlierf, M., Li, H., and Fernandez, J.M. (2004). The unfolding kinetics of ubiquitin captured with single-molecule force-clamp techniques. Proc Natl Acad Sci U S A *101*, 7299–7304.

Schlierf, M., and Rief, M. (2005). Temperature softening of a protein in single-molecule experiments. J Mol Biol *354*, 497–503.

Schoenauer, R., Bertoncini, P., Machaidze, G., Aebi, U., Perriard, J.-C., Hegner, M., and Agarkova, I. (2005). Myomesin is a molecular spring with adaptable elasticity. J Mol Biol *349*, 367–379.

Seddon, A.M., Curnow, P., and Booth, P.J. (2004). Membrane proteins, lipids and detergents: not just a soap opera. BiochimBiophysActa *1666*, 105–117.

Shao, Z., and Zhang, Y. (1996). Biological cryo atomic force microscopy: a brief review. Ultramicroscopy *66*, 141–152.

Sharma, A., Anderson, K., and Muller, D.J. (2005). Actin microridges characterized by laser scanning confocal and atomic force microscopy. FEBS Lett *579*, 2001–2008.

Shuster, T.A., Martin, F., and Nagy, A.K. (1996). Zinc causes an apparent increase in rhodopsin phosphorylation. Curr Eye Res *15*, 1019–1024.

Shuster, T.A., Nagy, A.K., Conly, D.C., and Farber, D.B. (1992). Direct zinc binding to purified rhodopsin and disc membranes. Biochem J *282* (Pt 1), 123–128.

Soekarjo, M., Eisenhawer, M., Kuhn, A., and Vogel, H. (1996). Thermodynamics of the membrane insertion process of the M13 procoat protein, a lipid bilayer traversing protein containing a leader sequence. Biochemistry *35*, 1232–1241.

Stahlberg, H., Muller, D.J., Suda, K., Fotiadis, D., Engel, A., Meier, T., Matthey, U., and Dimroth, P. (2001). Bacterial Na(+)-ATP synthase has an undecameric rotor. EMBO Rep *2*, 229–233.

Stöffler, D., Goldie, K.N., Feja, B., and Aebi, U. (1999). Calcium-mediated structural changes of native nuclear pore complexes monitored by time-lapse atomic force microscopy. J Mol Biol *287*, 741–752.

Stojanovic, A., Stitham, J., and Hwa, J. (2004). Critical role of transmembrane segment zinc binding in the structure and function of rhodopsin. J Biol Chem *279*, 35932–35941.

Struckmeier, J., Wahl, R., Leuschner, M., Nunes, J., Janovjak, H., Geisler, U., Hofmann, G., Jahnke, T., and Müller, D.J. (2008). Fully automated single-molecule force spectroscopy for screening applications. Nanotechnology *19*, 384020.

Swaminath, G., Lee, T.W., and Kobilka, B. (2003). Identification of an allosteric binding site for Zn2+ on the beta2 adrenergic receptor. J Biol Chem *278*, 352–356.

Taglicht, D., Padan, E., and Schuldiner, S. (1991). Overproduction and purification of a functional Na+/H+ antiporter coded by nhaA (ant) from Escherichia coli. J Biol Chem *266*, 11289–11294.

Thoma, J., Bosshart, P., Pfreundschuh, M., and Müller, D.J. (2012). Out but not in: the large transmembrane β-barrel protein FhuA unfolds but cannot refold via β-hairpins. Structure *20*, 2185-2190.

Thomson, N.H. (2005). Imaging the substructure of antibodies with tapping-mode AFM in air: the importance of a water layer on mica. J Microsc *217*, 193–199.

Torii, A., Sasaki, M., Hane, K., and Okuma, S. (1996). A method for determining the spring constant of cantilevers for atomic force microscopy. Meas Sci Technol *7*, 179–184.

Tskhovrebova, L., Trinick, J., Sleep, J.A., and Simmons, R.M. (1997). Elasticity and unfolding of single molecules of the giant muscle protein titin. Nature *387*, 308–312.

Uchihashi, T., Iino, R., Ando, T., and Noji, H. (2011). High-speed atomic force microscopy reveals rotary catalysis of rotorless F1-ATPase. Science *333*, 755-758.

Van den Berg, B., Clemons, W.M.J., Collinson, I., Modis, Y., Hartmann, E., Harrison, S.C., and Rapoport, T.A. (2004). X-ray structure of a protein-conducting channel. Nature *427*, 36–44.

Vendruscolo, M., and Dobson, C.M. (2005). A glimpse at the organization of the protein universe. Proc Natl Acad Sci U S A *102*, 5641–5642.

Viani, M.B., Pietrasanta, L.I., Thompson, J.B., Chand, A., Gebeshuber, I.C., Kindt, J.H., Richter, M., Hansma, H.G., and Hansma, P.K. (2000). Probing protein-protein interactions in real time. Nat Struct Biol *7*, 644–647.

Viani, M.B., Schäfer, T.E., Chand, A., Rief, M., Gaub, H., and Hansma, P.K. (1999a). Small cantilevers for force spectroscopy of single molecules. J Appl Phys *86*, 2258–2262.

Viani, M.B., Schäffer, T.E., Paloczi, G.T., Pietrasanta, L.I., Smith, B.L., Thompson, J.B., Richter, M., Rief, M., Gaub, H.E., Plaxco, K.W., et al. (1999b). Fast imaging and fast force spectroscopy of single biopolymers with a new atomic force microscope designed for small cantilevers. Rev Sci Ins *70*, 4300–4303.

Vinckier, A., Gervasoni, P., Zaugg, F., Ziegler, U., Lindner, P., Groscurth, P., Pluckthun, A., and Semenza, G. (1998). Atomic force microscopy detects changes in the interaction forces between GroEL and substrate proteins. Biophys J *74*, 3256–3263.

Wagner, O.I., Ascano, J., Tokito, M., Leterrier, J.F., Janmey, P.A., and Holzbaur, E.L. (2004). The interaction of neurofilaments with the microtubule motor cytoplasmic dynein. Mol Biol Cell *15*, 5092–5100.

Wieprecht, T., Beyermann, M., and Seelig, J. (2002). Thermodynamics of the coil-alpha-helix transition of amphipathic peptides in a membrane environment: the role of vesicle curvature. Biophys Chem *96*, 191–201.

Wiita, A.P., Ainavarapu, S.R., Huang, H.H., and Fernandez, J.M. (2006). Force-dependent chemical kinetics of disulfide bond reduction observed with single-molecule techniques. Proc Natl Acad Sci U S A *103*, 7222–7227.

Wiita, A.P., Perez-Jimenez, R., Walther, K.A., Grater, F., Berne, B.J., Holmgren, A., Sanchez-Ruiz, J.M., and Fernandez, J.M. (2007). Probing the chemistry of thioredoxin catalysis with force. Nature *450*, 124–127.

Wolynes, P.G., Onuchic, J.N., and Thirumalai, D. (1995). Navigating the folding routes. Science *267*, 1619–1620.

Woodside, M.T., Anthony, P.C., Behnke-Parks, W.M., Larizadeh, K., Herschlag, D., and Block, S.M. (2006). Direct measurement of the full, sequence-dependent folding landscape of a nucleic acid. Science *314*, 1001–1004.

Xie, K., Hessa, T., Seppala, S., Rapp, M., von Heijne, G., and Dalbey, R.E. (2007). Features of transmembrane segments that promote the lateral release from the translocase into the lipid phase. Biochemistry *46*, 15153–15161.

Ye, Y., Meyer, H.H., and Rapoport, T.A. (2001). The AAA ATPase Cdc48/p97 and its partners transport proteins from the ER into the cytosol. Nature *414*, 652–656.

Yu, J., Bippes, C.A., Hand, G.M., Muller, D.J., and Sosinsky, G.E. (2007). Aminosulfonate modulated pH-induced conformational changes in connexin26 hemichannels. J Biol Chem *282*, 8895–8904.

Zhong, Q., Inniss, D., Kjoller, K., and Elings, V.B. (1993). Fractured polymer silica fiber surface studied by tapping mode atomic-force microscopy. Surf Sci *290*, L688–L692.

Zocher, M., Fung, J.J., Kobilka, B.K., and Müller, D.J. (2012). Ligand-specific interactions modulate kinetic, energetic, and mechanical properties of the human β2 adrenergic receptor. Structure *20*, 1391-1402.

Zwanzig, R. (1988). Diffusion in a rough potential. Proc Natl Acad Sci U S A *85*, 2029–2030.

6

Biological Imaging Using X-Ray Free-Electron Lasers

STEFAN P. HAU-RIEGE

1. INTRODUCTION

X-ray diffraction using X-ray tubes as a photon source was critical in uncovering the structure of DNA in the 1950s (1). In the 1970s, with the advent of synchrotron light sources, enormous advances in structural biology were achieved through protein crystallography (2). More recently, a new revolutionary type of light source, the X-ray free-electron laser (XFEL), has become available. XFELs deliver light pulses with extreme physical characteristics. Typical pulse lengths range from 10 fs = 10^{-14} s to 300 fs, the wavelength is about 1 Å = 10^{-10} m, and the peak brightness is more than a factor of 10^9 larger than that of synchrotrons. The pulse lengths of XFELs are on the order of the oscillation period of bond stretching and bending in organic chemistry, and the wavelength is comparable to interatomic distances. Coupled with the enormous peak brightness, XFELs will enable imaging of small objects at the atomic scale with fs time resolution, maybe even individual or small clusters of molecules (3–5). This capability would make it possible to determine the structure of biological macromolecules that cannot be arranged in large arrays with translational symmetry (*crystals*) for X-ray crystallography. X-ray crystallography has been the basis for almost all biomolecular structure determination to this point. It fundamentally relies on the availability of large well-ordered crystals. However, since most biological macromolecules are very difficult to crystallize, large systematic gaps exist in the database of protein structures (6). An important example is the class of protein molecules that are responsible for cross-membrane communication, which is vital for drug development to cure human diseases. Since this class of proteins is particularly difficult to crystallize, their structures are mostly unknown today. Further, high-resolution imaging of nonrepetitive nonreproducible structures, such as living cells, has not been possible altogether. Beyond determining the static high-resolution structure of all these materials, the ultrashort pulses of XFELs will in addition enable time-resolved imaging and will provide the opportunity to track the dynamics of chemical reactions.

In this chapter, we will review the implementation, challenges, and promises of using XFELs for imaging the structure and dynamics of biological particles and molecules. We will first compare different imaging techniques including X-ray crystallography and coherent diffractive imaging of single macromolecules. We will then discuss their implementation at synchrotron light sources and, in more detail, at XFELs. We conclude this chapter with a summary of recent achievements, an outlook, and an overall summary.

2. X-RAY IMAGING TECHNIQUES

Our goal is to obtain the high-resolution three-dimensional structure of biological macromolecules and particles. X-ray photons with energies between 8 and 12 keV are particularly suitable as probes since they can easily penetrate biological systems, and their wavelength, on the order of 1 Å, is comparable to interatomic distances and is sufficiently short to achieve atomic resolution. When X-rays irradiate an object, they interact primarily the electrons in the system since the electron mass is much smaller than the mass of the atomic nuclei. X-rays can interact with matter in two primary ways: absorption and scattering. Absorbed X-rays possibly cause sample damage associated with ionization and heating. X-rays may also be scattered. If the energy of a scattered X-ray photon does not change, the process is called *elastic*, otherwise it is called *inelastic*. Elastically scattered X-rays are particularly useful since they have a well-defined phase relationship with respect to each other, and the intensity distribution of the scattered photons relates back to the electron density distribution in the sample.

In the simplest scenario, a plane wave of X-rays illuminates an object. After the object has interacted with the wave field, the radiation propagates to a 2D detector that measures its intensity distribution in the *far field*—that is, at distances at which the angular intensity distribution is independent of the distance from the object. The detector is typically protected by a beam stop or has a hole in the middle to prevent it from getting damaged by the incident light. The geometry of such an experiment is sketched in Figure 6.1. For sufficiently high X-ray energies, the Fourier transform of

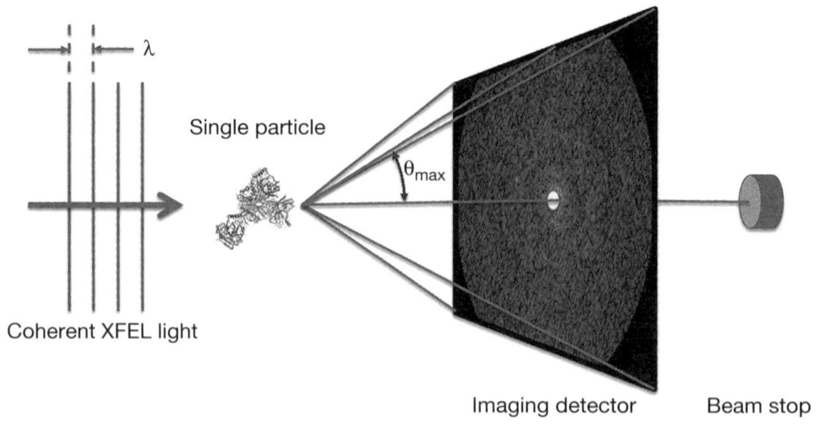

Figure 6.1 Geometry of a generic x-ray imaging experiment.

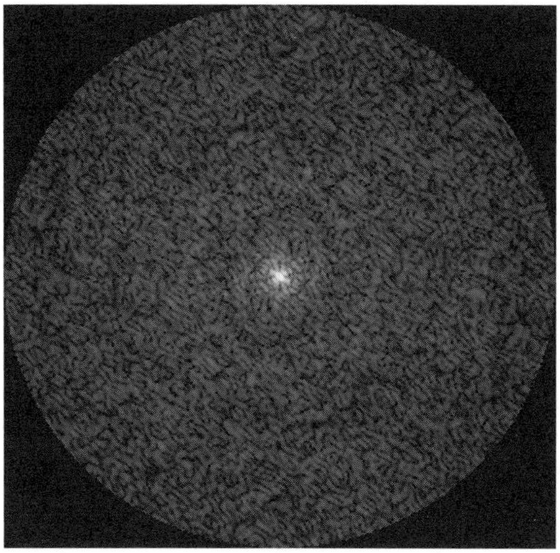

Figure 6.2 Diffraction pattern of an isolated molecule of GDPRAN-NTF2 COMPLEX.

the electron density in the sample is proportional to the electric field at the detector (7). Since usually only the intensity of the scattered radiation can easily be measured, referred to as the *molecular transform*, the phase information of the electric field is lost. Numerous *phasing algorithms* have been proposed to retrieve the phases of the electric field, in which case a simple inverse Fourier transform may provide an electron density map. Some of these algorithms are discussed below.

Figure 6.2 shows the diffraction pattern of an isolated molecule of GDPRAN-NTF2 COMPLEX (PDB code 1A2K), a particular molecule of interest. The center of the pattern corresponds to the straight-through beam that usually cannot be measured. We used a logarithmic intensity scale since the intensity drops quickly at larger distances from the center, corresponding to high-resolution structural information. In *coherent diffractive imaging*, the continuous diffraction patterns of isolated particles are measured, from which the electron density may be reconstructed. In *X-ray crystallography*, we measure diffraction data from objects with translational symmetry instead. In this case, the diffraction pattern is modulated by the symmetry of the object.

2.1. X-Ray Crystallography

We first consider a sample that is made up of identical unit cells arranged with translational and/or rotational symmetry, referred to as a *crystal*. In this case, the diffracted intensity is significantly increased in certain scattering directions; the recorded spots are called *Bragg peaks*. Bragg peaks occur when the coherent light that is elastically scattered by the unit cells adds constructively at the detector. If the number of cells in the crystal is N, the peak intensity may be up to N^2 larger than the intensity of a single cell, which, given the large number of cells in a crystal, is an enormous increase in signal. The width of a Bragg peak is proportional to $1/N$ and therefore the integrated intensity of a Bragg peak scales as N. In between the Bragg peaks, the intensity nearly vanishes.

If we record just the intensity of the Break peaks, as it is done in X-ray crystallography, we sample the diffraction pattern of an object at discrete points instead of recording the full pattern, as it is done in coherent diffractive imaging. According to sampling theory and, in particular, based on Shannon's information theorem (8), accurate reconstruction of a continuous function from a sampled set of discrete data requires sampling at a frequency that is larger than the *Nyquist* frequency f_{Nyq}, given by the inverse of the sample size. This assumes that both amplitudes and phases are recorded. If we sample below the Nyquist frequency, we *undersample* and information is lost, resulting in errors in the reconstruction. Sayre (9) pointed out that in X-ray crystallography, we always sample at the Nyquist frequency, and since with image plates or other detectors we typically measure only the intensities and not the phases of the electric field, we do not have enough information to reconstruct the electron density. This is referred to as the (*fundamental*) *phase problem in crystallography*. Further information or assumptions about the sample are required in order to obtain an image of the molecules.

For the cases when biological molecules can be crystallized into an extended, ordered array, X-ray crystallography has been proven to be a very powerful tool for structure determination. The phase problem can, at least in part, be overcome by making certain assumptions about the electron density or by determining phases through experimental measurements such as isomorphous replacement or using anomalous differences that occur near absorption edges. The resolution in X-ray crystallography is limited only by the wavelength, the quality of the crystal, and the size of the detector. However, it is often difficult to crystallize the molecule, so that direct single molecule imaging (coherent diffractive imaging) is desirable. This is discussed in the following section.

2.2. Coherent Diffractive Imaging

Coherent diffractive imaging techniques are used to image objects without the translational or rotational symmetry of a crystal (although these symmetries may be present as noncrystallographic symmetry and can be exploited during reconstruction). Instead of using an optical lens to re-interfere the light scattered from the object, as is done in conventional optics, we use computational phasing algorithms. Once we know the phases and amplitudes, we can back-propagate the X-ray wave from the detector to the exit surface of the sample and infer an image of the object. In the simplest case of a coherent plane wave irradiating the object, the achievable resolution is $d = \lambda/\sin\theta_{max}$, where θ_{max} is the highest scattering angle collected by the detector, and λ is the X-ray wavelength, see Figure 6.1. Since we are no longer limited to the Bragg peaks, we can record the continuous molecular transform at a spacing much finer than the Nyquist frequency f_{Nyq}. This concept is called *oversampling*. As pointed out by Sayre (9), the additional intensity information may compensate for the missing phase information. It is necessary to sample at least at twice the Nyquist frequency. Oversampling requires finer recording of the features in the diffraction pattern, and so better coherence of the incident beam is needed.

Numerous phasing algorithms have been developed for this purpose. A successful class of algorithms relies on using a Fourier transformation to iterate between the electron density in real space and the scattered electric field in reciprocal space, while applying appropriate constraints in both spaces, until a converging solution is reached. Such an algorithm was originally proposed by Gerchberg (10) and extended to finite-size objects by Fienup (11). These algorithms tend to be noise-stable and can tolerate the missing information near the center of a diffraction pattern where the beam stop protects the imaging detector from the direct beam; see Figure 6.1. Most modern algorithms are a variant of Fienup's *hybrid input-output* (HIO) algorithm, which consists of the following five steps:

1. Use the square root of the measured intensity as an estimate for the magnitude of the electric field, and make a guess for the phases (or simply use random phases).
2. Apply the inverse Fourier transformation to obtain an electron density.
3. Impose various constrains on the electron density, such as setting the electron density to zero outside a finite support volume, or setting the real part of the electron density toward zero in case it is negative. This positivity constraint may separate the correct and the conjugate phases.
4. Use the Fourier transformation to obtain the electric field distribution. Replace the electric field amplitudes by the measured values, but keep the phases.
5. Go to step 2.

This algorithm has been applied successfully to nonperiodic test samples (12) and to crystalline samples (13). As a variation, Oszlanyu and Suto (14) proposed a *charge flipping* algorithm, in which the sign of the charge density in regions which lie below a certain threshold is flipped. Elser (15) proposed another iterative phase retrieval algorithm that works well for poorly known supports. Spence et al. (16) proposed direct methods, and Hau-Riege et al. (17) proposed methods based on real-space conjugate-gradient minimization. Marchesini et al. (18) proposed the *shrink-wrap* algorithm in which the finite support is generated dynamically as part of the reconstruction algorithm. In this algorithm, a low-resolution version of the current guess is used as a new support. For larger objects, a finite illumination field can also provide effective constraints for reconstruction (19).

The scattered intensity from a nonperiodic sample is much weaker than from a crystal since the Bragg peaks are absent. In order to achieve sufficient signal-to-noise ratios, very bright X-ray sources are required, but the sample must withstand the greater X-ray exposure. If a protein is surrounded by vitreous ice, the number of scattered photons varies as $1/\theta^4$. Assuming a *conventional* damage limit of 160 photons / $Å^2$, then statistically significant signal levels can be obtained only at resolutions larger than 10 nm (20), which is typical for coherent diffraction on synchrotrons. To outdo this limitation and achieve atomic resolution, we need to overcome the conventional damage limit, which may be done using X-ray free-electron lasers, as described below.

2.3. Other Imaging Techniques

Numerous other X-ray imaging techniques have been suggested. Holographic techniques use reference objects that enhance the scattering from a weakly scattering object. Demonstrated techniques include Fourier-transform holography (21) and massively parallel holography (22). Ptychography is a scanning coherent diffraction microscopy technique in which the diffraction patterns of overlapping illuminations of large samples are measured. These data allow the determination of the complex object transmission function and of the wave field (23).

3. IMAGING USING SYNCHROTRONS

The advent of the first dedicated synchrotron light sources more than two decades ago triggered a revolution in the life sciences. With light intensities 50 to 100 times larger than rotating-anode X-ray sources, synchrotrons have been used to determine the structures of more than 60,000 biological macromolecules (24). Figure 6.3 shows the number of structures determined by the different analysis techniques as a function of time. More than 80 percent of today's newly determined structures of biological macromolcules are obtained using X-ray crystallography. Most remaining structures are determined using NMR spectroscopy, which is limited to relatively small molecules. Cryogenic electron microscopy is rapidly developing, and 1.9 Å resolutions have been achieved with 2D protein crystals (25), and 5.5 Å for asymmetrical ribosome (26).

Large-scale light sources producing high-brightness X-ray beams are typically classified into generations. First-generation light sources exploited the synchrotron radiation from electron storage rings that were originally built for high-energy physics applications. In the 1970s, these facilities were used to pioneer protein crystal diffraction (2). Second-generation light sources were optimized for generating synchrotron radiation, and much more effective equipment was available to generate and record diffraction patterns. Some second-generation light sources were retrofitted for insertion devices called *wigglers* and *undulators* that can be located in the straight sections of the synchrotron to

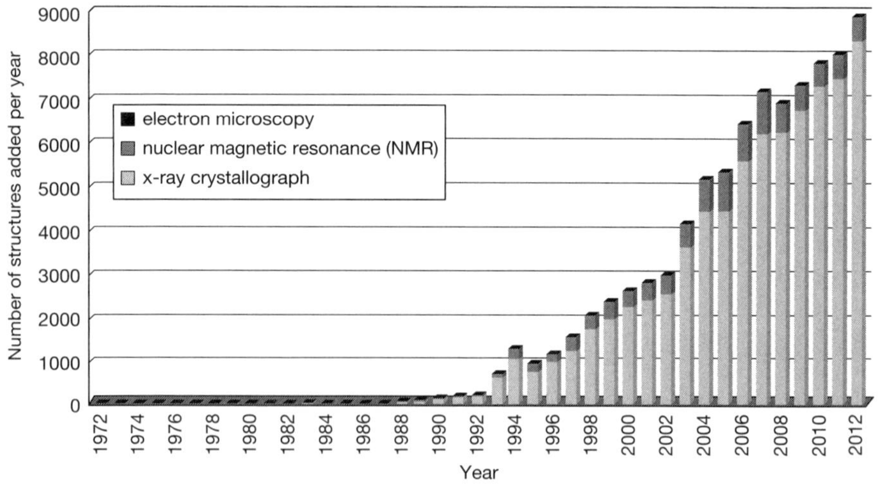

Figure 6.3 Number of structures added to the PDB data bank as a function of time for the three structure determination methods x-ray crystallography, electron microscopy, and nuclear magnetic resonance [6].

efficiently generate synchrotron radiation. Undulators are sets of magnets of alternating polarity and produce bright X-ray beams, which allowed using smaller crystals and lower protein concentrations, performing time-resolved studies, and achieving better spatial resolution. The third-generation light sources were optimized to accept these kinds of insertion devices. The high intensity of second and third generation synchrotron sources pushed researchers to develop cryogenic cooling techniques to minimize the deleterious effects of X-ray induced radiation damage.

In a synchrotron, groups of electrons, called electron bunches, move at nearly the speed of light for several kilometers on a closed path. A schematic of a synchrotron is shown in Figure 6.4 (a). In order to keep the electron bunches on their path, their direction must be periodically changed which is achieved by applying a magnetic field. Whenever the propagation direction of an electron bunch changes, the charges are accelerated, and, according to classical electrodynamics, emit radiation (27). The wavelength distribution of the synchrotron radiation depends on the velocity of electron bunch and the detailed nature of the acceleration. In comparison to XFELs, synchrotron radiation is of low intensity and incoherent, so that X-ray crystallography is the primary protein imaging technique of choice.

The achievable resolution in X-ray crystallography depends on the homogeneity of the sample, the crystal quality, the wavelength, and the detector size, and is further limited by radiation damage of the biological materials. For sufficiently large, high-quality crystals, atomic resolution may be achieved. We will now discuss the damage limit for biological materials imaged at synchrotrons. Since the pulse length at synchrotrons is much longer than that of free-electron lasers, we refer to the threshold fluence for the onset of damage as the *static* damage limit. For XFELs, we

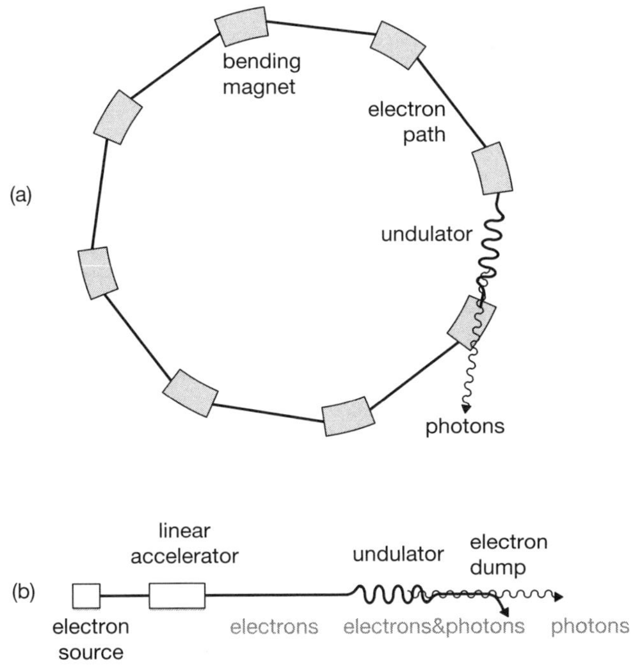

Figure 6.4 (a) Schematic of a synchrotron. Only one undulator insertion device is shown. (b) Schematic of an XFEL.

will later introduce a *dynamic* damage limit. While cryogenic cooling can slow sample deterioration in biological samples, morphological changes cannot be completely avoided. Based on calculations by Blake and Philips, showing that a single 8 keV electron disrupts about 70 bonds in a biological crystal (28), Henderson suggested that the damage limit for partial or complete breakdown of a protein crystal is 160 photons / $Å^2$ (29, 30). At higher fluences, the temperature rises, and since, typically, cells and protein crystals can tolerate a temperature increase of only a few degrees, the admissible fluence is very limited.

The resolution of living cells imaged using conventional continuous exposure is generally limited to about 10 nm, even when the samples are dried, chemically fixated, or cooled to the temperature of liquid nitrogen to avoid diffusion and mass loss (20). For a protein crystal, atomic resolution may be achieved since the Bragg peaks are very strong, and the dose is essentially shared among all the molecules in the crystal. Miao et al. (31) demonstrated two-dimensional coherent diffractive imaging on a nonperiodic, radiation-resistant specimen. In general, coherent diffractive imaging (CDI) at synchrotrons is limited to large objects since the peak flux is relatively small, the signal is not amplified due to Bragg diffraction, and the dose is limited by the static damage limit (32). To achieve atomic resolution in CDI, much larger peak brightnesses and shorter X-ray pulses are required. XFELs are a major step forward into this direction and so promise to revolutionize structural biology to an even larger extent than synchrotrons did.

4. IMAGING USING X-RAY FREE-ELECTRON LASERS

The pulse duration of bright X-ray sources has significantly decreased over the last decades. Synchrotrons are typically limited to pulse lengths longer than tens of picoseconds since the electron beam size increases with time due to the stochastic generation of synchrotron light. This effect can be reduced by using synchrotrons with larger radii, with the ultimate limit being a linear accelerator. The electron beam in a linear accelerator may be chirped so that the kinetic energy varies from head to tail, which may allow it to be compressed in length. Certain electron beam properties, such as the emittance described below, are solely determined by the properties of the source. The advent of high-brightness electron guns has made it possible to produce ever-smaller emittance beams. These have been utilized at the short-pulse photon source (SPPS) at SLAC, producing X-ray pulses with 10^7 photons and of 80 fs duration (33). The SPPS was an enormous improvement over synchrotrons. XFELs take it even further and produce pulses with a factor of 10^5 more photons per pulses than SPPS. Such a large number of photons could lead to substantial damage to the molecule that can manifest itself as increased temperature factors and missing side chains in the reconstructed biomacromolecule. However, the pulses are sufficiently short so that radiation damage mechanisms that occur on time scales longer than the pulse length are overcome.

4.1. X-Ray Free-Electron Lasers

X-ray free-electron lasers are considered fourth-generation light sources. They typically consist of an electron source, a linear accelerator, and a high-performance undulator. The low-gain-FEL concept was proposed by Madey (34), and its operation in the single-pass high-gain limit was proposed by Bonifacio et al. (35). The key features of XFELs are their large coherence with a *degeneracy parameter* (expectation value of the number of photons in the same quantum state (36)) of more than 10^9, a time resolution that is 10^3 to 10^4 times better than third-generation light sources, and a peak brightness that is 10^9 times larger than any other X-ray source. These output parameters make XFELs useful not only for analyzing structure and function of biological systems, but also for diffraction tomography, probing femtochemistry, studying atomic physics and warm dense matter, and, using extreme X-ray focusing, possibly exploring fundamental quantum electrodynamic effects, such as the boiling of the vacuum.

The first operational soft X-ray FEL was the FLASH facility at DESY in Hamburg, Germany (37), followed by the SPring-8 compact SASE source (SCSS) in Hyogo, Japan (38). At these facilities, a wavelength of 60 Å was achieved. This has been reduced to 41 Å more recently. In 2009 the Linac Coherent Light Source (LCLS) became operational at SLAC in the USA (39), generating photon pulses between 15 and 1.5 Å. LCLS is designed to deliver X-ray pulses with a total pulse energy of several mJ, which corresponds to 10^{12} to 10^{13} photons with the pulse length less than 100 fs and a repetition rate of 120Hz. Pulse energies up to 3.5 mJ and pulse lengths down to 10 fs have recently been demonstrated (40). A similar facility with slightly smaller pulse energies has become available in Japan in 2011 called the SPring-8 Ångstrom compact FEL (SACLA). Other facilities scheduled to become available are the European XFEL in 2017 (41) and the SwissFEL in 2016. These facilities will have similar output characteristics, except that the European XFEL will offer a substantially higher pulse repetition rate.

4.1.1. Functioning of an XFEL

In an XFEL, a photocathode gun generates a short, low-emittance electron bunch that is injected into a linear accelerator. The emittance is a measure for the angular content of the beam and characterizes how much the beam diverges upon propagation—low emittance is preferable. In the accelerator, the electron bunch is accelerated and compressed in length by bunch compressors. Bunch compressors are devices consisting of a series of magnets that reduce the length of an electron bunch with an imprinted energy dispersion. The very compact electron bunch is then injected into an undulator that is similar to the ones already used in synchrotrons. A schematic of a free-electron laser is shown in Figure 6.4 (b). The intense undulator radiation interacts with the electron bunch, causing it to "micro-bunch" into regions of the size of the radiation wavelength, as sketched in Figure 6.5. Once this is achieved, the electrons act coherently, leading to increased photon emission, and with that, increased micro-bunching. This nonlinear effect leads to an exponential increase in the radiation power down the length of the undulator until the electrons are fully micro-bunched and saturation is reached. The electrons in a synchrotron emit radiation incoherently, so that the radiation intensity is linear in the number of electrons N, whereas it is super-linear in an XFEL (up to N^2).

XFELs have been made possible by improved accelerator technology and low-emittance electron guns, providing dense and focused electron bunches, as well as improved undulator technology, allowing the fabrication of very long undulators. For example, the LCLS undulator is 100 m long. The output X-ray energy can be

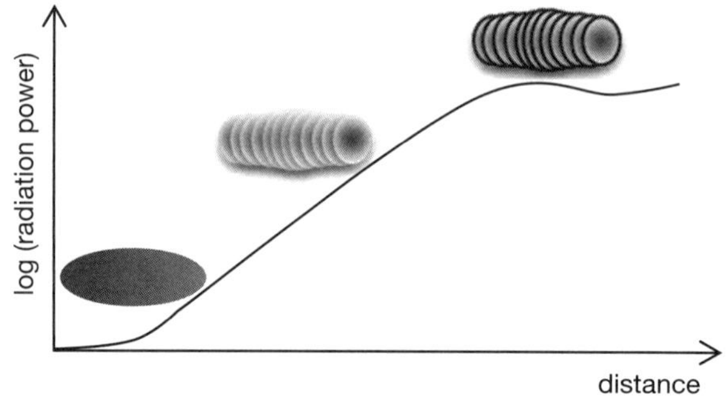

Figure 6.5 Micro-bunching and increase of radiation power along the undulator.

continuously tuned by altering the kinetic energy of the relativistic electron bunch. No mirrors or lasing cavities are needed, which is a major advantage over conventional lasers. Currently, most XFELs start the lasing action from noise and operate in the self-amplification-of-spontaneous-emission (SASE) mode, leading to a limited temporal coherence of about 0.3 fs at 8 keV. By seeding the lasing action, full temporal coherence has been achieved (42).

4.1.2. Conditioning and Detecting XFEL Radiation

The divergence of an XFEL beam is typically a few μrad (1 μrad = $180\times10^{-6}/\pi$ degrees), so that 100 m downstream from the exit of the undulator, the beam diameter increases by at least 100 μm. In order to achieve sufficient signal-to-noise ratios for single molecule imaging at atomic resolution, focal diameters of less than 100 nm are required (43). For protein nanocrystallography, a focal beam diameter of 1 μm is typically sufficient but the ideal case is presented by the focus being the size of the nanocrystal itself. Various means of achieving such focal sizes are being developed, for example X-ray zone plates, which are diffraction-based focusing optics (44), diffractive lenses (45), Kirkpatrick-Baez (KB) focusing mirror pairs, which are pairs of concave, grazing-incidence X-ray mirrors (46), and adaptive X-ray optics (47). Given the high intensity of even the unfocused beam, grazing-incidence mirrors as used in KB focusing optics may be preferable over normal-incidence optics to reduce the effect of radiation damage.

It would be beneficial to compress the beam in time (in the longitudinal spatial direction) in order to stay below the static and possibly the dynamic damage threshold. Methods to decrease the pulse length could be based on a chirp in the photon beam, which means that the photon energy varies along the length of the beam (48). Typical energy chirps that can be created in XFELs are on the order of a percent. By allowing only a certain energy window to be transmitted through one element in the optics, an effectively shorter pulse is created. A demonstrated method to reduce the XFEL pulse length is the spoiling of most of the electron beam except a time slice that will lase (40). The disadvantage of these *slicing* methods is that a substantial fraction of the beam energy is discarded. In order to preserve the total beam energy, it is more advantageous to use methods that *compress* the beam, such as dispersive optical elements in which the propagation time of photons through the optic depends on the energy or color of the light (49).

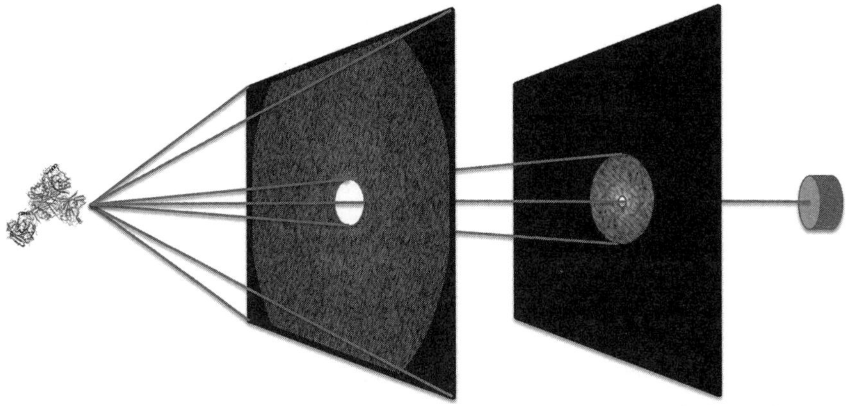

Figure 6.6 Geometry of a generic x-ray imaging experiment with two imaging detector in series to increase the dynamic range.

As shown in Figure 6.1, a two-dimensional diffraction pattern is recorded using an imaging detector, such as a charged-coupled device (CCD). The Cornell-SLAC Pixel Array Detector (CSPAD) is an example for a detector that can be read out at 120 Hz (50), matching the output frequency of LCLS. The optimal size of the detector depends on the distance of the sample to the detector, since that defines the angular acceptance range, and with that the image resolution. The number of pixels on the detector should be chosen depending on the size of the sample since, according to Shannon's theorem, the diffraction pattern of larger particles need to be sampled more finely. For single molecule imaging, the number of pixels on the detector should be in the range of 1024^2 to 4096^2. For smaller samples, the pixels can be binned into groups. A very large data stream is created, and dedicated infrastructure is needed to store and process all these data. In order to be able to record low-resolution data near the center of the diffraction pattern as well as high-resolution data at the edges, a dynamic range between 10^5 to 10^6 is desirable. The dynamic range of CCDs is usually only on the order of 10^3. To overcome this discrepancy it has been suggested to use two imaging detectors in series, as sketched in Figure 6.6, with the first detector having a hole in the middle, so that the high-intensity, low-resolution image is transmitted to the second detector (43). The CXI end station at LCLS has been designed accordingly (51).

4.2. Particle Injection

Our goal is to achieve atomic-resolution imaging of biological particles, for example in the form of single macromolecules or nanocrystals. Besides the sample itself, anything in the interaction region will also scatter X-rays, including a potential sample holder. It is therefore desirable to use spraying techniques to inject hydrated samples into the XFEL beam while maintaining high and sometimes ultra-high vacuum conditions. There are numerous challenges associated with such an approach. The particle density in the injected beam has to be sufficiently large so that the probability that the XFEL beam hits the particle is reasonably high, a 10–25 percent estimated hit rate is usually desired. Further, if the particles are aligned randomly, it is preferable for the XFEL beam to hit only a single particle per pulse in order to avoid interference effects. A difficulty is that the injected particle beam has to be pure and the molecules should be kept in their native conformation. For this, they should preferably be uncharged and fully hydrated. It has been found that a 20 percent hydration fraction, defined as the mass ratio of water and protein, is sufficient to conserve biological activity (52–53). It has further been shown that vitreous ice may also be an acceptable sample environment (30). Thermal and mechanical stresses should be avoided in order to keep the biological macromolecules or particles in native-like conformations (54–55). It would be desirable to inject the particles into the XFEL beam in a known orientation, but this is usually very difficult to achieve.

Material may be injected either as a continuous stream or in the form of droplets or particles. In either case, the amount of injected material should be kept to a minimum to maintain good vacuum conditions, reduce noise, minimize the amount of sample consumed, and alleviate the effect of the limited penetration depth of softer X-rays, if required. The lower limit for the diameter of a regular liquid jet (Rayleigh jet) is about 20 μm due to the onset of nozzle clogging. It was proposed (56) that the Rayleigh jet can be further compressed in diameter by the pressure of co-axially flowing gases, producing a gas-dynamic virtual nozzle (57). This kind of liquid microjet has been used in recent experiments to introduce protein nanocrystals into the LCLS beam, see for example (58) and (59). A microjet eventually transforms into a micro-droplet stream by Rayleigh breakup associated with the reduced surface energy of a stream of droplets compared to a continuous stream. The breakup can be induced by an acoustical disturbance, such as a piezoelectric actuator (60). So far, droplet sizes of about 1 μm have been demonstrated. Combining the technique with electrospray ionization leads to a reduced droplet size of 100 nm (60). It is desirable for the biomolecules and nanocrystals to remain hydrated until the XFEL irradiation to prevent minor structural changes in the micro droplet compared to the cell environment. In Reference (61), photosystem II microcrystals were injected into the LCLS X-ray beam in a liquid jet based on the electrohydrodynamic spraying of glycerol in vacuum (62) using the cone-jet mode (56).

Certain techniques developed in the field of mass spectrometry may provide alternatives to liquid-jet injection. Benner et al. (63) introduced an injection system that may shoot cells directly into the LCLS beam. The output of a charge-reduction nanoflow electrospray passes a pressure reducer and is then injected into an aerodynamic lens stack to increase the particle density. After passing a second pressure reducer, the pressure is reduced to 10^{-6} Torr and the particles have a velocity of 200 m/s. This system could, in principle, be used to inject small living cells, single virus particles, and single biomolecules. Note that it was found that picoplankton cells are still alive after passing through a vacuum chamber with a velocity of 200 m/s, that the cell morphology is similar, and that fluorescence is observed similar to native cells in solution. It has also been demonstrated that electrosprayed multicellular zebrafish embryos are still viable and develop normally (64). In a recent LCLS experiment on imaging whole viruses (65), an aerosol sample injector with an aerodynamic lens stack was used to deliver hydrated samples into the LCLS beam.

Some alternative sample delivering systems have been proposed, including the ability to select and store particle in ion-trapping devices. It has also been suggested that especially large samples may be embedded in a thin layer of vitreous ice, similar to techniques used in cryo-electron microscopy, which may provide the ability of controlled sample rotation.

4.3. Dynamic Damage Limits

Solem and Baldwin suggested that X-ray-laser flash-imaging microscopy could be used to overcome the radiation-induced degradation of images of living wet cells by obtaining useful structural information *before* the radiation destroys the sample (3). Hydrodynamic estimates have suggested that images with 10 nm resolution may be obtainable with picosecond pulses. Neutze et al. built on this idea and suggested that the static damage limit for single biological macromolecules at atomic resolution may be overcome by using femtosecond pulses from XFELs (5): The finite time duration required for damage to manifest itself gives rise to the concept of a dynamic damage limit. Simulations showed that the dynamic damage limit is several orders of magnitudes larger than the damage limits in conventional X-ray diffraction experiments (5).

Diffractive imaging using XFELs requires a detailed understanding of the interaction of the biological material with X-rays (66). We now consider the interaction of 12 keV X-rays, corresponding to a wavelength of about 1 Å, with the most abundant element in biological materials by weight, carbon. The majority (90 percent) of all photon-carbon interaction events are photoionizing, in which a photon is absorbed from a low-lying K orbital, and an electron with an energy of about 12 keV is emitted. This process produces a hollow atom with an unstable electron configuration and initiates a *damage cascade*: The hollow atom relaxes through the Auger process in which an electron from an outer shell falls into the hole, and another bound electron picks up the liberated energy and is ejected with a kinetic energy around 0.25 keV (67). For carbon ions, the Auger decay time is 11 fs, whereas for sulfur it is 1.3 fs (67). Fluorescence processes are not common for atoms with a low atomic charge number. Shake-up excitations, in which multiple ionization steps follow inner-shell ionizations, may also occur. On their way out of the molecule, both photoelectrons at 12 keV and Auger electrons at 0.25 keV may interact with other ions through multiple electron-impact ionization events, liberating further electrons, primarily from the valence shell, each with kinetic energies of tens of eV (68).

After photoionization, the elastic scattering of photons is the next common photon-carbon interaction event in 7–10 percent of all cases. During elastic scattering, the photon energy does not change. All scattered photons have a well-defined phase relationship with each other, and they form the signal in the diffraction pattern. Inelastic scattering, during which the photon transfers energy to an electron and also changes its phase, occurs in 3 percent of all cases. This process may also lead to ionization.

Early on in the pulse, the photo and Auger electrons may escape the molecule. Once the molecule is sufficiently positively charged, the lower-energy (and eventually the higher-energy) electrons are electrostatically trapped, which occurs more easily in large molecules. In larger particles, such as nanocrystals, nearly all of the electrons are trapped. Trapping of energetic electrons leads to an increase in the temperature of the trapped electrons, which, in turn, can lead to increased ionization, but also, in case of high ionization states, to enhanced three-body recombination.

Charging of the molecule due to loss of energetic electrons eventually leads to a Coulomb explosion of the particle. The large temperature of the trapped electron gas is associated with a high hydrostatic pressure inside the molecule, leading to its expansion due to hydrodynamic forces. Charge trapping also affects the expansion dynamics since it may slow the Coulomb expansion.

The damage processes in biological molecules have been modeled using various techniques, including molecular dynamics, hydrodynamics, and Monte Carlo methods (5, 69–71). It is generally agreed that pulses much shorter than 50 fs are required in order to overcome damage effects for typical XFEL conditions (5, 72). This has been roughly confirmed by recent XFEL experiments (73).

4.4. Protein Nanocrystallography

Protein nanocrystallography is a new type of imaging technique that uses sub-micrometer crystals for structure determination. Growing large, well-ordered protein crystals is often the bottleneck for structure determination, and this is particularly challenging for membrane proteins, which are the targets for half of all pharmaceutical drugs. As discussed above, sizeable crystals are needed for conventional crystallography to obtain high-resolution diffraction data since the exposure is limited by cumulative radiation damage. Sometimes smaller, high-quality crystals can be obtained much more easily, and diffraction patters from these single nanocrystals can be measured using XFELs. In a typical experiment, nanocrystals are injected into the XFEL beam at random orientation using a liquid-jet-based injector. The XFEL bandwidth is sufficiently broad so that a large number of Bragg peaks is observed, and the orientation of nanocrystals (in the laboratory frame) is determined by indexing the Bragg peaks using standard crystallographic methods. Unlike in conventional X-ray crystallography, the measured intensities are all partials (i.e., not integrated fully) since the sample is not continuously rotated to measure full Bragg peak intensities. Nanocrystallography was first demonstrated by Chapman et al. in 2009 shortly after the LCLS was opened to user operation (58). Protein nanocrystallography was used to determine the structure of a glycosylated propeptide to a resolution of 2.1 Å in early 2011 (74). A total of 4 million diffraction patterns were collected, which were then sorted for hits and subsequently indexed. The reconstructions were performed using a molecular replacement algorithm that starts from a known, similar structure. The crystals were grown in-vivo in cells and were a few μm by a fraction of a μm in size.

4.5. Coherent Diffractive Imaging Using XFELs

The enormous peak brightness and short pulses of XFELs offer the possibility for diffractive imaging of single particles at atomic resolution. The recognition of this opportunity is based on the realizations that (i) X-ray diffraction data of non-crystalline objects can be reconstructed without prior knowledge to yield a real-space image (9), as discussed in the previous section, and that (ii) femtosecond XFEL pulses may provide usable diffraction data even at fluences far beyond the static damage limit (5). In the long run, scattering from single macromolecules, viruses, nanocrystals, and nanoclusters of proteins may be recorded. Unlike in crystals, where resolution is closely linked to the quality of the crystal, resolution now depends on radiation intensity, pulse duration, wavelength, detector, and the extent of sample damage in the form of ionization and atomic motion during exposure, although resolution of the 3D reconstruction may well depend on sample homogeneity, similar to the case of crystallography.

In a likely scenario, copies of identical molecules are injected into the XFEL beam one-by-one in random, unknown orientations, and two-dimensional diffraction patterns are recorded with an imaging detector. Each recorded diffraction pattern from a single molecule is a section of the full three-dimensional diffraction pattern, sampled on a spherical shell, which is a part of the Ewald sphere. Even with the large peak brightness of the XFELs, these diffraction patterns will be very noisy since the molecules are small and scatter only weakly. It will be necessary to average over multiple diffraction patterns from identical samples in order to build up sufficient signal. The deceptively simple task of classifying the noisy diffraction patterns into groups of similar orientation turns out to be very difficult since the signal-to-noise ratio is so low. Assuming this problem can be solved and the diffraction patterns have been classified, they can be oriented with respect to each other by considering their intersection in three dimensions (75). The algorithms for the final step, phasing the three-dimensional diffraction pattern, are probably most advanced in comparison to other issues. However, these algorithms have been developed primarily for static objects and need to be tested for objects that undergo changes such as ionization and atomic motion during the pulse.

Coherent diffractive imaging has been demonstrated at the soft XFEL facility FLASH (37) in the wavelength range of 32 to 6 nm (76–77). In these experiments, it has been shown that imaging beyond the static damage limit is indeed possible. In the first diffractive imaging experiments at FLASH using a wavelength of 32 nm (76), stationary samples were heated to temperatures beyond 60,000K, even though the reconstructed images show no sign of damage during the pulse. Imaging was demonstrated at diffraction-limited resolution. In follow-up experiments, time-delay holography was used to follow the X-ray pulse-induced particle explosion with femtosecond time resolution (77). It was found that the experimentally measured shape evolution and explosion rates are in good agreement with predictions based on a hydrodynamic treatment of the explosion. Subsequently, the first single-particle imaging experiments have been demonstrated at FLASH (78). These experiments have demonstrated that XFELs can be used to follow processes with exquisite temporal and spatial resolution. A quantitative understanding is emerging about ultrafast processes in samples exposed to intense XFEL pulses. In another set of experiments performed at LCLS, the first diffractive imaging of a large virus was demonstrated (65).

4.5.1. Orientation and Classification

In the simplest scenario for diffractive imaging of macromolecules, the particles are injected in a random, unknown orientation. The signal in the diffraction pattern has to be sufficiently strong to allow us to determine the sample orientation, or at least to group them into classes of similar orientation, which sets the lower bound for the incoming photon fluence. Classification techniques for real-space images have been developed in the field of electron microscopy, but the signal in that case is three orders of magnitude larger. In the X-ray diffraction case, the measured diffraction pattern lies on a curved surface in reciprocal space and is a part of the Ewald sphere. When two diffraction patterns of different orientations are measured, they will intersect on a curve that runs through the origin. If the signal-to-noise ratio is sufficiently large, the relative angles of the projections may be determined.

When multiple diffraction patterns of different orientations are collected, it is often sufficient to classify them into groups of similar orientation, and only after the group averages have a sufficiently large signal-to-noise ratio are the group averages oriented. The per-shot dose needed for this technique has been estimated by Huldt et al. (75) and is significantly larger than the X-ray fluences that are currently available. The required fluences generally decrease with increasing particle size since larger particles scatter more photons. Fung et al. (79) have suggested a statistical method that combines the classification and orientation step, and possibly also the reconstruction. They have shown through simulations that even exceptionally weak diffraction patterns with on average only 0.01 photon per detector pixel at a resolution of 1.8 Å may still be assembled into a three-dimensional diffraction pattern. Their methodology exploits the correlation in the entire scattered photon ensemble and is based on the observation that for each pulse, the molecule has only three degrees of freedom, so that the diffraction patterns occupy a three-dimensional subset of all possible intensity measurements. Using Bayesian statistics, this subspace may be classified.

Instead of determining the orientation from the diffraction pattern, the molecular alignment may also be determined from the ion fragments given off by the sample explosion, even when the diffraction signal is weak (43). Three-dimensional time-of-flight spectrometer may be used to image the Coulomb explosion directly. Particularly useful may be the escape of heavy sulfur or intentionally placed marker atoms. This technique is especially useful for smaller molecules and may allow for a dramatic reduction in X-ray dose.

The classification and orientation problems may be circumvented by controlling the orientation of single molecules using nonresonant laser pulses (80). Laser alignment of small molecules is well established (81). Elliptically polarized laser light may be used to simultaneously align the molecule along three axes. An additional DC field may allow fixing the sense of the molecule, as well (82). If the alignment is not perfect, the diffraction pattern will be smeared out which affects the resolution of the reconstructed electron density (83). The authors of (83) have calculated that a four-degree alignment leads to a 1 nm resolution in relatively large molecules. Internal molecular motion complicates reconstruction further. Instead of aligning single molecules, it has also been suggested to align a gas of molecules (84), in which case the diffraction experiment may even be performed on a third generation synchrotron light source.

4.6. Comparison with Competing Techniques

It is not clear yet what the role of XFELs will be in the field of molecular structure determination that is currently dominated by synchrotron protein crystallography. It is conceivable that XFELs will find their place for specialty applications, for example to study rapid dynamics in molecules or to study single molecules. There is great hope that XFELs can overcome conventional damage mechanisms associated with continuous X-ray exposure, giving rise to the static damage limit. Currently, a major issue is access to the facilities. Whereas synchrotrons, by design, can support a large number of experiments simultaneously, the SACLA and LCLS facilities are much more sequential and allow typically only one experiment at a time. This will be increased to a couple in the future.

As an alternative to X-rays, electron beams are also promising probes for imaging single particles at atomic resolution. Comparing the signal per incident probe particle (x-ray or electron) with the amount of damage, electrons give more useful signal per damage event than X-rays (30). A useful figure of merit is the ratio of elastic scattering cross section and the product of the cross section of the inelastic interaction event and the energy transferred to the sample during this event (43). For a 300 keV electron, this ratio turns out to be about 1/(3×30eV), whereas for 10 keV X-rays, this ratio is about 1/(10×10 keV), so it is 1000 times worse for X-rays. For small samples, the photoelectrons escape, so the deposited energy drops from the energy of a photoelectron, about 10 keV, to the energy of an Auger electron, about 0.25 keV. In this case, X-rays are still twenty-five times worse than electrons.

On the other hand, X-rays offer unique advantages over electrons: X-rays can be focused to very small cross sectional areas, and they also may be time-compressed. For electrons, this is very difficult since the Coulomb forces do not allow compressing an electron bunch too much. Finally, electrons have a very limited penetration depth into biological materials, so that they can only be used for thin specimens. X-rays can penetrate much thicker biological samples: For a biological material with a density of 1 g/cm^3, the penetration depth at 8 keV is 1.9 mm. Further, XFELs offer the possibility to expose samples at room temperature.

5. OUTLOOK

5.1. First XFEL Bioimaging Experiments

The first hard XFEL bioimaging experiment was the demonstration of serial femtosecond X-ray protein nanocrystallography in 2009 (58). In this experiment, hydrated photosystem I crystals were injected in a continuous liquid jet into the LCLS beam (85), and high-angle diffraction patterns up to resolution of 8.5 Å were collected with a repetition rate of up to 30 Hz. The X-ray energy was 1.8 keV, corresponding to a wavelength of 6.9 Å. This work was immediately followed by an experiment at LCLS to image nonreproducible, non-crystalline single mimivirus particles with 32 nm resolution, also using X-rays with an energy of 1.8 keV (65). The first published atomic resolution diffraction of protein microcrystals (lysozyme) was performed at 9.4 keV, corresponding to a wavelength of 1.32 Å (59). In that experiment, several million diffraction patterns were collected with a rate of 120 Hz, and they were indexed to a resolution of 1.9 Å. Micrometer-sized membrane protein crystals grown in the lipidic sponge phase have been measured, and diffraction patterns to molecular resolution have been recorded (86). Simultaneous X-ray diffraction and X-ray emission spectroscopy of microcrystals of photosystem II at room temperature with a final resolution of 5.9 Å has been demonstrated by Kern et al. (61, 87). Finally, as described above, the first previously partially unknown atomic structure of a protein was determined in 2013 (74) with a resolution of 2.1 Å.

5.2. Future Opportunities at X-Ray Free-Electron Lasers

The short duration of XFEL pulses offers the opportunity to perform ultrafast imaging of static and evolving objects that may allow us to determine the time dependence of the dynamics and function giving an unprecedented opportunity for mechanistic inference. For example, it will be possible to image the assembly and disassembly of biomolecular complexes, the process of viral infections, and the charge separation and vibrational energy transfer following excitation. Going beyond nanocrystallography, it will be possible to image the structure of nanoclusters, subcellular organelles, whole cells, and single particles, such as biomacromolecules, with femtosecond time resolution. Combined with laser spectroscopy, studying photochemical reactions is possible.

Most X-ray diffraction techniques that have been developed at synchrotrons and rotating anode X-ray sources can be used at XFELs, including crystalline, powder, and surface diffraction, small-angle X-ray scattering (SAXS), as well as extended X-ray absorption fine structure (EXAFS) and near-edge X-ray absorption fine structure (NEXAFS) spectroscopy. As discussed above, ultimately, X-ray diffraction may be performed on single particles and biological macromolecules by introducing spraying techniques and methods from mass spectrometry. For such single particle applications, numerous techniques still need to be developed or further refined until coherent XFEL diffractive imaging could become widely available. Particularly challenging is the area of sample handling. Spraying techniques have to be developed that allow the selection and injection of samples and their mass selection at cryogenic temperatures and in high vacuum. Ultimately, we need the ability to inject native proteins, viruses, nanoclusters, cell organelles, and cellorganelles. Significant progress has been made in this area over the last decade. Improvements in detector performance are a necessary requisite for CDI to be able to reconstruct atomic resolution structures of nonperiodic particles. Finally, software and algorithms need to be developed to convert the enormous data stream of more than 100 (very noisy) diffraction patterns per second into a three-dimensional electron density. This may require the development of reconstruction techniques that take advantage of the intrinsic symmetry present in various macromolecules such as viruses.

6. SUMMARY

In summary, XFEL bioimaging promises to have a major impact on structural biology and the life sciences. XFEL pulses are sufficiently short to overcome damage processes associated with continuous X-ray exposure. The newly developed technique of nanocrystallography allows structure determination of proteins that do not crystallize to form large, well-ordered crystals. Nanocrystallography promises fast-turnaround structure determination while the sample stays at room temperature.

The future might hold single-molecule imaging, which could be achieved by combining XFEL's coherent, ultrashort, and bright X-rays pulses with oversampling phasing methods. Coherent diffractive imaging using XFELs is a newly developed technology that promises to enable ultrafast imaging of non-crystalline samples with unprecedented temporal *and* spatial resolution. If successful, we may gain an understanding of the structure-function relationship, which, for membrane proteins, would be an invaluable contribution to biochemical physiology and medicine.

ACKNOWLEDGMENTS

This work was performed under the auspices of the U.S. Department of Energy by Lawrence Livermore National Laboratory under Contract No. DE-AC52-07NA27344. The author would like to thank M. Hunter for useful discussions and suggestions for the manuscript, and J. Evans with some references on cryo-electron microscopy.

REFERENCES

(1) J.D. Watson and F.H.C. Crick, "Molecular structure of nucleic acid", Nature **171**, 737 (1953).
(2) J.C. Phillips, A. Wlodawer, M.M. Yevit, and K.O. Hodgson, "Applications of synchrotron radiation to protein crystallography: preliminary results", Proc. Natl. Acad. Sci. **73**, 128–132 (1976).
(3) J. Solem and G.C. Baldwin, "Microholography of living organisms", Science **218**, 229–235 (1982).
(4) L. N. Johnson and T. L. Blundell, "Introductory Overview", J. Synch. Rad. **6**, 813–815 (1999).
(5) R. Neutze, R. Wouts, D. van der Spoel, E. Weckert, and J. Hajdu, "Potential for biomolecular imaging with femtosecond X-ray pulses", Nature **406**, 752–757 (2000).
(6) H.M. Berman, J. Westbrook, Z. Feng, G. Gilliland, T.N. Bhat, H. Weissig, I.N. Shindyalov, P.E. Bourne, "The Protein Data Bank", Nucleic Acids Research **28**, 235–242 (2000).
(7) J. Goodman, "Introduction to Fourier Optics", Robert and Company Publishers, 3rd Edition (2004).
(8) C.E. Shannon, "A Mathematical Theory of Communication", Bell System Technical Journal **27**, 379–423 and 623–656 (1948).
(9) D. Sayre, "Some implications of a theorem due to Shannon", Acta Cryst. **5**, 843 (1952).
(10) R.W. Gerchberg and W.O. Saxton, "A practical algorithm for the determination of phase from image and diffraction plane pictures", Optik **35**, 237–246 (1972).
(11) J.R. Fienup, "Reconstruction of an object from the modulus of its Fourier transform", Opt. Lett. **3**, 27–29 (1978).
(12) J. Miao, P. Charalambous, J. Kirz, D. Sayre, "Extending the methodology of X-ray crystallography to allow imaging of micrometre-sized non-crystalline specimens", Nature **400**, 342 (1999).
(13) M.A. Pfeifer, G.J. Williams, I.A. Vartanyants, R. Harder, I.K. Robinson, "Three-dimensional mapping of a deformation field inside a nanocrystal", Nature **442**, 63 (2006).
(14) G. Oszlányi and A. Süto, "Ab initio structure solution by charge flipping", Acta Cryst. A **60**, 134–141 (2004).
(15) V. Elser, "Phase retrieval by iterated projections", J. Opt. Soc. Am. A **20**, 40–55 (2003).
(16) J.C.H. Spence, J.S. Wu, C. Giacovazzo, B. Carrozzini, G.L. Cascarano, and H.A. Padmore, "Solving non-periodic structures using direct methods: phasing diffuse scattering", Acta Cryst. A **59**, 255–261 (2003).
(17) S. P. Hau-Riege, H. Szoke, H. N. Chapman, A. Szoke, S. Marchesini, A. Noy, H. He, M. Howells, U. Weierstall and J. C. H. Spence, "SPEDEN: reconstructing single particles from their diffraction patterns", Acta Cryst. A **60**, 294–305 (2004).

(18) S. Marchesini.H. He, H. N. Chapman, S. P. Hau-Riege, A. Noy, M. R. Howells, U. Weierstall, and J. C. H. Spence, "X-ray image reconstruction from a diffraction pattern alone", Phys. Rev. B **68**, 140101(R) (2003).
(19) B. Abbey, K.A. Nugent, G.J. Williams, J.N. Clark, A.G. Peele, M.A. Pfeifer, M. de Jonge, and I. McNulty, Keyhole coherent diffractive imaging, Nat. Phys. **4**, 394 (2008).
(20) M.R. Howells, T. Beetz, H.N. Chapman, C. Cui, J.M. Holton, C.J. Jacobsen, J. Kirz, E. Lima, S. Marchesini, H. Miao, D. Sayre, D.A. Shapiro, J.C.H. Spence, D. Starodub, "An assessment of the resolution limitation due to radiation-damage in X-ray diffraction microscopy", J. Electr. Spectr. Rel. Phen. **170**, 4 (2009).
(21) S. Eisebitt, J. Luning, W.F. Schlotter, M. Lorgen, O. Hellwig, W. Eberhardt, J. Stohr, "Lensless imaging of magnetic nanostructures by X-ray spectro-holography", Nature **432**, 885 (2004).
(22) S. Marchesini, S. Boutet, A.E. Sakdinawat, M.J. Bogan, S. Bajt, A. Barty, H.N. Chapman, M. Frank, S.P. Hau-Riege, A. Szoke, C. Cui, D.A. Shapiro, M.R. Howells, J.C.H. Spence, J.W. Shaevitz, J.Y. Lee, J. Hajdu, M.M. Seibert, "Massively parallel X-ray holography", Nat. Photon. **2**, 560 (2008).
(23) P. Thibault, M. Dierolf, A. Menzel, O. Bunk, C. David, F. Pfeiffer, "High-Resolution Scanning X-ray Diffraction Microscopy", Science **321**, 379 (2008).
(24) H.M. Berman, K. Henrick, and H. Nakamura, "Announcing the worldwide Protein Data Bank", Nature Structural Biology **10**, 980 (2003)
(25) T. Gonen, Y. Cheng, P. Sliz, Y. Hiroaki, Y. Fujiyoshi, S.C. Harrison and T. Walz, "Lipid–protein interactions in double-layered two-dimensional AQP0 crystals", Nature **438**, 633 (2005).
(26) J.-P. Armache, A. Jarasch, A.M. Anger, E. Villa, T. Becker, S. Bhushan, F. Jossinet, M. Habeck, G. Dindar, S. Franckenberg, V. Marquez, T. Mielke, M. Thomm, O. Berninghausen, B. Beatrix, J. Söding, E. Westhof, D.N. Wilson, and R. Beckmann, "Cryo-EM structure and rRNA model of a translating eukaryotic 80S ribosome at 5.5-Å resolution", Proc. Nat. Acad. Sci. **107**, 19748 (2010).
(27) J.D. Jackson, "Classical Electrodynamics", 3rd Edition, Wiley (1998).
(28) C.C.F. Blake and D.C. Phillips, "Biological Effects of Ionizing Radiation at the Molecular Level", IAEA Symposium, Brno, Czechoslovakia, 183 (1962).
(29) R. Henderson, "Cryoprotection of protein crystals against radiation damage in electron and X-ray diffraction", Proc. R. Soc. London Sci. Ser. B **241**, 6–8 (1990).
(30) R. Henderson, "Realizing the potential of electron cryo-microscopy", Q. Rev. Biophys. **37**, 3–13 (2004).
(31) J. Miao, P. Charalambous, J. Kirz and D. Sayre, "Extending the methodology of X-ray crystallography to allow imaging of micrometre-sized non-crystalline specimens", Nature **400**, 342–344 (1999).
(32) H.N. Chapman, A. Barty, S. Marchesini, A. Noy, S.P. Hau-Riege, C. Cui, M.R. Howells, R.Rosen, H. He, J.C.H. Spence, U. Weierstall, T. Beetz, C. Jacobsen, and D. Shapiro, High-resolution ab initio three-dimensional x-ray diffraction microscopy, J. Opt. Soc. Am. A **23**, 1179–1200 (2006).
(33) M. Cornacchia, J. Arthur, L. Bentson, R. Carr, P. Emma, J. Galayda, P. Krejcik, I. Lindau, J. Safranek, J. Schmerge, J. Stohr, R. Tatchyn, and A. Wootton, "A Sub-Picosecond Photon Pulse Facility for SLAC", SLAC-PUB-8950, LCLS-TN-01-7, (2001).
(34) J.M. Madey, "Stimulated emission of Bremsstrahlung in a periodic magnetic field", J. Appl. Phys. **42**, 1906 (1971).

(35) R. Bonifacio, C. Pellegrini, L.M. Narducci, "Collective instabilities and high-gain regime in a free-electron laser", Opt. Comm. **50**, 373–378 (1984).
(36) L. Mandel and E. Wolf, "Optical coherence and quantum optics", Cambridge University Press, Cambridge (1995).
(37) K. Tiedtke, A. Azima, N. von Bargen, L. Bittner, S. Bonfigt, S. Düsterer, B. Faatz, U. Frühling, M. Gensch, Ch. Gerth, N. Guerassimova, U. Hahn, T. Hans, M. Hesse, K. Honkavaar, U. Jastrow, P. Juranic, S. Kapitzki, B. Keitel, T. Kracht, M. Kuhlmann, W.B. Li, M. Martins, T. Nuñez, E. Plönjes, H. Redlin, E. L. Saldin, E.A. Schneidmiller, J.R. Schneider, S. Schreiber, N. Stojanovic, F. Tavella, S. Toleikis, R. Treusch, H. Weigelt, M. Wellhöfer, H. Wabnitz, M.V. Yurkov and J. Feldhaus, "The soft x-ray free-electron laser FLASH at DESY: beamlines, diagnostics and end-stations", New J. Phys. **11**, 023029 (2009).
(38) T. Tanaka, T. Shintake (Eds.), SCSS X-FEL Conceptional Design Report, Riken Harima Institute, Hyogo, Japan, May 2005; http://www-xfel.spring8.or.jp.
(39) H. Winick, K. Bane, R. Boyce, J. Cobb, G. Loew, P. Morton, H. D. Nuhn, J. Paterson, P. Pianetta, T. Raubenheimer, C. Pellegrini, J. Rosenweiz, G. Ttravish, D. Prosnitz, E. T. Scharlemann, K. Halbach, K. J. Kim, R. Schlueter, M. Xie, R. Bonifacio, L. DeSalvo, and P. Pierini, "Short Wavelength FEL's using the SLAC Linac", Nucl. Instrum. Meth. A **347**, 199 (1994).
(40) Y. Ding, A. Brachmann, F.-J. Decker, D. Dowell, P. Emma, J. Frisch, S. Gilevich, G. Hays, Ph. Hering, Z. Huang, R. Iverson, H. Loos, A. Miahnahri, H.-D. Nuhn, D. Ratner, J. Turner, J. Welch, W. White, and J. Wu, "Measurements and Simulations of Ultralow Emittance and Ultrashort Electron Beams in the Linac Coherent Light Source", Phys. Rev. Lett. **102**, 254801 (2009).
(41) W. Decking, The European XFEL Project, in Brilliant Light in Life and Material Sciences, Springer, Netherlands (2007).
(42) J. Amann, W. Berg, V. Blank, F. Decker, Y. Ding, P. Emma, Y. Feng, J. Frisch, D. Fritz, J. Hastings, Z. Huang, J. Krzywinski, R. Lindberg, H. Loos, A. Lutman, H. Nuhn, D. Ratner, J. Rzepiela, D. Shu, Y. Shvyd'ko, S. Spampinati, S. Stoupin, S. Terentyev, E. Trakhtenberg, D. Walz, J. Welch, J. Wu, A. Zholents, D. Zhu, "Demonstration of self-seeding in a hard-X-ray free-electron laser", Nat. Photon. **6**, 693 (2012).
(43) J. Hajdu. K. Hodgson, J. Miao, D. van der Spoel, R. Neutze, C.V. Robinson, G. Faigel, C. Jacobsen, J. Kirz, D. Sayre, E. Weckert, G. Materlik, A. Szöke, "Structural Studies on Single Particles and Biomolecules", in "LCLS—The First Experiments", September 2000.
(44) E.A. Spiller, "Soft X-ray Optics", SPIE Press Monograph PM15 (1994).
(45) C. G. Schroer, O. Kurapova, J. Patommel, P. Boye, J. Feldkamp, B. Lengeler, M. Burghammer, C. Riekel, L. Vincze, A. van der Hart, and M. Küchler, "Hard x-ray nanoprobe based on refractive x-ray lenses", Appl. Phys. Lett. 87, 124103 (2005).
(46) P.Kirkpatrick and V.Baez, "Formation of Optical Images by X-Rays", J. Opt. Soc. Am. **38**, 766 (1948).
(47) H. Mimura, S. Handa, T. Kimura, H. Yumoto, D. Yamakawa, H. Yokoyama, S. Matsuyama, K. Inagaki, K. Yamamura, Y. Sano, K. Tamasaku, Y. Nishino, M. Yabashi, T. Ishikawa, K. Yamauchi, "Breaking the 10 nm barrier in hard-X-ray focusing", Nat. Phys. **6**, 122 (2010).
(48) V. Bharadwaj, A. Chao, M. Cornacchia, P. Emma, T. Kotseroglou, P. Krejcik, I. Lindau, H.-D. Nuhn, G. Stupakov, R. Tatchyn, R. Bionta, A. Toor, C. Pellegrini, "Ultrashort Optical Pulses in the Linac Coherent Light Source," SLAC-TN-05-036, LCLS-TN-00-8 (2000).

(49) H.N. Chapman and K.A. Nugent, "X-ray pulse compression using strained crystals", Opt. Comm. **205**, 351–359 (2002).

(50) P. Hart, S. Boutet, G. Carini, M. Dubrovin, B. Duda, D. Fritz, G. Haller, R. Herbst, S. Herrmann, C. Kenney, N. Kurita, H. Lemke, M. Messerschmidt, M. Nordby, J. Pines, D. Schafer, M. Swift, M. Weaver, G. Williams, D. Zhu, N. Van Bakel, J. Morse, "The CSPAD megapixel x-ray camera at LCLS", Proc. SPIE **8504**, X-Ray Free-Electron Lasers: Beam Diagnostics, Beamline Instrumentation, and Applications, 85040C (2012).

(51) S. Boutet and G.J. Williams, "The Coherent X-ray Imaging (CXI) instrument at the Linac Coherent Light Source (LCLS)", New J. Phys. **12**, 035024 (2010).

(52) R.V. Dunn and R.M. Daniel, "The use of gas-phase substrates to study enzyme catalysis at low hydration", Philos. Trans. R. Soc. Lond. B Biol Sci **359**, 1309–1320 (2004).

(53) V. Kurkal, R.M. Daniel, J.L. Finney, M. Tehei, R.V. Dunn, and J.C. Smith, "Enzyme activity and flexibility at very low hydration", Biophys. J. **89**, 1282–1287 (2005).

(54) E.I. Howard and R.E. Cachau, "Ink-jet printer heads for ultra-small-drop protein crystallography", Biotechniques **33**, 1302 (2002).

(55) G.M. Nishioka, A.A. Markey, C.K.Holloway, "Protein damage in drop-on-demand printers", J. Am. Chem. Soc. **126**, 16320–16321 (2004).

(56) A.M. Ganan-Calvo, "Cone-Jet Analytical Extension of Taylor's Electrostatic Solution and the Asymptotic Universal Scaling Laws in Electrospraying", Phys. Rev. Lett. **79**, 217 (1997).

(57) D.P. DePonte, U. Weierstall, K. Schmidt, J. Warner, D. Starodub, J.C.H. Spence and R.B. Doak, "Gas dynamic virtual nozzle for generation of microscopic droplet streams", J. Phys. D: Appl. Phys. **41**, 195505 (2008).

(58) H.N. Chapman, P. Fromme, A. Barty, T.A. White, R.A. Kirian, A. Aquila, M.S. Hunter, J. Schulz, D.P. DePonte, U. Weierstall, R.B. Doak, F.R.N.C. Maia, A.V. Martin, I. Schlichting, L. Lomb, N. Coppola, R.L. Shoeman, S.W. Epp, R. Hartmann, D. Rolles, A. Rudenko, L. Foucar, N. Kimmel, G. Weidenspointner, P. Holl, M. Liang, M. Barthelmess, C. Caleman, S. Boutet, M.J. Bogan, J. Krzywinski, C. Bostedt, S. Bajt, L. Gumprecht, B. Rudek, B. Erk, C. Schmidt, A. Homke, C. Reich, D. Pietschner, L. Struder, G. Hauser, H. Gorke, J. Ullrich, S. Herrmann, G. Schaller, F. Schopper, H. Soltau, K. Kuhnel, M. Messerschmidt, J.D. Bozek, S.P. Hau-Riege, M. Frank, C.Y. Hampton, R.G. Sierra, D. Starodub, G.J. Williams, J. Hajdu, N. Timneanu, M.M. Seibert, J. Andreasson, A. Rocker, O. Jonsson, M. Svenda, S. Stern, K. Nass, R. Andritschke, C. Schroter, F. Krasniqi, M. Bott, K.E. Schmidt, X. Wang, I. Grotjohann, J.M. Holton, T.R.M. Barends, R. Neutze, S. Marchesini, R. Fromme, S. Schorb, D. Rupp, M. Adolph, T. Gorkhover, I. Andersson, H. Hirsemann, G. Potdevin, H. Graafsma, B. Nilsson, J.C.H. Spence, "Femtosecond X-ray protein nanocrystallography", Nature **470**, 73 (2011).

(59) S. Boutet, L. Lomb, G.J. Williams, T.R.M. Barends, A. Aquila, R.B. Doak, U. Weierstall, D.P. DePonte, J. Steinbrener, R.L. Shoeman, M. Messerschmidt, A. Barty, T.A. White, S. Kassemeyer, R.A. Kirian, M.M. Seibert, P.A. Montanez, C. Kenney, R. Herbst, P. Hart, J. Pines, G. Haller, S.M. Gruner, H.T. Philipp, M.W. Tate, M. Hromalik, L.J. Koerner, N. van Bakel, J. Morse, W. Ghonsalves, D. Arnlund, M.J. Bogan, C. Caleman, R. Fromme, C.Y. Hampton, M.S. Hunter, L.C. Johansson, G. Katona, C. Kupitz, M. Liang, A.V. Martin, K. Nass, L. Redecke, F. Stellato, N. Timneanu, D. Wang, N.A. Zatsepin, D. Schafer, J. Defever, R. Neutze, P. Fromme,

J.C.H. Spence, H.N. Chapman, I. Schlichting, "High-Resolution Protein Structure Determination by Serial Femtosecond Crystallography", Science **337**, 362 (2012).
(60) U. Weierstall, H. Chapman, R. Doak, J. Spence, D. Starodub, D. Shapiro, P. Kennedy, J. Warner, G. Hembree, P. Fromme, "Droplet streams for serial crystallography of proteins", Exp. in Fluids **44**, 675–689 (2008).
(61) J. Kern, R. Alonso-Mori, R. Tran, J. Hattne, R.J. Gildea, N. Echols, C. Gloeckner, J. Hellmich, H. Laksmono, R.G. Sierra, B. Lassalle-Kaiser, S. Koroidov, A. Lampe, G. Han, S. Gul, D. DiFiore, D. Milathianaki, A.R. Fry, A. Miahnahri, D.W. Schafer, M. Messerschmidt, M.M. Seibert, J.E. Koglin, D. Sokaras, T. Weng, J. Sellberg, M.J. Latimer, R.W. Grosse-Kunstleve, P.H. Zwart, W.E. White, P. Glatzel, P.D. Adams, M.J. Bogan, G.J. Williams, S. Boutet, J. Messinger, A. Zouni, N.K. Sauter, V.K. Yachandra, U. Bergmann, J. Yano, "Simultaneous Femtosecond X-ray Spectroscopy and Diffraction of Photosystem II at Room Temperature", Science **340**, 491 (2013).
(62) B.K. Ku and S.S. Kim SS, "Electrohydrodynamic spraying characteristics of glycerol solutions in vacuum", J. Electrostat. **57**, 109 (2003).
(63) W.H. Benner, M.J. Bogan, U. Rohner, S. Boutet, B. Woods and M. Frank, "Non-Destructive Characterization and Alignment of Aerodynamically Focused Particle Beams Using Single Particle Charge Detection", J. Aerosol Sci. **39**, 917–928 (2008).
(64) J.D.W. Clarke and S.N. Jayasinghe, "Bio-electrosprayed multicellular zebrafish embryos are viable and develop normally", Biomed. Mater. **3**, 011001 (2008).
(65) M.M. Seibert, T. Ekeberg, F.R.N.C. Maia, M. Svenda, J. Andreasson, O. Jonsson, D. Odic, B. Iwan, A. Rocker, D. Westphal, M. Hantke, D.P. DePonte, A. Barty, J. Schulz, L. Gumprecht, N. Coppola, A. Aquila, M. Liang, T.A. White, A. Martin, C. Caleman, S. Stern, C. Abergel, V. Seltzer, J. Claverie, C. Bostedt, J.D. Bozek, S. Boutet, A.A. Miahnahri, M. Messerschmidt, J. Krzywinski, G. Williams, K.O. Hodgson, M.J. Bogan, C.Y. Hampton, R.G. Sierra, D. Starodub, I. Andersson, S. Bajt, M. Barthelmess, J.C.H. Spence, P. Fromme, U. Weierstall, R. Kirian, M. Hunter, R.B. Doak, S. Marchesini, S.P. Hau-Riege, M. Frank, R.L. Shoeman, L. Lomb, S.W. Epp, R. Hartmann, D. Rolles, A. Rudenko, C. Schmidt, L. Foucar, N. Kimmel, P. Holl, B. Rudek, B. Erk, A. Homke, C. Reich, D. Pietschner, G. Weidenspointner, L. Struder, G. Hauser, H. Gorke, J. Ullrich, I. Schlichting, S. Herrmann, G. Schaller, F. Schopper, H. Soltau, K. Kuhnel, R. Andritschke, C. Schroter, F. Krasniqi, M. Bott, S. Schorb, D. Rupp, M. Adolph, T. Gorkhover, H. Hirsemann, G. Potdevin, H. Graafsma, B. Nilsson, H.N. Chapman, J. Hajdu, "Single mimivirus particles intercepted and imaged with an X-ray laser", Nature **470**, 78 (2011).
(66) S.P. Hau-Riege, "High-intensity x-rays—interaction with matter", Wiley-VCH, Weinheim (2011).
(67) E. J. McGuire, "K-Shell Auger Transition Rates and Fluorescence Yields for Elements Be-Ar", Phys. Rev. **185**, 1–6 (1969).
(68) Y.-K. Kim and M. E. Rudd, "Binary-encounter-dipole model for electron-impact ionization", Phys. Rev. A **50**, 3954–3967 (1983).
(69) Z. Jurek, G. Faigel, and M. Tegze, "Dynamics in a cluster under the influence of intense femtosecond hard X-ray pulses", European Physical Journal D **29**, 217–229 (2004).
(70) M. Bergh, N. Tîmneanu, and D. van der Spoel, "Model for the dynamics of a water cluster in an x-ray free electron laser beam", Phys. Rev. E **70**, 051904 (2004)
(71) S.P. Hau-Riege, R.A. London, and A. Szoke, Dynamics of X-Ray Irradiated Biological Molecules, Phys. Rev. E **69**, 051906 (2004).

(72) S.P. Hau-Riege, R.A. London, G. Huldt, and H.N. Chapman, "Pulse requirements for x-ray diffractive imaging of single biological molecules", Phys. Rev. E **71**, 061919 (2005).

(73) A. Barty, C. Caleman, A. Aquila, N. Timneanu, L. Lomb, T.A. White, J. Andreasson, D. Arnlund, S. Bajt, T.R.M. Barends, M. Barthelmess, M.J. Bogan, C. Bostedt, J.D. Bozek, R. Coffee, N. Coppola, J. Davidsson, D.P. DePonte, R.B. Doak, T. Ekeberg, V. Elser, S.W. Epp, B. Erk, H. Fleckenstein, L. Foucar, P. Fromme, H. Graafsma, L. Gumprecht, J. Hajdu, C.Y. Hampton, R. Hartmann, A. Hartmann, G. Hauser, H. Hirsemann, P. Holl, M.S. Hunter, L. Johansson, S. Kassemeyer, N. Kimmel, R.A. Kirian, M. Liang, F.R.N.C. Maia, E. Malmerberg, S. Marchesini, A.V. Martin, K. Nass, R. Neutze, C. Reich, D. Rolles, B. Rudek, A. Rudenko, H. Scott, I. Schlichting, J. Schulz, M.M. Seibert, R.L. Shoeman, R.G. Sierra, H. Soltau, J.C.H. Spence, F. Stellato, S. Stern, L. Struder, J. Ullrich, X. Wang, G. Weidenspointner, U. Weierstall, C.B. Wunderer, H.N. Chapman, "Self-terminating diffraction gates femtosecond X-ray nanocrystallography measurements", Nat. Photon. **6**, 35 (2012).

(74) L. Redecke, K. Nass, D.P. DePonte, T.A. White, D. Rehders, A. Barty, F. Stellato, M. Liang, T.R. Barends, S. Boutet, G.J. Williams, M. Messerschmidt, M.M. Seibert, A. Aquila, D. Arnlund, S. Bajt, T. Barth, M.J. Bogan, C. Caleman, T. Chao, R.B. Doak, H. Fleckenstein, M. Frank, R. Fromme, L. Galli, I. Grotjohann, M.S. Hunter, L.C. Johansson, S. Kassemeyer, G. Katona, R.A. Kirian, R. Koopmann, C. Kupitz, L. Lomb, A.V. Martin, S. Mogk, R. Neutze, R.L. Shoeman, J. Steinbrener, N. Timneanu, D. Wang, U. Weierstall, N.A. Zatsepin, J.C.H. Spence, P. Fromme, I. Schlichting, M. Duszenko, C. Betzel, H.N. Chapman, "Natively Inhibited Trypanosoma brucei Cathepsin B Structure Determined by Using an X-ray Laser", Science **339**, 227 (2013).

(75) G. Huldt G, A. Szoke, and J. Hajdu, "Diffractive imaging of single particles and biomolecules", J. Struct. Biol. **144**, 219–227 (2003).

(76) H. N. Chapman, A. Barty, M. J. Bogan, S. Boutet, M. Frank, S. P. Hau-Riege, S. Marchesini, B. W. Woods, S. Bajt, W. H. Benner, R. A. London, E. Plönjes, M. Kuhlmann, R. Treusch, S. Düsterer, T. Tschentscher, J. R. Schneider, E. Spiller, T. Möller, C. Bostedt, M. Hoener, D. A. Shapiro, K. O. Hodgson, D. van der Spoel, F. Burmeister, M. Bergh, C. Caleman, G. Huldt, M. M. Seibert, F. R. N. C. Maia, R. W. Lee, A. Szöke, N. Timneanu and J. Hajdu, "Femtosecond diffractive imaging with a soft-X-ray free-electron laser", Nature Phys. **2**, 839–843 (2006).

(77) H. N. Chapman, S. P. Hau-Riege, M. J. Bogan, S. Bajt, A. Barty, S. Boutet, S. Marchesini, M. Frank, B. W. Woods, W. H. Benner, R. A. London, U. Rohner, A. Szöke, E. Spiller, T. Möller, C. Bostedt, D. A. Shapiro, M. Kuhlmann, R. Treusch, E. Plönjes, F. Burmeister, M. Bergh, C. Caleman, G. Huldt, M. M. Seibert, and J. Hajdu, "Femtosecond time-delay X-ray holography", Nature **448**, 676–679 (2007).

(78) M. J. Bogan, W. H. Benner, S. Boutet, U. Rohner, M. Frank, A. Barty, M. M. Seibert, F. Maia, S. Marchesini, S. Bajt, B. Woods, V. Riot, S. P. Hau-Riege, M. Svenda, E. Marklund, E. Spiller, J. Hajdu, H. N. Chapman, "Single Particle X-Ray Diffractive Imaging", Nano. Lett. **8**, 310–316 (2008).

(79) R. Fung, V. Shneerson, D.K. Saldin, and A. Ourmazd, "Structure from fleeting illumination of faint spinning objects in flight", Nature Phys. **5**, 64–67 (2009).

(80) J.S. Wu and J.C.H. Spence, "Phasing diffraction data from a stream of hydrated proteins", J. Opt. Soc. Am. **22**, 1453–1459 (2005).

(81) H. Stapelfeldt and T.Seideman, "Colloqium: aligning molecules with strong laser pulses", Rev. Mod. Phys. **75**, 543–557 (2003).

(82) H. Sakai, S. Minemoto, H. Nanjo, H. Tanji, and T. Suzuki, "Controlling the orientation of polar molecules with combined electrostatic and pulsed nonresonant laser fields", Phys. Rev. Lett. **90**, 083001 (2003).

(83) J.C.H. Spence, K. Schmidt, J.S. Wu, G. Hembree, U. Weierstall, B. Doak, and P. Fromme, "Diffraction and imaging from a beam of laser-aligned proteins: resolution limits", Acta. Cryst. A **61**, 237–245 (2005).

(84) D. Starodub, R.B. Doak, K. Schmidt, U. Weierstall, J.S. Wu, J.C.H Spence, M. Howells, M. Marcus, D. Shapiro, A. Barty, H.N. Chapman, "Damped and thermal motion of laser-aligned hydrated macromolecule beams for diffraction", J. Chem. Phys. **123**, 244304–244304-7 (2005).

(85) M. S. Hunter and P. Fromme, "Toward structure determination using membrane-protein nanocrystals and microcrystals", Methods **55**, 387 (2011).

(86) L.C. Johansson, D. Arnlund, T.A. White, G. Katona, D.P. DePonte, U. Weierstall, R.B. Doak, R.L. Shoeman, L. Lomb, E. Malmerberg, J. Davidsson, K. Nass, M. Liang, J. Andreasson, A. Aquila, S. Bajt, M. Barthelmess, A. Barty, M.J. Bogan, C. Bostedt, J.D. Bozek, C. Caleman, R. Coffee, N. Coppola, T. Ekeberg, S.W. Epp, B. Erk, H. Fleckenstein, L. Foucar, H. Graafsma, L. Gumprecht, J. Hajdu, C.Y. Hampton, R. Hartmann, A. Hartmann, G. Hauser, H. Hirsemann, P. Holl, M.S. Hunter, S. Kassemeyer, N. Kimmel, R.A. Kirian, F.R.N.C. Maia, S. Marchesini, A.V. Martin, C. Reich, D. Rolles, B. Rudek, A. Rudenko, I. Schlichting, J. Schulz, M.M. Seibert, R.G. Sierra, H. Soltau, D. Starodub, F. Stellato, S. Stern, L. Struder, N. Timneanu, J. Ullrich, W.Y. Wahlgren, X. Wang, G. Weidenspointner, C. Wunderer, P. Fromme, H.N. Chapman, J.C.H. Spence, R. Neutze, "Lipidic phase membrane protein serial femtosecond crystallography", Nat. Meth. **9**, 263 (2012).

(87) J. Kern, R. Alonso-Mori, J. Hellmich, R. Tran, J. Hattne, H. Laksmono, C. Gloeckner, N. Echols, R.G. Sierra, J. Sellberg, B. Lassalle-Kaiser, R.J. Gildea, P. Glatzel, R.W. Grosse-Kunstleve, M.J. Latimer, T.A. McQueen, D. DiFiore, A.R. Fry, M. Messerschmidt, A. Miahnahri, D.W. Schafer, M.M. Seibert, D. Sokaras, T. Weng, P.H. Zwart, W.E. White, P.D. Adams, M.J. Bogan, S. Boutet, G.J. Williams, J. Messinger, N.K. Sauter, A. Zouni, U. Bergmann, J. Yano, V.K. Yachandra, "Room temperature femtosecond X-ray diffraction of photosystem II microcrystals", Proceedings of the National Academy of Sciences **109**, 9721 (2012).

PART II

Imaging Cellular and Tissue Architecture

How It All Fits Together to Sustain Life

Introduction to Part II

MANFRED AUER, NATALIA PINZÓN, AND GARY C. HOWARD ■

In the first section of the book, we focused on the three-dimensional structure of the macromolecular components and on mechanistic insights into their individual functions. In the second section, we pay tribute to cellular function. Understanding cell processes requires an understanding of the functions of each individual macromolecular component and also of their spatiotemporal organization—their location in time and space—and their interaction partners, and their relative positions to one another and to subcellular compartments. One cannot determine the workflow of an industrial manufacturing plant by just looking at the tools. This is especially true if most of our information is at the level of screwdrivers and drill presses.

Just like in a manufacturing plant, many of basic functional tools (enzymatic activities) are integrated into larger machines (i.e., macromolecular machines/assemblies) that carry out a set of different functions. In an industrial setting, such as in an automobile manufacturing plant, the workflow is typically organized around a certain subcomponent of the car (e.g., the chassis, engine, transmission). Each subcomponent is, in itself, highly complex. It is manufactured independently from the others and then combined to form the final product: the car. Likewise, different regions or compartments in the cell produce different "components" that altogether enable life. Determining the production rates, the controls on each of these processes, and how they interact to sustain life lies at the heart of systems biology.

To fully understand living organisms (e.g., metabolism, signaling, and cellular responses to internal or external signals), we must determine (1) the three-dimensional architecture of cellular scenes, including organelle, cell, and tissue two- and three-dimensional organization, (2) the spatiotemporal localization of target proteins within organelles, cells, and tissues, and (3) the molecular and macromolecular composition, including ions, metabolites and other small molecules, as well as macromolecules (e.g., oligo- and polysaccharides, nucleic acids, genes, transcripts, and regulatory (macro)molecules).

The techniques in this section provide static and/or dynamic information at different sizes and resolution scales. Each has its advantages and disadvantages. None is superior to the others. All will be needed if we try to paint a holistic picture of cellular and tissue activity, physiology, and pathogenesis.

One may argue that most of the techniques covered by the first section (free-electron diffraction is the one exception) are mature and have delivered outstanding results for many years now. Hence, improvements in these techniques now likely will lie in automation, higher throughput, and for X-ray diffraction and NMR, expansion to larger macromolecular assemblies. In contrast, the technologies of this second section are much less mature, and therefore, one may expect significant improvements in coming years. Most importantly, we are just at the beginning of an era in which these approaches will be combined to obtain compositional and architectural information of the system under investigation.

We believe that insights obtained by these approaches—when integrated at the specimen or model level across multiple scales and resolution ranged and when combined with the knowledge obtained from the imaging approaches described in section one—will result in a holistic systems biology understanding of underlying biological processes.

ELECTRON (MICROSCOPY) TOMOGRAPHY AND SERIAL BLOCK SURFACE SCANNING ELECTRON MICROSCOPY IMAGING

Transmission electron microscopy (EM) can be applied to a large range of biological samples, such as individual proteins (electron crystallography) protein complexes (single particle analysis), and more complex cellular sceneries (electron tomography). All of these approaches, except electron tomography, rely on implicit (exploiting symmetry) or explicit averaging of projection views of the three-dimensional sample, followed by reconstruction into a three-dimensional volume that must be interpreted by model building, either through threading of an amino acid sequence through the density map or often through docking of known higher-resolution structural models into the density maps provided by EM.

In the case of tomographic three-dimensional imaging, a large number of projection views of the same object viewed under a range of known tilt angles are recorded, aligned, and reconstructed into a three-dimensional volume. It is typically carried out with whole-mount samples of less than 1 μm thickness (e.g., bacteria, viruses, organelles, thin cell edges, or large macromolecular complexes). Larger objects require sectioning either in a frozen-hydrated state or after embedding into resin. Since no averaging is performed, it is hard to estimate what resolution can be achieved, but geometrical considerations and limitations of the dose that will be tolerated by a biological specimen put its upper limit at 2–3 nm for an object of ~100–150 nm thickness.

The strengths of EM are that literally anything can be visualized and that this approach bridges the gap between the higher resolution structural techniques (featured in the first section of the book) and the lower resolution techniques, some of which are featured in this section. Apart from some practical shortcomings, such as the data anisotropy along the direction of the beam (due to the limited tilt range), and the high level of shot noise and reconstruction artifacts, the challenge one faces in interpreting such cellular volumes is that the macromolecular composition of this depicted unique cellular volume is rather ill-defined. Furthermore, the three-dimensional ultrastructure is often complex and further complicated by macromolecular crowding, with proteins and protein complexes literally touching one another. While large complexes, such as the ribosome, can be identified, smaller complexes are not distinct enough to allow direct detection. It is therefore of great interest to combine fluorescence and electron microscopy imaging and to correlate the two imaging modes to detect specific macromolecules in their ultrastructural context. Where possible, subvolumes can be averaged, resulting in improved density maps. One of the main downsides of transmission EM tomography is that only small volumes can be studied, far less than 1 percent of a typical eukaryotic cell. While montaging and serial tomography approaches are useful to increase the volume examined, these approaches are tedious and often result in detectable gaps between the data sets, complicating quantitative analysis.

In its two forms, focused ion beam scanning (S)EM and serial block face SEM, serial block surface SEM imaging overcomes the problem of inadequate statistical sampling by imaging large volumes with tens of microns in X, Y, and Z. Its resolution is slightly lower than transmission EM but good enough to detect the presence of ultrastructural cellular landmarks, such as large complexes and organelles, and therefore to quantify cellular architecture.

SOFT X-RAY TOMOGRAPHY

Soft X-ray tomography (SXT) enjoys all of the advantages of tomographic imaging. For example, all kinds of samples can be visualized, and it doesn't require metal staining or dehydration or resin embedding. Instead, SXT can be performed on samples in their frozen state. Contrast in SXT is derived from the differences in density and biochemical composition between the various structures inside the cell. The resolution is isotropic as the specimen can be rotated 360° around one axis.

One key advantage over electron tomography is that much larger volumes (e.g., ~10-μm-thick samples, about the width of a prototypic eukaryotic cell), can be examined in their native liquid (although frozen) state due to the weaker interaction of X-rays with matter. Another advantage is that SXT data collection is rather fast, allowing a significant number of samples to be imaged in three dimensions with higher throughput.

A disadvantage of SXT is its limited resolution. It is one to two orders of magnitude lower than EM imaging. SXT delivers slightly better resolution than optical microscopy but, unlike optical approaches, shows all cellular features at least those that display a contrast in the soft X-ray range, such as organelles, and not just the ones that contain a reporter molecule or label.

SXT suffers from the same shortcomings as electron tomography, namely a lack of adequate certainty on the molecular and macromolecular composition of the volume under investigation. Since the resolution is limited, the challenges arising from macromolecular crowding and overall complexity are less of an issue, but would become one if the resolution could be significantly improved.

The combination of SXT with cryogenic fluorescence optical microscopy is promising, as fluorescence reporter molecules such genetically encoded fluorescence fusion proteins can be imaged in their respective subcellular architectural context. While the location of many proteins has been mapped by fluorescence microscopy, the combination with ultrastructural landmarks may yield novel insight into these biological processes.

HIGH-RESOLUTION SINGLE-MOLECULE FLUORESCENCE MICROSCOPY

Fluorescence light microscopy has been one of the most widely used imaging technologies in modern cell biology. For the last two decades, scientists have recorded various localization and co-localization patterns, and the dynamics of their changes, using genetically encoded fluorescence proteins. These have developed a model for cells that looks more or less like the idealized cartoon depiction found in cell biology textbooks. From such observations, scientists have deduced mechanistic insights about biological processes and sometimes claimed the interaction of two proteins because their localization pattern partially or mostly overlapped. While this deduction may be correct for some proteins, others simply happened to be in the same subcellular neighborhood without ever encountering one another.

The smallest spot that can be achieved by focusing lenses is limited by diffraction. The limit of diffraction becomes—once aberrations, light scattering, and poor signal-to-noise ratios are no longer obstacles—the factor that keeps light microscopy from resolving biological phenomena below the 200nm range.

Higher resolution fluorescence imaging was desirable, but how could the diffraction limit of light be overcome?

STED microscopy was the first method to break through the hard "limit" on spatial resolution and, thus, opened the door to super-resolution light techniques. The basic idea of STED is to reduce the size of the focal spot and hence improve spatial resolution, by actively inhibiting the fluorescence on the edge of the focal spot, by use of a second laser.

Other super-resolution approaches rely on signals from different fluorophores that are detected sequentially to map a location with high precision. Hence, a combination of advances in laser, detector, and probe technology are combined with novel data acquisition protocols for precision mapping.

Another approach has been to use interference patterns of light for illumination of the sample, typically increasing the resolution by a factor of two to four compared with conventional fluorescence confocal 3D imaging. While all these novel fluorescence imaging approaches are currently far from routine, they are beginning to be incorporated into commercial setups and are expected to be more widely accessible to the biological community.

MASS SPECTROMETRY IMAGING

The remaining three chapters mark a fundamental shift from previous chapters. While all chapters so far have dealt with three-dimensional architectural organization (e.g., at the level of proteins and complexes, cells and tissues), we now enter the territory of label-free chemical imaging. Electron and soft X-ray tomography require prior knowledge of the macromolecular composition of the three-dimensional volume under investigation, through prior localization studies or correlative reporter-tagged fusion protein fluorescence imaging studies. However, such knowledge is often restricted to a small number of proteins, with other macromolecules and small molecules necessarily being ignored.

Mass spectrometry imaging (MSI) overcomes the need for expensive and time-consuming labeling of a very small number of proteins and hence is a powerful and versatile method of mapping the spatial distribution of (small) proteins and also metabolites, and biomarkers based on their molecular masses. Two variations are used on biological samples: secondary ion mass spectrometry (SIMS) and matrix-assisted laser desorption/ionization (MALDI) imaging. For example, MALDI links the chemical specificity of materials by mass spectrometry with positional information. Depending on the size of the laser used, it can show the distribution of many ion signals of up to 20 kDa. The resolution in selected applications can be as high as about 1 μm, and hence not too far from the resolution obtained by fluorescence imaging. Some bottlenecks exist: some MSI imaging approaches require the sample to be in a vacuum; in a three-dimensional volume it is not completely clear where the MS signal comes from; and often one is simply overwhelmed with the spectrum of signals. While we can easily detect what we expect to find, identifying unknown components is significantly more difficult and sometimes impossible. Despite these challenges, MSI has undergone enormous improvement in recent years and, unlike many other techniques, can detect small-molecule metabolites, small proteins, lipids, nucleic acids, and carbohydrates.

SYNCHOTRON FTIR IMAGING

Fourier transform infrared (FTIR) imaging, like MSI and Raman imaging, yields a spectrum for every pixel raster-imaged. Infrared spectroscopy is based on the selective absorption at each frequency, depending on the vibrational states (in dipole moment with vibration) of the chemical components of the biological material. In the case of the FTIR, a polychromatic infrared beam is raster-imaged repeatedly over the sample, while systematically varying the frequency composition of the incoming beam through interferometry and detecting the infrared absorption spectrum. Using extensive mathematical processing (Fourier transform), the absorption spectrum for each wavelength can be determined and the spectra scrutinized for different spectral signatures to reveal the presence of certain (macro) molecules. Its XY resolution is about 1 μm and hence comparable to other optical approaches, but the conventional infrared signal strength is very low. Synchotron FTIR overcomes this issue by utilizing the high flux (and hence high signal) of the synchotron.

RAMAN IMAGING

Raman imaging exploits the inelastic scattering of incident photons in the infrared range due to polarizability of the chemical component, resulting in an energy shift relative to the frequency of the laser used as the source of radiation. Like infrared imaging, the resulting spectrum is typically very weak since only one in 10 million photons that hit the sample undergo Raman scattering. Raman spectra often display sharp bands specific for certain chemical configurations. One challenge for Raman imaging is a fluorescent signal that can overwhelm the Raman signal, making chemical interpretation impossible. Unlike infrared, Raman imaging is insensitive to water in the sample, making it an ideal choice for biological material. It requires very little (if any) sample preparation, but does require very flat surfaces for imaging. The resolution is similar to other optical techniques, and this approach has been used for a variety of biological samples.

AUTOMATED MICROSCOPIC IMAGING AND SURVIVAL STATISTICS

In this technique, imaging is linked with statistical analysis of a biologically relevant outcome, namely the survival of the cells. By monitoring individual cells over their lifetime in a high-throughput and fully automated manner, multiple disease-relevant endpoints can be tracked in live neurons at once, including survival, morphology, levels/localization/aggregation state of disease-causing proteins, mitochondrial trafficking and morphology, ubiquitin-proteasome function, autophagy/mitophagy flux, axonal trafficking, and energy metabolism. While conventional systems base comparisons on single snapshots, the automated microscope system collects data on large numbers of individual cells over extended periods. In fact, only eight cells per well allowed the prediction of the overall outcome of a particular treatment or perturbation with about 90 percent accuracy. As a result, this method is significantly more sensitive in its ability to directly measure biological processes in individual live cells and to distinguish changes that are pathogenic from those that might be incidental to the disease process or even coping responses to disease.

SUMMARY

In this section, we have a division between architectural and chemical information through the different biophysical imaging and chemical imaging techniques. Obviously, a combination of architectural, compositional, and chemical imaging is superior to architectural and chemical imaging alone, and hence, current and future efforts aim at combining such approaches to gain a systems biology insight into biological processes.

Organizational Imaging and Reporter Molecule Localization

	EM tomography	FIB/SEM; SBF/SEM	Soft X-Ray Tomography	Wide field Epifluorescence Microscopy	Confocal Fluorescence Microscopy	2-Photon Fluorescence Microscopy	Structured Illumination Microscopy (SIM)	PALM/STORM & related	STED
Lateral Resolution	3–10 nm	~10 nm	~100 nm (25–50 nm may be possible)	200–300 nm	200–300 nm	250–400 nm	100–200 nm	20–50 nm	20–50 nm
Axial Resolution	10–15 nm	~15 nm (FIB) 30–50 nm (SBF), ~10 nm possible	~100 nm (25–50 nm may be possible)	Multiple microns 800 nm with deconvolution	800 nm	800 nm	400 nm	50–100 nm in TIRF mode, otherwise not really functional	800 nm
Penetration Depth	100–500 nm 1 micron possible	100+ microns	Typically 10 microns	100 microns	100 microns	300 microns	100 nm in TIRF mode, up to 5 microns in pseudo-TIRF mode	10 microns	10 microns
Radiation type	Electron beam	Electron beam	Soft X-ray Photons	Visible light Photons	Visible light Photons, laser	Visible light Photons, laser	Visible light Photons, laser	Visible light Photons, laser	Visible light Photons, laser
Sample preparation	HPF/FS resin Cryo-sectioning cryo-whole mounts	Staining, resin embedding and sectioning	Chemical fixation or cryo-immobilization	Fluorescently labeled affinity probe or genetically encoded fluorescent reporter molecule required, imaging of live cells possible. Some tissues are difficult to image or require "clearing" reagents For PALM/STORM/STED live imaging only possible if processes are slow, fixed samples better, requires photoactivatable, photoswitchable probes					

(*Continued*)

Type of object	Micromolecular complexes, organelles, cells, tissues	Cells and tissues	Individual cells (bacterial, yeast, mammalian cell culture cells)	Organelles, cells, tissues					
Dynamic or static	Static	Static	Static	Static &dynamic				Mostly static, dynamic possible	
Data type	Density maps of all components in entire volume			Monochromatic or multicolor distribution of reporter molecules					
Advantages/ challanges	Visualizes intact cells, cell extracts, provides cellular contexts at macromolecular resolution Analysis is complex due to crowdedness Molecular identity is missing	Visualizes intact tissues at macromolecular complex resolution, large volume coverage Data sets can be of extremely high complicty	Compared to EM higher through-put, but cell architecture at (modest) organelle resolution, requires access to synchotron radiation	Fast, easy data collection, very high throughput, automatable, allows time-lapse imaging, but axial resolution poor, diffraction—limited resolution	Realtively fast data collection with reasonable throughput, axial resolution improved but still not great, diffraction limited resolution Cellular context is missing, only see what is tagged	Similar to confocal imaging but better penetration depth	Expands resolution by a factor of about two, with the need of any specialized dyes, relatively easy to use, easily compatible with live-cell imaging	High lateral resolution but requires specific photoactivatable/ photoswitchable reporter molecules, data acquisition is complex, optimal for thin objects, no cellular context	High lateral resolution, reasonable axial resolution, faster than PALM allowing more dynamic information, but only image tagged molecules, no cellular context

	SIMS	NIMS	MALDI	DESI	SR-FTIR	Raman Imaging
Lateral Resolution	50–100 nm	~15 microns	~10 microns	~100 microns	1 micron	~ 200–400 nm
Axial Resolution	2D surface, 3D possible	only molecule on 2D surface	2D surface	2D surface	Several microns	~1 micron
Energy type for ionization/ radiation type	ion beam, electric field		laser, electric field	charged solvent, electric field	infrared lasers	UV, VIS or NIR lasers
Sample preparation	dried samples	adsorption to matrix			little preparation, label free, near native	
Area/size of object	tens of microns	millimeters	~100 microns	millimeters	~100 microns	~100 microns
Data type	ion spectra, mass<500 Da	ion spectra, wide mass range (10–2,000 Da)	ion spectra, mass>500 Da	ion spectra, wide mass range	absorption spectra of macromolecules	scattering spectra of macromolecules

(Continued)

Advantages/ challenges	—excellent spatial resolution —minimal sample preparation —imaging under vacuum —long acquisition times for large areas —not suitable for most organic molecules due to extensive fragmentation —typically used for elemental mapping —destructive	—wide mass range —medium resolution —requires sample preparation —imaging under vacuum —detects only molecules in contact with the matrix surface —destructive	-fast and sensitive —medium resolution —detects only larger ions —requires sample preparation -imaging under vacuum —destructive	—fast, high throughput — ambient conditions, soft ionization (little fragmentation) — little sample preparation —low substrate requirements —poor spatial resolution (100–250 microns) —destructive	—non invasive, dynamic, compositional imaging —Spectral noise from water in living cells, hence elaborate sample preparation needed —cannot visualize subcellular components —long acquisition time of large areas —requires synchrotron facility —nondestructive	-noninvasive dynamic compositional imaging —minimal sample preparation —high specificity; chemical and molecular information —native state —insensitive to water —versatile —analysis of multiplexed data —requires flat surfaces —weak signal swamped by fluorescent background —potentialphotodamage —long acquisition time —low SNR —nondestructive

ABBREVIATIONS: SIMS, secondary ion mass spectroscopy; NIMS, nanostructure-initiator mass spectrometry; MALDI, Matrix-assisted laser desorption/ ionization; DESI, desorption electrospray ionization mass spectrometry; SR-FTIR, synchrotron radiation-Fourier transform infrared.

7

Bridging the Resolution Gap

Electron Tomography and Advanced Three-Dimensional SEM Approaches for Cellular Volumes

MANFRED AUER

ELECTRON MICROSCOPY—EXAMINING CELLS AND TISSUES

Seeing Is Believing—Visualizing the Inner Workings of Cells

In the seventeenth century, the English natural philosopher Robert Hooke contributed to the scientific revolution when he used a device for augmenting our sense of sight though the optical magnification of biological objects too small to be seen by the naked eye. Often referred to as the founding father of modern microscopy, he coined the term "cell" to describe what he considered the fundamental unit of life, and published *Micrographia*, a book depicting the astonishing details of life as observed through the lenses of his hand-crafted, leather gold-tooled microscope. Immediately, scientists and the interested public were intrigued by the degree of order that one could observe at scales beyond the finesse of human craft. To this day biologists in general, and microscopists in particular, revere the beauty of tissue and cellular organization with its remarkable degree of order and organization that is aesthetically pleasing and hence reassuring to the observer. The determination of structure always had a profound impact on our understanding of function, and it is therefore not too surprising that the discipline of structural biology, which emerged out of anatomy, aimed to dissect and visualize form and function with ever finer tools down to the molecular and atomic level. Early on, microscopic analysis of cells revealed the presence of organelles, and hence a compartmentalization of cellular function. The tools of the twentieth century allowed biochemistry to distinguish that certain parts of the metabolic pathway were spatially separated, but for the longest time it was still assumed that a cell in good approximation was a second-order reaction vessel, a bag of enzymes, where random encounters of substrates with their respective enzymes led to reaction products, and hence the focus of biology then was on studying individual proteins and their respective active site chemistry. Only recently it became appreciated that most if not all processes in cells were likely to be carried out by protein complexes, containing ten or more individual proteins that work together to ensure efficient execution of their respective tasks (Alberts 1998). In other words, cellular function now could be viewed as the spatial and temporal patterns of 3D cellular organization and activity of such macromolecular machines. As a result, the system as a whole became the target of investigation. Such insight, however, made it necessary to develop new approaches for the direct visualization of such macromolecular machines in their native cellular context in order to study such patterns at the macromolecular level, and hence to understand function in the context of macromolecular and cellular organization.

The invention of the electron microscope, arguably one of the most important technical developments in cellular biology in the twentieth century, revealed that cells were highly organized into subcellular compartments. This was followed by the discovery of new membrane-encased organelles, filaments that constitute the cytoskeleton including actin bundles, microtubules and intermediate filaments, as well as of intracellular, as well as cell-cell and cell-matrix junctional complexes, all of which appeared to be well-defined in size and shape. Among such junctional complexes are tight junctions, which seal the luminal space of epithelial tissues and therefore prevent the diffusion of ions and other molecules between luminal and basolateral space. Another well-defined example for crucial cell-cell adhesion complexes are the desmosomes, whose apparent role is to provide strong adhesion between cells, therefore ensuring the structural stability of the respective tissues. Detailed ultrastructural analysis of these two adhesion complex systems revealed the preferential association of such junctional complexes with the actin cytoskeleton or the network of intermediate filaments, respectively, and hence it became clear that cellular function wasn't just compartmentalized into organelles, but also was highly organized within the cytoplasm, and that it was probably safe to assume that all these processes are compositionally, spatially, and temporarily tightly regulated.

While clearly powerful and suggestive, all these detailed ultrastructural images of cells were static in nature since the electron microscope requires lengthy and invasive sample preparation techniques incompatible with live-cell imaging. Optical imaging approaches had dramatically improved by the late twentieth century, allowing phase-contrast live cell imaging and hence the study of the dynamic behavior of cells. However, conventional optical approaches are inherently diffraction-limited, hence limiting the resolution to ~200–400 nm and therefore one to two magnitudes above molecular resolution. In addition, just like in electron microscopy (EM), phase contrast images alone could not discriminate between the different molecules and hence had limited success for the study of cellular physiology. Direct visualization of specific dynamic cellular processes in live cells became possible only with the development of genetically encoded fluorophores, such as realized in the jellyfish-derived Green Fluorescent Protein (GFP) and its spectrally shifted derivatives. The fact that reporter molecules could now be genetically linked to the protein of interest allowed in vivo dynamic visualization of a large variety of cellular processes, a technique so fundamental that its development was rewarded with the Nobel Prize in Chemistry in 2008.

As a consequence, at the end of the twentieth century, EM imaging had temporarily taken a back-seat in cell biological analysis, in part to some misconceptions about EM, which at that time had a reputation of being prone to a vast variety of sample preparation artifacts, time-consuming and tedious in nature and only accessible to a small community of experts. Since by then we all "knew" (or at least we thought so) what the ultrastructure of a typical cell looks like, we did not need to go through the trouble to determine it for every sample, but settled for a "one-fits-all" generic typical cellular ultrastructure, and instead simply assumed the presence of functional complexes through fluorescence detection of surrogate proteins.

However, with rapid cryo-immobilization approaches that were developed in the later part of the twentieth century, artifact-free cellular preservation became possible. More recently, cell biologists have started to realize that fluorescent surrogates can be inadequate, hence requiring direct ultrastructural validation of functional complexes, rather than just the presence of constituting proteins. As a result, the early twenty-first century has witnessed a renewed interest in electron microscopy, including in the recently developed 3D EM imaging approach also known as electron tomography.

Paradigm Shift in Cell Biology: From Bags of Enzymes to Ordered Assembly of Macromolecular Machines

The paradigm shift from individual proteins to protein complexes also meant that such large molecular machines needed to be analyzed, both as biochemically isolated and purified multiprotein complexes as well as in their native cellular context. Electron microscopy had been used successfully since the middle of the twentieth century for the structure determination of biological complexes typically exploiting crystallographic and non-crystallographic symmetry (Auer 2000, Henderson 2004), including icosahedral symmetry for the study of viruses (Baker et al. 1999), helical symmetry for cytoskeletal filaments (Stewart 1988, Egelman 2007) and 2D crystalline symmetry for membrane proteins that would arrange in a 2D crystalline pattern (Glaeser 2007). Exploiting symmetry allowed implicit averaging of identical macromolecules in defined orientations, each of which is related to its neighbors by a respective symmetry operator. Particles without inherent symmetry needed to be aligned and explicitly averaged, a process that is covered in detail in the single-particle EM chapter of this book (see also Nogales & Grigorieff 2001). Single-particle averaging involves imaging tens of thousands of identical particles, ideally in random orientation, and subsequent alignment and classification of their projection views, followed by determination of their respective exact orientation in order to allow averaging for 3D reconstruction. Common to all of the approaches that implicitly or explicitly rely on averaging is that the biological object under study is assumed to be homogeneous and to a certain degree identical to one another in composition, size, and shape, thus often requiring stringent biochemical purification procedures. As an alternative to biochemical isolation of macromolecular complexes and therefore destruction of cellular context, molecular machines can be imaged in situ. This approach is particularly powerful in cases where the macromolecular machine is too rare or too fragile to be isolated, or where function of such macromolecular machines depends on cellular context. Electron tomography allows the visualization of both intact cells and crude cell extracts, followed by *in-silico* selection of suitable particles, for example employing segmentation and template matching approaches. Electron tomography can also be performed on isolated individual large particles such as viruses and can yield starting models for higher resolution structural analysis using single particle cryo-EM.

Some of the fundamental differences between the electron tomographic study of cellular volumes and isolated macromolecular complexes lies in the level of certainty with respect to the compositional identity, the crowdedness of the cellular volume and its consequences for visualization and quantitative analysis, as well as the overwhelming wealth of information, making it often necessary to extract the features of interest in order to comprehend complexity of the molecular scenery.

Tools for Analyzing Architecture and Function

Studying the inner workings of cells requires the study of 3D architecture at different levels of resolution and complexity, including the 3D atomic and molecular structure of individual proteins, the 3D architecture of macromolecular complexes that are composed of those invididual proteins, as well as the 3D organization of such molecular machines in cells and tissues, where in concert with their cellular surrounding they perform their respective task to sustain cellular physiology.

The atomic and molecular level is usually best addressed by classical structural biology approaches such as X-ray crystallography of well-suited 3D crystals, electron crystallography of well-suited 2D crystals and helical assemblies, NMR spectroscopy, and to some limited degree atomic force microscopy of 2D crystals.

Supramolecular complexes are typically the territory of single-particle negative-stain—and cryo-electron microscopy, although substantial success has been accomplished by Small-Angle X-ray Scattering (SAXS) (Putnam et al. 2007) and X-ray crystallography (Chiu et al. 2006, Mueller et al. 2009), provided that the complex is stable and homogeneous enough for high quality 3D crystals. A hybrid approach has been taken, where either the single particle cryo-EM data serves as a starting model for X-ray diffraction pattern phasing, or where density maps obtained by cryo-EM under different experimental conditions can be interpreted by using the atomic model obtained by X-ray crystallography. The use of such hybrid approaches for data reconstruction and/or data interpretation has increased over the last decade with often impressive results (Lindert et al. 2009).

The area of cellular imaging is best covered by electron tomography, with its core strength being that, unlike all other typical structure determination methods, it does not rely on the presence of identical objects and is hence ideally suited in instances where the cellular context is crucial to function. Electron tomography can visualize a large number of interesting biological cellular systems (Auer 2000, Baumeister 2002, McIntosh et al. 2005, Lucic et al. 2005, McDonald & Auer 2006, Frey et al. 2006, Nickell et al 2006, Downing et al. 2007), including signaling complexes that require the interplay between extracellular, transmembrane, and intracellular macromolecular complexes (Auer et al. 2008).

All the approaches discussed so far are "true" 3D methods that result in true 3D density maps that can be interpreted through model building and quantitative analysis. In addition, there are some pseudo-3D approaches, which are often confused with real 3D approaches as they provide a 3D impression by projecting a 3D object onto the 2D plane of the recording device such as film or CCD cameras. One such approach is freeze-etch scanning electron microscopy (SEM) which has provided to date some of the most stunning images of the cytoskeleton and other cellular features. However, like atomic force microscopy (AFM), SEM is inherently limited to surfaces and hence is often of limited value to the study of the inner workings of cells, but can be very useful for the study of intact cell surfaces or otherwise prepared surface views of cellular volumes.

The Good, the Bad, and the Ugly: The Power and Limitations of Electron Microscope Imaging

Since any direct imaging technique is ultimately diffraction-limited, the wavelength used for structure determination must be comparable to the desired resolution, and the only electromagnetic spectral window suitable for imaging at molecular dimensions is X-ray. Strictly speaking, X-ray crystallography is not an imaging technique, as it produces diffraction patterns—the Fourier Transform of an image, with the phase information being lost upon photon detection. However, a model of the original atomic distribution can be obtained through clever introduction of site-specific heavy metals that allow phasing of the diffraction pattern followed by model building and subsequent iterative refinement of the model comparing model-based calculated and experimentally observed diffraction patterns. Unlike the visible part of the electromagnetic spectrum, X-rays are difficult to focus, and hence X-ray microscopy of biological specimens has only recently been explored (Le Gros et al. 2005). While X-ray microscopy is not fundamentally limited in resolution, the current practical resolution limit is several tens of nanometers, and hence it will be interesting to see whether the full potential of this relatively young discipline for molecular-resolution imaging of cellular processes can technically be realized in the near future.

Apart from the electromagnetic spectrum, both electrons and neutrons can be used for imaging, but the interaction of neutrons with matter is weak, therefore rendering neutron scattering isn't suitable for most biological imaging applications.

Electron microscopy provides an interesting alternative to photons in the study of biological matter, with wavelengths well below a nanometer, and despite relative shortcomings of electromagnetic lenses, accelerated electrons are well suited for structural analysis of proteins, macromolecular machines, and entire cells. However, due to the strong interaction of electrons with matter, a few millimeters of air are sufficient to stop the electron beam completely. As a consequence, imaging must occur in a high vacuum, which poses problems to hydrated biological specimens, which simply would dry out under such conditions. In addition, samples are required to range in thickness up to a micron, but typically are chosen to be closer to 100–200 nm, depending on microscope acceleration voltage.

As a consequence, all isolated macromolecular complexes and some bacteria can be studied as frozen-hydrated whole-mount samples, whereas most eukaryotic cells and basically all tissues need to be sectioned to allow beam penetration and to minimize multiple scattering events, which would degrade image quality. The downside with sectioning is that for most specimens it is hard to control the exact orientation and hence the features of interest may not be present as a whole in any of the sections, and their position relative to the rest of the cell can be hard to assess.

The most fundamental limitation in electron microscopy imaging is probably the dose that can be tolerated by a biological specimen before radiation damage renders the biological sample under investigation useless for structural analysis. Radiation damage cannot be avoided, but its effect can be mitigated through lose-dose and low-temperature imaging.

SAMPLE PREPARATION

Sample preparation is arguably the most crucial of all steps towards imaging of the cellular macromolecular inventory and its respective 3D architecture, and despite its importance it is rarely discussed in detail. This section will aim to address the various choices one has for sample preparation and explain the rationale behind the various protocols. An introduction to biological sample preparation can be found in references (McDonald & Auer 2006, Vanhecke et al. 2008, Maunsbach 1998). Unlike in optical microscopy, imaging of biological samples in the electron microscope requires sophisticated sample preparation protocols. First, as has been discussed in the previous section, the strong interaction of electrons with matter requires typical biological samples to be thinner than about one micrometer, and hence for typical samples to be cut into thin sections, although in few selected cases thicker specimen can be imaged using high-voltage electron microscopes. Second samples need to be imaged in a vacuum, posing a significant challenge for aqueous solution containing cells, which would instantaneously dry out in the microscope's vacuum. Third, besides elastic scattering events there exist inelastic scattering events and as a net result there is deposition of energy into the biological sample under study. This deposition of energy leads to the breakage of chemical bonds and the formation of radicals that catalyze further bond breakages and therefore the disintegration of the macromolecules, also known as radiation damage (Bozzolla & Russel 1999).

Depending on the dimensions and nature of the biological sample, specimens either can be viewed as whole-mounts, as is the case for individual proteins, protein complexes, other macromolecular assemblies, and bacterial cells, or need to be sectioned as is the case for eukaryotic cells, tissues, and microbial communities. Whole-mount samples can be either examined in their unstained frozen-hydrated state using cryo-EM low-dose imaging, or visualized with enhanced contrast after heavy metal staining (usually osmium, uranium, and/or lead), which results in either positive or negative staining. Stained biological samples can withstand a higher electron dose and lead to increased contrast, which facilitates visualization. However, heavy metal staining bears the potential of structural reorganization of the macromolecules and the danger of preferential staining, thus altering the structure of interest.

Conventional Protocols

As discussed above, the challenges one faces in EM are the need for thin specimen, the necessity to image in vacuum as well as radiation damage caused by the electron beam. The conventional solution is to embed biological specimens in an epoxy or other "plastic" resin, which can be hardened, and sectioned to a thickness of

50–100 nm. Exposure of the specimen to heavy-metal salts (typically osmium, uranium, and lead) during the sample preparation process generates contrast and allows subsequent visualization in the electron microscope. Such resin-sections are stable in the vacuum of the electron microscope and often yield sufficient contrast for straightforward ultrastructural interpretation. However, resins are not water-miscible, and hence require a dehydration step, where cellular water is being replaced by a polar organic solvent (such as ethanol) in order to allow subsequent resin-infiltration and resin polymerization. It is the dehydration step that can be considered the most detrimental of all sample preparation steps. Stripping proteins and other macromolecules of their hydration shell promotes denaturation, likely through conformational changes that expose hydrophobic regions, which leads to the aggregation artifacts typically associated with conventionally prepared samples. To reduce the effect of the solvent exchange, biological samples are typically exposed to primary fixatives, typically glutaraldehyde, and to secondary fixatives such as osmiumtetroxide and uranylacetate prior to dehydration. The idea is that the fixatives will physically connect the network of macromolecules into a giant gel-like structure, which in turn should resist aggregation during dehydration. These chemicals penetrate the sample by diffusion, which results in rapid fixation for objects thinner than a few micrometers in the shortest dimension, but may take several hours for larger pieces of tissues or organisms that contain natural diffusion barriers, including animal exocytoskeletons, protective layers around embryos or plant cell walls, therefore resulting in slow cellular death and sometimes gross alterations of cell architecture, such as autolysis. Even thorough aldehyde-fixation will only cross-link certain protein residues, but not nucleic acids, carbohydrate, or lipids, hence extended exposure to solvents in the dehydration, especially when performed at room temperature, and resin-infiltration process can lead to extraction of unfixed macromolecular components, often generating the impression of sparse protein distribution within cells (Small 1981). Nothing can be further from the truth as well-preserved samples display a very dense matrix of proteins that literally touch each other. The extraction from and relocation of proteins within the cells, together with vastly altered membrane structures that is most likely due to a cellular stress response of dying cells, have given conventional EM sample preparation a bad reputation. Fortunately, there exist sophisticated methods that can overcome most if not all of these artifacts. Interestingly, some investigators still prefer the artifacts associated with conventional protocols over the more densely populated, complex, and low-contrast images that result from cryogenic approaches of sample preparation.

Cryogenic Approaches of Sample Preparation

Unlike conventional sample preparation approaches, ultrarapid freezing immobilizes all cellular molecules within milliseconds, therefore "fixing" all molecules in their respective position and avoiding any fixation artifacts (Gilkey & Staehelin 1986, Studer et al. 2008). Such cryo-immobilized samples can now be either cryo-sectioned or subjected to a process called freeze-substitution that is designed to minimize any dehydration artifacts. Dehydration of frozen-hydrated specimen occurs over the course of several days at dry-ice temperature, using either acetone or methanol as the organic solvent, both of which are liquid at this temperature. While the detailed mechanisms remain unclear, it was empirically found that at such low temperature the macromolecules are shielded from the effect of changes in their hydration state. It is presumed that at these low temperatures macromolecules have a limited ability to undergo drastic conformational changes in response to the changes of the solvent, therefore avoiding protein denaturation and subsequent aggregation (Kellenberger 1987). Once the solvent is exchanged, macromolecules seem stable in the new solvent environment: they can be warmed up to room temperature and resin-infiltrated without apparent changes to the ultrastructure. Typically, fixatives, such as glutaraldehyde and osmium tetroxide or uranyl acetate, are dissolved in the organic solvent, but it is assumed that they chemically fix the samples only after the solvent exchange has taken place and when the sample is being slowly warmed up, with glutaraldehyde and osmium tetroxide beginning to be chemically reactive at –50°C and –30°C, respectively (Steinbrecht & Müller 1987). Therefore, at the temperature that macromolecules can undergo conformational changes the cells are already fixed.

Freeze-substituted samples can be warmed up to room temperature and subsequently infiltrated with commercially available epoxy-araldite or LRWhite resins, for morphological or immunolabeling studies, respectively. Alternatively, cells can be infiltrated with methacrylate resins (such as Lowicryl HM20 or K4M), and polymerized by UV irradiation at –25°C to –70°C. Infiltration at room temperature can be accelerated through the use of a suitable specialized microwave oven.

Embedding in resin, followed by chemical, thermal, or UV-radiation polymerization, results in blocks that can be sectioned just like conventionally prepared samples. One of the caveats of freeze-substitution is that membranes often appear less well pronounced and almost missing. Such short-comings can be avoided by deliberately retaining trace amounts of water throughout sample preparation (Buser & Walther 2008) or adding tannic acid to the freeze-substitution mixture (Jimenez et al. 2009).

High-Pressure Freezing—the Promise of Artifact-Free Specimen Fixation

The main objective of successful rapid freezing is the vitrification of the biological specimen, meaning the avoidance of ice crystal formation, which threatens to damage the cell's ultrastructure. Therefore, the heat needs to be conducted away rapidly from the specimen, while at the same time high pressure discourages ice crystal formation due to the lower density of ice compared to liquid water. For very thin objects, less than about one micron in thickness, vitrification can be accomplished by plunge freezing into liquid ethane. Most eukarotic cells and tissues require either slam freezing onto a cooled metal block (Sitte et al. 1987) or high pressure freezing, resulting in tens or hundreds of microns of vitrified samples, respectively. Often cryo-protectants, such as glycerol, are added to mitigate ice-crystal formation, which is recognizable as freezing damage, but these cryo protectants can create their own set of ultrastructural artifacts (Gilkey & Staehelin 1986) and can be a challenge for osmotically sensitive systems (Triffo et al. 2008). If live samples cannot be frozen directly, one can freeze samples that have been chemically fixed prior to freezing, ideally through the use of a specialized microwave oven, which allows vast acceleration of penetration of the fixative solution, and minimizes fixation artifacts (Leong & Sormunen 1998, Giberson & Demaree 1999).

The type of freezing depends on specimen dimensions. As discussed above, objects thinner than a few microns can be vitrified by plunge-freezing in liquid ethane, some 10 micrometers of ice-crystal-free freezing can be accomplished by slam-freezing where the specimen is rammed into a liquid nitrogen-cooled copper block. Anything thicker than some 10 micrometers needs to be high-pressure frozen, with sample dimensions of up to a thickness of 200–300 micrometers. While the idea of freezing at about 2,000 atmospheres of pressure has been around for half a century (Moor & Riele 1968), the first commercial machines became only available in the mid-1980s, and interest from the biological community only started to pick up by the mid-1990s, but recently has been rapidly growing, along with the interest for 3D ultrastructural imaging by tomography. For a more thorough discussion on the theory and practice of high pressure freezing, the interested reader is referred to Moor (1987). In short, the high pressure building up just prior to freezing discourages ice-crystal formation due to the density differences of water and ice. Figure 7.1 summarizes the choices one has for sample preparation and shows representative images of bacteria illustrating the differences in preservation between conventionally processed samples, high-pressure frozen (HPF) and freeze-substituted and resin-embedded samples, cryo-sectioned samples, and frozen-hydrated whole mounts. HPF samples retain the organization of the cytoplasm with few if any visible distortions or artifacts, both for resin-embedded samples as well as cryo-sectioned samples. Since investigators will spend weeks and months collecting and analyzing the tomograms, preservation of the specimen as closely as possible to the native state is crucial.

Whole-Mount and Cryo-Section Tomography

Thin objects that were prepared by plunge-freezing of isolated macromolecular machines on holey-carbon coated grids are typically imaged as whole-mounts in their frozen-hydrated state. Like in most other cryo-EM techniques thickness of the ice in the holes is crucial for sample contrast, where too thick of an ice thickness leads to undetectable contrast, whereas samples too thin may alter macromolecular organization due to air-water interface and flattening artifacts. Thicker objects prepared by high-pressure freezing can be sectioned in their frozen-hydrated state, but this technique, also known as cryo-sectioning or cryo-EM of vitreous sections, is still under constant development with very few labs worldwide having mastered the art of cryo-sectioning (Zhang et al. 2004, Al-Amoudi et al. 2004, Pierson et al. 2009). While contrast is inherently low, as expected for unstained samples, application of sufficient defocus during data collection can reveal the position and shape of large cellular machines. While the TEM imaging of frozen-hydrated specimen is widely accepted as the gold standard, low signal-to-noise and the fact that area cannot be prescreened (in order to not waste precious electron dose) limit its use for a large number of biological questions. Nevertheless, it is the method of choice for all whole-mount specimens thinner than some 500–1000 nm. Apart from sectioning of resin-embedded samples, cryo-sectioning avoids sample preparation artifacts but is not well suited for specimens that require extensive screening and/or need to be viewed in a particular orientation. Also, if localization of particular macromolecules is the goal of the study, cryo-EM of vitreous sections is severely limited, as lengthy labeling protocols require the thawing of the specimen and thus chemical fixation.

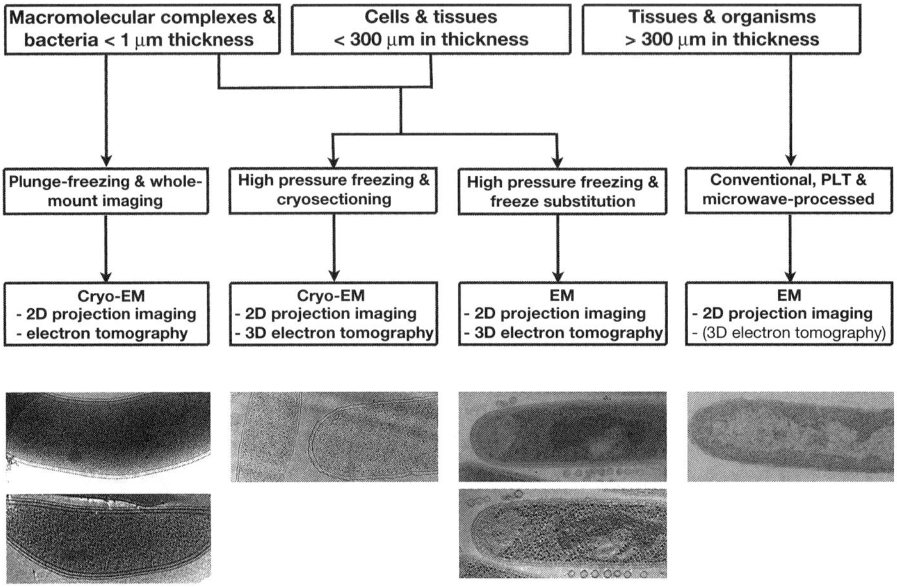

Figure 7.1 Sample preparation and imaging of biological specimen. Choice of sample preparation depends on the characteristics of the samples, including specimen thickness and complexity. Specimens are either viewed as whole-mounts or as sections, resulting in either 2D projection images (top) or slices through 3D (bottom) imaging approaches. Representative images of comparable bacterial samples are shown: Left: *Caulobacter crescentus* as whole-mount of ~500 nm thickness (top) and 1.2 nm thick slice through 3D reconstruction (courtesy of Drs. Luis Comolli and Kenneth H. Downing, Lawrence Berkeley National Laboratory). Second from left: projections view of 30 nm thick cryo-section of *Escherichia coli* (courtesy of Dr. Sriram Subramaniam, National Institute of Health, Zhang et al. 2004). Note that the amount of details visible is comparable to slices of the 3D reconstructions of whole-mount tomograms. Second from right: 100 nm thick resin-embedded section imaged in 2D projection view (top) and 1 nm slice through 3D reconstruction of *Myxococcus xanthus* biofilms (Palsdottir et al. 2008). Right: 2D projection view of 100 nm thick resin-embedded section of *Desulfovibrio vulgaris*. Note excellent preservation of high-pressure frozen, freeze-substituted sample compared to microwave assisted or conventional processing.

The Tokuyasu approach (Tokuyasu 1973, Webster et al. 2008, Peters et al. 2006) of cutting frozen sections that were fixed and sucrose-infiltrated prior to freezing to avoid ice crystal formation, has been successfully used for immuno-EM labeling but often results in poor ultrastructure.

Alternative Approaches—the Illusion of 3D

Other approaches that have been taken for the macromolecular and ultrastructural analysis of biological specimens include rotary shadowing and freeze-etch electron microscopy, respectively. For an overview on these techniques the interested reader is referred to Heuser (1980). Rotary shadowing has been extensively used in the past and reveals the approximate shape of macromolecules, but it remained a 2D projection of a 3D molecules and, at the low resolution it provides, it is therefore of limited value. Freeze-etching and subsequent SEM analysis has delivered stunning images of cells and, while being a 2D photograph of a 3D object, has probably been the most influential in shaping our understanding of the cell's interior. Our brains can generate a 3D impression from the perspective, light and shadows of the images, but the 2D nature of the images fail to provide us with quantifiable information of the 3D organization of the macromolecular machines. A true alternative to TEM tomography is the approach developed very recently of focused ion beam scanning electron microscopy (FIB/SEM), which combines TEM sample preparation with block-surface SEM imaging, and delivers 3D data at a resolution comparable to electron tomography, but unlike conventional tomography approaches it allows samples to be studied continuously for tens of microns in depths.

ELECTRON TOMOGRAPHY—MOLECULAR MACHINES, CELLS, AND TISSUES IN 3D

Imaging Pleiomorphic Objects—Introducton to Tomography

The roots of electron tomography reach back to the 1960s when Walter Hoppe proposed that pleimorphic objects which do not possess symmetry could be studied by tomographic imaging similar to the then-newly developed Computed Axial Tomography (CAT scan) medical imaging technique. However, the electron dose then necessary to acquire the necessary number of views was prohibitive. Hence it took almost thirty years of technical development before tomograms of cells and macromolecules became meaningful. Electron tomography has seen dramatic developments over the last two decades illustrated by several selected reviews (Frank 1992, Baumeister et al. 1999, Downing et al. 2007). Years of pioneering work both on the hardware as well as the software side have resulted in commercially available setups for semi-automated data collection. The stability and accuracy of electron microscope stages, as well as the sensitivity of detection devices such as CCD have been significantly improved, along with software for low-dose imaging and automated data acquisition, minimizing the electron dose that is spent on tracking and focusing the image (Koster et al. 1992). It is the total dose acceptable for tomographic imaging that ultimately prevents biologists from getting to subnanometer resolution, and while the total dose can be distributed among the individual images of the tilt series, without loss of signal-to-noise ratio in the final 3D reconstruction, the reduction of exposure for every image that can limit the ability to align the images, and by noise in the detector (McEwen et al. 1995). Resin-embedded and stained samples can tolerate a higher electron dose, however one concern has been that the fidelity of the final reconstruction is limited by the faithfulness of the stain distribution.

We are now in an age where semi-automated data collection is no longer a bottleneck, and where progress is limited by the clarity of the original question to be addressed, adequate sample preservation as well as data inspection and analysis tools, as will be discussed in more detail below.

One of the obsessions in the field of structural biology, which electron tomography can be considered being part of, is the question of resolution. In the case of diffraction techniques and cryo-EM single particle analysis, the resolution limit can be determined as the frequency shell, which contains sufficiently consistent high-quality information. However, since electron tomography is performed often on unique cellular volumes, implicit or explicit means of averaging cannot be applied, although a variety of approaches for estimating the resolution have been proposed (Penczek 2002, Unser et al. 2005). Based on geometrical considerations we can estimate the maximum theoretical resolution as a function of sample thickness "D" and number of different views "n" being taken in a tilt series. Assuming the images are spaced evenly throughout 180° the resolution limit is ~πD/n. For example, for a 100 nm thin section, and 150 images recorded at 1° increments the resolution in xy plane is 2–3 nm. The actual quality and fidelity of the reconstructed volume will also depend on the noise level, conformational alterations occurring during sample preparation, the amount of radiation damage encountered during data collection, and reconstruction artifacts stemming from imperfections of image frame alignment and the interpolation algorithms used for 3D reconstruction.

Data Collection—Getting the Raw Data

The principle of electron tomography is similar to CAT scan medical imaging, with the difference that the microscope stays stationary and the EM specimen is being rotated and imaged at different orientations (Figure 7.2). Projection images at various tilt angles are aligned before 3D reconstruction, typically through weighted back projection. Image alignment is often aided by fiducial gold markers adhered to the support film, although fiducial-free alignment approaches have been reported (Winkler & Taylor 2006). Data sets are collected by tilting the specimen around the stage tilt axis typically from plus 75 degrees to minus 75 degrees at 1 degree increments, although such high tilts may not always be achievable due to grid or stage geometry limitations. Higher tilt angles may not be practicable as images may be obstructed by grid bars, and/or may be limited in quality by the thickness of the specimen at such tilt angles. As a result, there is a missing wedge of information resulting in resolution anisotropy along the axis of the electron beam. The missing wedge can be reduced to a missing pyramid, if the region is imaged again while being tilted perpendicular to the first axis, typically by rotating the grid by 90 degrees prior to collection of the second data set (Mastronarde 1997). Such dual-axis data sets are more informative and suffer from less distortion, but require the exact area to be found again after grid rotation, and the total dose to be spread over double the number of images, which for cryo-specimen can be problematic. For resin-embedded samples, on the other hand, collection of dual-axis data sets is significantly easier to accomplish. Hence with typical CCD setups of 2048 × 2048 pixels, at a magnification corresponding to a detector pixel size of 0.5–1 nm, the field of view is typically between ~1 μm and ~2 μm, which is sufficient to cover most macromolecular machines but may only reveal a small proportion of eukaryotic cell systems. Laterally overlapping regions can be imaged in 3D resulting in montages of tomograms, which allow larger "cellular real estate" to be covered, but such montages are somewhat restricted to resin-embedded samples as during data collection typically an area larger than recorded by CCDs will be illuminated. If serial sections can be obtained, individual tomograms or larger montages of corresponding regions can be recorded for each section, resulting in serial tomograms. Montaging and serial tomography in

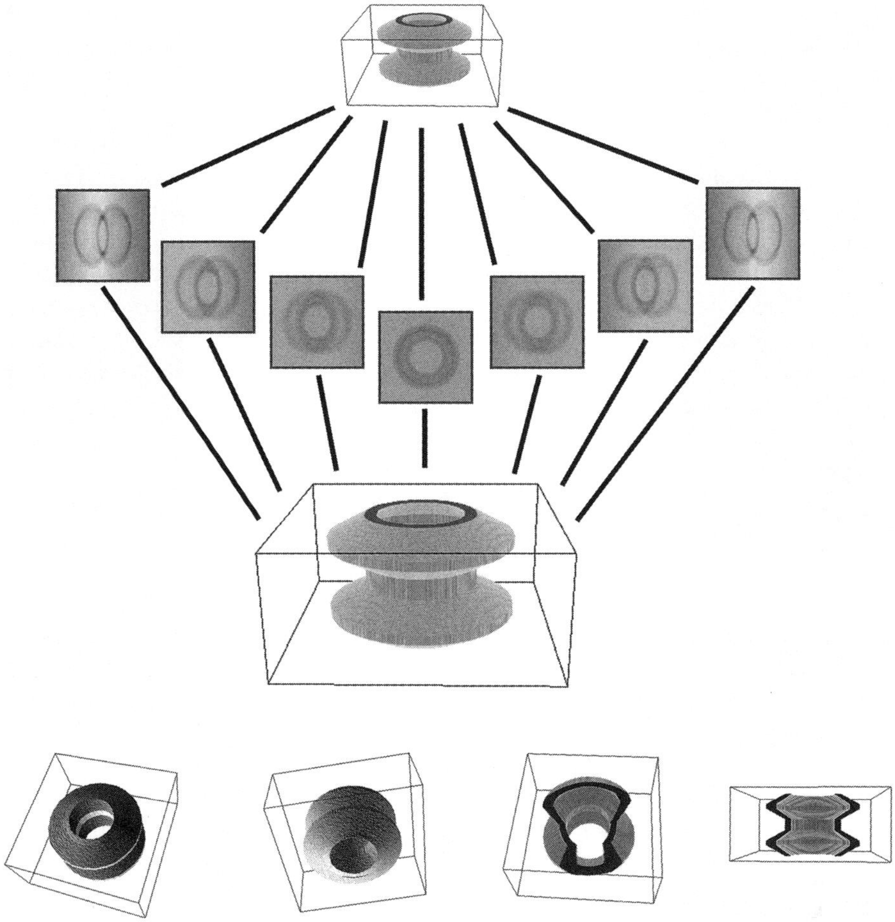

Figure 7.2 Schematic view of tomographic data collection, 3D reconstruction, and 3D visualization. Rotation of an object in the electron microscope leads to projection views of the objects at different tilt angles. After alignment of the images of the tilt series, the objected can be reconstructed in three dimensions through a process called weighted back projection. 3D reconstructions can then be visualized with suitable computer graphics programs, allowing different views of the objects including slices in arbitrary orientations, allowing a deeper understanding of its architecture and form.

principle allow one to reconstruct entire cellular 3D volumes at high resolution, but to accomplish this one needs to acquire extremely large data sets, that with current technology are nearly impossible to be displayed and analyzed, let alone mined for features of interest (Marsh 2007).

3D Reconstruction—from Views to Volumes

Once data sets (single axis or dual axis) have been collected, raw image frames need to be properly adjusted for brightness differences and aligned, typically with the help of fiducial gold markers. Precise tracking of the gold particles helps refine the geometrical parameters of data collection and hence minimize possible alignment artifacts, before the aligned images and their respective tilt angle are recombined into a 3D volume, a process typically accomplished through the so-called weighted back-projection algorithm. A number of packages are available for tilt series alignment and 3D volume reconstruction (Kremer et al. 1996, Frank et al. 1996, Chen et al. 1996, Nickel et al. 2005, Winkler & Taylor 2006, Zheng et al. 2007). Alternative schemes, including the computationally costly algebraic reconstruction technique (ART) algorithms have been proposed and realized (Marabini et al. 1998), but so far they have not been demonstrated to significantly improve the 3D reconstruction, although they are believed to be particularly powerful in cases where data is under-sampled. However, it remains to be seen whether the full potential of ART can be exploited for electron tomography. No matter what algorithms are used, the end result of tomographic data collection is a 3D file, represented as a stack of two 2D images. While for many people obtaining 3D reconstruction seems as an end-point, it is really just the beginning for the quest of interpreting these often very complex data sets.

DATA VISUALIZATION & ANALYSIS

Data Visualization and Exploration—Let's Just Take a Look

While data collection and 3D reconstruction are becoming routine tools, the further development of visualization tools is crucial for the study of increasingly larger and more complex 3D volumes. The result of tomographic imaging is a 3D density map that can be rotated and viewed from any angle and sliced at intervals defined by the detector pixel resolution (see Figure 7.2). Such 3D volumes can be remarkably nontrivial to display in part due to their enormous size of several gigabytes, which is beyond the rendering capabilities of most interactive visualization programs. Moreover, the data sets are often noisy, very crowded with protein complexes literally touching each other, and complex in nature, making inspection and analysis a challenging task.

While slice-by-slice inspection may serve as a first pass to look through the data, the 3D nature of the tomographic data is better appreciated by 3D rendering approaches, such as volume or surface rendering. Volume rendering retains the volume data and sums it from back to front with appropriate weights (ray casting). The information of each volume element (voxel) is incorporated, resulting in a transparent, ghost-like impression of the electron density. Surface rendering creates the illusion of depth by the removal of hidden surfaces, as well as the use of perspective, lighting options, shading, color, and texturing effects. Such 3D scenes are more familiar to our brain and therefore often preferred for display of the 3D data. However, surface rendering shows the object only at a certain chosen density threshold, and is often not very helpful for highly complicated objects with internal cavities. For a photorealistic impression, ray tracing is the method of choice, but it is computationally very expensive, since it follows the paths of each individual light ray as it interacts with the surface of the object. Ray tracing cannot be achieved in real-time and is therefore not suited for initial data inspection but used for data presentation and in publications.

Stereo display techniques as well as a looped video animation of the rendered object, particularly when rotated around an axis, can substantially ease the understanding of structural detail. Physical models, although out of fashion, are a powerful alternative to computer graphics display, and even complicated models can—in principle—be produced by stereolithography.

The study of cellular volumes resembles the exploration of uncharted territory, and while the individual protein structures that constitute the complex may be known, one often cannot anticipate what the 3D architecture of complex cellular machines will look like. Real-time rendering of tomographic volumes is therefore key for interactive data exploration and has recently been implemented using multi-resolution interactive geometry (Bajaj et al. 2003). Further developments towards a cellular "flight simulator" are needed, which would allow the 3D navigation through the densities, zooming in and out in scale from individual proteins to the whole cell, as well as interactive, semi-automated fitting of candidate protein structures to the density envelopes, and validation of the fit. Such a virtual-reality-based program would be of enormous value to the analysis of tomographic data.

Quantitative Data Analysis—Beyond the Pretty Pictures

Tomographic data sets suffer from a variety of shortcomings, including reconstruction artifacts, resolution anisotropy, poor signal-to-noise ratio, high density, and often complex 3D arrangement of macromolecules.

Noise Reduction

The commonly found poor signal-to noise ratio stems from a variety of sources, including uneven stain distribution, low-dose imaging, detector noise, radiation damage, and imperfections in the 3D reconstruction process. High noise levels complicate the visualization of the volume and automated feature extraction. Since averaging techniques most of the time cannot be applied to unique cellular 3D volumes, other methods for signal enhancement are required. A variety of noise-reduction algorithms have been proposed for tomographic volumes including iterative median filtering (van der Heide 2007), bilateral filtering (Jiang et al. 2003), and nonlinear anisotropic diffusion filtering (Frangakis et al. 2001).

Feature Extraction

Tomograms of cellular volumes contain thousands of individual proteins, which are typically densely packed and organized into cellular machines. This macromolecular crowding complicates the analysis as straightforward segmentation routines perform poorly in these crowded environments. Our senses perform not optimally when presented with overwhelming levels of detail, and we must therefore reduce the complexity of the scenery by extracting the features of interest to allow biological interpretation. Segmentation methods can electronically extract multi-protein complexes from their cellular surrounding, allowing an unobstructed view into the machinery's architectural organization. Figure 7.3 shows two examples of manually

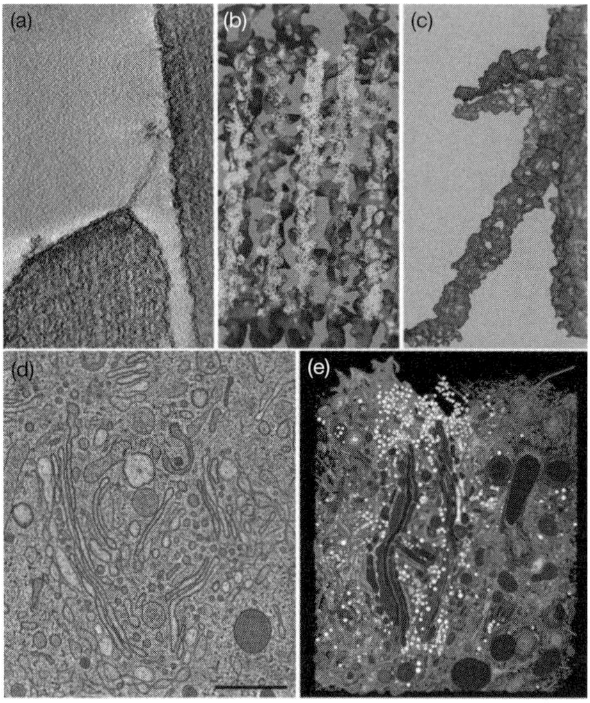

Figure 7.3 Feature extraction/segmentation leading to structure-based macromolecular and cellular models. Examples of inner ear hair cell (a–c) and pancreatic cell tomography (d–e). (a) Single-pixel (0.8 nm) thick image slice of 3D reconstruction of inner ear hair cells displaying inter-stereocilia tip links of 160 nm length and actin-bundles (Auer et al. 2008). (b) close-up view of the stereocilia core with an actin bundle-model docked into the 3D electron density map, note actin-filament spacing of 12 nm fits well with observed density map. (c) 3D rendering of tip link illustrating its complex architecture that was not obvious from slice-by-slice inspection of the raw tomograms, tip link length amounts to ~160 nm. (d) A magnified view from part of a single pixel-thick image slice extracted from the 315 slices that together comprised the 3.1 × 3.2 × 1.2 μm^3 tomographic volume of a pancreatic cell line. Each compartment was manually segmented in the conventional manner using IMOD (Kremer et al. 1996). (e) Contour sets defining each compartment were meshed for 3D visualization and quantitative analysis. The membranes of the Golgi complex are displayed in the context of all other neighboring organelles, vesicles, ribosomes, and microtubules in the region. Color-coding: Golgi cisternae (cis-trans): light blue, pink, cherry, green, dark blue, gold, red; endoplasmic reticulum (yellow), membrane-bound ribosomes (blue), free ribosomes (orange), microtubules (bright green), insulin granules (blue), clathrin-negative vesicles (white), clathrin-positive compartments/vesicles (red), clathrin-negative compartments/vesicles (purple), and mitochondria (dark green). Scale bar, 500 nm. Reproduced from data originally published in Marsh et al. 2001 (courtesy of Brad Marsh, Institute for Molecular Bioscience, The University of Queensland, Brisbane, Australia).

segmented macromolecular complexes (Auer et al. 2008) and cell organelles (Marsh et al. 2001), respectively, illustrating the amount of information that is gained upon feature extraction and false-color representation.

While manual segmentation is still widely used, automated segmentation methods are desirable due to the labor-intensive and somewhat subjective nature of manual segmentation. Robust automated segmentation algorithms based on eigenvector analysis (Frangakis & Hegerl 2002), density thresholds, for instance watershed immersion (Volkmann 2002), and density gradients, such as boundary segmentation (Bajaj et al. 2003), have been proposed and allow fast, reproducible, and more objective segmentation of cellular volumes.

Data Interpretation—Making Sense of the Data

The goal of all data visualization and segmentation efforts is to answer the biological questions that prompted the imaging efforts. We therefore need to comprehend the 3D volumes, for instance through simplification and abstraction of the data, followed by biological interpretation.

Data Abstraction

One way of simplifying complex 3D volumes is to turn them into simple geometric models, such as sticks and balls, just like a cartoon is a simplified version of the real world. Skeletonization approaches have been developed (Bajaj et al. 2003) and facilitate quantitative analysis of the object by allowing one to obtain geometric information about features of interest, including dimensions, distances, and angles.

Data Annotation

While not yet realized, it would be desirable to interactively replace complex 3D shapes with actual or idealized models of candidate proteins as we go along exploring the tomographic volumes. For example, actin filaments could be represented by 8 nm diameter thin rods, microtubules by 25 nm diameter hollow cylinders, and membrane with 5 nm thin straight or curved planes. Molecular sceneries can thus be annotated either using an alphanumerical description or by simply substituting with a database-derived model or simply an idealized simple geometrical object.

Gradual replacement of actual features with a cartoonlike version would simplify the crowded scenery and allow one to focus on the macromolecular features that have not yet been identified while keeping an abstract representation of their surroundings. As an additional benefit by mixing actual density and vector-based geometrical objects, one may be able to display an increasing size of the tomographic volume, since simplified geometrical objects require less random access memory than complex shapes.

Model Docking

Model building has long been the key in interpreting biological structures, and has led to a level of insight that was not available from observing the obtained density distribution alone. If the exact molecular makeup of a complex is known, and the resolution is sufficient, protein structures can be fitted into the density maps, either manually using interactive 3D graphics or semi-automatically.

At 2–3 nm resolution the shape of macromolecular machines reveals quarternary structure and in some cases allows the determination of the protein's presence and orientation at least for complexes that have a sufficiently distinct shape (Grimm et al. 1998).

Template Matching

Entire 3D volumes can be searched with structural templates, as has been successfully demonstrated for the thermasome and proteasome in phantom cells (Frangakis et al. 2002) or ribosomes in intact cells (Ortiz et al. 2006), yet given the low signal-to-noise ratio and the macromolecular crowding it remains to be seen whether template matching can be extended to the more typical case of smaller proteins and protein complexes that may not display sufficient features for unambiguous docking. While so far template matching has been applied only to cryo-tomography data, there is no reason why the same approach should not also work for data sets from resin-embedded samples, which display a higher signal-to-noise ratio. One difficulty that arises when examining cellular volumes, particularly in applying template matching, is that unlike purified, well-characterized macromolecular complexes, each cellular machine may be unique in its exact composition and 3D architecture, even though all target complexes carry out the same function. Aside from the possibility that each machine may be in a different state of its mechanical cycle, multiple protein complex architectural solutions can be consistent with accomplishing a particular physiological task, and therefore architectural flexibility and redundancy in design may play an important role in the robustness of cellular processes. In other words, while few houses in any given neighborhood are exactly identical, they all contain windows, doors, walls, and a roof, and therefore are all designed to provide shelter, independent of their exact 3D architecture. As an example may serve the mechano-electrical transduction and adaptation machinery in hair cells, whose tomographic analysis indicates some multiplicity in design. Although the overall structures display some similarities, they do not seem to be identical (unpublished observation), thereby limiting the usefulness of classical cross-correlation approaches. Instead, for complex cellular machineries, we may need to resort to pattern-recognition approaches and principal components analysis to detect which aspects of the design are conserved and which aspects can be varied without loss of function.

Motif Averaging

Repeating motifs, if they can be identified in the cellular volume, can be superimposed and averaged, resulting in a higher fidelity of the density map and a higher confidence with respect to model docking. The insect flight muscle was one of the first and still is one of the most impressive examples where motif averaging has been successfully applied and has led to fundamental new insight into the conformational changes of myosin when engaged with actin-filaments (Taylor et al. 1999), but in the meantime other investigators have used motif averaging to improve tomographic density maps (Winkler 2007, Sui & Downing 2006, Schmid & Booth 2008). In addition to automated template matching, atomic models have been successfully docked into tomographic densities showing that a hybrid approach with high-resolution models can be docked into low-resolution envelopes provided by EM, hence bridging the gap between individual proteins and cellular complexes (Volkmann 2009, Lindert et al. 2009). The insect flight muscle also demonstrates that rapid-frozen, stained, and resin-embedded specimens may be an excellent alternative to cryo-sectioning.

Exploiting Symmetry

Repeating motifs can be superimposed and averaged, resulting in a higher fidelity of the density map and a higher confidence with respect to model docking. A particular form of motif averaging is to detect and then impose symmetry on the respective multiprotein complex, which often results in an aesthetically more pleasing and hopefully more correct macromolecular structure.

It is inherently difficult to validate the 3D structures of macromolecular machines obtained from tomograms, unless there is a sufficient number of examples that lend themselves suitable for single-particle averaging, or unless the structure of isolated machines has been determined independently and a similar protein composition and conformation can be safely assumed in vivo, as is the case for ribosomes (Brandt et al. 2008). Often the analysis of tomograms leads to new, unexpected hypotheses, which then need to be experimentally tested. In the end, interpretation of the data usually requires substantial expert knowledge with respect to the biological problem, and hence is best performed by the biologist who posed the problem and provided the specimen in the first place.

Data Storage, Annotation, and Presentation

In addition to visualization, segmentation, and quantitative analysis of molecular machines, storage of large cellular data sets in searchable databases as well as convenient annotation and storage of metadata need to be addressed. Most approaches rely on manual segmentation of small volumes, and the resulting analysis typically ignores the contextual information available in the tomogram. A tomographical volume, although at molecular resolution, often represents only a small cellular volume. It would be desirable to combine this approach with light microscopy, which can survey larger areas. Image registration then becomes important, because light microscopy and EM data need to be correlated. Correlative microscopy typically relies on markers that are visible by fluorescence and EM. Tag-based fluorescent labeling and subsequent photoconversion approaches as well as quantum dot-based labeling may allow larger areas, such as an entire tissue sample, to be surveyed; the regions of interest can then be imaged in more detail by electron tomography. With automated EM imaging, large areas of tissue samples can be imaged at molecular resolution and tiled together into gigantic montages. Although this scenario is not routine yet, a dramatic example of the potential is provided in a report on mapping membrane and cytoskeletal structures throughout an entire eukaryotic cell (Marsh 2007). This work has not approached the ultimate resolution limit but is already providing vast amounts of information. Large-area imaging of resin-embedded sections would overcome one of the dilemmas of correlative microscopy, namely that increased resolution is traded for a narrow field of view. Automation that will greatly improve throughput is key to routinely achieving this goal.

COMPLIMENTARY AND ALTERNATIVE APPROACHES

If the target protein cannot be identified by shape comparison alone, its molecular identity needs to be established by other means, such as through the use of labeling approaches. Traditionally, high-resolution localization of macromolecular components has been accomplished by immunogold labeling approaches, where electron-dense gold clusters are conjugated to antibodies specific for the target macromolecule. However, with labeling efficiencies typically ranging from 1–5 percent (Reedy et al. 1994), traditional immuno-affinity-based approaches are limited to abundant proteins and do not fully reflect their distribution. Moreover, for each epitope, a suitable antibody must be found and characterized, and the protocol must then be optimized to reduce nonspecific binding and epitope antigenicity retention, which can be a rather tedious endeavor and sometimes may fail altogether.

Candidate Identification—the Quest for the "GFP" of EM

Due to the complexity of a cell and its unknown exact molecular composition, the localization of a candidate protein molecule within the tomographic volume is not a straightforward task. In other words, cellular volumes do not come annotated. In some cases, the cellular location and association with easily identifiable features, such as actin filaments, as well as the size and shape of the densities may give clues to the identity of a protein, but even a good fit of a likely candidate protein to the density in itself is no proof.

What is needed is a widely applicable reporter system that recognizes the protein of interest, ideally through a genetically encoded tag. This reporter system must provide sufficient contrast to be unambiguously detected in the electron microscope, be small enough to ensure access to the protein, be stable throughout the sample preparation and imaging, show a high affinity and be specific for its target in order to avoid false-negative and false-positive signals, respectively. Some progress towards this goal has been made (reviewed in Giepmans et al. 2006), but this is one of the least developed areas in modern cellular electron microscopy.

An ideal method would be based on a genetically encoded tag that is recognized by an electron-dense label or by a fluorphore-conjugated ligand that can be converted into an electron-dense form. One of the major advantages of such a universal tagging and labeling system is that the labeling protocol only would have to be worked out once and would then be applicable to all biological systems, provided the tag-modified protein is expressed correctly and the protein remains functional.

Methods that rely on fluorescent moieties that can be observed in the light microscope and then photoconverted into an electron-dense signal by photooxidation would fall in this category, whether the reporter fluorophore was a protein (Grabenbauer et al. 2005) or a small organic compound that binds to a genetically encoded tag. In the presence of diaminobenzidine (DAB), these can be photoconverted into an osmiophilic precipitate, which is localized within ~3 nm around the sites of fluorescence and can be detected after osmium staining of DAB deposits (Gaietta et al. 2002).

In the FlAsH/ReAsH system a rather small genetically encoded tag (typically some 11–12 amino acids) is recognized by a nontoxic membrane-permeable dye, resulting in a steep increase in fluorescence. The use of the fluorescence photooxidation technique allows correlative microscopy, meaning that a system can be first studied by epifluorescence or confocal microscopy and subsequently processed for EM analysis as was so elegantly shown for gap junction proteins in cultured cells (Gaietta et al. 2002). Despite its promise, neither FlAsH/ReAsH—nor SNAP-labeling have yet been widely adopted, possibly due to a lack of robustness of the labeling process, including problems encountered with specificity and sensitivity of the labeling and also the photoconversion process. One of the downsides of utilizing a fluorescence tag is that it must be visualized in the electron microscope by means of photoconversion, which produces a "fuzzy" osmiophilic diaminobezidin precipiate.

Quantum (Q-) dots are also of great interest to optical imaging, since all Q-dots are excited by the same short wavelength band, but yet emit at different longer wavelengths, allowing simultaneous detection of several reporter molecules. Q-dots are bright and do not fade under the optical light beam. Incidentally, Q-dots can also be recognized in the electron microscope through their difference in size and shape (Giepmans et al. 2005). However, Q-dots are typically bound to either immuno- or other affinity probes, and they require penetration of the cellular volume prior to TEM sample preparation or successful labeling of the section surface, which may result in poor ultrastructural preservation or low labeling efficacy, respectively.

Yet another system of interest is the Nickel-nitrilotriacetic acid (Ni-NTA) nanogold, which has been shown to bind to purified hexahistidine-tagged protein containing complexes (Hainfeld et al. 1999) but still needs to be validated for intact cells or sectioned material. Currently, efforts are underway to use this labeling approach on tissue-sections, which would turn this approach into a viable alternative to immunolabeling, combining the advantage of a well-defined, high-contrast label with the universal labeling of a genetically encoded tag, which compared to many other systems would be very small (six amino acids) and in addition such hexahistidine tags have been extensively used for biochemical purification of functional proteins. One of the advantages of the covalently bound gold is that it is readily recognized due to its strong scattering properties, unlike DAB-based photoconversion products and certain Q-dots, which can be mistaken for ribosomes or other stained large protein complexes.

Correlative LM/EM Imaging

Optical microscopy is limited in resolution and often provides only information about the localization of a reporter molecule, yet electron microscopy can provide ultrastructural context and macromolecular shape information but in most cases does not allow identification of individual proteins. Hence both approaches visualize different aspects of biological systems, and when combined can yield both compositional and architectural information. Correlative imaging relies on the ability of the reporter molecule to be detected by the different methods, such as through the photoconversion of the fluorescence reporter molecule into an electron dense precipitate, the emission spectrum/shape correlation of Q-dots, or the fluorescence preservation in sectioned material. While the combination of fluorescence imaging with electron microscopy has so far been the predominant, although relatively rare, application of correlative imaging, the concept can be readily expanded to other forms of advanced imaging, including Raman microscopy or mass spectrometry-based techniques such as nanoSIMS, and possibly others, such as MRI or PET imaging.

Superresolution Optical Imaging Approaches—Potential and Limitations

One of the most exciting developments of optical microscopy imaging over the recent years is the development of super-resolution microscopy, most notably structured illumination (Gustafsson et al. 1999), Photo-Activated Localization Microscopy

(PALM) (Betzig et al. 2006), and Stochastic Optical Reconstruction microscopy (STORM) (Huang et al. 2008a, 2008b). Structured illumination builds on the use of Moire patterns in the illumination that allows collection of higher frequency space data, and therefore the reconstruction of the image beyond the diffraction limit of light. PALM and STORM rely on the stochastic activation of a very small number of all fluorophores at low illumination intensity conditions. Instead of imaging all fluorophores simultaneously, only a small fraction is activated at any given time, resulting in well resolved spots of individual fluorophores, whose precise position can be determined by deconvolution with their respective point spread function. This procedure is repeated for many cycles until all fluorophores have been activated and/or bleached. Some of the newer genetically encoded fluorescent tags allow the wavelength-dependent on- and off-switching of the fluorophores. Some caveats are that while positioning in x and y can reach some 10–20 nm, positioning in z (along the beam) is currently about 1 or 2 magnitudes worse, but to some extent can be improved by using either total internal reflection fluorescene (TIRF) imaging or confocal microscopy setups. While optical sectioning remains challenging, one could image physically sectioned material at some 50–70 nm resolution, followed by, for instance, PALM analysis. One of the challenges that then arises is to retain the photoswitchable properties of the fluorophore, which to date has not been accomplished, but seems feasible. There is another caveat with super-resolution imaging: while relatively precise measurements can be made with respect to numbers and distances between all fluorophores, the images resemble the star-lit night sky. Hence it is often invaluable to combine this technology with electron microscopy, which can then provide ultrastructural landmarks that help interpretation of the fluorophore position mapping. Not too surprisingly, it would seem that correlative PALM/EM imaging of a sample, for instance sections or whole-mounts, would be most desirable, but requires the development of protocols for preservation of the PALM signals in sections suitable for electron microscopy.

Serial Section Imaging and Array Tomography

While electron tomography is very valuable for the precise 3D description of macromolecular complexes, there are biological questions that can be answered by physical sectioning of the specimen and collecting high-resolution 2D images of adjacent sections, also known as serial sectioning imaging. The z-resolution is limited by the physical section thickness size and is usually worse by about one order of magnitude when compared to tomograms of individual sections. Serial sectioning can be combined with tomography, also known as serial section tomography, but often it is sufficient to obtain a coarse 3D reconstruction of cellular volumes that allows one to cover relatively quickly a reasonably thick volume, allowing one to image entire cells and parts of tissue. However, one needs to be careful on aligning the serial sections into a 3D volume; due to the mechanical deformations encountered during cutting of a thin slice of a resin block, 2D images often need to be slightly morphed (e.g., rotation by a few degrees or changes in magnifications by a few percent from one serial section to another), rendering this approach suitable for large cellular organelles but not well suited for macromolecular complexes.

When combining serial section EM imaging with on-section sequential immunolabeling, one can localize a variety of epitopes using sequential fluorescence microscopy followed by bleaching of each secondary-conjugated fluorophor, and superimposition of all the fluorescence data onto the cellular ultrastructure (Micheva & Smith 2007), allowing for coarse 3D mapping of epitopes with respect to defined coarse 3D ultrastructure.

Novel Advanced 3D SEM Imaging Approaches

One of the most exciting developments in EM imaging over the last few years has been the repeated focused ion beam (FIB) milling of specimen blocks followed by repeated SEM imaging of the block surface (Hildebrand et al. 2009). This method, also referred to as FIB/SEM, allows one to follow the 3D ultrastructure of resin-embedded samples for tens of micrometers in the z-direction at a resolution of some 3–5 nm in xy and some 10–15 nm in z-direction, comparable to conventional electron tomography approaches. Successful imaging, however, depends on sufficient specimen contrast, which is attainable in high-pressure frozen, freeze-substituted heavy-atom-treated and resin-embedded specimen blocks but unlikely to be achieved for unstained frozen-hydrated samples. While serial section tomography—that is, the tomographic reconstruction of serial sections—is feasible, serial sectioning followed by serial tomography can be tricky for a large number of samples and physical sectioning may lead to loss of material and undesirable surface properties of sections (including thinning due to radiation damage), which renders the top and bottom of tomographic reconstructions often less reliable and hence may result in gaps of information. A somewhat competing, but in many ways complementing, technique to FIB/SEM is the so-called Serial Block-Face SEM, which like FIB/SEM uses the backscatter electrons to image the block surface but instead of an ion beam uses a ultramicrotome inside the chamber of the SEM to allow for repeated trimming and imaging of the block surface (Denk & Horstmann 2004). Its z-resolution depends on how thin a section can be cut. While routine ultramicrotoming would result in z-steps of 50–100 nm, it appears to be possible to cut much thinner sections somewhat comparable to what is feasible with an ion beam, possibly due to the fact that the section does need to be recovered, allowing smaller section thickness. However, this approach requires specialized knives and optimal conditions for ultrathin sectioning beyond the 50 nm thickness, and is as of yet not routinely available. While it remains to be seen whether SBF/SEM or FIB/SEM can achieve the resolution required to distinguish shapes of macromolecular machines, it appears particularly strong where uninterrupted deep z-direction coverage at 10–15 nm resolution is desirable,such as when studying microbial communities or investigating complex 3D eukaryotic organelles or membrane systems.

FUTURE: INTEGRATED IMAGING

One of the themes that emerged in all disciplines over the recent years was to further improve sample preparation and automate data collection; each discipline continues to strive for brighter colors, more contrast, faster data acquisition, and in some ways global dominance. While all that is certainly of value, it doesn't address the core issue underlying current biological imaging, namely that different techniques collect fundamentally different information. For instance, fluorescence microscopy gathers precise temporal and spatial information about a specific reporter molecules, but usually fails to provide context, whereas electron microscopy provides rich ultrastructural and therefore contextual information, but often fails to identify specific macromolecules. Likewise, Raman imaging and mass spectrometry imaging, while currently remaining resolution-challenged, provide compositional information, but fail to visualize 3D organization of these molecular and macromolecular components. Moreover, fluorescently labeled surrogate marker proteins can be misleading, since an optical signal may reveal the presence of a protein, such as the zonula adherents junction protein ZO-1, but does not guarantee the existence of a functional tight junction, hence the use of fluorescence signal of ZO-1 as an argument for apical-basolateral cell polarity could be fundamentally flawed (unpublished observation).

It would therefore seem best to combine the various imaging modalities and to superimpose the image information to gain complimentary and therefore fundamentally more comprehensive insight into the system under investigation. Another challenge that arises in imaging is that one typically needs to decide between imaging a large area at low magnification, allowing an overview and cell or tissue context but without enough detail, or a small area at higher magnification, which allows one to focus in on the details with the caveat that contextual information is typically lost. At the time of imaging it is often not completely obvious which part of the "cellular real estate" will be of most interest in the analysis and one may only find out later that it would have been most informative to also image an adjacent area. Instead, one either shoots fairly randomly, assuming that one obtains an unbiased sampling or one shoots on the first area that seems to look great and/or confirms expectations. Few groups routinely take wide-field, high-resolution data sets to accomplish statistics sampling and high-resolution detail, but advances in automated data collection have allowed tiling of the data into large composite images and volume that can be scrutinized not unlike "Google Maps," allowing one to zoom in and out and to move around. Multiscale imaging may be necessary for the needle-in-a-haystack problem of rare cellular events that can only be appreciated by careful analysis, and hence only after the data has been collected. Also, one may need to first establish the history of a molecular event, such as by live-cell optical imaging, to discriminate the area of interest for further in-depth higher resolution analysis. The benefit of multiscale imaging becomes obvious to anybody who has ever tried to orient herself/himself in a city or region she/he is unfamiliar with: While high-resolution details such as street names are most useful to find one's exact position, one really often needs to see major landmarks or know what city is nearby in order to navigate with confidence.

Like "Google Maps," multimodal imaging promises to allow different kinds of information to be integrated, such as maps of roads, railroads, and municipal buildings overlaid onto satellite images and traffic information combined with street view photographs and information about nearby shopping opportunities. Having all this information combined in one map that can be explored in real-time, for instance at every point being connected to other databases, is a dream of any cell biologist and very timely. Imagine you could walk around your cell regions of interest, at every point appreciate their architectural beauty and elegance, recognize their macromolecular composition, and link both architecture and function back to databases providing real-time information about these structures and molecules. Wouldn't we all enter a new era of holistic understanding of the inner workings of life in a cell?

So what is holding us back? It turns out that there is no fundamental limitation, merely technical hiccups that need to and can be overcome, and a willingness of researchers of the different imaging fields to come together and work towards integration of the different imaging modalities. Some hurdles that need to be overcome along the way are the following: 1) we need to find sample preparations that are compatible with the various imaging approaches; 2) we need to find reporter systems that can be visualized in different modalities, for instance fluorescence probes that can be visualized in combination with electron microscopy; 3) we need to collect large data sets particularly at the high-resolution end in order to cover the same area, ideally from the same identical sample or near-identical samples; 4) we need to register the data to allow precise overlay of the structures, such as through the use of cross-platform fiducial markers; 5) we need to develop tools for appropriate visualization and exploration of the data, similar to Google Maps in 2D, but even more urgently needed in 3D. 3D visualization and exploration of complex data sets is presumably one of the biggest hurdles for in-depth understanding, and might require virtual-reality-based immersive visualization settings that are continue to be refined and improved for the video-game-community; 6) we need to segment the data to extract features of interest, allowing us to focus on what is important and to conduct quantitative analysis; 7) we need to simplify data replacing pixel-based densities by simplified models wherever possible to reduce complexity and allow larger and larger volumes to be explored; 8) we need to be able to interactively annotate the data and query databases in real time to examine the richness of the data while exploring the volume; 9) we need to make our findings accessible to collaborators, to other scientists in the world that may use the same datasets for their own questions, and the interested public at large,for example by turning our most precious findings into blockbuster movies that illustrate the elegance of nature; and last but not least, 10) we must interpret the data as a team, with members coming from different backgrounds, such as biologists, mechanical engineers, computer scientists, in order to fully exploit the information hidden in design and composition.

While the challenges on integrated multiscale multimodal imaging are numerous, they are not fundamental in nature, yet difficult to overcome. However, its promise is a more fundamental and holistic insight into the molecular events underlying cellular properties responsible for physipological function and dysfunction of cells and tissues, as well as microbial biofilms. Such insight would allow us to understand the molecular events of pathogenesis for the many diseases humans are threatened with or reveal the molecular and cellular strategies with which microbes perform their extraordinary functions, such as the breakdown of lignocellulose or bioremediation, which in turn may help in the quest for biomass-based transportation fuels or efficient cleanup of toxic sites.

Cellular tomography in conjunction with tag-based protein localization, as well as correlative imaging across platforms and scales, provides the link between the high-resolution work of classical structural biology and the worlds of cellular biology, and possibly will reign in an era of systems microscopy.

ACKNOWLEDGEMENTS

I would like to thank my colleague Kenneth H. Downing for spirited discussions on the past, present, and future of electron microscopy imaging and for carefully reading this book chapter and making valuable suggestions. I would like to acknowledge image material for Figure 7.1 depicting whole-mount cryo-EM and cryo-tomography as well as cryo-sections kindly provided by Drs. Luis Comolli and Kenneth H. Downing, Lawrence Berkeley Laboratory, Berkeley, CA, as well as

Dr. Sriram Subramaniam, NIH, Bethesda, DC, respectively. Images for Figures 7.3d, 7.3e are courtesy of Brad Marsh, Institute for Molecular Bioscience, The University of Queensland, Brisbane, Australia.

REFERENCES

Al-Amoudi A, Norlen LP, Dubochet JJ (2004) Cryo-electron microscopy of vitreous sections of native biological cells and tissues, *J. Struct. Biol.*148: 131–135.

Alberts B (1998) The cell as a collection of protein machines: preparing the next generation of molecular biologists, *Cell* **92**: 291–294.

Auer M (2000) Three-dimensional electron cryo-microscopy as a powerful structural tool in molecular medicine. *J. Mol. Med.* **78**: 191–202.

Auer M, Koster AJ, Ziese U, Bajaj C, Volkmann N, Wang DN, Hudspeth AJ (2008) Three-dimensional architecture of hair-bundle linkages revealed by electron-microscopic tomography, *J. Assoc. Res. Otolaryngol.* **9**: 215–224

Bajaj C, Yu Z & Auer M (2003) Volumetric feature extraction and visualization of tomographic molecular imaging. *J. Struct. Biol.* **144**: 132–143

Baker TS, Olson NH, Fuller SD (1999) Adding the third dimension to virus life cycles: three-dimensional reconstruction of icosahedral viruses from cryo-electron micrographs, *Microbiol. Mol. Biol. Rev.* **63**: 862–892

Baumeister W, Grimm R, Walz J. (1999) Electron tomography of molecules and cells, *Trends Cell Biol.* **9**: 81–85

Baumeister W. (2002) Electron tomography: towards visualizing the molecular organization of the cytoplasm, *Curr. Opin. Struct. Biol.* **12**: 679–684.

Betzig E, Patterson GH, Sougrat R, Lindwasser OW, Olenych S, Bonifacino JS, Davidson MW, Lippincott-Schwartz J, Hess HF (2006) Imaging intracellular fluorescent proteins at nanometer resolution, *Science* **313**:1642–1645.

Bozzolla, J.J. and L.D. Russell (1999) Electron Microscopy, 2nd ed. Jones & Bartlett, Sudbury, MA.

Brandt F, Etchells SA, Ortiz JO, Elcock AH, Hartl FU, Baumeister W (2009) The native 3D organization of bacterial polysomes, *Cell,* **136**: 261–271

Buser C, Walther P (2008) Freeze-substitution: the addition of water to polar solvents enhances the retention of structure and acts at temperatures around –60 degrees C, *J. Microsc* **230**: 268–277

Chen H, Hughes DD, Chan TA, Sedat JW, Agard DA (1996) IVE (Image Visualization Environment): a software platform for all three-dimensional microscopy applications, *J. Struct. Biol.* **116**: 56–60.

Chiu W, Baker ML, Almo SC (2006) Structural biology of cellular machines, *Trends Cell Biol.,* **16**: 144–150

Denk W, Horstmann H (2004) Serial block-face scanning electron microscopy to reconstruct three-dimensional tissue nanostructure, *PLoS Biol.* **2**: e329

Downing KH, Sui H, Auer M (2007) Electron tomography: a 3D view of the subcellular world, *Anal. Chem.* **79**: 7949–7957

Egelman EH (2007) Single-particle reconstruction from EM images of helical filaments, Curr. *Opin. Struct. Biol.,* **17**: 556–561

Frangakis AS Stoschek A, Hegerl R (2001) Wavelet transform filtering and non-linear anisotropic diffusion assessed for signal reconstruction performance on multidimensional biomedical data, *IEEE Trans Biomed Eng,* **48**: 213–222

Frangakis AS, Böhm J, Förster F, Nickell S, Nicastro D, Typke D, Hegerl R, Baumeister W (2002) Identification of Macromolecular complexes in cryoelectron tomograms of phantom cells, *Proc. Natl. Acad. Sci. U.S.A.* **99**: 14153–14158

Frangakis AS, Hegerl R (2002) Segmentation of two—and three-dimensional data from electron microscopy using eigenvector analysis, *J. Struct. Biol.* **138**: 105–113

Frank J (1992) Electron tomography, Plenum Press, New York

Frank J, Radermacher M, Penczek P, Zhu J, Li Y, Ladjadj M, Leith A (1996) SPIDER and WEB: processing and visualization of images in 3D electron microscopy and related fields, *J. Struct. Biol.* **116**: 190–199

Frey TG, Perkins GA; Ellisman MH (2006) Electron tomography of membrane-bound cellular organelles, *Annu. Rev. Biophys. Biomol. Struct.* **35**: 199–224.

Gaietta G, Deerinck TJ, Adams SR, Bouwer J, Tour O, Laird DW, Sosinsky GE, Tsien RY, Ellisman MH. (2002) Multicolor and electron microscopic imaging of connexin trafficking. *Science* **296**: 503–507.

Giepmans BN, Deerinck TJ, Smarr BL, Jones YZ, Ellisman MH (2005) Correlated light and electron microscopic imaging of multiple endogenous proteins using Quantum dots, Nat. Methods 2: 743–749.

Giepmans BN, Adams SR, Ellisman MH, Tsien RY (2006) The fluorescent toolbox for assessing protein localization and function, *Science* **312**: 217–224

Gilkey, JC, Staehelin. LA (1986) Advances in ultrarapid freezing for the preservation of cellular ultrastructure. *J. Electron Microsc. Tech.* **3**: 177–210.

Giberson RT, Demaree Rs Jr (1999) Microwave processing techniques for electron microscopy: a four-hour protocol, *Methods Mol. Biol.* **117**: 145–158

Glaeser RM (2007) Electron Crystallography of Biological Macromolecules, Oxford University Press, USA, ISBN 0195088717

Grabenbauer M, Geerts WJ, Fernandez-Rodriguez J, Henger A, Koster AJ, Nilsson T (2005) Correlative microscopy and electron tomography of GPF through photooxidation, *Nat. Methods* **2**: 857–862

Grimm R, Singh H, Rachel R, Typke D, Zillig W, Baumeister W (1998) Electron tomography of ice-embedded prokaryotic cells, *Biophys. J.* **74**: 1031–1042.

Gustafsson MG, Agard DA, Sedat JW (1999) I5M: 3D widefield light microscopy with better than 100 nm axial resolution, *J. Microsc.* **195**: 10–16.

Hainfeld JF, Liu W, Halsey CM, Freimuth P, Powell RD (1999) Ni-NTA-gold clusters target His-tagged proteins, *J. Struct. Biol.***127**: 185–198.

Henderson R (2004) Realizing the potential of electron cryo-microscopy, *Q. Rev. Biophys.*, **37**: 3–13.

Heuser, JE (1980). Quick-freeze, deep-etch method of preparing samples for 3-D electron microscopy. *Trends in Biochemical Sciences* **6**: 64–68.

Hildebrand M, Kim S, Shi D, Scott K, Subramaniam S. (2009) 3D imaging of diatoms with ion-abrasion scanning electron microscopy, *J. Struct. Biol.* **166**: 316–328.

Huang B, Wang W, Bates M, Zhuang X (2008) Three-dimensional super-resolution imaging by stochastic optical reconstruction microscopy, *Science* **319**:810–813.

Huang B, Jones SA, Brandenburg B, Zhuang (2008) Whole-cell 3D STORM reveals interactions between cellular structures with nanometer-scale resolution, *Nature Methods* **5**: 1047–1052.

Jiang W, Baker ML, Wu Q, Bajaj C, Chiu W (2003) Applications of a bilateral denoising filter in biological electron microscopy, *J. Struct. Biol.* **144**: 114–122.

Jiminez N, Vocking K, van Donselaar EG, Humbel BM, Post JA, Verkleij AJ (2009) Tannic acid-mediated osmium impregnation after freeze-substitution: a strategy to enhance membrane contrast for electron tomography, *J. Struct. Biol.* **166**: 103–108.

Kellenberger E (1987) The response of biological macromolecules and supramolecular structures to the physics of specimen cryopreparation, 35–63. In R.A. Steinbrecht and K. Zierold (Eds.), Cryotechniques in Biological Electron Microscopy. Springer-Verlag, Berlin.

Koster AJ, Chen H, Sedat JW, Agard DA (1992) Automated microscopy for electron tomography, *Ultramicroscopy,* **46**: 207–228.

Kremer JR, Mastronarde DN, McIntosh, JR (1996) Computer visualization of three-dimensional image data using IMOD, *J. Struct. Biol.* **116**: 71–76.

Le Gros MA, McDermott G, Larabell CA (2005) X-ray tomography of whole cells, *Curr. Opin. Struct. Biol.* **15**: 593–600.

Leong As, Sormunen RT (1998) Microwave procedures for electron microscopy and resin-embedded sections, *Micron* **29**: 397–409.

Lindert S, Stewart PL, Meiler J (2009) Hybrid approaches: applying computational methods in cryo-electron microscopy, *Curr. Op. Struct. Biol.* **19**: 218–225.

Lucic V, Förster F, Baumeister W (2005). Structural studies by electron tomography: from cells to molecules, *Annu. Rev. Biochem.* **74**: 833–865.

Marabini R, Herman GT, Carazo JM (1998) 3D reconstruction in electron microscopy using ART with smooth spherically symmetric volume elements (blobs), *Ultramicroscopy* **72**: 53–65.

Marsh BJ, Mastronarde DN, Buttle KF, Howell KE, McIntosh JR (2001) Organellar relationship in the Golgi region of the pancreatic beta-cell-line HIT-T15, visualized by high-resolution electron tomography, *Proc. Natl. Acad. Sci. USA* **98**: 2399–2406.

Marsh BJ (2007) Reconstructing mammalian membrane architecture by large area cellular tomography, *Methods Cell Biol.* **79**: 193–220.

Mastronarde DN (1997) Dual-axis tomography: an approach with alignment methods that preserve resolution, *J. Struct. Biol.* **120**: 343–352.

Maunsbach AB (1998) Biomedical electron microscopy, Academic Press, ISBN 0124806104

McDonald KL & Auer M (2006) High-pressure freezing, cellular tomography, and structural cell biology, *Biotechniques* **41**: 137–141.

McEwen BF, Downing KH, Glaeser RM (1995) The relevance of dose-fractionation in tomography or radiation-sensitive specimens, *Ultramicroscopy* **60**: 357–373.

McIntosh JR, Nicastro D, Mastronarde D (2005) New views of cells in 3-D: an introduction to electron tomography, *Trends Cell Biol.* **15**:43–51.

Micheva KD, Smith SJ (2007) Array tomography: a new tool for imaging the molecular architecture and ultrastructure of neural circuits, *Neuron* **55**: 25–36.

Moor H, Riehle U (1968). Snap-freezing under high pressure: a new fixation technique for freeze-etching. *Proc.* **4***th, Eur. Reg. Conf. Electron Microsc.* **2**: 33–34.

Moor H (1987) Theory and practice of high pressure freezing, p. 175–191. In R.A. Steinbrecht and K. Zierold (Eds.) Cryotechniques in Biological Electron Microscopy. Springer-Verlag, Berlin.

Mueller M, Jenni S, Ban N (2009) Strategies for crystallization and structure determination of very large macromolecular assemblies, *Curr. Opin. Struct. Biol.* **17**: 572–579.

Nickell S, Förster F, LInaroudis A, Net WD,Beck F, Hegerl R, Baumeister W, Plitzko JM (2005) TOM software toolbox: acquisition and analysis for electron tomography, *J. Struct. Biol.* **149**: 227–234.

Nickell S, Kofler C, Leis AP, Baumeister W (2006) A visual approach to proteomics, *Nat. Rev. Mol. Cell Biol.* **7**: 225–230.

Nogales E, Grigorieff N (2001). Molecular Machines: putting the pieces together, *J. Cell Biol.* **152**: F1-F10.

Ortiz JO, Förster F, Kürner J, Linaroudis AA, Baumeister W (2006) Mapping 70S ribosomes in intact cells by cryoelectron tomography and pattern recognition, *J. Struct. Biol.* **156**: 334–341.

Palsdottir H, Remis JP, Schaudinn C, O'Toole E, Lux R, Shi W, McDonald KL, Costerton JW, Auer M (2009) Three-dimensional macromolecular organization of cryo-fixed Myxococcus Xanthus biofilms as revealed by electron microscopic tomography, *J. Bacteriol.* **191**: 2077–2082.

Penczek PA (2002) Three-dimensional spectral signal-to-noise ratio for a class of reconstruction algorithms, *J. Struct. Biol.* **138**: 34–46.

Peters PJ, Bos E, Griekspoor A (2006) *Cryo-immunogold electron microscopy, Curr. Protoc. Cell Biol.* Chapter 4, Unit 4.7.

Pierson J, Sani M, Tomova C, Godsave S, Peters PJ (2009) Towards visualization of nanomachines in their native cellular environment, Histochem. *Cell Biol.* **132**: 253–262.

Putnam CD, Hammel M, Hura GL, Tainer JA (2007) X-ray resolution scattering (SAXS) combined with crystallography and computation: defining accurate macromolecular structures, conformations and assemblies in solution, *Q. Rev Biophys* **4**: 191–385.

Reedy MC, Reedy MK, Leonard KR, Bullard B(1994) Gold/Fab immuno electron microscopy localization of troponin H and troponin T in Lethocerus flight muscle, *J. Mol. Biol.* **239**: 52–67.

Schmid & Booth (2008) Method for aligning and for averaging 3D volumes with missing data, *J. Struct. Biol.* **161**: 243–248.

Sitte, H., L. Edelmann, and K. Neumann. (1987) Cryofixation without preatreatment at ambient pressure, p. 87–113. In R.A. Steinbrecht and K. Zierold (Eds.), Cryotechniques in Biological Electron Microscopy. Springer-Verlag, Berlin.

Small, J.V. (1981) Organization of actin in the leading edge of cultured cells: influence of osmium tetroxide and dehydration on the ultrastructure of actin meshworks. *J. Cell Biol.* **91**: 695–205.

Steinbrecht RA, Müller M (1987) Freeze substitution and freeze drying, p. 149–172. In R.A. Steinbrecht and K. Zierold (Eds.), Cryotechniques in Biological Electron Microscopy, Springer-Verlag, Berlin.

Stewart M (1998) Computer image processing of electron micrographs of biological structures with helical symmetry, *J. Electron Microsc. Tech.* **9**: 325–358.

Studer D, Humbel BM, Chiquet M (2007) Electron microscopy of high pressure frozen samples: bridging the gap between cellular ultrastructure and atomic resolution, Histochem. *Cell. Biol.* **130**: 877–889.

Sui H, Downing KH (2006) Molecular architecture of axonemal microtubule doublets revealed by cryo-electron tomography, *Nature* **442**: 475–478

Taylor KA, Schmitz H, Reedy MC, Goldman YE, Franzini-Armstrong C, Sasaki H, Tregear RT, Poole K, Lucaveche C, Edwards RJ, Chen LF, Winkler H, Reedy MK (1999) Tomographic 3D reconstruction of quick-frozen, Ca2+-activated contracting insect flight muscle, *Cell* **99**: 421–431.

Tokuyasu KT (1973) A technique for ultracryotomy of cell suspensions and tissues, *J. Cell Biol.* **57**: 551–565.

Triffo WJ, Palsdottir H, McDonald KL, Lee JK, Inman JL, Bissell MJ, Raphael RM, Auer M (2008) Controlled microaspiration for high-pressure freezing: a new method for ultrastructural preservation of fragile and sparse tissues for TEM and electron tomography, *J. Microsc.* **230**: 278–287.

Unser M, Sorzano CO, Thevenaz P, Jonic S, El-Bez C, De Carlo S, Conway JF, Trus BL (2005) Spectral signal-to-noise ratio and resolution assessment of 3D reconstructions, *J. Struct. Biol.* **149**: 243–255.

Van der Heide P, Xu XP, Marsh BJ, Hanein D, Volkmann N (2007) Efficient automatic noise reduction of electron tomographic reconstructions based on iterative median filtering, *J. Struct. Biol.* **158**: 196–204.

Vanhecke D, Graber W, Studer D (2008) Close-to-native ultrastructural preservation by high pressure freezing, *Methods Cell Biol.* **88**: 151–164.

Volkmann N (2002) A novel three-dimensional variant of the watershed transform for segmentation of electron density maps. *J. Struct. Biol.* **138**: 123–129.

Volkmann N (2009) Confidence intervals for fitting of atomic models into low-resolution densities, *Acta Crystallogr D Biol Crystallogr,* **65**: 679–689.

Webster P, Schwarz H, Griffiths G (2008) Preparation of cells and tissues for immuno EM, *Methods Cell Biol.* **88**: 45–58.

Winkler H, Taylor KA (2006) Accurate marker-free alignment with simultaneous geometry determination and reconstruction of tilt series in electron tomography, *Ultramicroscopy* **106**: 240–254.

Winkler H (2007) 3D reconstruction and processing of volumetric data in cryo-electron tomography, *J. Struct. Biol.* **157**: 126–137.

Zhang P, Bos E, Heymann J, Gnaegi H, Kessel M, Peters PJ, Subramaniam (2004) Direct visualization of receptor arrays in frozen-hydrated sections and plunge-frozen specimens of E. coli engineered to overproduce the chemotaxis receptor Tsr, *J. Microsc.* **216**: 76–83.

Zheng SQ, Keszthelyi B, Branlund E, Lyle JM, Braunfeld MB, Sedat JW, Agard DA (2007) UCSF tomography: an integrated software suite for real-time electron microscopic tomographic data collection, alignment and reconstruction, *J. Struct. Biol.* **157**: 138–147.

8

Correlated Soft X-Ray Tomography and Cryo-Light Microscopy

ELIZABETH A. SMITH, BERTRAND P. CINQUIN, GERRY MCDERMOTT, MARK A. LE GROS, AND CAROLYN A. LARABELL

INTRODUCTION

Biological processes occur over an enormous range of scale (Subramaniam 2005). As a consequence, no single imaging technique is capable of visualizing all biological processes and structures. Instead an array of techniques must be used, with each having characteristics that allow it to image optimally within limits in terms of specimen sizes, spatial resolution, and type of information that can be visualized. Of these characteristics, specimen illumination is the most significant, and the one that determines the overall capabilities of a modality. For example, bright-field light microscopy is best suited to imaging specimens that are a few microns or larger where the required spatial resolution is few hundred nanometers (Cowhig 1974). If higher spatial resolution views of the specimen are needed then sources of illumination—such as electrons or X-rays—must be used. In most instances satisfactorily answering a biological question requires imaging data from a number of complementary imaging modalities with, for example, molecular localization data in conjunction with visualization of cellular structures. In this chapter we will describe the principles behind an imaging technique that uses soft X-rays for specimen illumination. In particular, we will discuss the soft X-ray tomography (SXT) and the application of this technique to visualizing the detailed architecture inside cells with high three-dimensional spatial resolution. We will also describe cryogenic fluorescence microscopy, a complementary imaging technique to provide information on the position of tagged molecules within a cell imaged by SXT. Using this combination of modalities provides deep insights into the location of particular biochemical processes and their cellular and phenotypic effects.

SOFT X-RAY TOMOGRAPHY—HISTORY AND EVOLUTION OF THE TECHNIQUE

SXT has now matured and become an established player in the area of cellular imaging (Larabell and Nugent 2010). The development of SXT satisfies a long-standing need for an imaging technique capable of visualizing intact, fully hydrated cells—including eukaryotic cells—without the need to chemically fix the specimen or pre-treat it with contrast enhancing agents. SXT imaging uses specimens that have been cryo-preserved prior to data collection. The specimens are, therefore, imaged in a "near-native state," that is, the ideal scenario for visualizing any biological system (Le Gros, McDermott et al. 2005; McDermott, Le Gros et al. 2009). SXT has an additional advantage of being quantitative. Image contrast in SXT is generated by the differential attenuation of soft X-rays by the specimen according to the Beer-Lambert law, and is therefore a function of chemical species and thickness (McDermott, Le Gros et al. 2009; Larabell and Nugent 2010). In SXT the specimen is illuminated with X-rays from a region of the spectrum known as the "water window" (between 2.3–4.4 nm; 280, 530 eV~2.4nm; 517 eV). X-ray photons in this energy range are attenuated an order of magnitude more strongly by biomolecules than by water. In terms of a biological specimen this leads to the production of images with excellent natural contrast.

SXT can produce full-field images of a specimen with 50 nm isotropic spatial resolution, a big improvement over the resolution that is possible with visible light microscopy (typically ~250nm in x,y and ~750nm along z, respectively). SXT has much greater specimen penetration than is possible with electron microscopy (~15 μm vs ~500 nm) (Henke, Gullikson et al. 1993; Kirz, Jacobsen et al. 1995; Subramaniam 2005; Lucic, Leis et al. 2008). Specimen throughput is another positive attribute when SXT is compared with other high-resolution cellular imaging techniques (Larabell and Nugent 2010). An SXT data set required to isotropically reconstruct a 15 × 15 × 15 μm volume can be collected in three to five minutes (McDermott, Le Gros et al. 2009). Depending on their size, multiple cells can be contained in this field of view, from one large eukaryotic cell to a handful of yeast (McDermott, Le Gros et al. 2009) to tens or even hundreds of small prokaryotic cells. In concert with the relatively short data collection time, this means large numbers of cells can be imaged in a very short space of time. Accordingly, it is now routine to image and analyze several hundred reconstructions of bacterial or yeast cells to study phenotypic changes that take place during processes such as the cell cycle, when the cell responds to specific environmental factors (Uchida, McDermott et al. 2009), or as a function of key events during the cell cycle (Uchida, Sun et al. 2011).

Whilst it's true that SXT can provide unprecedented insights into cell structure, a complete understanding of cellular function requires the integration and interpretation of many types of data, including data from multiple imaging modalities (Le Gros, McDermott et al. 2009; McDermott, Le Gros et al. 2009; Larabell and Nugent 2010). As a consequence, much effort has been, and will continue to be, directed towards the establishment of new and/or improved methods to sequentially image a specimen with two or more modalities. Of course, it is most productive if the modalities use different sources of illumination. Rapid progress has been made towards correlating SXT and light microscopy imaging. In particular, we have worked to correlate SXT with high numerical-aperture cryogenic light microscopy (Le Gros, McDermott et al. 2009; McDermott, Le Gros et al. 2009). Used in conjunction with SXT, this new modality allows a cryo-preserved specimen to be sequentially imaged with soft X-rays and light (full-field and fluorescence). In practice, the X-ray image gives an accurate description of the subcellular architecture, with light-based imaging providing information on the location of fluorescent-tagged molecules within the cell (Le Gros, McDermott et al. 2009; McDermott, Le Gros et al. 2009; Larabell and Nugent 2010). These two pieces of data can be overlaid to form a single, information-rich image that contains powerful insights into both the cellular structure and the local environment of fluorescently labeled molecules (Le Gros, McDermott et al. 2009). This information can also be integrated with molecular biology, genetics, and computer modeling to increase our understanding of basic cell biology, or contribute to translational research, for example by visualizing the effect of candidate drug molecules on the subcellular structural phenotype of the cell. As a result SXT, especially in conjunction with correlated fluorescence imaging, is now widely recognized as being a powerful new imaging tool for biological and biomedical research (Larabell and Nugent 2010).

SOFT X-RAY TOMOGRAPHY: PRINCIPLES

In SXT, the specimen is imaged using a soft X-ray microscope (Le Gros, McDermott et al. 2005). In common with all microscopes the instrument produces two-dimensional projection images of the portion of the specimen that is within the depth of focus of the microscope (Spitta 1907). Most cells have a thickness greater than a few hundred nanometers and are therefore too complex to be usefully imaged in two dimensions, since in projection, structures inside the specimen are confusingly superimposed on top of each other (Derosier and Klug 1968). With tomography, the specimen is imaged from many angles, from which a three-dimensional, volumetric representation of the specimen, called a tomographic reconstruction, can be determined. Fortunately, tomography is a straightforward, commonly used, and a very well-established technique (Natterer 1986). To allow calculation of a tomographic reconstruction, a number of projection images must be collected at angular intervals around a rotation axis. This data set then contains all of the information necessary to calculate a three-dimensional representation of the specimen. This is the principle behind medical CT (computed tomography). In a CT scanner, hard X-rays (very short wavelength) are used to take a series of two-dimensional images of portions of a human body. In the CT instruments, the X-ray source and the detector, rotate around the stationary specimen, the patient. These projection images are then used to compute a reconstruction of the internal structures of the body with a spatial resolution of a millimeter or better. The basic concepts behind CT are used in SXT to image cell-sized specimens at significantly higher resolution (better than 50 nm) (Larabell and Nugent 2010). In SXT, the X-ray source and detector remain fixed, and the specimen is rotated inside the soft X-ray microscope to obtain a series of projection images (these are generally referred to as a "tilt" or "rotation" series) (Le Gros, McDermott et al. 2005). All soft X-ray microscopes capable of imaging biological specimens follow similar design principles, which we shall now describe.

Soft X-Ray Microscopes

The first X-ray microscope was developed by Paul Kirkpatrick and Albert Baez in the latter half of the 1940s (Kirkpatrick and Baez 1948). Their microscope used grazing-incidence reflective optics (termed K-B optics) to focus the X-rays onto the specimen. This was a landmark development for X-ray science and materials research, as witnessed by the fact that the optical configuration Kirkpatrick and Baez developed remains in widespread use to this day (Ortega, Bohic et al. 2004; Matsuyama, Mimura et al. 2006). However, this type of microscope didn't produce any meaningful biological images. This didn't occur until much more recently, when Fresnel Zone Plate microscopes were installed at third generation synchrotron light sources. These microscopes were the first indication of the real potential of soft X-ray microscopy for cellular imaging (Schmahl, Rudolph et al. 1996; Attwood 1999). In these instruments the specimen is illuminated with X-rays with energies that lie within the "water window" region of the X-ray spectrum. Almost immediately, these microscopes began to produce clear projection images of biological specimens (Schneider 1998; Meyer-Ilse, Hamamoto et al. 2001). The structural information contained in many of these images could be readily interpreted based on data from light- and electron-based microscopes, but with greater spatial resolution than the former technique, and unlike the latter, from intact, unstained cells (Meyer-Ilse, Hamamoto et al. 2001; Larabell and Le Gros 2004; Le Gros, McDermott et al. 2005). These results sparked resurgence in the development of soft X-ray microscopy specifically for biological imaging.

A soft X-ray microscope is similar to a simple full-field light microscope in that a condenser lens focuses the X-ray illumination onto the specimen and an objective lens refocuses unabsorbed X-rays onto a detector (Andersen, Harteneck et al. 1999). For the condenser lens, soft X-ray microscopes typically use a Fresnel zone plate optic. However, a glass capillary waveguide may also be used. Both types of condenser have inherent advantages and disadvantages (Ice 1997). To avoid confusion, and eliminate discussion that is not particularly germane to the thrust of this chapter, we will restrict our discussion to soft X-ray microscope designs that employ Fresnel zone plate condensers. Fresnel zone plates contain radially symmetric rings, known as Fresnel zones (Andersen, Harteneck et al. 1999; Attwood 1999) that alternate between being opaque and transparent towards X-ray photons. In operation, soft X-rays diffract around the opaque zones (Attwood 1999). The zones are spaced so that the diffracted light constructively interferes at the desired focus. For the condenser zone plate, the focal point is typically the specimen. The geometric condition to do so is that both opaque and transparent zones have the same area. In zone plate optics, the rings get thinner and more closely packed as they move outwards from the center point, until the outermost "zone" is reached (Andersen, Harteneck et al. 1999). To a first approximation, the spacing between adjacent zones of absorbing material in the objective zone plate defines the maximum spatial resolution obtainable from that particular microscope's optical system (Andersen, Harteneck et al. 1999; Chao, Harteneck et al. 2005). To date, SXT has been has been carried out with objective zone plates that produce a spatial resolution of 25 to 50 nm (McDermott, Le Gros et al. 2009; Larabell and Nugent 2010). However, lens designers and fabricators have now produced lenses that have a spatial resolution better than 15 nm (Chao, Harteneck et al. 2005). Installing one of these lenses into an existing soft X-ray microscope would be relatively trivial, however, since as the spatial resolution of the lens increases, the depth of field decreases (McDermott, Le Gros et al. 2009). Consequently, the latest generation of zone plates has a depth of field that is shallower than the thickness of eukaryotic cells. Overcoming this shortfall will require the development of techniques such as a combination of deconvolution with tomography. This work is now in progress (personal communication).

SOURCES OF SOFT X-RAYS

With currently available technologies, synchrotrons remain the optimal source of the illuminating photons for a soft X-ray microscope (Kirz, Jacobsen et al. 1995; Attwood, Chao et al. 2006; Larabell and Nugent 2010). Synchrotrons produce incredibly intense beams of soft X-rays that can be readily collimated and focused onto very small specimens (Attwood, Chao et al. 2006). The soft X-ray microscope XM-2 is located at the Advanced Light Source of Lawrence Berkeley National Laboratory, California. This facility is the brightest source of soft X-rays in the world. Soft X-ray microscopes can use illumination from ports on "bend magnet" sections of the synchrotron lattice. This makes the construction of a soft X-ray microscope a relatively easy and cost-effective proposition compared with instruments that require ports on insertion devices, such as wigglers or undulators.

Despite their advantages, access to experimental time at a synchrotron source is limited. There are relatively few of such facilities in the world, and all are heavily oversubscribed. To address this potential weakness much work is also underway on the development of "table top" X-ray sources (de Groot, Johansson et al. 2003; Hertz, Tuohimaa et al. 2007; Stollberg, Pokorny et al. 2007; Tuohimaa, Ewald et al. 2008; Hertz, von Hofsten et al. 2012). Small, laboratory-sized X-ray microscopes are now beginning to appear on the market. These instruments will soon make it feasible for individual research groups to have in-house SXT capabilities, rather than having to compete for experimental time at an oversubscribed synchrotron facility.

Soft X-Ray Imaging in Three Dimensions

In addition to a soft X-ray microscope, carrying out three-dimensional tomography on biological specimens requires a stage for cryogenically fixed specimens to be rotated (Le Gros, McDermott et al. 2005). Imaging a typical cell by soft X-ray tomography requires the collection of many projection images at angular increments over a 180° range. For a 3 μm-thick cell to be imaged with 50 nm isotropic resolution, this means collecting 180 projection images at 1° rotational increments. (Le Gros, McDermott et al. 2005; McDermott, Le Gros et al. 2009). Consequently, there are two challenges to be overcome. First, all biological specimens get damaged when they are exposed to intense light, whether this is intense UV illumination in a fluorescence microscope or photons in a soft X-ray microscope. Since tomography requires repeated exposure of the specimen to a fairly intense beam of X-rays this could quickly lead to a significant cumulative radiation dose, and therefore open the prospect of radiation damage, which in turn results in the generation of experimentally induced artifacts. However, it has been long accepted and repeatedly demonstrated that radiation damage can be mitigated by cooling the specimen to cryogenic temperatures whilst it is being imaged. In practice, if a biological specimen is cooled to liquid nitrogen temperature, a thousand soft X-ray microscope images can be collected with no apparent signs of radiation damage (Le Gros, McDermott et al. 2005;

McDermott, Le Gros et al. 2009). The second issue is one of structures moving or changing inside the specimen during the time it takes to collect a number of exposures. With a high spatial resolution technique, such as soft X-ray microscopy, even relatively small movements inside the cell during data collection have a profound impact on the fidelity of the calculated tomogram. Therefore, the specimen must be fixed prior to being imaged. There are two possible methods for doing this: by either using chemical fixatives or cryogenic immobilization (Weiss, Schneider et al. 2000). Chemical fixation is potentially very damaging to cellular structures (Leis, Rockel et al. 2009) and frequently seen to cause the collapse of highly solvated structures, such as vacuoles, or the swelling of others organelles, such as mitochondria, leading to concerns about other possible damage to fine structures (Leis, Rockel et al. 2009). Consequently, cryogenic fixation is considered by far to be the better option; this method is virtually instantaneous and has been seen to retain fine structural details inside the cell (Leis, Rockel et al. 2009). Accordingly, cryofixation is the principle method used in SXT as a means of minimizing artifacts due to internal movements during data collection, and/or radiation damage (Jensen and Briegel 2007; McDermott, Le Gros et al. 2009; Larabell and Nugent 2010). As we will discuss below, cryo-fixation has another major additional advantage in terms of correlated imaging; fluorescent labels have a much longer working lifetime when imaged at low temperature (Moerner and Orrit 1999; Thompson, Larson et al. 2002; Le Gros, 2009 #2840).

Specimen Preparation

Specimen preparation is a very important factor in determining the final quality and fidelity of any biological imaging technique. Artifacts generated by specimen preparation protocols can't be mitigated or eliminated post facto; they can only be accounted for. For SXT the specimen is flash cooled to cryogenic temperatures prior to imaging (Weiss 2000; Weiss, Schneider et al. 2001; Le Gros, McDermott et al. 2005; McDermott, Le Gros et al. 2009; Larabell and Nugent 2010). Typically, the specimen is mounted in an appropriate holder and rapidly cooled to liquid nitrogen temperature by either moving the holder into a stream of cold helium gas or rapidly plunging the specimen into a cryogen, such as liquid propane (Le Gros, McDermott et al. 2005). Either method cools the specimen quickly enough to promote the formation of amorphous rather than crystalline ice. This means that the structure of the ice inside the cell remains similar to that of the liquid before freezing. Consequently, the specimen is not damaged by the formation of crystalline ice.

Specimen Mounting

Prior to being imaged, specimens must obviously be mounted in a suitable holder. There are a number of possible options and holder configurations, all of which meet the criteria of mechanical stability during data collection. There is one further consideration, freedom of rotation. Cylindrical holders offer 360° of rotation around a central axis (Le Gros, McDermott et al. 2005; McDermott, Le Gros et al. 2009). Specimens sitting on top of a flat holder have a maximum rotation that can be used, which leads to a "missing wedge" of data (McDermott, Le Gros et al. 2009). For this important reason alone cylindrical holders are the preferred option. In practice they can easily be manufactured from inexpensive thin-walled capillary tubes.

CAPILLARY SPECIMEN HOLDERS

Before being loaded with cells, thin glass micro capillaries are heated and pulled to form an extended narrow tip using a micropipette puller. The tip region is typically 1 mm long with a diameter of 4 to 10 μm. This configuration combines mechanical stability with a long viewing length, along which cells can be imaged (Le Gros, McDermott et al. 2005). Multiple fields of view can be imaged along the length of the capillary by simply translating the holder in the microscope. Cells, suspended in a suitable medium, are easily loaded into the capillary using a regular pipette. Typically, capillaries are loaded with 1 μL of suspension containing between 10,000 and 100,000 cells. For most cell types capillary force is sufficient to draw the cells into the tapered region of the capillary to be imaged. In some instances, for example if the cells are particularly "sticky" or irregularly shaped, it may be necessary to use an adapted centrifuge to move the cells in to the narrow bore of the capillary (Dilworth, Parkinson et al. 2012). In general, the total time needed to mount cells in a capillary holder is very short, just several minutes, and except in very rare cases where the cells are particularly fragile, not considered long enough for damage to occur.

Large Cells on Flat Membranes

Although desirable, capillary specimen holders are impractical for mounting large eukaryotic cells or tissue specimens. The force required to move large, fragile cells into the tube would likely cause mechanical damage, plus the cylindrical geometry of a capillary is not a good match for relatively flat adherent cells. Instead, specimens such as adherent cells are mounted on 100 nm-thick silicon nitride membranes embedded in a silicon support frame (Dilworth, Parkinson et al. 2012). Cells can then be directly cultured onto this type of specimen holder by suspending it in the growth flask, or by adding cells using a pipette and then removing excess buffer using filter paper.

Fiducial Markers

The series of projection images contained in a data set are first aligned to a common axis of rotation prior to calculation of a three-dimensional tomographic reconstruction. There are a number of computational methods available for aligning projection images; these can be manual, automatic, or semi-automatic techniques. In all cases, the alignment of the images is quicker and more accurate if the specimen contains readily identifiable structural features, known as fiducial markers. Therefore, it is common practice to coat the outside of specimen holders with 60 nm or 100 nm gold nano-particles. These electron dense particles strongly attenuate the transmission of soft X-rays, and appear as vivid features in the projection image; hence they function as excellent fiducial markers (Stollberg, Guttmann et al. 2007). Fiducials can be identified manually, or recognized in the projection images by a computer algorithm (Jing and Sachs 1991; Amat, Moussavi et al. 2007). In either case, tracking a population of fiducials between adjacent images allows the rotation axis, the relative rotation, and any movements due to instrumental instabilities to be precisely determined (Jing and Sachs 1991; Amat, Moussavi et al. 2007). The alignment of adjacent images in a rotation series has a significant bearing on the amount of noise and artifacts seen in calculated reconstructions, therefore significant care is taken to ensure that good quality fiducial markers are present. That said, much progress has been made recently on the development of algorithms for "fiducial free" alignment (Winkler and Taylor 2006; Parkinson, Knoechel et al. 2012); these methods are improving and coming so close to being sufficiently accurate that the need to add fiducials will be completely eliminated in the near future.

Data Collection and Processing

A detailed discussion of the methods used to process data is out of the scope of this chapter; instead a brief summary will be given to provide a flavor of the process. In practice, SXT data processing is done using software packages developed and refined for electron tomography (for example IMOD (Kremer, Mastronarde et al. 1996)). It is important to stress again that with capillary specimen holders there is no limit to the angular rotation through which data can be collected (Le Gros, McDermott et al. 2005). The use of capillaries allows data to be collected at any angle through a full 360° of rotation. In other words, even though images are collected using a single "tilt" axis, there is no missing data as would be the case if a single axis (or even a double axis) electron microscopy cryostage was used (Le Gros, McDermott et al. 2005). This completeness of data also greatly improves the quality of the calculated tomogram by minimizing noise (Le Gros, McDermott et al. 2005). In conjunction with the excellent contrast inherent in soft X-ray microscope images, this leads to the generation of high fidelity SXT volumetric reconstructions.

An alternative approach for imaging extended adherent mammalian cells is implemented at the soft X-ray microscope at the Helmholtz Zentrum Berlin. Here, an electron tomographic cryo-tilt stage is implemented into the microscope. In recent work, Schneider and colleagues used data from this microscope to calculate three-dimensional cellular reconstructions and to identify the plasma and nuclear membrane, vesicles, lysosomes, mitochondria, and nuclear components (Schneider, Guttmann et al. 2010). The methods they describe are applicable to a wide range of adherent and non-adherent cells.

As with any three-dimensional imaging technique, once cells have been reconstructed they can be "segmented" (Leis, Rockel et al. 2009; McDermott, Le Gros et al. 2009). This is the process whereby regions containing a specific feature are isolated and typically bound by a surface. Organelles are the most commonly segmented objects inside a cell. Segmentation in SXT is guided by the measured Linear Absorption Coefficient (LAC) (Le Gros, McDermott et al. 2005; Parkinson, McDermott et al. 2008; Leis, Rockel et al. 2009; McDermott, Le Gros et al. 2009; Uchida, McDermott et al. 2009; Larabell and Nugent 2010; Uchida, Sun et al.

2011). The contrast in SXT is derived from differences in density and biochemical composition between the various structures inside the cell. For example, densely packed droplets of lipids absorb soft X-rays strongly and have correspondingly high LAC values, whereas vacuoles are highly solvated and contain a comparatively lower density of bio-molecules, consequently, these organelles are seen to have low LAC values. In this way it is a straightforward process to identify and isolate the major organelles in a reconstruction of a cell. Organelles such as the mitochondria, nucleus, nucleolus, vacuoles, and lysosomes all have characteristic LAC values (that hold constant between cells, and in some cases between markedly different cell types). This makes their identification straightforward in a SXT reconstruction of a cell. Segmentation can be carried out manually, or by employing software algorithms that automate the process. In a recent study, Uchida and co-workers compared organelle volumes from reconstructed cells that had been segmented both manually and automatically (Uchida, McDermott et al. 2009). They found the correlation between the volumes to be remarkably close, with only some minor variations. This is an important advance, given that data can be obtained in large quantities; in contrast, segmentation would be a major bottleneck if it remained a manual process. Currently, it takes an experienced person hours or days to segment a single unicellular eukaryote, such as yeast. Automated algorithms can achieve the same results in seconds which allows obtaining quantitative information on cellular structures from a statistically significant number of cells a fast, routine procedure.

SOFT X-RAY TOMOGRAPHY: EXAMPLES

As stated above, SXT can be used to image virtually all cell types of interest to biologists, from simple bacteria, to yeast, algae, and higher order eukaryotes. See Figures 8.1–4 for examples. A particular strength of SXT is the capability of imaging intact, fully hydrated eukaryotic cells (Larabell and Nugent 2010). See Figure 8.3 for an SXT reconstruction and segmentation of an intact T-cell. Understandably, given the strong interest in this data, this is the area where most effort has been focused. The ground was first broken with the report that the algae *Chlamydomonas reinhardtii* had been imaged in exquisite detail using SXT (Weiss 2000; Weiss, Schneider et al. 2000; Weiss, Schneider et al. 2001). This was a key piece of work in the field, and even today, this work remains an excellent demonstration of the power of SXT. Unfortunately, the cryo-rotation stage they used was unwieldy and labor intensive, making data collection very time consuming. As a consequence, there we no further reports of SXT of biological specimens during the lifetime of this particular microscope.

It took many more years of development before Larabell and colleagues carried out a number of studies on both fission and budding yeasts, as shown in Figure 8.4 (Larabell and Le Gros 2004; Gu, Etkin et al. 2007; Parkinson, McDermott et al. 2008). This work was carried out using a significantly more automated cryo-rotation stage mounted on XM-1, a general-purpose soft X-ray microscope at the Advanced Light Source, Berkeley. Since XM-1 was primarily designed for magnetism and materials research, it did not have the optimal optical configuration for biological imaging. However, the cellular images it produced were of sufficient quality to prove that SXT

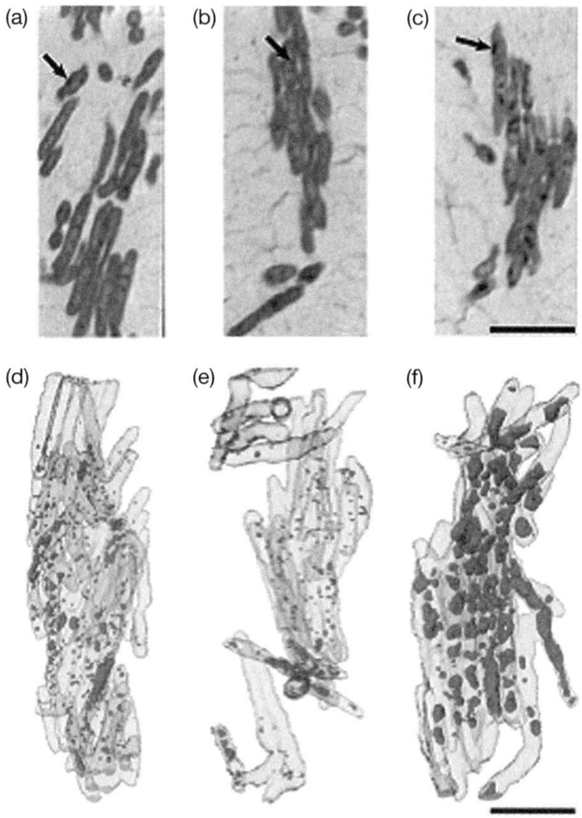

Figure 8.1 Orthoslices from three-dimensional tomographic reconstructions (A–C) and segmented volumes (D–F) of *Mycobacterium smegmatis* during early log phase growth (A, D), late log phase growth (B, E), and stationary phase (C, F). The lipid bodies (arrows a–c), which are highly absorbing (linear absorption coefficient, LAC, 0.71 μm^{-1}), are easily distinguished from the surrounding organic material (LAC, 0.38 μm^{-1}). Lipid bodies were segmented and color-coded yellow (D–F) to show the changes in relative concentration under different conditions. Scale bars (A–C) = 3 μm, (D–F) = 2 μm. (Reproduced from McDermott, Fox et al. 2012).

could be developed as a useful, high-throughput biological imaging technique. For example, unlike in the initial work of Weiss et al., these later publications contained data from not one, but many tens of cells (Larabell and Le Gros 2004; Gu, Etkin et al. 2007; Parkinson, McDermott et al. 2008).

With the confidence provided by these results a new soft X-ray microscope, fully dedicated to biological imaging, was constructed at the ALS. Once commissioned this new microscope, designated XM-2, quickly surpassed all of the previous work on other soft X-ray microscopes in terms of quality and quantity of usable data. In addition to having an optical system specifically tailored for imaging biological specimens, XM-2 operates with the latest generation of cryo-rotation stages (McDermott, Le Gros et al. 2009). This combination of factors resulted in an enormous improvement in image quality and fidelity compared to data from previous instruments. This

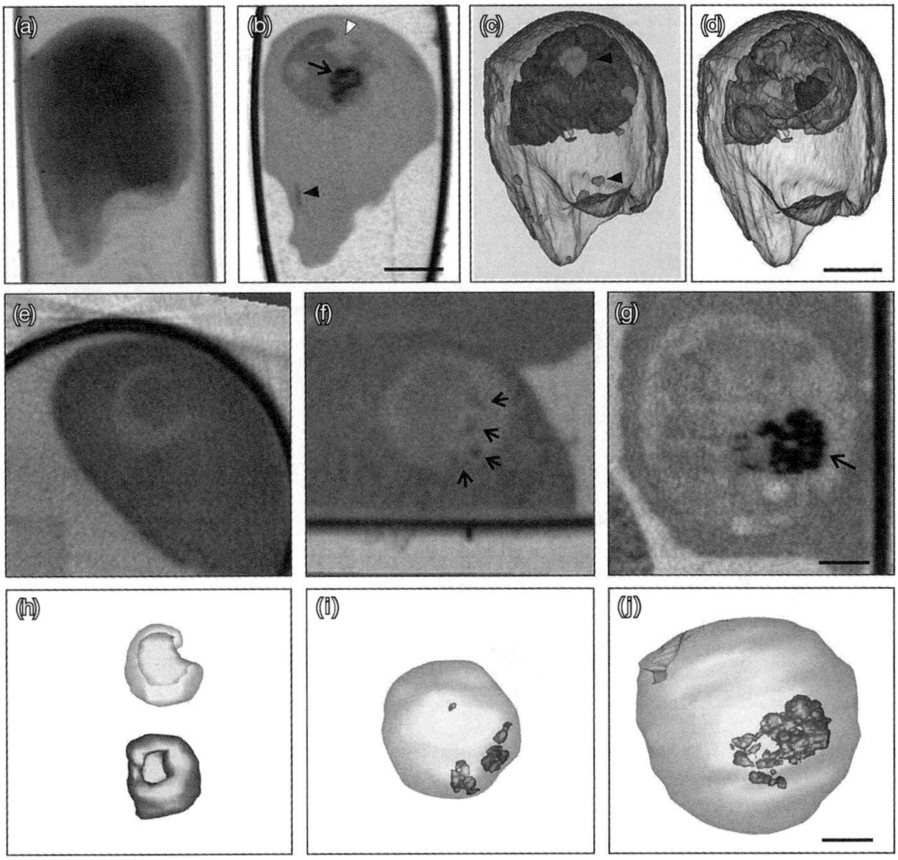

Figure 8.2 X-ray tomography of intact *P. falciparum*-infected RBCs Parasite-infected RBCs were mounted in glass capillaries then snap frozen and imaged using the XM-2 transmission x-ray microscope during rotation through 180°. (A) A projection view of a trophozoite-infected RBC in which the outline of the host cell and the parasite are visible. (B,E–G) Virtual sections through tomograms generated for parasites at different stages of development. Invaginations are observed in the surface of some of the trophozoites (B, white arrowhead). X-ray-dense hemozoin crystals are observed in more mature stage parasites (B,F,G, arrows). Membrane features are observed in the RBC cytoplasm (B, black arrowhead). (C,D) The RBC surface is rendered in translucent pink and the parasite surface in solid (C) or translucent (D) brown. Membrane features in the RBC cytoplasm are rendered in red (C) and hemozoin is depicted in purple (D). (H–I) Rendered models of an early ring (H, side and top views), a late ring (I) and a mid trophozoite (J) showing the parasite surface (translucent beige) and the hemozoin crystals (gold). The cupped parasite surface is illustrated in solid gold in H (bottom object). Scale bar, A–D 2 μm; E–J 1 μm. (Reproduced from Hanssen, Knoechel et al. 2011). See insert for color version.

fact is especially evident when cells are imaged mounted in capillary specimen holders. As discussed above, the lack of restriction on angular rotation eliminates the so-called "missing wedge" of data that can't be avoided using flat specimen holders, and generates reconstructions with significantly lower levels of noise and artifact (Le Gros, McDermott et al. 2005; McDermott, Le Gros et al. 2009).

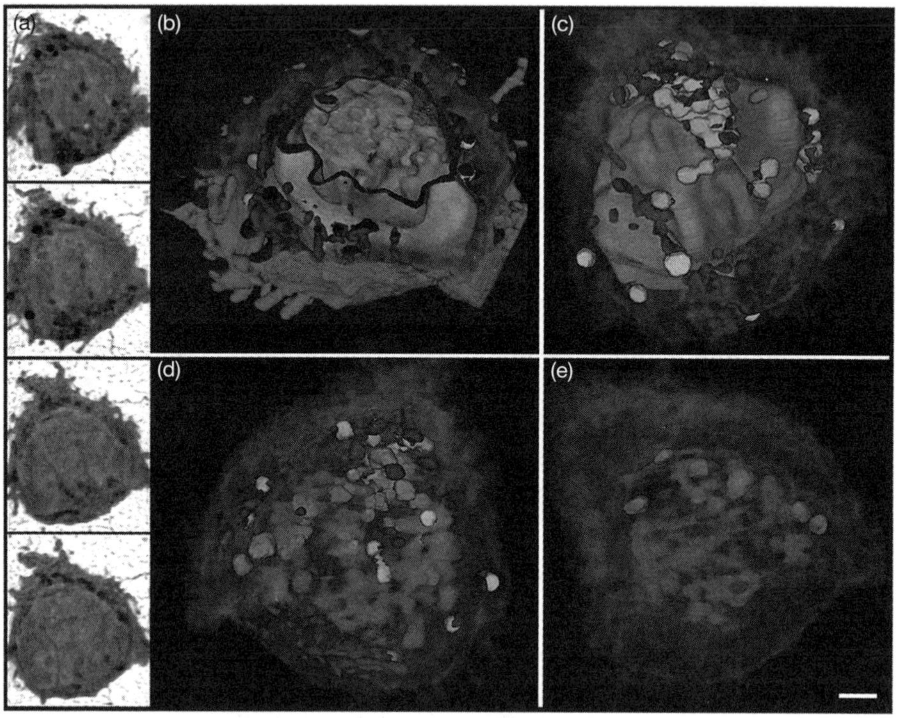

Figure 8.3 Soft X-ray tomography of a lymphocyte (T-cell). (A) Four orthoslices from the tomographic reconstruction of a cryo-immobilized T-cell. (B–E) Segmented volumes color-coded for identification of internal structures. (B) Reconstruction of a cryo-immobilized T-cell. (B) Cut-away view showing several filopodial extensions on the cell surface (blue) and the internal cytoplasm (purple) containing organelles such as mitochondria (magenta) and highly absorbing vesicles (yellow). A cut-away view of the nuclear envelope (cyan) reveals the chromatin (salmon) and nuclear bodies (green) within. (C) The cell surface has been removed, showing the 3D organization of cytoplasm and organelles with respect to the intact nuclear envelope, which is highly folded. (D) View of the cell interior in which the nuclear envelope has been made transparent revealing the 3D organization of the chromatin with respect to cytoplasmic organelles. (E) View of the cell showing the 3D organization of chromatin and nuclear bodies with respect to cytosol. Scale bar, 1.0 mm. (Reproduced from Larabell and Nugent 2010). See insert for color version.

LOCALIZING MOLECULES IN A SXT RECONSTRUCTION OF A CELL

SXT can provide exquisite insights into the subcellular structure, and give us unique, quantitative information on the subcellular phenotype. However, a more complete understanding of cellular behavior also requires knowledge of the location of specific molecules. Particularly if the molecules perform different functions depending on their cellular location, or if the molecule needs to interact with specific partner molecules to become active, methods for localizing molecules directly in the context of a SXT reconstruction of a cell would be of high value.

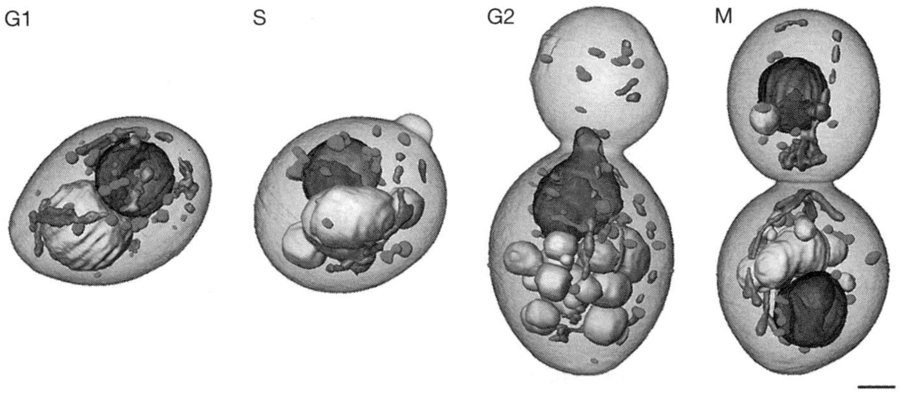

Figure 8.4 Soft X-ray tomography of rapidly frozen Saccharomyces cerevisiae cells imaged at each phase of cell cycle – G1, S, M, and G2. Organelles are color-coded as follows: blue, nucleus; orange, nucleolus; gray, mitochondria; ivory, vacuoles; green, lipid bodies. Scale bar, 2 μm (reproduced from Larabell and Nugent 2010)

Initial efforts were directed towards using immuno-labeling methods that had previously been very successful when applied to electron microscopy/tomography. In these methods the molecule of interest is immuno-labeled with an electron dense tag (Meyer-Ilse, Hamamoto et al. 2001) or with a molecule that contains an absorption edge located within the "water window" (Ashcroft, Gu et al. 2008). The basic principle behind this method is simple; the label attenuates soft X-rays significantly more strongly that any biological features, allowing it to be unambiguously identified in a soft X-ray microscope image. Unfortunately, immuno-labeling methods have significant disadvantages. Most seriously, the specimen must be chemically fixed and incubated with membrane permeabilizing agents, such as detergents or organic solvents, to allow ingress of large antibodies into the interior of the cell (Leis, Rockel et al. 2009). Clearly, disrupting the integrity of cellular membranes potentially results in proteins and other molecules leaking out of organelles, or the cell itself, and will likely cause a varying amount of damage to cellular substructures (Leis, Rockel et al. 2009). Moreover, once inside the cell the antibody may have very limited access to the target epitope-containing molecule. Therefore, immuno-labeling is a less than optimal solution to the problem (McDermott, Le Gros et al. 2009). A much better option for localizing molecules is to use genetically encoded fluorescence labels, rather than antibodies, to localize molecules of interest, and correlate this data directly with SXT images. Fortunately, the methods and instruments needed to perform this technique have now been developed. Le Gros and colleagues recently developed the world's first high-numerical aperture cryogenic light microscope capable of being operated at (Le Gros, McDermott et al. 2009) liquid nitrogen temperature. Integration of this light-based modality with the soft X-ray microscope allows the *same* specimen to be imaged sequentially with these two very different modalities (Larabell and Nugent 2010). In the case of XM-2, both imaging techniques are carried out on the same instrument after a simple translation of the specimen holder.

Localizing molecules by correlated fluorescence and X-ray imaging has many distinct and obvious advantages. There is a wealth of readily available fluorescence data in the literature. In addition, most bio-labs have a fluorescent microscope and routinely utilize the power of genetically encoded fluorescence tags, such as Green Fluorescent Protein and its many derivatives, in their day-to-day research efforts (Tsien 2003; Tsien 2005; Giepmans, Adams et al. 2006). Consequently, there is a wealth of existing knowledge on molecular localization based on this technique. However, information on the environment surrounding the labeled molecule is lacking, and it is often difficult, if not impossible, to delineate the precise cellular location of a molecule based on fluorescence data alone. Fundamentally important questions frequently remain unanswered, such as "is the molecule inside or outside the organelle?" This can only be answered by localizing the molecule in the context of a high-spatial resolution reconstruction. There is a secondary advantage to being able to carry out correlated light and SXT imaging. The fluorescence data can resolve ambiguities over the assignment of a segmented region in a reconstruction, as shown in the correlated images of *Schizosaccharomyces pombe* in Figure 8.5. This occurs when functionally different organelles are similar in size and have similar LAC values, or in cases when the objects have similar LAC values as neighboring cellular contents.

As an aside, the development of high numerical aperture cryogenic light microscopy is a significant development in its own right, and is not just a useful imaging adjunct to SXT. It has been long understood that cryo-cooling significantly enhances their fluorescence working lifetimes (Moerner and Orrit, 1999). Consequently, cryogenic immersion microscopy meets a long-standing need in biological and biomedical imaging by providing a means of imaging fluorescent probes that ordinarily have short working lifetimes. Although low temperature microscopes have long been used to image frozen cells these instruments all operated in air, therefore they could only use lenses with relatively low numerical apertures. The development of a cryogenic immersion lens allows frozen specimens to be imaged at high resolution by using index-matched cryogens. This is a very important factor, since it opens up the door for high or even "super resolution" fluorescent imaging on specimens containing labels with greatly enhanced working lifetimes.

Over the years, significant progress has been made towards enhancing genetically encodable fluorescent proteins derived from jellyfish and corals (Dickson, Cubitt et al. 1997; Zhang, Campbell et al. 2002; Tsien 2003; Shaner, Steinbach et al. 2005; Tsien 2005; Giepmans, Adams et al. 2006; Shu, Shaner et al. 2006; Shaner, Lin et al. 2008). Most of this work has been aimed at increasing the range of available colors, reducing oligomerization of the molecules to produce physically smaller tags, and so on. This work has been fruitful and now genetically encoded tags are a highly useful, versatile research tool. However, there has been limited success in extending the fluorescent working lifetime of these molecules (Ai et al. 2006; Shaner et al. 2008). The highest possible spatial resolution determination of a well-isolated fluorescent signal single molecule, and therefore the precision with which relative protein locations can be determined, is dependent on the number of photons collected from the fluorophore (Schermelleh, Carlton et al. 2008). By cryogenically cooling the specimen, and therefore increasing the number of photons that can be collected before the fluorophore beaches, in principle it becomes possible to collect very high-resolution data on the location of specific molecules inside the cell. Consequently, the addition of high numerical cryogenic light microscopy into a soft X-ray microscope is the combination of two very powerful, unique techniques. As a result the data obtained by correlated imaging is enormously rich and informative.

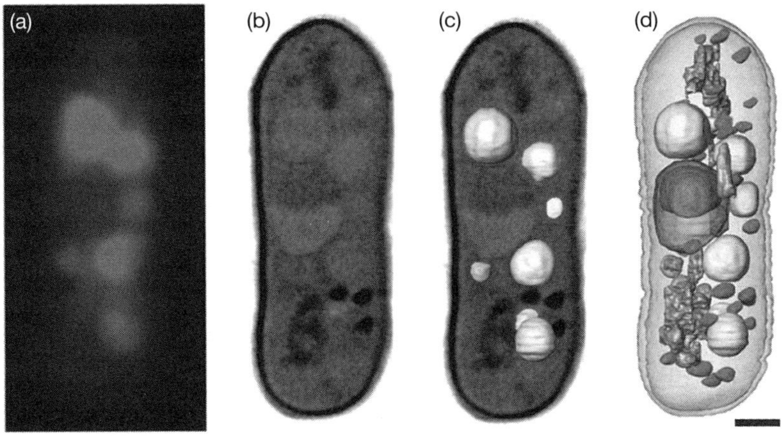

Figure 8.5 Correlated soft X-ray tomography and cryo-light imaging (wide-field fluorescence) of Schizosaccharomyces pombe. (a) The vacuoles were fluorescently labeled and imaged by cryo-light microscopy. (b, c) Slices through the volumetric reconstruction calculated from soft X-ray tomography data, with the vacuoles shown as segmented volumes in panel c. The segmented vacuoles correlate closely with the locations determined from cryo-light microscopy. (d) The same cell after the major organelles have been segmented. Scale bar = 1 μm. (Reproduced from McDermott, Le Gros et al. 2012)

SUMMARY AND FUTURE PROSPECTS

Soft X-ray tomography is a unique imaging technique, capable of giving us unprecedented insights into the structure of cells and tissue specimens that are held in a near-native state (Larabell and Nugent 2010). The combination of high spatial resolution, together with a contrast mechanism that reveals subtle subcellular features, makes SXT an excellent cellular-level imaging tool. Work to date using this modality only hints at the spectrum of cell types amenable to this kind of imaging. In the future, SXT will be employed on an ever-expanding range of the specimen types, including cells whose dimensions and complexities push the limits of the technique. This is particularly true for complex, high-order eukaryotic cells.

The spatial resolution that can be achieved with SXT depends on the optical system installed in the microscope. In the past, these nanofabricated objects were both expensive and difficult to manufacture. However, the revolution in nanofabrication technology has made production of Fresnel lenses efficient and comparatively low in cost. Moreover, it is now possible to manufacture such optics with increasingly finer structural features. This is enormously important in terms of spatial resolution that can be obtained using SXT, since the width between the outer zones in a Fresnel optic determines the maximum spatial resolution that can be achieved in a soft X-ray microscope. State-of-the-art soft X-ray Fresnel lenses have been demonstrated to be capable of resolving test objects with 12 nm features (Chao, Kim et al. 2009). Rapid progress is being made on the development of methods that combine tomography with through-focus deconvolution techniques. This work is in the early stages of evolution, but builds on a very firm theoretical and experimental foundation generated by many workers in electron and light microscopy. Once completed, this combination of methods will allow ultra-high-resolution imaging to be carried out on thick specimens in the near-native state. Consequently, use of high-resolution zone plates with shallow depth of field will undoubtedly become a common feature in SXT in the near future.

In the future, the most exiting technological developments will be in the continued development of methods for correlated cryo-fluorescence and X-ray imaging. This is truly innovative in terms of instrumentation and potential impact on biomedical research. Using the new cryo-light microscope in tandem with soft X-ray tomography it will be possible to localize the position of fluorescently tagged molecules in the context of the framework produced by a high-resolution reconstruction of the cell. This prospect has already generated enormous community interest. In closing, we feel confident in saying soft X-ray tomography has finally come of age, and has emerged as a mainstream technique with wide applicability in areas as diverse as drug discovery, biomedical research, and basic cell biology.

REFERENCES

Ai HW, Henderson JN, Remington SJ, Campbell RE (2006) Directed evolution of a monomeric, bright and photostable version of Clavularia cyan fluorescent protein: structural characterization and applications in fluorescence imaging. *Biochem. J.* ***400:*** 531–540.

Amat, F., F. Moussavi, et al. (2007). "Markov random field based automatic image alignment for electron tomography." *J Struct Biol* ***161***: 260–275.

Andersen, E. H., B. Harteneck, et al., Eds. (1999). *Nanofabrication of X-ray ZonePlates with the Nanowriter Electron-Beam Lithography System.* X-ray Microscopy VI. Berkeley, CA, American Institute of Physics.

Ashcroft, J. M., W. Gu, et al. (2008). "TiO2 nanoparticles as a soft X-ray molecular probe." *CHEM COMMUN* ***21***: 2471–2473.

Attwood, D., W. Chao, et al. (2006). "Imaging at High Spatial Resolution: Soft X-Ray Microscopy to 15 nm." *Journal of Biomedical Nanotechnology* ***2***(2): 75–78.

Attwood, D. T. (1999). *Soft x-rays and extreme ultraviolet radioation: principles and applications.* Cambridge, New York, Cambridge Uhiversity Press.

Chao, W., B. D. Harteneck, et al. (2005). "Soft X-ray microscopy at a spatial resolution better than 15 nm." *Nature* ***435***(7046): 1210–1213.

Chao, W., J. Kim, et al. (2009). "Demonstration of 12 nm Resolution Fresnel Zone Plate Lens based Soft X-ray Microscopy." *Optics Express* ***17***(20): 17669–17677.

Cowhig, J. (1974). *The world under the microscope.* New York, Bounty Books.

de Groot, J., G. A. Johansson, et al. (2003). "Improved liquid-jet laser-plasma source for X-ray microscopy." *Journal de Physique IV* ***104***: 121–122.

Derosier, D. J. and A. Klug (1968). "Reconstruction of 3 Dimensional Structures from Electron Micrographs." *Nature* ***217***(5124): ***130***-&.

Dickson, R. M., A. B. Cubitt, et al. (1997). "On/off blinking and switching behaviour of single molecules of green fluorescent protein." *Nature* ***388***(6640): 355–358.

Dilworth, Y., Parkinson, L. R. E., Gerry McDermott, Mark A. Le Gros, Rosanne M. Boudreau, and Carolyn A. Larabell (2012). *Nanoimaging cells using soft x-ray tomography.* USA Humana Press (Springer publishing group).

Giepmans, B. N., S. R. Adams, et al. (2006). "The fluorescent toolbox for assessing protein location and function." *Science* ***312***(5771): 217–224.

Gu, W. W., L. D. Etkin, et al. (2007). "X-ray tomography of Schizosaccharomyces pombe." *Differentiation* ***75***(6): 529–535.

Hanssen, E., C. Knoechel, et al. (2011). "Cryo transmission X-ray imaging of the malaria parasite, P. falciparum." *Journal of Structural Biology* ***173***(1): 161–168.

Henke, B. L., E. M. Gullikson, et al. (1993). "X-Ray Interactions—Photoabsorption, Scattering, Transmission, and Reflection at E=50-30,000 Ev, Z=1-92." *Atomic Data and Nuclear Data Tables* ***54***(2): 181–342.

Hertz, H. M., T. Tuohimaa, et al. (2007). "Phase-contrast x-ray imaging with a liquid-metal-jet-anode microfocus source." *Applied Physics Letters* ***91***(7).

Hertz, H. M., O. von Hofsten, et al. (2012). "Laboratory cryo soft X-ray microscopy." *Journal of Structural Biology* ***177***(2): 267–272.

Ice, G. E. (1997). "Microbeam-forming methods for synchrotron radiation." *X-Ray Spectrometry* ***26***(6): 315–326.

Jensen, G. J. and A. Briegel (2007). "How electron cryotomography is opening a new window onto prokaryotic ultrastructure." *Curr Opin Struct Biol* ***17***(2): 260–267.

Jing, Z. Q. and F. Sachs (1991). "Alignment of tomographic projections using an incomplete set of fiducial markers." *Ultramicroscopy* ***35***(1): 37–43.

Kirkpatrick, P. and A. V. Baez (1948). "Formation of optical images by x-rays." *Journal Of The Optical Society Of America* ***38***(9): 766–774.

Kirz, J., C. Jacobsen, et al. (1995). "Soft x-ray microscopes and their biological applications." *Quarterly reviews of biophysics* ***28***(1): 33–130.

Kremer, J. R., D. N. Mastronarde, et al. (1996). "Computer visualization of three-dimensional image data using IMOD." *Journal of Structural Biology* ***116***(1): 71–76.

Larabell, C. A. and M. A. Le Gros (2004). "X-ray tomography generates 3-D reconstructions of the yeast, Saccharomyces cerevisiae, at 60-nm resolution." *Molecular Biology of the Cell* ***15***(3): 957–962.

Larabell, C. A. and K. A. Nugent (2010). "Imaging cellular architecture with X-rays." *Current Opinion in Structural Biology* ***20***(5): 623–631.

Le Gros, M. A., G. McDermott, et al. (2005). "X-ray tomography of whole cells." *Curr Opin Struct Biol* ***15***(5): 593–600.

Le Gros, M. A., G. McDermott, et al. (2009). "High-aperture cryogenic light microscopy." *J Microsc* ***235***(1): 1–8.

Leis, A., B. Rockel, et al. (2009). "Visualizing cells at the nanoscale." *Trends Biochem Sci* ***34***(2): 60–70.

Lucic, V., A. Leis, et al. (2008). "Cryo-electron tomography of cells: Connecting structure and function." *Histochemistry and Cell Biology* ***130***(2): 185–196.

Matsuyama, S., H. Mimura, et al. (2006). "Development of scanning x-ray fluorescence microscope with spatial resolution of 30 nm using Kirkpatrick-Baez mirror optics." *Review Of Scientific Instruments* ***77***(10): 103102.

McDermott, G., D. M. Fox, et al. (2012). "Visualizing and quantifying cell phenotype using soft X-ray tomography." *Bioessays* ***34***(4): 320–327.

McDermott, G., M. A. Le Gros, et al. (2009). "Soft X-ray tomography and cryogenic light microscopy: the cool combination in cellular imaging." *Trends Cell Biol* ***19***(11): 587–595.

McDermott, G., M. A. Le Gros, et al. (2012). "Visualizing Cell Architecture and Molecular Location Using Soft X-Ray Tomography and Correlated Cryo-Light Microscopy." *Annual Review of Physical Chemistry* ***63***(1): 225–239.

Meyer-Ilse, W., D. Hamamoto, et al. (2001). "High resolution protein localization using soft X-ray microscopy." *J Microsc* ***201***(Pt 3): 395–403.

Moerner, W. E. and M. Orrit (1999). "Illuminating single molecules in condensed matter." *Science* ***283***(5408): 1670–1676.

Natterer, F. (1986). *The Mathematics of Computerized Tomography*. New York, NY, Wiley.

Ortega, R., S. Bohic, et al. (2004). "Microchemical element imaging of yeast and human cells using synchrotron X-ray microprobe with Kirkpatrick-Baez optics." *Analytical chemistry* ***76***(2): 309–314.

Parkinson, D. Y., C. Knoechel, et al. (2012). "Automatic alignment and reconstruction of images for soft X-ray tomography." *Journal of Structural Biology* ***177***(2): 259–266.

Parkinson, D. Y., G. McDermott, et al. (2008). "Quantitative 3-D imaging of eukaryotic cells using soft X-ray tomography." *J Struct Biol* ***162***(3): 380–386.

Schermelleh, L., P. M. Carlton, et al. (2008). "Subdiffraction multicolor imaging of the nuclear periphery with 3D structured illumination microscopy." *Science* ***320***(5881): 1332–1336.

Schmahl, G., D. Rudolph, et al. (1996). "X-ray microscopy." *Naturwissenschaften* ***83***(2): 61–70.

Schneider, G. (1998). "Influence of roughness and interdiffusion in zone plate structures on the diffraction efficiency described by coupled wave theory." *Applied Physics Letters* ***73***(5): 599–601.

Schneider, G., P. Guttmann, et al. (2010). "Three-dimensional cellular ultrastructure resolved by X-ray microscopy." *Nature Methods* *7*(12): 985–U116.

Shaner, N. C., M. Z. Lin, et al. (2008). "Improving the photostability of bright monomeric orange and red fluorescent proteins." *Nat Methods* ***5***(6): 545–551.

Shaner, N. C., P. A. Steinbach, et al. (2005). "A guide to choosing fluorescent proteins." *Nat Methods* ***2***(12): 905–909.

Shu, X., N. C. Shaner, et al. (2006). "Novel chromophores and buried charges control color in mFruits." *Biochemistry* ***45***(32): 9639–9647.

Spitta, E. J. (1907). *Microscopy—the Construction, theory and Use of the Microscope*. New York, NY, E.P. Dutton.

Stollberg, H., P. Guttmann, et al. (2007). "Size-selective colloidal-gold localization in transmission X-ray microscopy." *Journal of Microscopy-Oxford* ***225***(1): 80–87.

Stollberg, H., M. Pokorny, et al. (2007). "A vacuum-compatible wet-specimen chamber for compact X-ray microscopy." *Journal of Microscopy-Oxford* ***226***(1): 71–73.

Subramaniam, S. (2005). "Bridging the imaging gap: visualizing subcellular architecture with electron tomography." *Current Opinion in Microbiology* ***8***(3): 316–322.

Thompson, R. E., D. R. Larson, et al. (2002). "Precise nanometer localization analysis for individual fluorescent probes." *Biophys J* ***82***(5): 2775–2783.

Tsien, R. Y. (2003). "Imagining imaging's future." *Nat Rev Mol Cell Biol* ***4*** (Suppl): SS16–21.

Tsien, R. Y. (2005). "Building and breeding molecules to spy on cells and tumors." *FEBS Lett* ***579***(4): 927–932.

Tuohimaa, T., J. Ewald, et al. (2008). "A microfocus x-ray source based on a nonmetal liquid-jet anode." *Applied Physics Letters* ***92***(23).

Uchida, M., G. McDermott, et al. (2009). "Soft X-ray tomography of phenotypic switching and the cellular response to antifungal peptoids in Candida albicans." *Proceedings of the National Academy of Sciences of the United States of America* ***106***(46): 19375–19380.

Uchida, M., Y. Sun, et al. (2011). "Quantitative analysis of yeast internal architecture using soft X-ray tomography." *Yeast* ***28***(3): 227–236.

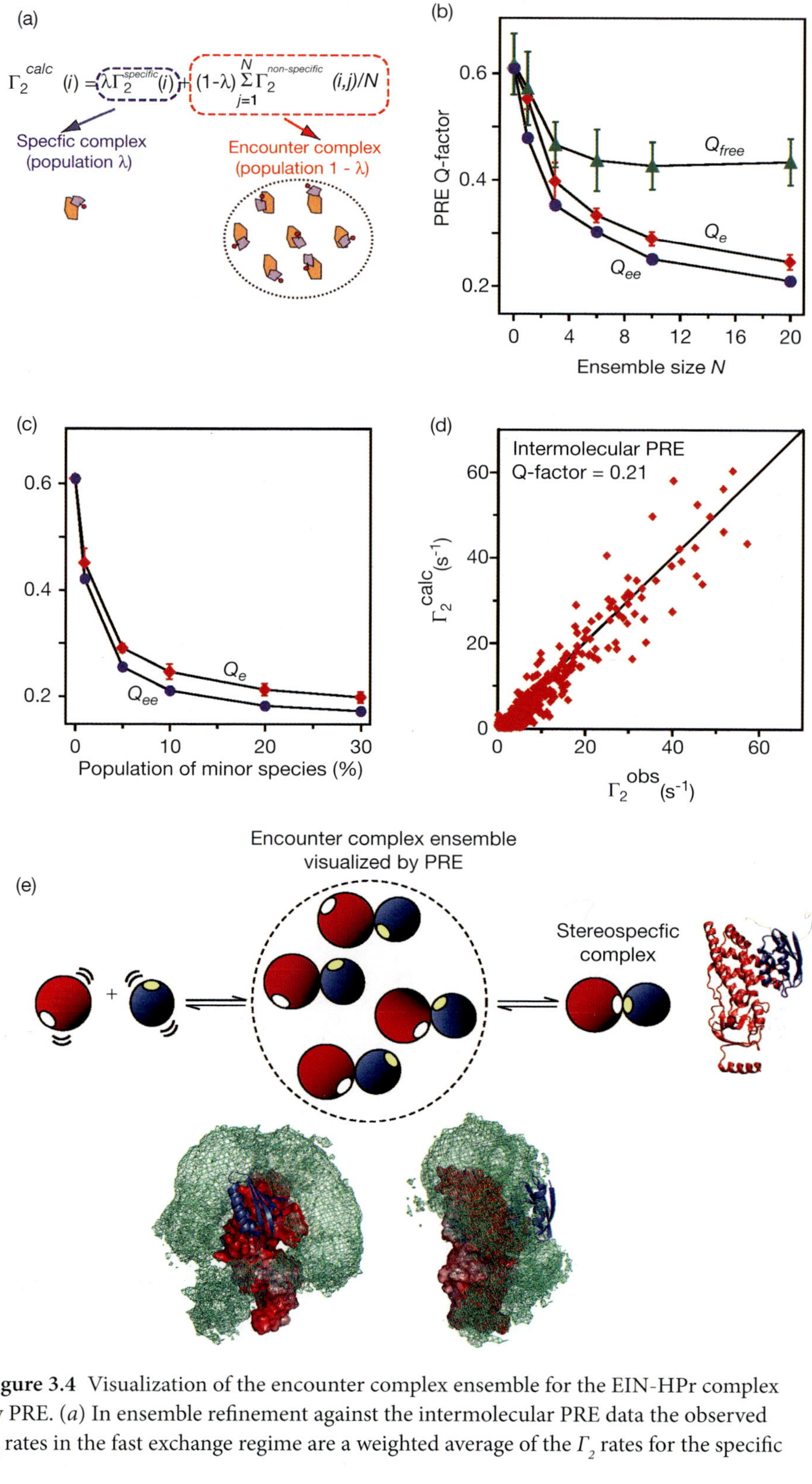

Figure 3.4 Visualization of the encounter complex ensemble for the EIN-HPr complex by PRE. (*a*) In ensemble refinement against the intermolecular PRE data the observed Γ_2 rates in the fast exchange regime are a weighted average of the Γ_2 rates for the specific

complex and an encounter complex ensemble comprising *N* species. (*b*) Dependence of working (Q_e and Q_{ee}) and complete cross-validated (Q_{free}) Q-factors on ensemble size *N*. (Q_e is the average Q-factor <Q> for all 100 calculated ensembles, and Q_{ee} is the ensemble of ensembles average Q-factor). (*c*) Dependence of working Q factors on population of the encounter complex ensemble. (*d*) Correlation between observed and calculated Γ_2 rates obtained with a population of 10 percent for the encounter complex species represented by an ensemble of size $N = 20$. (*e*) Schematic of the reaction pathway involved in stereospecific protein-protein complex formation, with two views of a reweighted atomic probability density map (63) illustrating the distribution of HPr molecules on the surface of EIN that make up the ensemble of encounter complexes. The encounter complex probability map (light green mesh plotted at a threshold of 20 percent maximum) is calculated from 100 independent calculations of ensemble size $N = 20$ at a population of 10 percent; the molecular surface of EIN is color coded by electrostatic potential ($\pm$ 8 kT); and the location of HPr in the steterospecific complex is shown as a blue ribbon. A ribbon diagram of the stereospecific complex with EIN in red and HPr in blue is also shown on the right hand side of the panel. Adapted from (17).

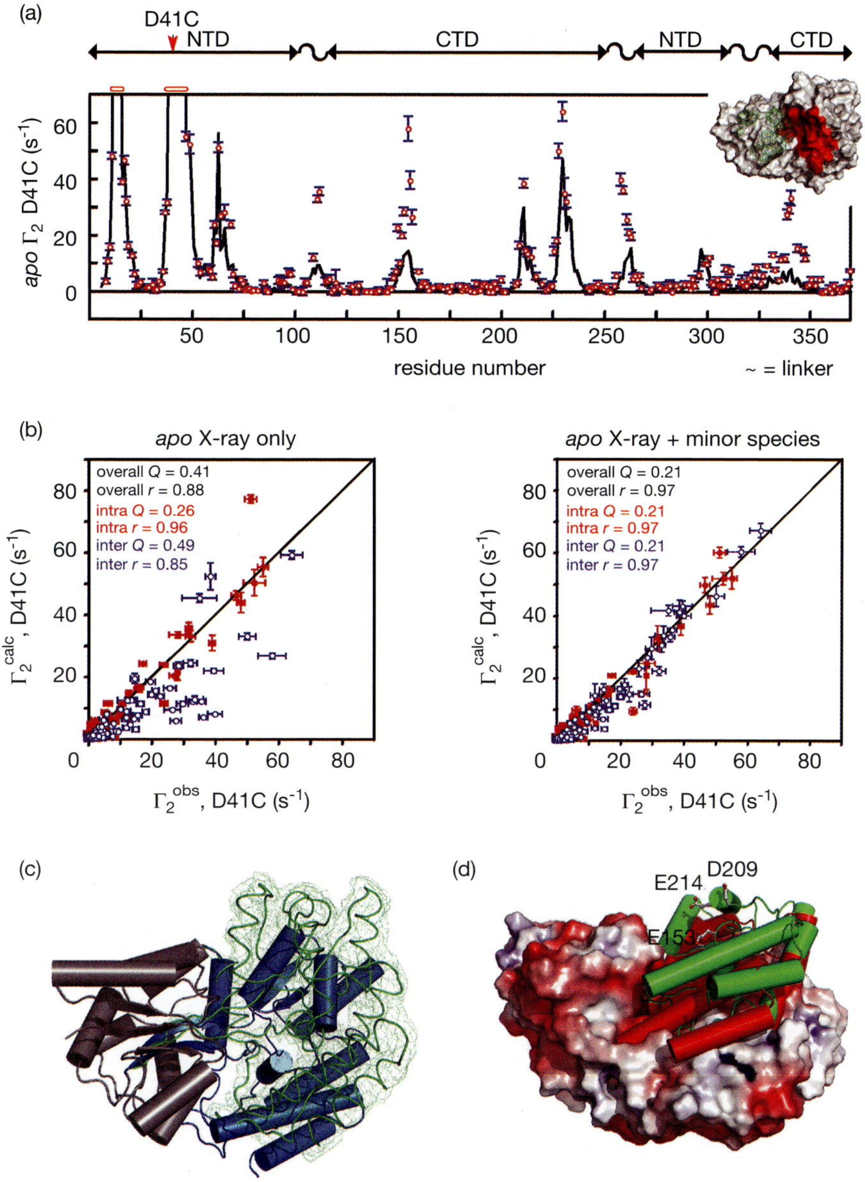

Figure 3.5 Open-to-partially closed transition in *apo* MBP. (*a*) Comparison of observed PRE profiles (red circles) with back-calculated values obtained by ensemble paramagnetic probe refinement against the intermolecular PRE data for the nitroxide spin-label at D41C (black line). (*b*) Comparison of observed versus calculated Γ_2 rates for the D41C data obtained with the X-ray *apo* MBP structure alone (*left panel*) and upon inclusion of a minor species at an occupancy of 5 percent (*right panel*). (*c*) Equilibrium mixture of the major open (blue cylinder) and minor partially-closed (green smoothed backbone trace with reweighted backbone atomic probability map shown as a light green mesh) forms of the CTD of *apo* MBP with the NTDs of the two species superimposed. (*d*) Structural comparison of the CTD of the minor partially closed state of *apo* MBP (green cylinders) and *holo* MBP (red cylinders) with the open form of *apo* MBP shown as a molecular surface color-coded according to electrostatic potential. Adapted from (18).

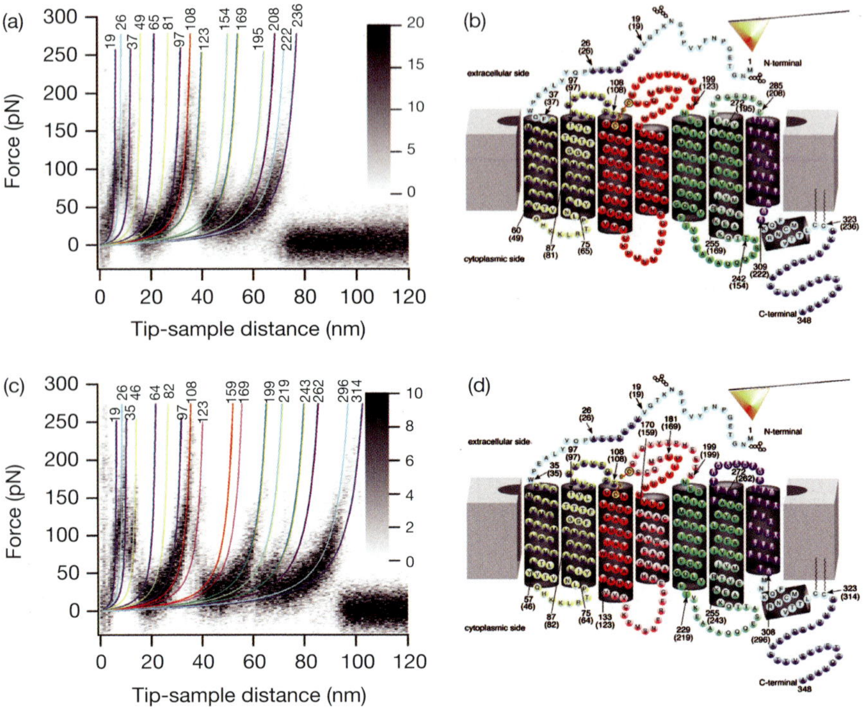

Figure 5.13 Single-molecule force spectroscopy curves recorded on native ROS disc membranes. (A) The superimposition of several force curves ($n = 42$) enhanced common features among the curves. Major (black) and minor peaks (grey shaded) were fitted using the WLC model as above. (B) Secondary structure of rhodopsin mapped with structural segments that constitute the unfolding barriers in SMFS. (C) The superimposition of several F-D curves ($n = 25$) enhanced the common features among the curves. Major (black) and minor peaks (grey shaded) were fitted using the WLC model as above. The analysis of force curves revealed the strengths and locations of molecular interactions established within rhodopsin (D). (D) Secondary structure of rhodopsin mapped with structural segments observed by SMFS. Arrows locate the starting and the end of each structural segment. Numbers in brackets are the amino acid values obtained from WLC fitting and the numbers without brackets are the corresponding residue number in the rhodopsin sequence. Equally colored WLC fits of force peaks (A and C) and structural segments correlate to each other.

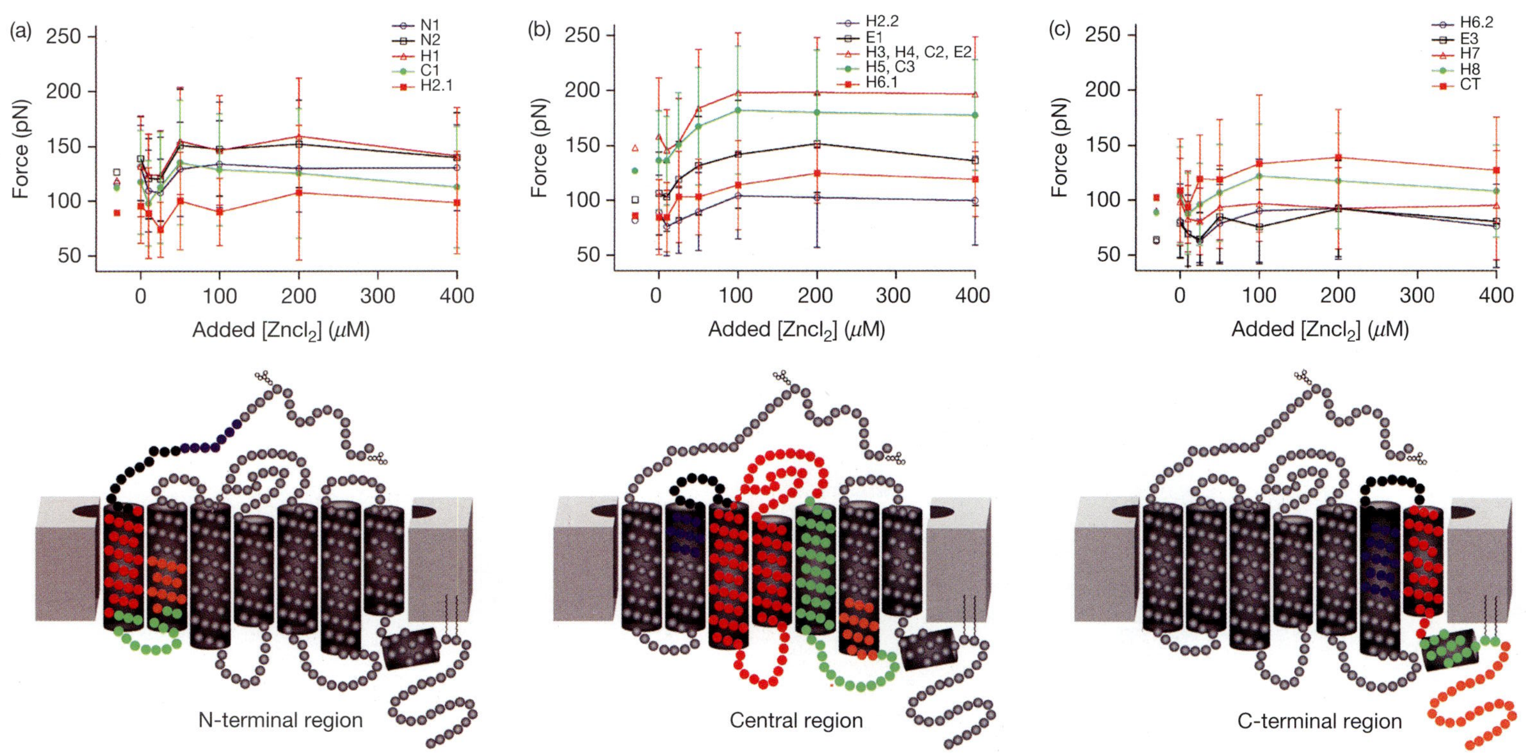

Figure 5.15 Rhodopsin unfolding forces depend on $ZnCl_2$ concentrations. SMFS was performed at $ZnCl_2$ concentrations of 0–400 μM, and the average unfolding force was determined for each structural segment (A–C). Data points below 0 μM added $ZnCl_2$ on the *x*-axis denote average unfolding forces from experiments done in the presence of 1 mM EDTA. The colors of the titration curves correspond to the structural segments shown in the lower panel. Error bars are the standard deviations of the unfolding forces of each structural segment at different Zn^{2+} concentrations.

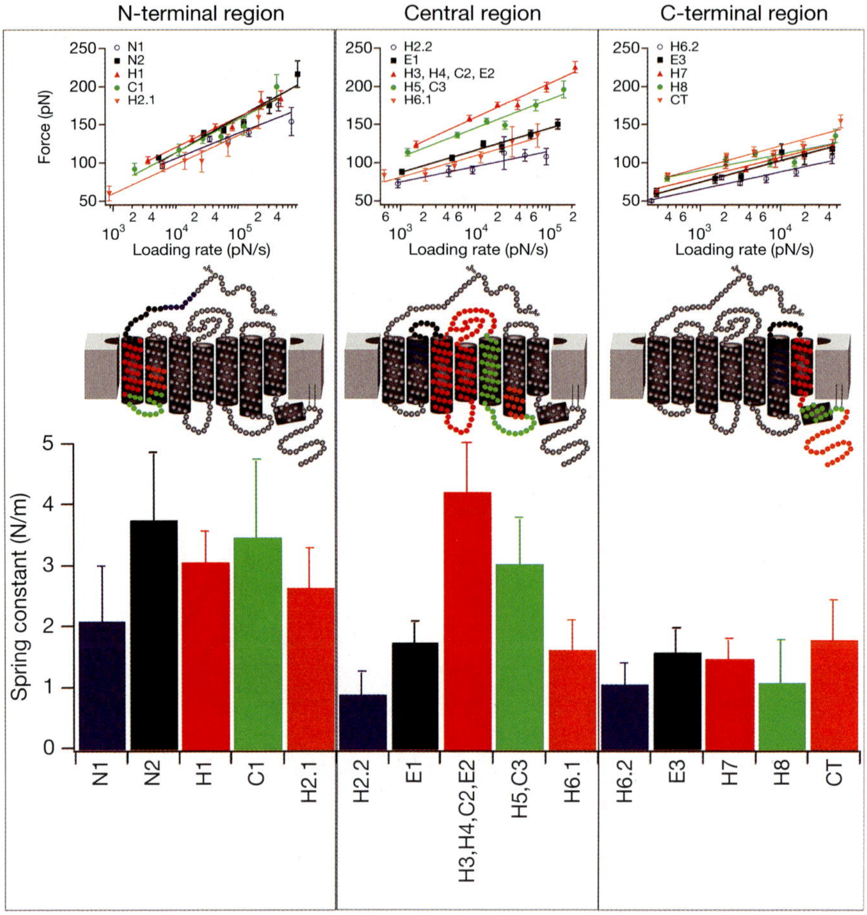

Figure 5.18 Rigidity of structural segments of rhodopsin as determined by DFS. Rigidity of structural segments (lower panel) was estimated by calculating the spring constant, κ, from x_u and $\Delta G_u^{\ddagger}$ values derived from DFS experiments (upper panel). The structural segment, H2.2, in helix II is the most flexible region ($\kappa = 0.9$ N/m), and the structural segment, H3,H4,C2,E2, constituted of helices III and IV with their connecting cytoplasmic loop CII and the extracellular loop EII is the most rigid region ($\kappa = 4.2$ N/m). Colors of fits to the DFS data (upper panel) and bars in the graph (lower panel) correspond to those of structural segments outlined in the secondary structures (middle panel).

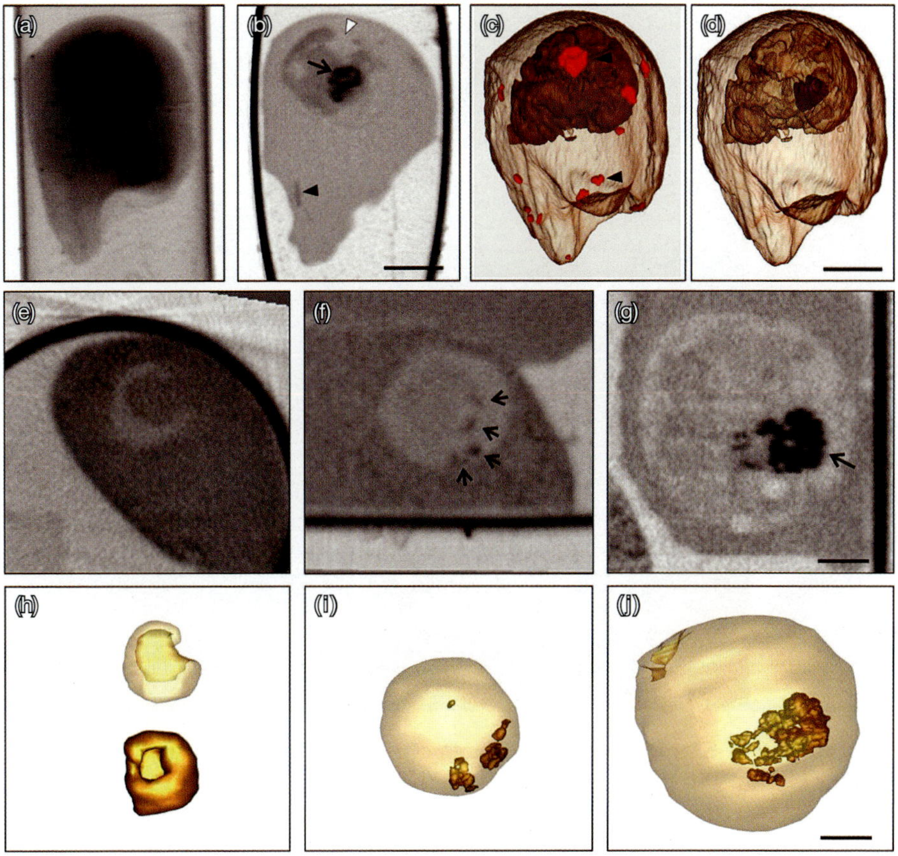

Figure 8.2 X-ray tomography of intact P. falciparum-infected RBCs Parasite-infected RBCs were mounted in glass capillaries then snap frozen and imaged using the XM-2 transmission x-ray microscope during rotation through 180°. (A) A projection view of a trophozoite-infected RBC in which the outline of the host cell and the parasite are visible. (B,E–G) Virtual sections through tomograms generated for parasites at different stages of development. Invaginations are observed in the surface of some of the trophozoites (B, white arrowhead). X-ray-dense hemozoin crystals are observed in more mature stage parasites (B,F,G, arrows). Membrane features are observed in the RBC cytoplasm (B, black arrowhead). (C,D) The RBC surface is rendered in translucent pink and the parasite surface in solid (C) or translucent (D) brown. Membrane features in the RBC cytoplasm are rendered in red (C) and hemozoin is depicted in purple (D). (H–I) Rendered models of an early ring (H, side and top views), a late ring (I) and a mid trophozoite (J) showing the parasite surface (translucent beige) and the hemozoin crystals (gold). The cupped parasite surface is illustrated in solid gold in H (bottom object). Scale bar, A–D 2 μm; E–J 1 μm. (Reproduced from Hanssen, Knoechel et al. 2011).

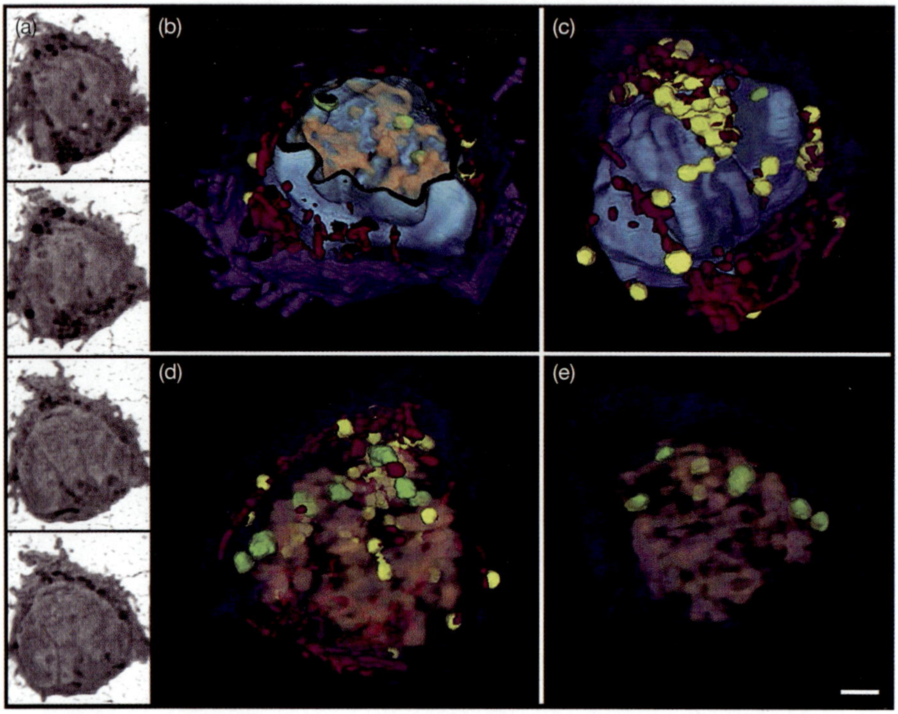

Figure 8.3 Soft X-ray tomography of a lymphocyte (T-cell). (A) Four orthoslices from the tomographic reconstruction of a cryo-immobilized T-cell. (B–E) Segmented volumes color-coded for identification of internal structures. (B) Reconstruction of a cryo-immobilized T-cell. (B) Cut-away view showing several filopodial extensions on the cell surface (blue) and the internal cytoplasm (purple) containing organelles such as mitochondria (magenta) and highly absorbing vesicles (yellow). A cut-away view of the nuclear envelope (cyan) reveals the chromatin (salmon) and nuclear bodies (green) within. (C) The cell surface has been removed, showing the 3D organization of cytoplasm and organelles with respect to the intact nuclear envelope, which is highly folded. (D) View of the cell interior in which the nuclear envelope has been made transparent revealing the 3D organization of the chromatin with respect to cytoplasmic organelles. (E) View of the cell showing the 3D organization of chromatin and nuclear bodies with respect to cytosol. Scale bar, 1.0 mm. (Reproduced from Larabell and Nugent 2010).

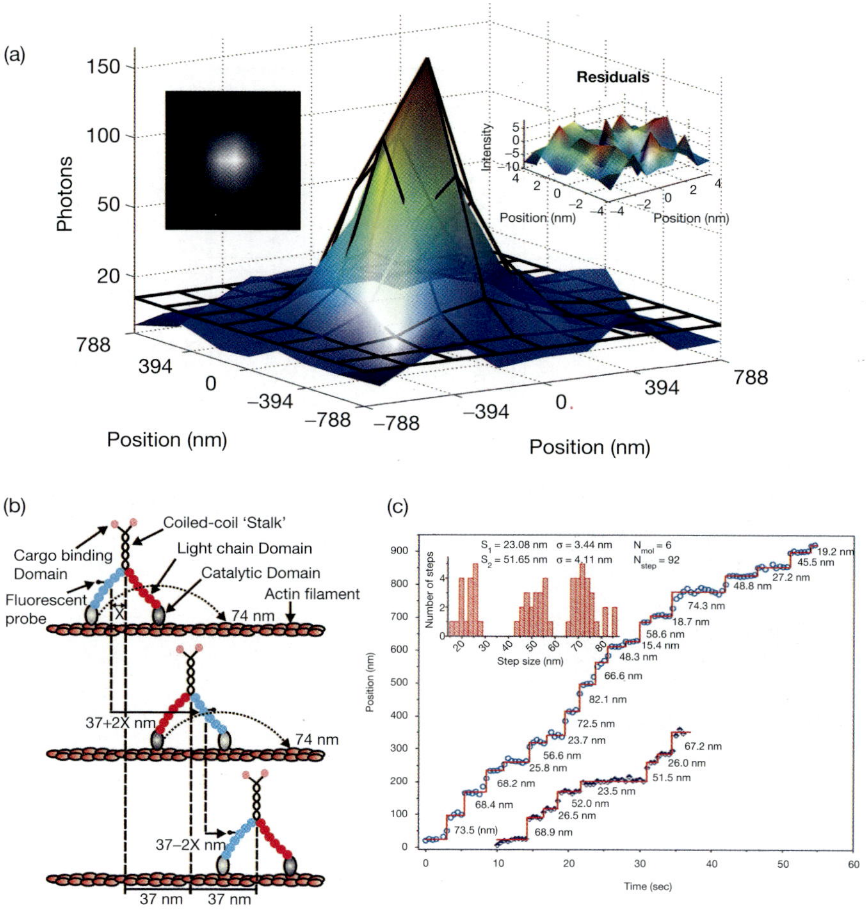

Figure 9.2 FIONA and its application. (a) CCD image of a single Cy5 labeled RNA (left inset) and its corresponding PSF fit to a 2D Gaussian. Residuals from the fit, which depict the deviation of data points from the fitted curve, are also shown (right inset). (b) The hand-over-hand translocation model of myosin V as evident from single molecule step traces of myosin V labeled with bifunctional rhodamine on one of the twelve calmodulin light chains (47). On average, the protein alternates between 52 nm and 23 nm steps as it moves. This stepping pattern suggests that X, the distance along the direction of motion between the center of the coiled-coil stalk and labeled calmodulin, is 7.5 nm, i.e., the third calmodulin from the top is labeled. Sometimes, the speed of myosin V translocation is much faster than the frame rate of image acquisition, due to which, some 52 nm and 23 nm steps combine to yield ~74 nm apparent steps. A histogram of six myosin V molecules taking a total of 92 steps, representing the frequency of occurrence of specific step-sizes, is shown in the inset. (Panel c reproduced with permission from (47)).

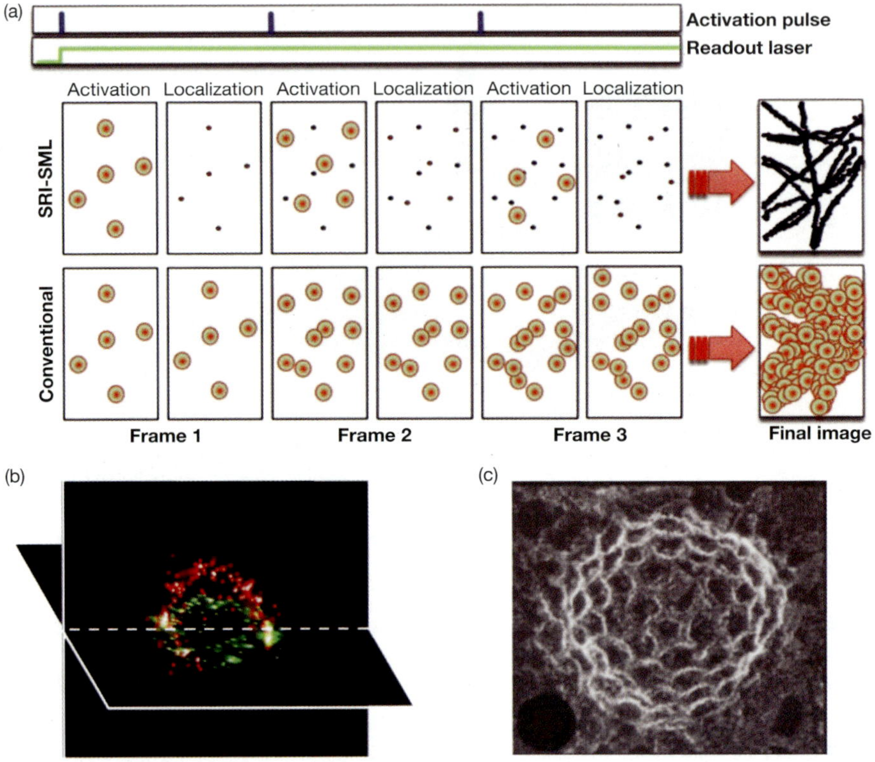

Figure 9.3 Comparison of SRI-SML imaging with conventional fluorescence and electron microscopy. (a) SRI-SML images have a much higher resolution compared to conventional fluorescence microscope images. This resolution enhancement is particularly striking in the case of microtubule filaments. As illustrated, due to overlapping of diffraction-limited spots, conventional imaging cannot resolve single filaments or capture the intricate criss-crossing pattern clearly observed in SRI-SML images. Note that, unlike SRI-SML, conventional imaging only requires one frame per image, thus the final image is not built up upon individual activation events. Individual frames are illustrated here merely for the sake of comparison. (b) and (c) Comparison of images of clathrin coated pits obtained using 3D STORM and electron microscopy (Reproduced with permission from (98, 133)).

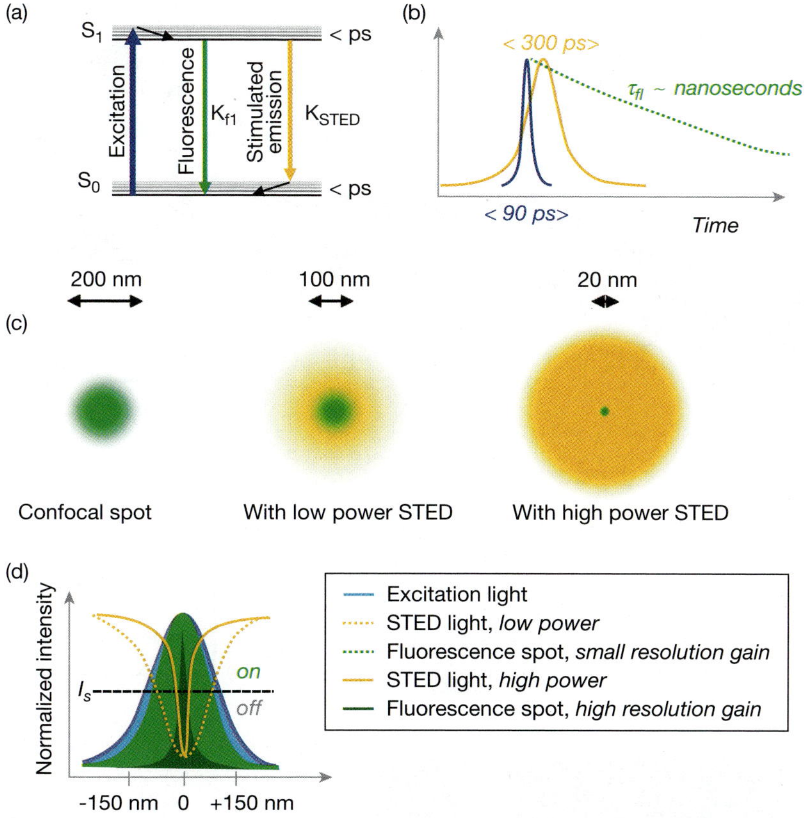

Figure 10.1 The principle behind STED microscopy. (A) Jablonksi diagram of the transitions of a fluorescent molecule, S_0 is the ground state, S_1 is the first excited singlet state, each with multiple vibrational sublevels. Absorption of a blue photon excites the dye molecule to S_1, from where it returns back to the ground state S_0 either via spontaneous emission of a green photon (fluorescence) or via stimulated emission by an orange photon. In fact, stimulated emission sends the molecules to an elevated (vibrational) sublevel of the ground state, from which they rapidly (in under a picosecond) decay to the final ground state. (B) The rationale behind using pulsed laser sources: After excitation by a brief pulse of blue laser light (90 picoseconds) the dye molecules normally relax back to the ground state via spontaneous emission of a green photon. The average time the molecules spend in the excited state is characterized by the lifetime (τfl), which is on the order of a few nanoseconds. If the STED pulse is timed so that it is delivered well within the lifetime τfl the fluorescence can be effectively inhibited. Provided the STED pulse is intense enough, most molecules will return to the ground state via stimulated emission instead of via spontaneous emission (fluorescence). Because of the short lifetime of the excited state, it makes sense to use synchronously pulsed lasers for STED microscopy, instead of continuous wave (cw) lasers, where the excitation and STED light is not synchronized. (C) Confining the fluorescence in space: the size of the green spot dictates the spatial resolution for a confocal microscope (*left*). The orange annulus (or "doughnut") is the focal intensity distribution of the STED laser. While it begins to suppress the fluorescence on the outer edge of the green spot for moderate STED laser intensities (*middle*), it drastically confines it at high STED laser power (*right*), leading in theory to a singularity of molecular size. (D) Normalized intensity profiles of the blue excitation laser, the STED laser at low (*dotted line*) and high (*solid line*) intensity, and the fluorescence distribution, before (*light green*) and after (*dark green*) strong STED suppression.

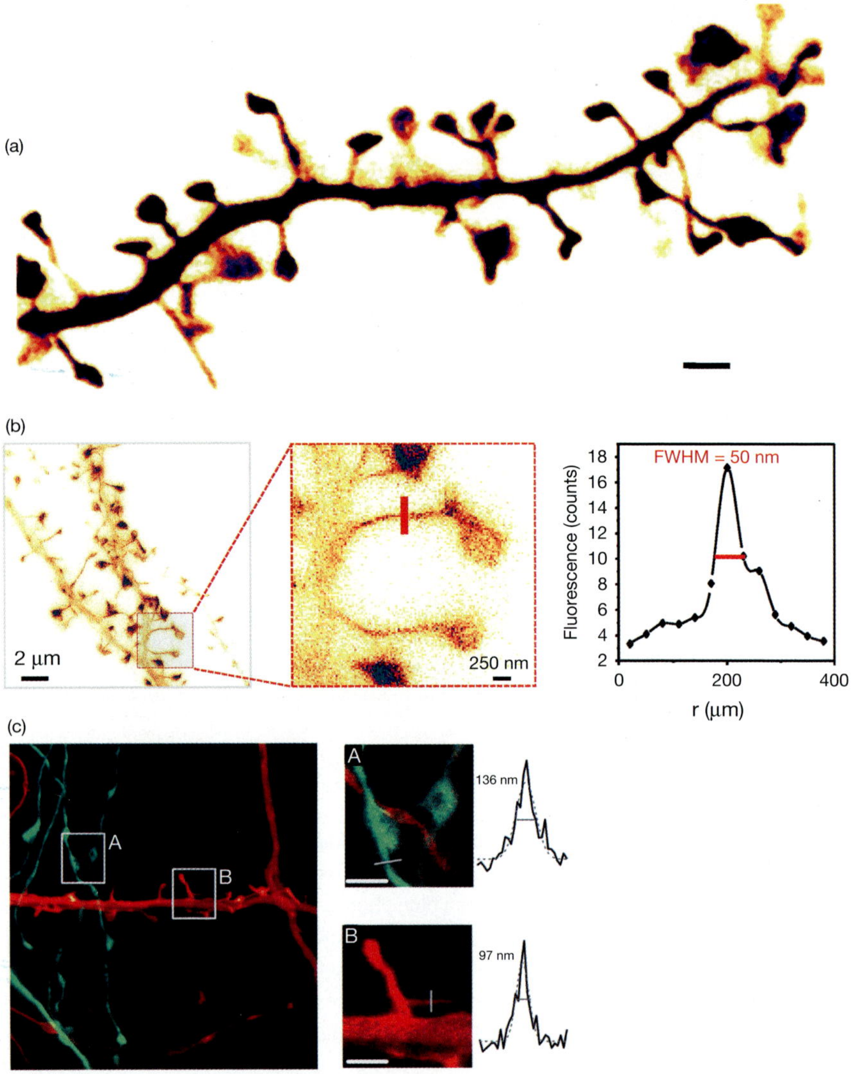

Figure 10.3 (A) Maximum intensity projection of a stretch of dendrite of a pyramidal neuron filled with YFP based on a 3D stack of STED images (scale bar, 1 micron). (B) Imaging the cytoskeleton of synapses at the nanoscale: Distribution of the cytoskeletal protein actin inside a dendritic spine imaged by STED using YFP-*lifeact* as label for actin. (reprinted from Urban et al, *Biophys J*, 2011). (C) Two-color STED microscopy of axons and dendrites using fluorescent proteins (GFP, YFP) and organic dyes (Alexa Fluor 488, Calcein-green) as volume markers (scale bar, 1 micron). (reprinted from Tonnesen et al., *Biophys J*, 2011)

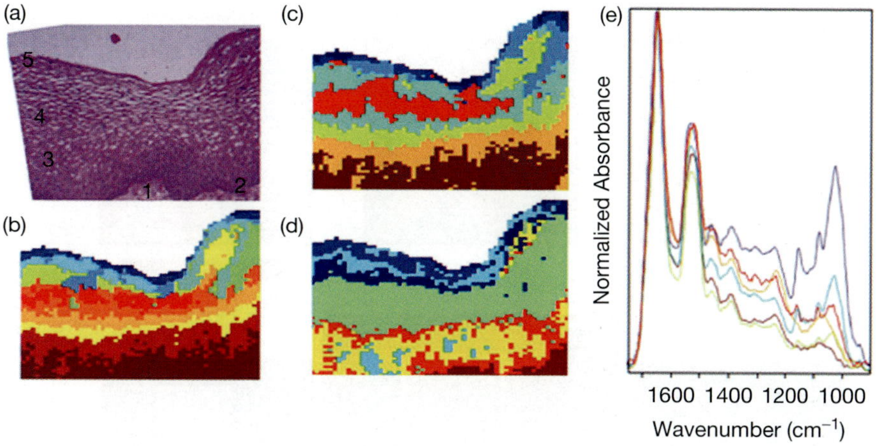

Figure 12.5 Sensitivity of different mid-infrared spectral regions as a diagnostic tool. (A) Photomicrograph of a section of cervical squamous epithelium from patient diagnosed with LSIL. The section was stained (H&E) after FTIR data acquisition. 1: Connective tissue; 2: basal layer; 3: parabasal layer; 4: intermediate layer; and 5: superficial layer. (B) 9-cluster maps using the entire mid-IR spectral range (1800–800 cm^{-1}). (C) 9-cluster maps using the "protein region" (1740–1470 cm^{-1}). (D) 9-cluster maps using the "glycogen/nucleic acid region" (1200– 1000 cm^{-1}). (E) Mean cluster spectra from six of the seven clusters color coded to enable correlation with 2C. Note: the individual colors in panels B–D are assigned by the algorithm for each data set and are not transferable between separate analyses (35).

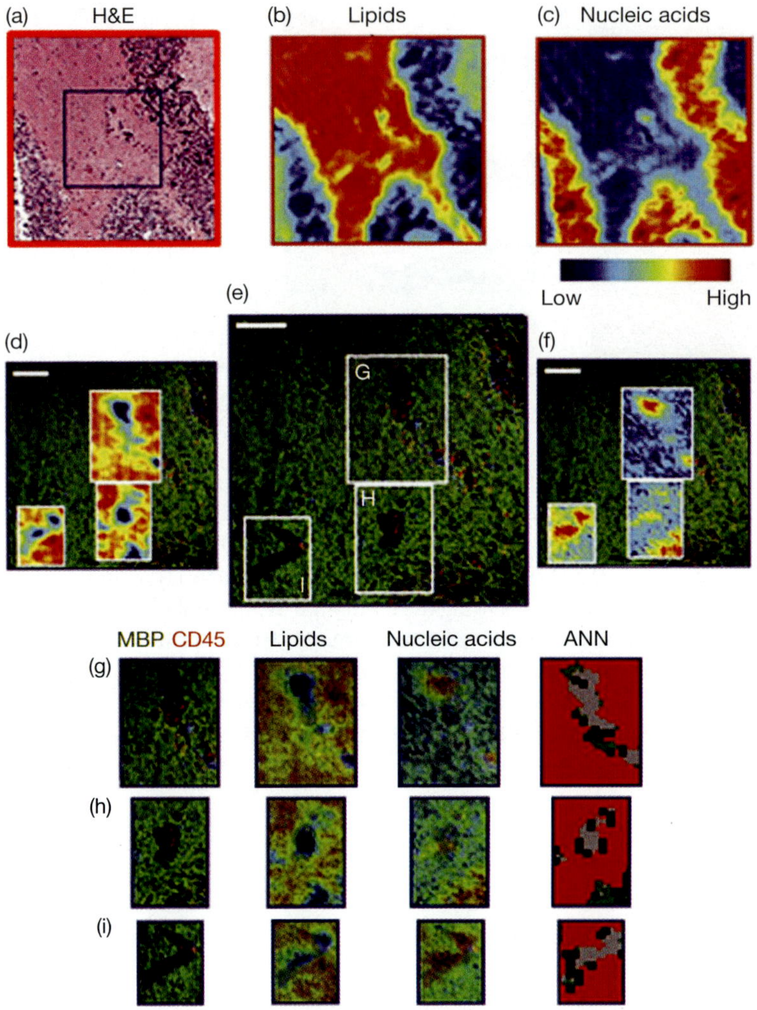

Figure 12.9 Synchrotron-based chemical mapping detects early experimental autoimmune encephalomyelitis (EAE). (A) Adjacent cerebellum section stained with H&E; blue square is the area studied by infrared FPA imaging; white square indicates area examined using the synchrotron FTIR. (B, C) FPA images of cerebellum showing the relative concentration of lipid and nucleic acid derived from FPA spectral data. Scale bar = 100 μm. (D–F) Areas interrogated with the synchrotron light source overlaid onto areas stained with immunofluorescence of 56 percent optical opacity. Note that the infrared analysis showing myelin organization and cellular infiltrate following staining with an anti-MBP antibody (green) and DAPI (blue). Corroboration of the pathological nature of the microlesion provided by ANN classification (G–I). These areas are classified in a similar manner to that seen in the advance stages of EAE; red classifies the cerebellum white matter, green classifies the granular layer, gray classifies the lesion spectra, and black pixels are unclassified spectra. Scale bar = 50 μm (88).

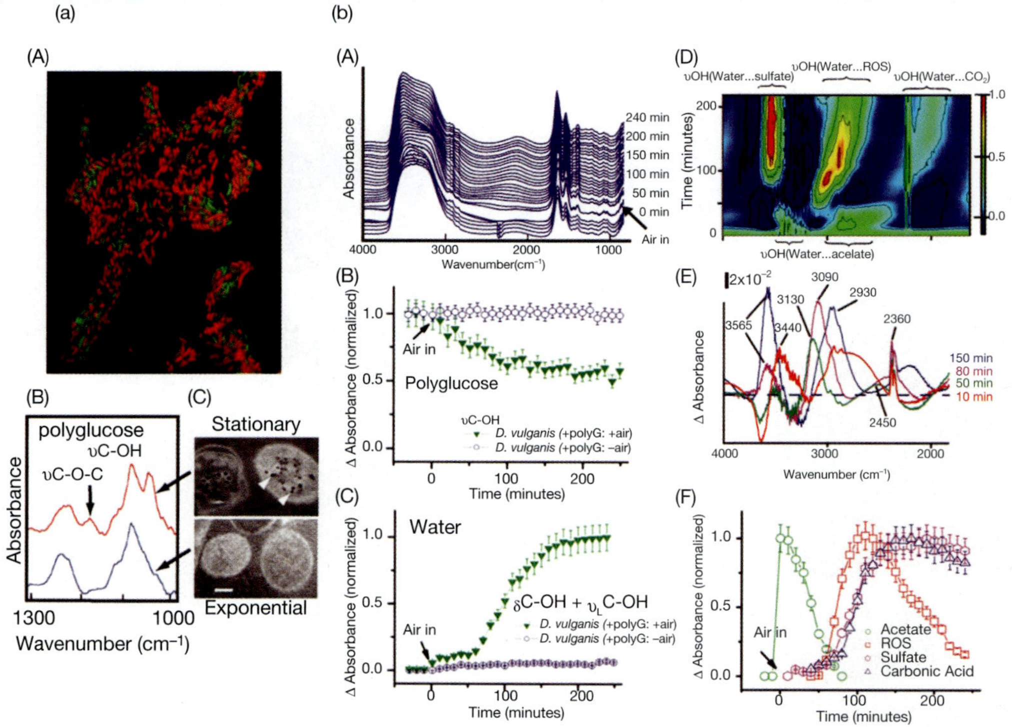

Figure 12.15 (A) Survivor sulfate reducing bacteria. (a) Viability of SRB after exposing to moist air for eight hours. Live (green) and dead (red) were assessed using reagents in LIVE/DEAD *Bac*Light Bacterial Viability kit. (b) Typical infrared absorption spectra of stationary-phase (red) and exponential-phase (blue) *D. vulgaris*. (c) Transmission electron microscopy (TEM) images of thin sections post-stained by the periodic acid thiosemicarbazide-osmium (PATO) method show intracellular polyglucose granules in stationary-phase but not exponential-phase *D. vulgaris*. (B) Oxidative Stress and adaptation. (a)Typical real-time SR-FTIR spectra of *D. vulgaris* during the transition from an anaerobic to aerobic environment. Since all spectra are derived using air as a reference, the abrupt change in the spectral feature at ~2348 cm^{-1} is associated with the presence of atmospheric CO_2. Typical time-course of infrared intensity (normalized by the maximum value) of (b) polyglucose content and (c) water. (d) Time versus frequency contour plot of SR-FTIR time-difference spectra in the hydride–OH dominated stretch region. (e) Snapshots of time-difference spectra for selected different time points. The dashed line marks zero difference absorbance. (F) Typical time-course of infrared intensity (normalized by the maximum value) of H-bonded species (115).

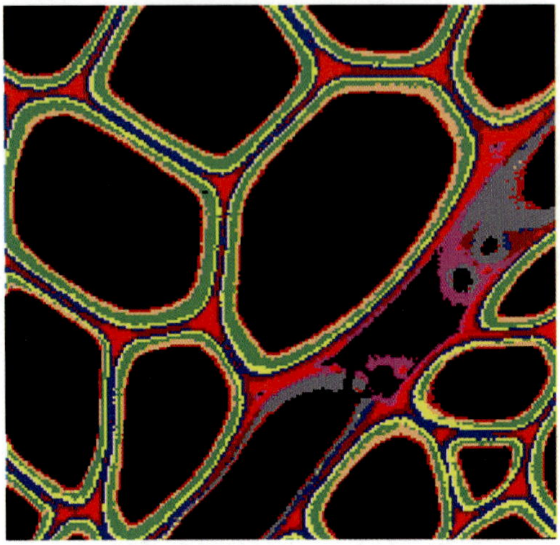

Figure 13.9 Reconstructed image of a xylem cross-section of poplar wood obtained by preprocessing and k-means cluster analysis.

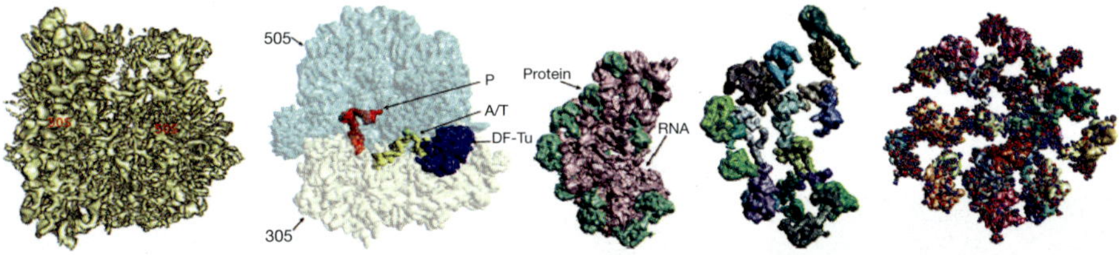

Figure 15.11 Segmentation of a 6.4 Å map for the 70S ribosome (72). (left to right) Original 3D map; segmentation into five primary subunits: 30S, 50S, EF-Tu, P site, and A/T site; further segmentation of the 30S subunit into protein and RNA parts; fine segmentation of the 30S proteins; segmentation of the 50S proteins after fitting with the associated X-ray structure.

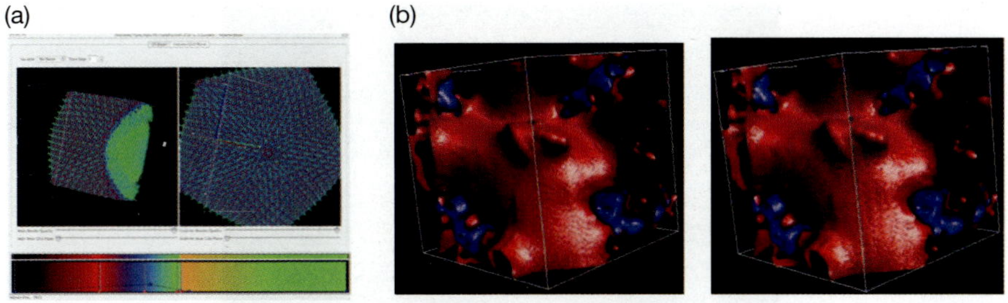

Figure 15.14 High-performance quantitative visualization in *VolumeRover*. (a). *VolumeRover* rendering of the chillo iridescent virus (EMDB 1580) where the right panel shows a low-resolution rendering of the virus and the left panel shows the rendering of the portion of the density map which is inside a moving rover. (b) Comparison of uncompressed (left) and compressed (right) volume rendering shows no visible difference despite 41:1 compression ratio (155).

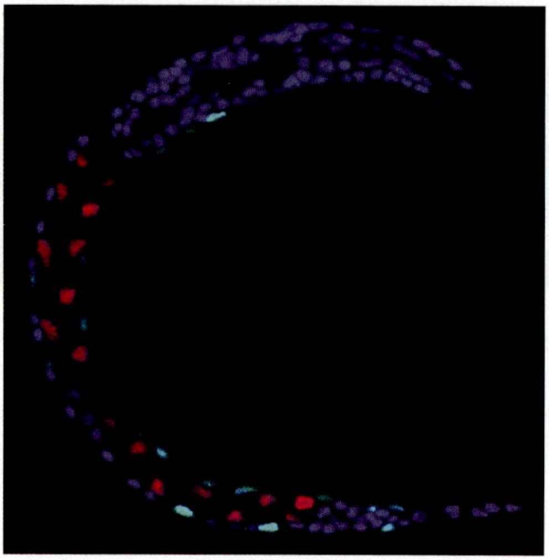

Figure 16.1 A typical confocal laser scanning microscopic image of *C. elegans* (L1 stage), where different nuclei are specifically labeled using fluorescent molecules. Blue: DNA-binding DAPI staining that labels all 558 nuclei of the animal. Green: enhanced Green Fluorescent Protein (eGFP) labeling of 81 body wall muscle cells and 1 depressor muscle cell. Red: mCherry labeling for specific cells for gene expression analysis.

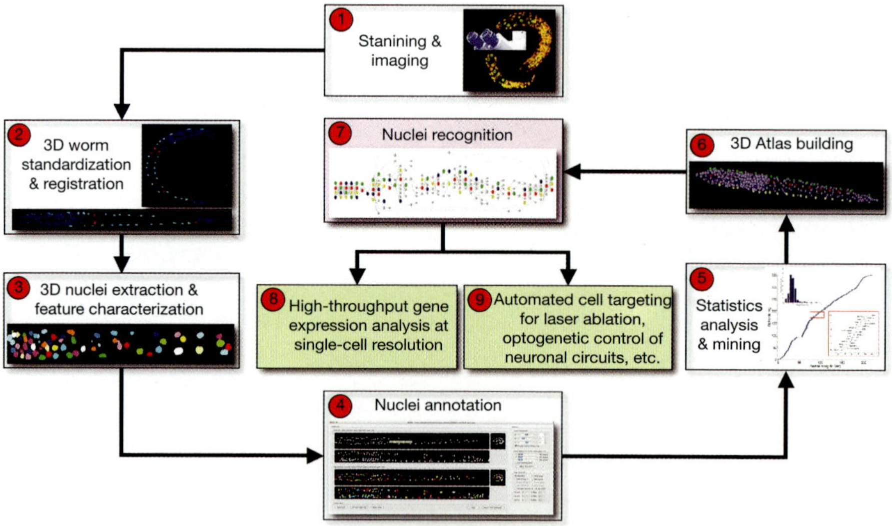

Figure 16.2 The major steps (boxes ◆~⌘) in building a 3D digital cell atlas of L1 larval stage of C. elegans and the applications (boxes ❀~❝) of this atlas.

order	cell name	mean location	neighboring nuclei	spacing	size	shape	...
332	INDRI	X: 81.52 ± 2.04 Y: 2.66 ± 0.46 Z: 1.84 ± 0.33	HYP7ABPRAAPPPP, INDL1, BWMDR11, HYP7ABARPAAPPP, BWMDL10	3.82 ± 1.14 3.84 ± 0.93 4.35 ± 0.76 4.50 ± 1.29 4.67 ± 1.09	9005 ± 993 (px) 1.5005 ± 0.71 (μm)	Aspect ratio 1.52 ± 0.21 ...	...
...	...	...	...	...	...	...	...

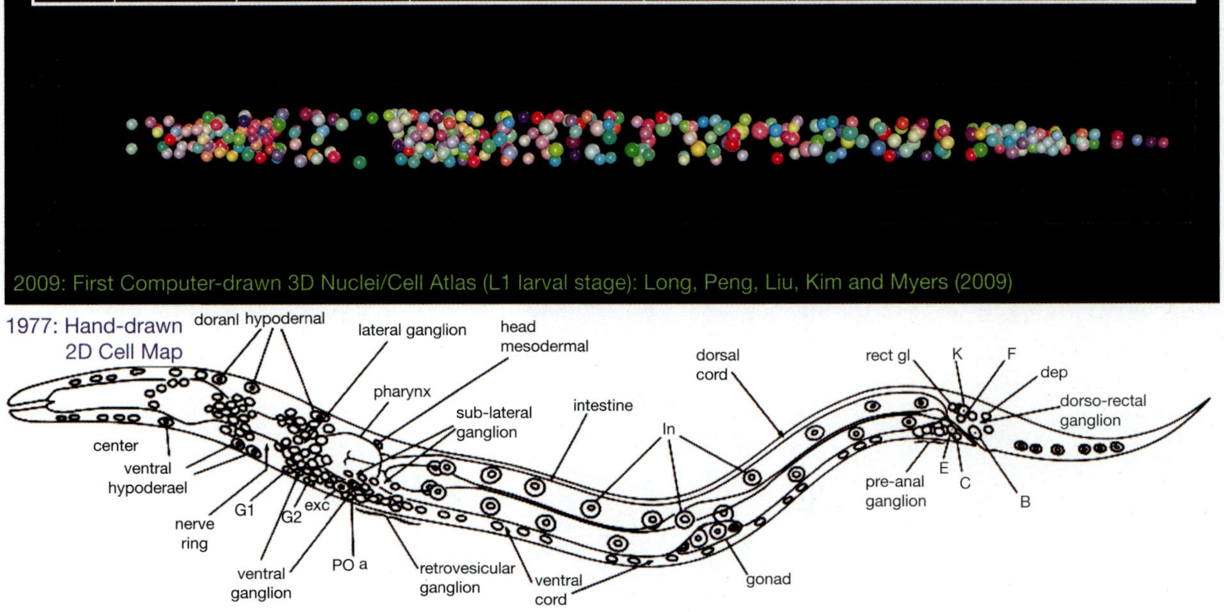

Figure 16.3 Comparison of the 3D digital cell atlas of *C. elegans* (Long, et al, 2009) and an earlier 2D version (Sulston and Horvitz, 1977). In the 3D atlas, each colored sphere corresponds to one nucleus; also for clear visualization, only a subset of nuclei are displayed.

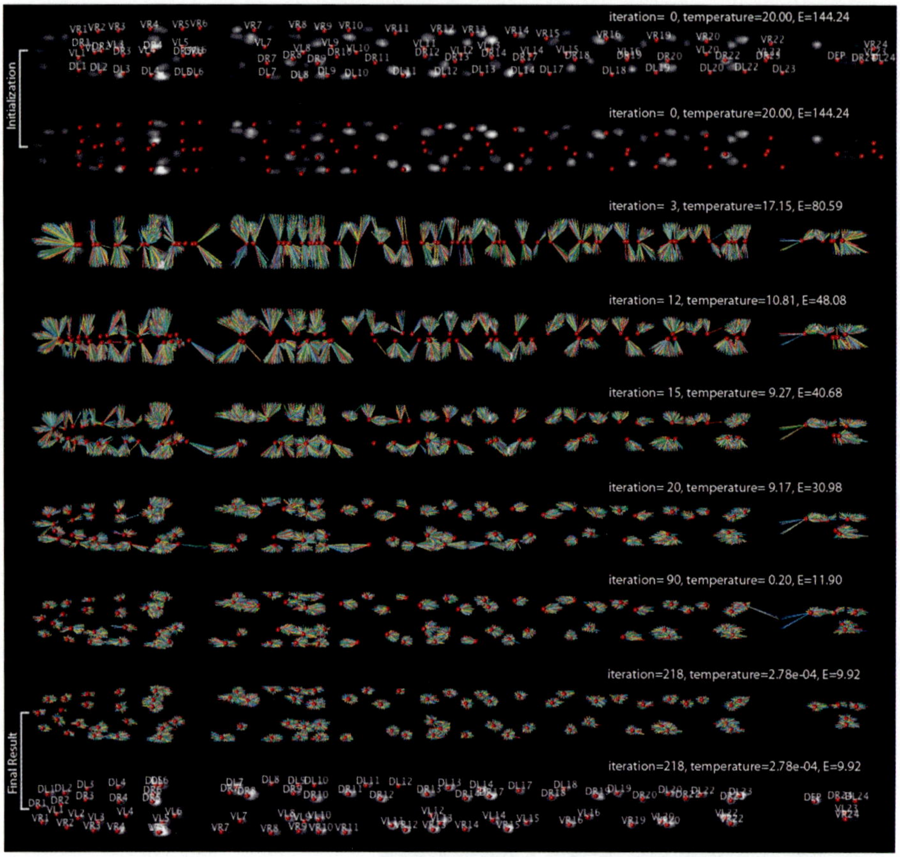

Figure 16.4 Simultaneous cell recognition and segmentation. In each image, the red spheres show the deformed atlas; the segmented pixels and their corresponding cells in the atlas are connected by lines with different colors. The first row shows the original image overlaid with the initial atlas; other rows show the results of several intermediate steps of the iterative optimization, with which the atlas of cells (body wall muscle cells shown here) deform to the optimal locations and the foreground image voxels are automatically classified to (and thus segmented) each of these cells. The energy values shown are normalized using the total number of image foreground voxels. When we produced this figure, the image intensity was enhanced for better visibility. The surface rendering of the segmented regions can be seen in Figure 16.5.

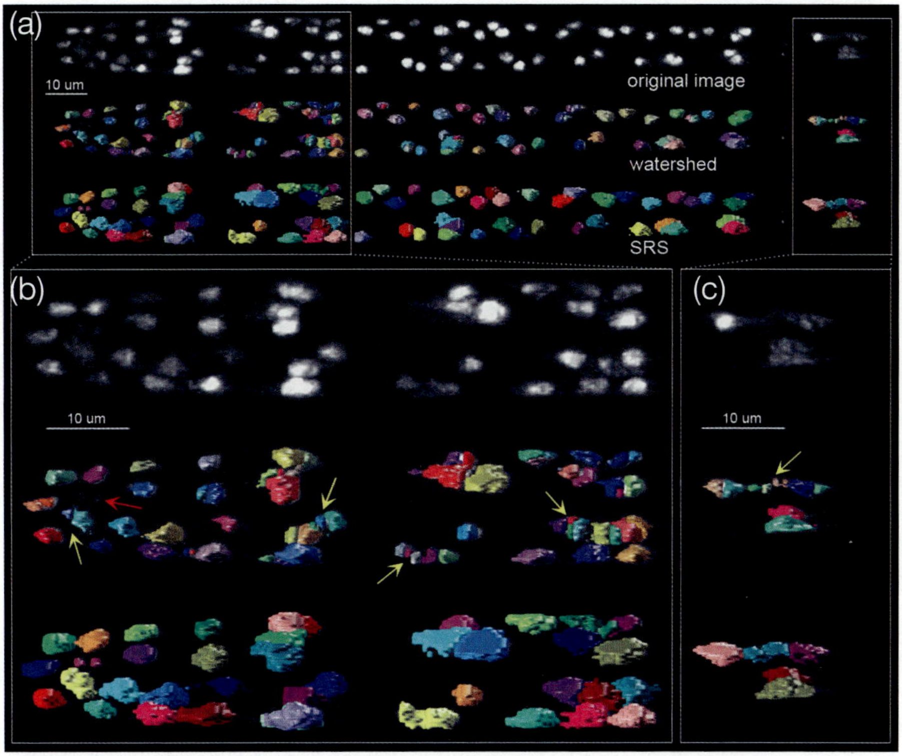

Figure 16.5 Comparison of the segmentation results of watershed and SRS. The same test image in Figure 16.4 was used. The color-surface objects indicate the different segmentation regions. In (c), yellow arrow: over-segmentation; red arrow: under-segmentation.

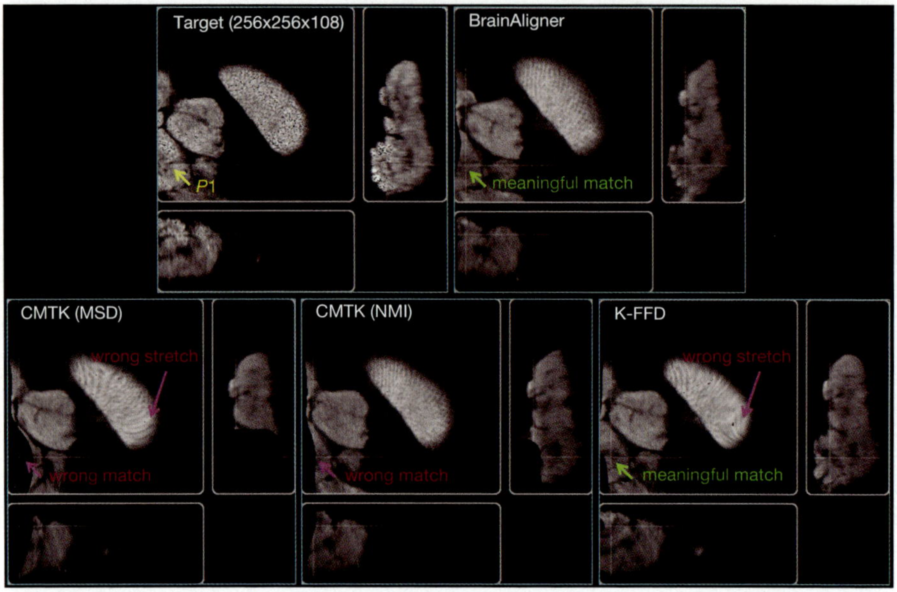

Figure 16.8 Comparison of the aligned images produced by BrainAligner and several previous methods. (a) and (b) show two typical locations we checked the aligned images. Compared to the target image, BrainAligner was able to produce smoothly aligned subject image. The irregular stretching of the warped image is one of the main problems of the previous alignments (38; Qu and Peng, unpublished data). Irregular stretching is due to wrong matching at various locations.

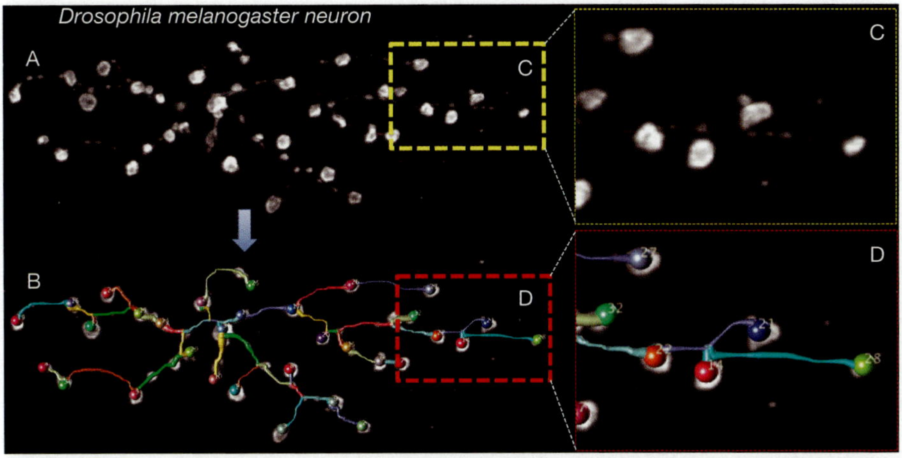

Figure 16.9 A challenging case of fruit fly neuron for tracing. This neuron has broken and often fuzzy structures and low single-to-noise ratio. (A) A single lamina neuron of fruit fly (courtesy of G. Rubin lab) along with the 3D reconstruction. In the input image (A) and (C) the fruit fly neuron has broken and punctuated neurites. We applied the V3D-Neuron to trace the neuron morphology (B) and (D) and visualize as a color-coded neurite model.

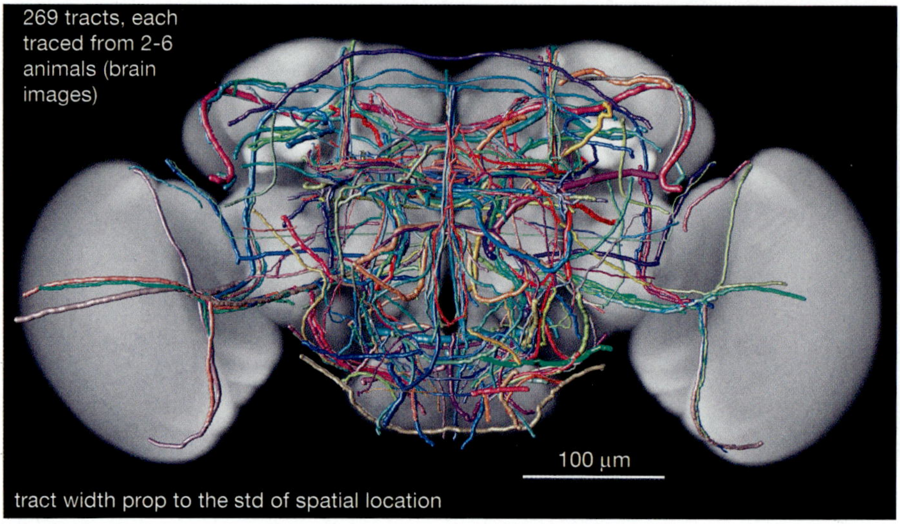

Figure 16.10 A 3D atlas of neurite tracts reconstructed from aligned GAL4 patterns. The neurote tracts were reconstructed from a number of brain images, which were all non-rigidly aligned in 3D to best match a standard target brain (Peng, Chung, et al, 2011). The neurite tracts are randomly color-coded for better visualization. The width of each tract equals the respective spatial deviation.

Weiss, D. (2000). *Computed Tomography Based on Cryo X-ray Microscopic Images of Unsectioned Biological Specimens*. Ph.D., Georg-August University of Göttingen.

Weiss, D., G. Schneider, et al. (2000). "Computed tomography of cryogenic biological specimens based on X-ray microscopic images." *Ultramicroscopy* ***84***(3–4): 185–197.

Weiss, D., G. Schneider, et al. (2001). "Tomographic imaging of biological specimens with the cryo transmission X-ray microscope." *Nuclear Instruments & Methods in Physics Research Section A-Accelerators Spectrometers Detectors & Associated Equipment* ***467***(Part 2): 1308–1311.

Winkler, H. and K. A. Taylor (2006). "Accurate marker-free alignment with simultaneous geometry determination and reconstruction of tilt series in electron tomography." *Ultramicroscopy* ***106***(3): 240–254.

Zhang, J., R. E. Campbell, et al. (2002). "Creating New Fluorescent Probes for Cell Biology." *Nat Rev Mol Cell Biol* **3**(12): 906–918.

9

Breaking Abbe's Law

Super-Accuracy and Super-Resolution Fluorescence Microscopy based on Single Molecule Detection

SETHURAMASUNDARAM PITCHIAYA,
JOHN R. ANDROSAVICH, AND NILS G. WALTER

INTRODUCTION

The cell is a complex mixture of single molecules that function in a highly organized and concerted manner to forge life. Conversely, each molecule within the cell is, in principle, a local reporter on the nature and dynamics of its surroundings. Yet studies of biomolecules, such as probing their structure, measuring their reaction kinetics, or determining their intracellular distribution, have mostly been conducted on the ensemble level, reporting the "average" structure, mechanism, or cellular organization. Important information is hidden within this average, as a seemingly homogenous population of molecules can, in fact, be composed of surprising heterogeneity. In addition, the flow of genetic information is spatially and temporally organized through the transport of single mRNA molecules to discrete regions in an organ (1) or even a single cell (2, 3) to synthesize proteins in the region where they are needed most. For example, as seen by live-cell fluorescence microscopy, a significant fraction of the mRNA encoding the cytoskeletal protein actin localizes to the leading edge of the cell. Upon tracking the diffusion of single actin mRNA molecules, multidirectional movements with net displacement vectors pointing towards the leading edge of the cell were observed, thus rationalizing the average cellular organization of the mRNA (4).

Single-molecule fluorescence microscopy (SMFM) generally reveals rare events and transient species or intermediates on a reaction pathway, like those found in protein (5, 6) or RNA folding landscapes (7, 8), without the need to perturb or synchronize the molecules of the system to generate a signal. Such information is easily lost by averaging in regular ensemble measurements. According to the ergodic theorem, the time average of any observable obtained from visualizing a single molecule is equivalent to the population average, provided the molecule is visualized over a sufficiently long period of time. Yet, single-molecule studies have also shown that some RNA and protein enzymes exhibit so-called static heterogeneity with conformationally and functionally distinguishable molecular species that only slowly exchange on an experimental time scale (7, 9–12). The types of information obtained from SMFM thus provide unprecedented insights into the functional mechanisms, dynamic properties, and subcellular localization of biomolecules, with only limited perturbation of the underlying biology.

Two major aspects of microscopy, accuracy and resolution, define the quality of imaging in a quantitative manner. On one hand, accuracy describes the radius of the circle (or sphere in the case of three-dimensional localization) within which it is certain that an object is localized (i.e., the error associated with localizing an object). On the other hand, resolution is the distance below which two closely located objects cannot be distinguished as separated from one another. In 1873 Ernst Abbe observed the failure of a lens-based optical microscope to distinguish two features closer than half of the wavelength of the illumination light on the lateral plane. This imposes a theoretical limit on the resolution of fluorescence microscopy of 200–300 nm (using visible, ~500-nm illumination light). This relationship is more accurately described by:

$$\Delta d = \frac{\lambda}{2n\sin\alpha} \tag{1}$$

where Δd is the distance between two closely located spots, λ is the wavelength of light used for illumination, n is the refractive index of the medium in which the lens is working, and α is the half-angle of the maximum cone of light that enters or exits the lens. The obtainable axial resolution as described by:

$$\Delta z = \frac{\lambda}{2n\sin^2\alpha} \tag{2}$$

is almost threefold higher than the resolution limit in the lateral dimension (~500–600 nm).

For over a century, Abbe's law (also referred to as "Rayleigh's resolution limit") was considered a fundamentally limiting factor in intracellular imaging. This problem is exacerbated for single molecules: their intracellular density is often much higher than that commonly used in SMFM, leading to severe overlap of their images. Given that a typical human cell contains ~10^{10} protein molecules (13) of ~2.5-nm radius and the cell by itself has a volume of ~$4\text{x}10^{12}$ nm^3, a conservative estimate of ~20 percent of the total cell volume is filled with just proteins. Factoring in other molecules, such as lipids, nucleic acids, inorganic ions, and small metabolites, the density of molecules, in general, within a cell is astoundingly high. Even in such a crowded environment biomolecules often diffuse at speeds of several nanometers per second, frequently colliding with other molecules and changing direction. Studying single molecules under such conditions necessitates the need for higher accuracy and resolution than that obtained from conventional optical microscopy.

This chapter will provide an overview of the currently available ultra-high-resolution microscopy methods that can be used to visualize single molecules labeled with fluorescent probes in vitro and in vivo (the latter mostly in live cultured cells). These methods can be categorized as: i.) methods pertaining to localization of a few single molecules with nanometer-scale precision (super-accuracy imaging (14)); and ii.)

methods resolving many single molecules within a densely populated sample by localizing only a sparse subset of them at a time through repeated super-accuracy imaging (often termed super-resolution imaging (14)). Other single molecule techniques, such as manipulation by optical tweezers or imaging by atomic force microscopy (AFM), are beyond the scope of this chapter. A detailed description of AFM can be found in an accompanying book chapter and a broad overview of all single-molecule methods can be found in several recent books and reviews (15–17).

EXPERIMENTAL DESIGN IN SINGLE-MOLECULE FLUORESCENCE MICROSCOPY

The sensitivity required to detect single molecules is achieved by carefully selecting the sample concentration, instrumentation, and probe used to label the molecule. The concentration of molecules should be kept low enough, in the nanomolar to picomolar range, to distinguish one molecule from another. While controlling sample concentration in vitro is relatively easy, it presents a major challenge in vivo. Another important parameter to be optimized in SMFM is the signal-to-noise ratio (SNR) (i.e., the ratio of useful signal (photons) emitted by single molecules and unwanted, spurious background signal (noise)). Noise can arise from impurities in the sample, from fluorescent molecules in the sample that are not in focus, or from shot noise (dark counts) of the detector. From its definition, it is clear that having a high SNR is preferred, but several factors can markedly affect this value (as discussed below). A whole arsenal of solutions, in the form of improved optical configurations (illumination methods, light sources, and detectors) and fluorescent probes, is currently available and constantly being refined to overcome low SNR.

MODES OF ILLUMINATION

Schemes to illuminate a sample can be primarily divided into two categories: far-field illumination and near-field illumination. In far-field imaging, the distance between the objective lens and sample is at least an order of magnitude greater than the wavelength of light used. Conversely, in near-field imaging, the sample is separated from the illuminating focusing lens or fiber optics by a distance much smaller than the illumination wavelength. Additionally, the diameter of the lens or fiber is smaller than the wavelength of light used (typically ~80 nm) (18, 19). Such a configuration ensures that the light interacts with the sample before effects of diffraction come into play, leading to higher resolution. Near-field illumination requires very flat samples for the lens or fiber optics to scan over it, making it less useful for intracellular imaging. In the following, we focus on some of the more popular far-field illumination schemes used in SMFM (Figure 9.1).

Wide-Field Epi-Illumination

In wide-field epi-illumination schemes, light passes through the objective's optical axis, illuminating all molecules in the light path. Although a specific plane of the

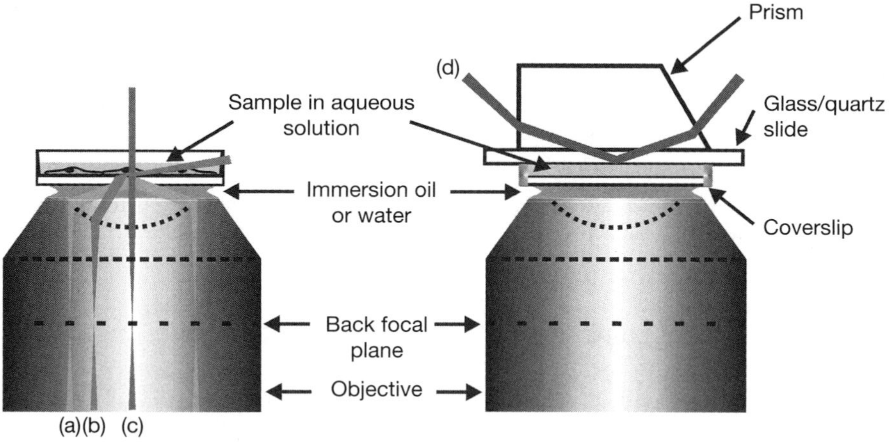

Figure 9.1 Wide-field illumination methods. In live cell imaging, typically, OTIRF (a), HILO (b), or wide-field epi-illumination (c) is used. By simply moving the excitation light away from the optical axis, one can switch from wide-field epi-illumination to HILO or OTIRF. In all three cases, the incident light and the fluorescent light pass through the same objective. By contrast, in PTIRF (d) the incident light is directed to the sample through a prism. The fluorescent light passes through the depth of the sample and the coverslip before being collected by the objective. As the sample has to be sandwiched between the slide and the coverslip, PTIRF is not suitable for live cell imaging, which demands frequent media change and precise control of O_2/CO_2 gas.

sample can be focused, molecules present in other focal planes also get excited, contributing significant background. This illumination method is suitable for SMFM only when bright molecules are sparsely distributed, such that background contribution from out-of-focus molecules is negligible.

Total Internal Reflection (TIR) Illumination

TIR was first used in 1981 by Daniel Axelrod in conjunction with fluorescence (TIR*F*) and fluorescence microscopy (TIR*FM*) to image the lipid probe 3,3′-dioc ta-decyclindocarbocyanine (diI) in human skin fibroblasts (20). When light is transmitted through a medium of higher refractive index (n_i; e.g., glass) into one of lower refractive index (n_r; e.g., water) at incident angles (θ_i) greater than the critical angle given by $\theta_c = \sin^{-1}(n_r/n_i)$ (Snell's Law), light is totally reflected within the first medium. However, because of its wave properties, some light will be transmitted into the medium of lower refractive index in the form of a "standing" or "evanescent" wave. The intensity of this light field decreases exponentially from the interphase, establishing only a thin lamina of ~150-nm depth within the aqueous sample, effectively reducing background fluorescence from outside regions. TIR can be achieved by illumination through either a high numerical aperture (NA) objective (objective-type TIRFM, or OTIRFM) or a quartz prism that couples the incident light into the glass surface (prism-type TIRFM, or PTIRFM). In OTIRFM, the incident laser light is focused onto the back focal plane (BFP) of the objective so that it is away from and parallel to the optical axis, thus θ_c is achieved without needing an inclined light source. Since the incident light and the fluorescent light are focused by the same objective, scattering of laser light within the objective leads to some background in the image. PTIRFM avoids this background, but poses geometric constraints due to the greater challenge to couple the incident light at the right angle and the fact that both sides of the sample need to be optically accessed. TIRFM is especially useful in imaging surface immobilized molecules or, in the case of (intra)cellular imaging, the basal plasma membrane and ~100 nm of the adjacent cytoplasm.

Highly Inclined and Laminated Optical Sheet (HILO) Illumination

Although TIRFM provides image quality greatly improved over that obtained by traditional epi-illumination, it is restricted to imaging molecules near an optically transparent surface. HILO microscopy (HILOM), also termed variable angle epi-fluorescence microscopy (VAEM) or near-TIRFM, overcomes this limitation by greater penetration into the sample without significantly compromising the signal-to-noise ratio (SNR) (21, 22). Similar to OTIRF, the light source is focused onto the BFP of the objective but at a smaller radial distance from the optical axis so that the incident angle does not exceed the critical angle. This way, light is refracted into the sample at high inclination from the optical axis, thus illuminating a small lamina within the sample, the width of which is dependent on the diameter of the field of view (FoV) and the angle of incidence. The laminar width of the beam is maintained by using a field stop, essentially decreasing the FoV, to prevent excessive divergence of the incident beam. Due to the transmission of the refracted beam through the objective into the sample, as in epi- and narrow-field illumination, any movement of the objective along the optical axis results in focusing on a different plane in the sample along the z-dimension, enabling imaging in 3D (21, 22). The excitation beam path for epi-illumination, OTIRFM and HILOM are relatively similar so that a single microscope can be used to implement all three schemes.

Narrow-Field Epi-Illumination

In narrow-field epi-illumination, light is focused onto a small volume element within the sample, thus exciting only molecules in the excitation volume and not those outside of it. Confocal laser scanning microscopes work by this principle, wherein focusing of light to a specific volume is achieved by introducing pinholes (or apertures) in the excitation and emission paths. Higher resolution as compared to wide-field illumination can be achieved by decreasing the width of the pinholes, but this comes at the cost of lower photon collection. The method also suffers from slow sample readout rates as scanning an entire specimen occurs at one volume element at a time. A significant improvement in the scanning rate can be achieved by using spinning-disc or Nipkow-disc confocal microscopy. A disc consisting of several microlenses spins at typically 3,500 rpm to create virtual pinholes for confocal detection (23), essentially exciting several distinct regions of the sample simultaneously.

LIGHT SOURCES AND DETECTORS

Lasers are preferred over lamps for high-resolution optical microscopy because of their monochromatic, coherent, and collimated light. Laser power should be high enough to excite all molecules optimally without damaging the sample. The wavelength of light spans anywhere from the far ultra-violet (UV) to the near-infra-red (NIR) region of the electro-magnetic spectrum (350–1000 nm), depending on the nature of the fluorescent probe and sample used. However, wavelengths from the visible part of the spectrum (450–680 nm) are used most frequently, mainly for two reasons. First, the transmission properties of available optics are best in the visible range, and second, most well-documented fluorescent probes (fluorophores) have their spectral properties in this region of the electromagnetic spectrum. Moreover, UV light is not suitable for imaging living cells: it induces DNA damage and sometimes even apoptosis (24). IR lasers are slowly gaining popularity, however, because they scatter the least, thus significantly improving sample penetration depth.

Charge coupled devices (CCDs) and avalanche photodiodes (APDs) or photomultiplier tubes (PMTs) are most commonly used in SMFM, and depending on the type of imaging performed, one or the other is preferred. In wide-field illumination, CCDs are the detectors of choice, whereas for narrow-field illumination where photons from only a small region of the entire sample are collected, APDs or PMTs are used. CCDs contain a large matrix of photosensitive elements (or pixels) that simultaneously collect photons over the entire detector surface. Photon flux information at each pixel is stored as an electronic charge (photoelectron) that is first converted to an analog voltage by an amplifier, then digitized. Intensified (I-)CCDs and electron-multiplying (EM-)CCDs amplify the number of photoelectrons through an external amplifier and on-chip, respectively, such that the digital number is (marginally) linearly dependent on the number of photons collected, making intensity values semi-quantitative at the (sub-)pixel level. APDs and PMTs are point detectors sensitive enough to detect and accurately count single photons, but have to be used in laser scanning microscopes to generate images of two-dimensional samples, limiting their time resolution. The time resolution obtainable with CCDs is primarily limited by the frame rate (number of frames or images per unit time), whereas the spatial resolution is dependent on the number of pixels and the quantum efficiency (i.e., the fraction of detected versus collected photons). A typical CCD used in SMFM has 512 × 512 pixels. While CCDs with higher numbers of pixels are available (e.g., 1024 × 1024), they are often less sensitive due to lower quantum yields.

PROBES

The sample or the single molecule(s) under inquiry must be conjugated to a (fluorescent) probe or carry strong inherent fluorescence to be visualized. Synthetic dyes (Cy dyes, Alexa dyes, various forms of rhodamine dyes), fluorescent beads, quantum dots, and fluorescent proteins (FPs) are among the common fluorescent probes used in SMFM. FPs are the mainstay of intracellular fluorescence microscopy. The cloning and expression of fluorescent proteins fused to cellular targets have become routine practices. However, this method of labeling comes at the cost that the protein fusion must be expressed exogenously, often resulting in overexpression, compared to endogenous copies. Using inducible expression systems (25) is one of the many possible tricks that can be employed to mitigate overexpression. Non-fluorescent beads, especially ones of large size, can be imaged with high precision even using microscopes with simple optics, making them an attractive alternative to synthetic dyes and FPs in single particle tracking (SPT). Conjugating biomolecules to beads and to a certain extent FPs, however, comes with the caveat that the attachment of a bulky load could skew the molecule's function, localization, and/or diffusion. The development of various bio-orthogonal labeling strategies (26) and fluorogenic photo-affinity probes (27, 28) has made the specific labeling of single molecules with small fluorescent probes a viable option, even in living cells. A more detailed description and comparison of conjugation chemistries and the different organic dyes and FPs used in SMFM can be found in several reviews (29–31).

A fluorophore suitable for SMFM should have high brightness, favorable photophysical properties, and sufficient inertness so that the label does not interfere with the function of the molecule to be tagged. The first two criteria have a direct impact on the localization accuracy of single molecules. Brightness, a value of photon output that is calculated as the product of a fluorophore's extinction coefficient and quantum yield, should be high enough to delineate signal from noise. Undesirable photophysical processes, such as intersystem crossing (triplet state excitation) and photobleaching, markedly affect the quality and length of a single-molecule recording. Although the precise mechanism is unclear, molecular oxygen is thought to be primarily responsible for photobleaching via photo-oxidation of the fluorophore. Several chemical agents (e.g., cyclooctatetraene (COT), trolox, and 4-nitobenzyl alcohol (32)) and enzymatic oxygen scavenging systems (e.g., those containing glucoseoxidase and catalase (9) or protocatechiuc acid and protocatechuate-3,4-dioxygenase (33)) can be used to prevent fast photobleaching and quench triplet state excitation. Attaching multiple fluorophores per molecule can also increase brightness and fluorescence longevity, but it also increases the risk of rendering the molecule nonfunctional.

Imaging living biological samples presents a unique set of problems, including phototoxicity and autofluorescence. Fluorophores, in their excited states, react with molecular oxygen in the cell, resulting in the accumulation of phototoxic free radicals that can compromise subcellular compartments or even the entire cell's livelihood (34). Thus, a balance has to be struck between the excitation laser power (and wavelength) used and the time for which the sample is illuminated, especially for long time-lapse experiments. In addition, naturally fluorescent molecules present inside cells, such as NADH, FADH, and heme, absorb visible light and fluoresce, contributing to a high background level of autofluorescence during imaging. One way to circumvent autofluorescence and simultaneously enhance photostability of FPs is by using cell culture media that do not contain any fluorescent molecules, especially vitamins (35). Arguably the best solution is, however, to use fluorophore probes that absorb light in the far-red visible or NIR part of the spectrum where cellular components show minimal absorption (36, 37).

IMAGING SINGLE MOLECULES WITH NANOMETER SCALE LOCALIZATION ACCURACY

Even before the advent of single-molecule detection methods, there was a widespread interest in tracking the movement or diffusion of molecules, especially membrane proteins and motor proteins, in solution and in living cells. Ensemble methods to calculate molecular velocities and diffusion coefficients have existed in the form of fluorescence correlation spectroscopy, or FCS (38), and fluorescence recovery after photobleaching, or FRAP (39). Tracking the movement of individual molecules with high spatio-temporal resolution seemed like a logical extension.

During the late 1980s, three different groups independently tracked the displacement of single beads over time by high-speed video microscopy both in vitro (40) and within living cells (41, 42) with ~10–100 nm lateral accuracy and sub-second temporal resolution. Around the same time, the optical absorption spectrum of single pentacene molecules was recorded in p-terphenyl crystalline matrices at liquid-helium temperature (43). Single-molecule fluorescence detection of cryo-cooled pentacene molecules soon followed (44, 45). In an effort to enhance resolution and extend single molecule detection to the third dimension, Van Oijen et al. (46) used a combination of confocal microscopy and spectrally selective imaging (SSI) to localize single pentacene molecules embedded in p-terphenyl crystals. SSI involves sequentially exciting individual pentacene molecules of distinct absorption maxima (caused by matrix effects), so that each molecule can be localized independent of the presence of another within a densely populated, otherwise un-resolvable region. The abilities to track beads with high accuracy and to visualize single fluorescent molecules with high resolution, together with a continual expansion of the available repertoire of optical instrumentation, have converged over the past two decades into what is now termed super-accuracy and super-resolution imaging.

A single fluorescent probe (or a cluster of closely positioned probes) can be considered to be a point source of emitted light whose image is, due to diffraction, spread over a much larger area on the detector than the actual size of the object. Such an image, referred to as a diffraction limited spot, appears as an airy disc, composed of a central bright spot surrounded by concentric rings of diminishing intensities. The intensity distribution of such an image can be mathematically described by what is termed a point spread function (PSF). By approximating the PSF with a simple two-dimensional (2D) Gaussian function, the intensity maximum, that is, the center of the Gaussian curve, can be localized with an accuracy similar in magnitude to the size of the fluorescent emitter, effectively breaking the diffraction barrier and Abbe's law.

FLUORESCENCE IMAGING WITH ONE NANOMETER ACCURACY (FIONA)

FIONA is a super-accuracy method that is used to localize and track single, typically fluorescently labeled, molecules with nanometer accuracy in the x-y plane (47). Fluorescence from single emitters is detected by a CCD and the intensity is plotted as a function of the spatial coordinates x and y (Figure 9.2a). The resulting PSF is fit with a 2D Gaussian function:

$$P_g(x,y;z_0,A,x_0,y_0,s_x,s_y) = z_0 + A\exp\left[-\frac{1}{2}\left[\left(\frac{(x-x_0)^2}{s_x}\right)+\left(\frac{(y-y_0)^2}{s_y}\right)\right]\right] \quad (3)$$

where z_0 is a constant term due to background fluorescence, A is the amplitude or maximal signal, x_0 and y_0 are the coordinates of the center of the distribution, and s_x and s_y are the standard deviations along the axes. The full width at half-maximum (FWHM) of the Gaussian depicts the variance of the PSF, and the (central) peak pinpoints the mean (Figure 9.2a). The variance divided by the number of photons collected results in the most simplistic form of the standard error of the mean (s.e.m.), or the precision by which the molecule can be localized (48, 49):

$$\left\langle (\Delta x)^2 \right\rangle = \frac{s^2}{N} \tag{4}$$

where Δx is the error in localization, $\langle(\Delta x)^2\rangle$ is the s.e.m., s^2 is the variance and N is the number of photons. The s.e.m. also depends on other important parameters, including the detector noise, background noise, and the effective pixel size (49):

$$\left\langle (\Delta x)^2 \right\rangle = \frac{s^2 + \left(\frac{a^2}{12}\right)}{N} + \frac{8\pi s^4 b^2}{a^2 N^2} \tag{5}$$

where a is the effective pixel size, which is simply the CCD pixel size divided by the overall magnification used, and b represents noise of any form. This equation shows that greater localization accuracy can be achieved by increasing the number of photons collected, decreasing the noise, and/or reducing the effective pixel size. High (1.2–1.65) NA objectives and back-thinned CCDs are typically used to collect as many photons as possible with minimal detector noise. Decreasing the effective pixel size requires a corresponding increase in magnification, which in turn reduces the number of photons collected per unit square area of the detector, significantly compromising the SNR. Therefore, a localization precision of 1–2 nm can be achieved under ideal conditions by collecting 5,000–10,000 photons using a CCD with effective pixel size of 80–120 nm. The same principles can be applied to each frame of a movie (or series of multiple frames), tracking the movement of a single molecule with ~1-nm accuracy, hence the name FIONA. Depending on the frame rate at which the movie is recorded, a temporal resolution of up to 1 ms can be achieved (47).

Perhaps one of the most convincing proof-of-concepts for FIONA, and a spearhead for the broader adoption of single molecule techniques, originated from the investigation of motor protein movements. The mechanism by which motor proteins translocate on a cytoskeletal "path" was under debate for several years. Myosin V is a dimeric-motor protein that contains a catalytic ("head") domain, a light chain "arm" that is composed of six calmodulins, and a cargo binding domain on each of its monomers that is held together by a coiled-coil "stalk" (Figure 9.2b). One study (50) postulated that kinesin, another motor protein, moves over microtubules by an "inchworm"-type mechanism, where one of the arms always leads, while another study (51) hypothesized a hand-over-hand mechanism, in which the two heads alternately move past one another. Yildiz et al. (47) resorted to FIONA to resolve the controversy. They used a myosin-light-chain-domain labeling strategy similar to that used by Goldman and coworkers (51). Rhodamine-labeled calmodulin was added to myosin V monomers so that each dimeric motor protein contained a single fluorophore (i.e., one fluorophore per two arms). The labeled arm is expected to take alternative 37-2X nm and 37+2X nm steps if it translocates by a hand-over-hand mechanism, where X is the distance of the fluorophore from the stalk. For a labeled calmodulin closest to the head, X would be ~18.5 nm, and hence, the motor should take alternating 74 nm and 0 nm steps. Alternatively, if the motor moves by an inchworm type mechanism, the step size should be a constant ~37 nm. Not only could the authors convincingly show that the protein moved via a hand-over-hand mechanism by plotting the position of the fluorophore with time, they could also distinguish between molecules that had the labeled calmodulin at different positions with respect to the stalk (Figure 9.2c). Similarly, FIONA has also been used to track the path taken by kinesin as it "walks" over microtubules in vitro (52).

MULTI-LABEL IMAGING AND PARTICLE TRACKING

Single-Molecule High-Resolution Imaging with Photobleaching (SHRImP) and Nanometer Localized Multiple Single Molecules (NALMS)

In an effort to extend the use of FIONA to densely populated samples in which the PSFs overlap, both Paul Selvin's group (53) and Norbert Scherer's group (54) developed a method that exploits the quantized photobleaching behavior of single fluorophores to perform FIONA localization, and termed it SHRImP and NALMS, respectively.

The principle of SHRImP/NALMS can be explained by considering a sample that contains two molecules A and B whose PSFs overlap ($PSF_A + PSF_B$). If both fluorophores photobleach, the signal reduction in a plot of intensity over time will typically be in the form of two consecutive steps, each step representing the photobleaching of a single fluorophore, since it is highly unlikely for multiple fluorophores to photobleach at the same time. This stepwise photobleaching can be used to count the number of fluorophores with overlapping PSFs (55, 56). Once A has photobleached, the image only contains PSF_B, which can now be localized with high accuracy using FIONA. The resultant image can then be subtracted from the original image ($PSF_A + PSF_B$) to derive PSF_A, another single PSF to which the FIONA can be applied, thus localizing both molecules with high accuracy in different images of a movie. The same concept can be applied to a sample that contains more than two overlapping PSFs, enabling localization of molecules that are separated by as low as ~10 nm (53, 54).

Gordon et al. (53) used SHRImP to resolve double-stranded (ds) DNA molecules each labeled with one Cy3 at both 5′ ends. Upon imaging immobilized biotin-tagged DNA molecules of different lengths (30–50 bp) with TIRF, they deduced the expected length for each molecule of 3.4Å per base pair. Qu et al. (54) used NALMS to arrive at a similar conclusion, but in addition to using short dsDNA also used long λ-DNA hybridized to multiple Cy3-PNA conjugates to resolve up to ~20 serially overlapping PSFs, each separated by ~7 nm.

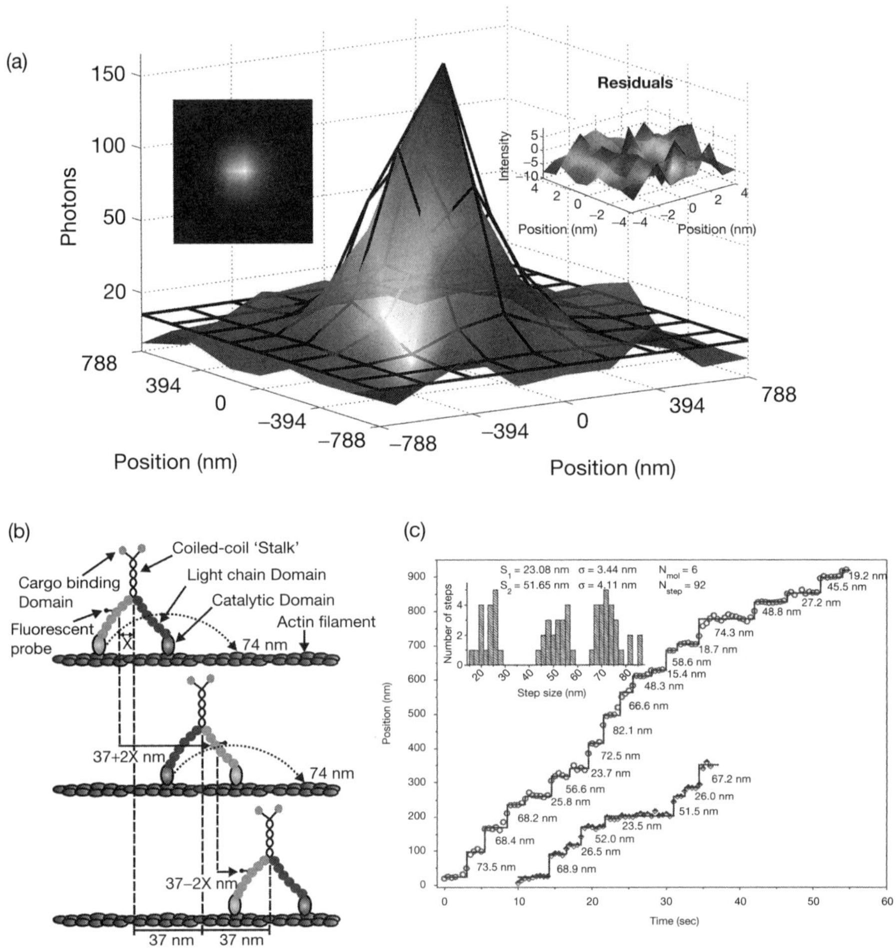

Figure 9.2 FIONA and its application. (a) CCD image of a single Cy5 labeled RNA (left inset) and its corresponding PSF fit to a 2D Gaussian. Residuals from the fit, which depict the deviation of data points from the fitted curve, are also shown (right inset). (b) The hand-over-hand translocation model of myosin V as evident from single molecule step traces of myosin V labeled with bifunctional rhodamine on one of the twelve calmodulin light chains (47). On average, the protein alternates between 52 nm and 23 nm steps as it moves. This stepping pattern suggests that X, the distance along the direction of motion between the center of the coiled-coil stalk and labeled calmodulin, is 7.5 nm, i.e., the third calmodulin from the top is labeled. Sometimes, the speed of myosin V translocation is much faster than the frame rate of image acquisition, due to which, some 52 nm and 23 nm steps combine to yield ~74 nm apparent steps. A histogram of six myosin V molecules taking a total of 92 steps, representing the frequency of occurrence of specific step-sizes, is shown in the inset. (Panel c reproduced with permission from (47)). See insert for color version.

Single-Molecule High-Resolution Colocalization (SHREC)

SHREC resolves overlapping PSFs originating from molecules labeled with spectrally distinct fluorescent dyes. Unlike SHRImP/NALMS, SHREC does not depend on photobleaching of fluorescent probes so that the observation window for imaging can be significantly extended (57). However, it requires the precise mapping of images from channels of distinct spectral range (color) onto one another, which is nontrivial due to chromatic aberrations of the optics. Spudich and coworkers therefore developed a mapping calibration for SHREC that uses fiduciary markers to measure the registration between the color detection channels. To this end, surface immobilized beads of sub-diffraction (100 nm) diameter labeled with both Cy3 and Cy5 (fiduciary markers) were imaged at such a low density that only a single bead was visible per FoV. This bead was then moved in 0.5-μm steps in the pattern of a 12×26 grid with a piezoelectric stage. The resultant 312 images were stacked to create a single image containing 614 regularly spaced PSFs (312 each in the Cy3 and Cy5 channels). Each PSFs in each channel was independently fit with a 2D Gaussian function and localized by a FIONA-like algorithm. Combining all locations thus determined, a local weighted mean transformation (58) was calculated to map any point in the Cy5 channel precisely onto its corresponding point in the Cy3 channel. The resultant map is used as a template to localize single dual-labeled molecules.

To benchmark their SHREC technique, the authors used Cy3/Cy5 doubly labeled dsDNA similar to that used in the SHRImP and NALMS studies and obtained very similar results (57). They then went on to use SHREC to concomitantly visualize the movement of both legs of myosin V in real-time. To this end, they exchanged the native calmodulines with a mixture of either Cy3- or Cy5 labeled calmodulins so that, on occasion, the two legs of a myosin would carry each a different fluorophore. As expected, they observed leg step sizes of ~74 nm, shifted by ~37 nm between the two color channels (57). More recently, Singer and coworkers have used a similar super-registration approach to track the transport of β-actin mRNAs through the nuclear pore complex (NPC). By achieving an unprecedented ~20-nm lateral accuracy at ~20-ms time resolution in living cells, they observed and measured the kinetics of a fast three-step mRNA transport process taking no more than 200-ms total time (59).

Of note, the multi-fluorophore super-accuracy imaging methods SHRImP, NALMS, and SHREC are closely related to and likely inspired the super-resolution imaging techniques, such as STORM and PALM discussed below. The primary goal of the former methods is to localize and track single or few molecules over time with high precision, whereas the latter tools aim to resolve ~10^3 times higher densities of single molecules over consecutive imaging cycles where only few molecules are turned on at any given time, accumulating data for a super-resolution image over time.

LOCALIZATION AND TRACKING OF SINGLE MOLECULES IN 3D

Initial efforts in SPT involved systems that were either two-dimensional in nature, such as labeled lipid molecules in the basal membrane of a cell, or the information obtained were 2D projections of objects moving in 3D. Verkman and coworkers developed a unique method for 3D SPT by introducing a cylindrical lens in the optical path of an epi-fluorescence microscope (60). The PSF of a particle in focus then is circular in shape and becomes larger and more ellipsoidal when the particle moves out of the focal plane. The major axis of the ellipsoid formed when the particle is above the focal plane is perpendicular to that formed when the particle is below the focal plane. Using this method, the authors localized red fluorescent beads of 100-nm diameter in living cultured cells with ~5-nm lateral and ~12-nm axial accuracy at ~500-ms time resolution (60).

Another powerful method for tracking particles in 3D is off-focus or defocused imaging. Developed by Florin and coworkers this method is based on the principle that the diameter and pattern of a PSF varies as the object is defocused (61). Particles that are slightly out-of-focus have a larger PSF diameter than when in focus and those even further from the focal plane give rise to a complex ring patterned PSF. The distance of the particle from the image plane can be deduced from the intensity pattern of the rings (i.e., the number, diameter, and relative intensity of rings). Schutz et al. used this method to record the trajectory of single potassium channels labeled with Cy5-tagged hongotoxin ligands in living T-lymphocyte cells (62). Although the method yielded lower axial accuracy than bead tracking (~40 nm versus ~3 nm), it was the first report of 3D SPT of molecules tagged with small fluorescent probes. Toprak et al. (63) used a variation of the above method to track phagocytosed beads with ~2.7-nm lateral accuracy and ~3.7-nm accuracy along the optical axis at 50-ms time resolution. In this technique, termed bifocal imaging, only 30 percent of the emitted light is focused, leaving the other 70 percent defocused, both of which are then split onto different regions of a CCD camera. The focused and defocused PSFs of a single particle are first mapped onto one another and then used to record the trajectory of the particle in the x-y plane and the z-axis, respectively.

Another approach for 3D SPT is based on the relative displacement of a pair of split images of the PSF onto distinct regions of the CCD. Appropriately named *Parallax* by Goldman and coworkers, either a wedge prism (64) or two closely spaced, parallel mirrors (65) are introduced at the back-focal-plane of the objective to view a fluorescent particle from two sides of the optical axis of the objective. With this optical configuration, the image of a particle moving in only the x-y plane is equally displaced in the x and y directions on either region of the CCD. By contrast, movements along the z-direction are reflected by the two images moving towards or away from each other in the x- (64) or y-direction (65). The authors localized fluorescent particles with ~3-nm accuracy along the optical axis, including glucose-transporter containing vesicles in living adipocytes and single myosins walking along actin filaments (65, 66).

Finally, Moerner and coworkers recently introduced a so-called 4*f* imaging system composed of two achromatic lenses and a reflective liquid crystal phase-only spatial light modulator that convolves the microscope image with a double-helical PSF (67, 68). As a result, the camera image obtained of a single fluorescent particle shows two spots whose relative x,y-positions will twist around each other as the z-position of the tracked particle changes (like the two backbones of a dsDNA viewed along the double-helical axis). The authors were able to localize brightly fluorescent 200-nm beads with ~5–10-nm precision in the z-dimension and weaker emitters with up to ~30-nm precision (68).

SUPER-RESOLUTION IMAGING BASED ON SINGLE-MOLECULE LOCALIZATION

The emergence of single-molecule techniques, such as FIONA, demonstrated the ability to break the classical optical diffraction barrier and provide super-accuracy resolution in determining a single fluorescent molecule's localization. Since a conventional wide-field image of a cell containing many fluorescent molecules can be thought of as a distribution of diffraction limited spots, it should then be possible to extend FIONA from a single-molecule "super-accuracy" approach to a method for obtaining the bigger picture: complete "super-resolution" images of macromolecular structures and assemblies.

The main challenge in applying FIONA to super-resolution imaging of cells is that FIONA requires well-isolated point sources. As the signals from multiple points begin to overlap, the diffraction pattern becomes distorted and the model for fitting the PSF begins to break down, causing decreased localization accuracy. For this reason, FIONA has mostly been applied to single molecule studies where the concentration of the molecule of interest can be kept low. If the goal, however, is to image a subcellular structure, then that particular structure must be labeled with a minimum number of point sources to construct an image with enough information to define its shape and contours. To prevent under-sampling and ensure accurate and complete imaging, the Nyquist-Shannon theorem postulates that the sampling interval must be at least half that of the desired resolution (69). Thus, to obtain an image with 50-nm lateral resolution, the labeling point sources must not be more than 25 nm apart in the x,y-plane. This translates into a high labeling density of at least $6.4 \times 10^3/\mu m^2$ or ≥ 200 points per diffraction-limited area (when imaging a fluorophore with emission wavelength $\lambda = 550$ nm with an objective lens of NA = 1.45).

In the first half of the 2000s, several groups worked to extend FIONA to samples with more than one label, which led to the development of SHREC and SHRImP/NALMS described above. These techniques resolve two dyes separated by < 10 nm—a considerable improvement, yet still not able to handle the densities required for subcellular super-resolution imaging. Around the same time, photoswitchable proteins and dyes were undergoing rapid development. In response to certain wavelengths of light, these fluorescent proteins undergo conformational changes that alter their fluorescence properties—turning them "on" or "off," or causing a shift in the spectral profile. The first photoswitchable protein discovered, Kaede, was cloned from the coral *Trachyphyllia geoffroyi* and irreversibly converts from a green to a red fluorescent protein upon irradiation with UV or blue light (70). Several other fluorescent proteins have since been engineered using high-throughput mutagenesis to have similar photoreactive properties. In addition, a photoswitchable small molecule dye formed between the cyanine dyes Cy3 and Cy5 was reported (71). These new fluorescent markers were immediately recognized for their potential in cell, organelle, and protein tracking experiments using existing imaging strategies (72). However, it took four years after the discovery of Kaede to realize the full potential of photoswitchable probes in super-resolution imaging.

STORM/PALM/F-PALM

In 2006 three groups independently published methods that take advantage of photoswitchable probes for super-resolution imaging based on single molecule localization (SRI-SML) termed, respectively, stochastic optical reconstruction microscopy (STORM) (73), photoactivation light microscopy (PALM) (74), and fluorescent photoactivation light microscopy (F-PALM) (75). Although different in name, these methods all rely on the same basic principle: By sparsely activating or "turning on" only a small subset of the total number of photoswitchable probes in any given video frame, the *effective* labeling density for that frame becomes sufficiently low to allow each activated probe to be localized with FIONA. Applying this concept, image acquisition proceeds in an iterative manner where the series of events—activation, readout, inactivation—is continuously repeated until enough molecules are localized to fulfill the Nyquist criterion for the desired resolution (Figure 9.3a). The final image is then reconstructed or mapped by super-positioning the coordinates of the probes localized for each frame onto the same reconstructed image. A Gaussian peak represents each particle whose width represents the uncertainty in localization. Therefore, the resolution of the final image is determined by only two factors: the localization accuracy from FIONA and the density of localized probes.

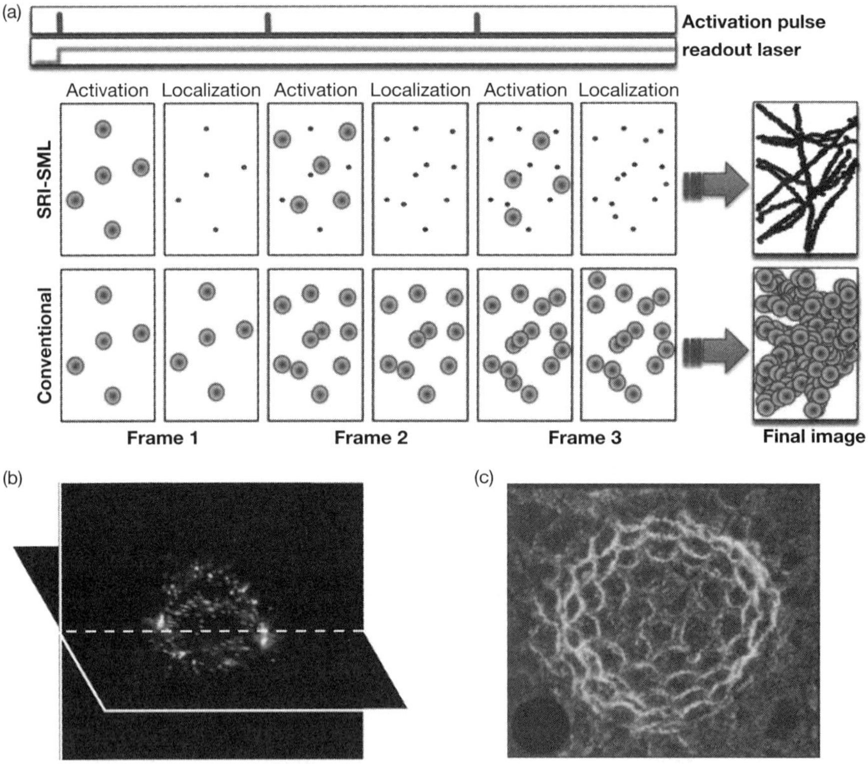

Figure 9.3 Comparison of SRI-SML imaging with conventional fluorescence and electron microscopy. (a) SRI-SML images have a much higher resolution compared to conventional fluorescence microscope images. This resolution enhancement is particularly striking in the case of microtubule filaments. As illustrated, due to overlapping of diffraction-limited spots, conventional imaging cannot resolve single filaments or capture the intricate criss-crossing pattern clearly observed in SRI-SML images. Note that, unlike SRI-SML, conventional imaging only requires one frame per image, thus the final image is not built up upon individual activation events. Individual frames are illustrated here merely for the sake of comparison. (b) and (c) Comparison of images of clathrin coated pits obtained using 3D STORM and electron microscopy (Reproduced with permission from (98, 133)). See insert for color version.

Unlike "hardware-based" super-resolution techniques (see final section below), SRI-SML does not require custom microscope components, lending itself to retrofitting of existing equipment. The basic setup consists of an inverted epi-fluorescence microscope equipped with a high NA objective lens, scientific grade EM-CCD camera, and a pair of continuous wave lasers. Often an objective-type total internal reflection (TIRF) or near-TIRF/HiLo illumination configuration is used to reduce background. Two lasers are typically used to drive the acquisition cycle—a weak short-wavelength laser for activation and a higher power longer-wavelength laser for readout and inactivation. The lasers are either sequentially alternated with a longer duration for readout/inactivation, or the activation laser is pulsed at appropriate intervals while the longer-wavelength laser continuously illuminates the sample. Each fluorophore exhibits unique switching kinetics, but in general, laser induced switching rates are faster than spontaneous rates and are linearly dependent on laser intensity (71). The activation laser intensity is optimized to activate the maximum number of molecules per frame without overlap, usually ~1 per μm^2 area (76). Since acquisition rates are largely limited by the off-rate of switching, the laser used for readout and inactivation is often set to a high intensity. The frame rate of the EM-CCD camera is often synchronized with the switching events, wherein the exposure times for each frame match the average fluorescence on-time of activated probes. Typical frame rates are on the timescale of 30–100 ms (a first-order approximation for the frame rates required for a particular fluorophore can be found in reference (76)). To simplify the acquisition process, asynchronous camera schemes (i.e., protocols that do not match the camera acquisition time to the duration of photon bursts) have also been demonstrated (77, 78).

The first studies with STORM/PALM/F-PALM produced striking images at a resolution nearly an order of magnitude greater than the diffraction limit. Zhuang and coworkers imaged individual RecA DNA binding proteins bound to plasmid DNA with ~ 40-nm x,y-resolution by indirect immunofluorescence wherein the secondary antibody was labeled with a reversibly switchable synthetic dye (73). Hess and colleagues achieved similar resolution when imaging lysosomal transmembrane protein CD63 fused with Kaede in Cos-7 cell cryosections (74). With this improved resolution, it was possible to distinguish closely spaced lysosomal particles that may have been interacting or undergoing transformation from late endosomes. In comparison, standard wide-field fluorescence images of the same sample showed these closely spaced endosomes as a single uniform object. The extraordinary imaging power of these techniques was further demonstrated with several other targets, including the cytochrome-C oxidase import sequence localized within the mitochondrial matrix, vinculin and actin proteins located at focal adhesion regions of fox lung fibroblast cells, and the HIV1 Gag protein within intact Cos-7 cells (74).

Since photoswitching mediated SRI-SML is a single-molecule technique, it can be used to derive quantitative data on the spatial distribution of dense molecular assemblies. Single molecule localizations enabled Owen et al. (79) to compile 2D clustering maps of proteins in the cell membrane. Using a statistical analysis of adjacent particles, the authors found that a truncated form of the tyrosine kinase Lck, fused to the photoactivatable protein tdEos, heterogeneously distributes throughout the membrane with 58 percent of molecules localizing in clusters of various sizes, ranging between 20–190 nm in radius. Another membrane bound tyrosine kinase, Src, has a different distribution pattern when expressed in the same cells. Similar protein counting experiments have been demonstrated using immunofluorescence staining of mitochondrial proteins (79). In addition to positional information, data on the orientations of molecules can also be derived. Gould et al. (80) determined the anisotropies of single actin and hemagglutinin proteins in fixed fibroblast cells by introducing a polarizing beam splitter in the detection path. Refinement of this approach could one day allow detection of the binding states of large populations of single molecules in cells. The marriage of SRI-SML with other single-molecule techniques, such as fluorescence resonance energy transfer (FRET), represents the current vanguard of technical development.

Switchable Probes for SRI-SML

The ideal probes for SRI-SML have photophysical attributes that maximize SNR. First and most importantly, probes should have a high photon output in the activated fluorescence on-state. Ideally, this output should occur rapidly and the on-state should be short-lived to decrease the necessary acquisition time. Second, the on-state should be several thousand-fold brighter than the off-state. Probes with a high contrast ratio between states will have the least amount of background. Third, the spontaneous interconversion rates between states should be much slower than the laser-controlled conversion rates to ensure that the number of activated probes in a given frame can be kept sufficiently low in a user-controlled manner.

A number of photoactivatable probes are currently available and fall within two categories: synthetic dyes and genetically encoded fusion proteins. Under current convention, the distinguishing factor between STORM- and PALM-type techniques is the category of probe used; STORM utilizes synthetic dyes while PALM utilizes fusion proteins. Both types of probes have several strengths and weaknesses, and neither is considered to be perfect. The best type of probe for any given experiment must be chosen on a case-by-case basis, depending on the molecular target, cell type, and overall goals of the experiment. Here, we juxtapose these two categories of photoactivatable probes in the context of SRI-SML and briefly highlight a few of the most commonly used probes under each category.

Synthetic Dyes

The most common synthetic switches are composed of cyanine dyes, a popular family of fluorescent dyes noted for their brightness due to large extinction coefficients. The original switchable synthetic dye used in STORM was composed of an activator Cy3 molecule and reporter Cy5 molecule (73). With this switch-pair, Cy5 emission is induced with red laser light and subsequently inactivated spontaneously by the same laser. Activation is induced using green laser light, which excites Cy3 and causes an energy transfer loosely analogous to FRET, which activates Cy5 but does not directly induce Cy5 emission in the absence of red laser light. Switching is reversible and can occur up to 1000 times before permanent photobleaching, enabling multiple localizations for each probe.

The exact mechanism of photoswitching is ill understood. High-resolution mass spectrometry suggests that Cy5 is switched into a reversible dark state by formation of a thiol-dye adduct produced by primary thiols, a required buffer component for switching, covalently bonding to the polymethine bridge of Cy5 (80). The role of

Cy3 in activation is less clear, and in fact, switching has been demonstrated in the absence of an activator dye, albeit at a much slower rate (30, 81). Cy3 can also be replaced by other fluorophores to serve as activators. Since the highest activation rates are induced with light at a wavelength corresponding to the maximum absorption wavelength of the activator, these alternative switch-pairs are convenient tools in multi-label SRI-SML (see below).

Typically, synthetic dyes must be conjugated to antibodies that bind to specific cellular proteins. Initially, antibodies for immunofluorescence were labeled with Cy3-Cy5 pairs by reacting amine-modified antibodies with dye mixtures containing amine-reactive Cy5 and an excess of Cy3 to increase the likelihood of each Cy5 label being closely adjacent to at least one activating Cy3 molecule (73). This process was later streamlined by directly conjugating Cy3 and Cy5 together, using commercially available starting materials (82). At ultra-high resolutions, one must consider the size of the fluorescent probe; greater distances between the probe and the target will reduce resolution. Despite the fact that synthetic dyes are >30-fold lower molecular weight than fluorescent proteins, the need to use randomly labeled antibodies as an intermediary considerably increases the "effective" size of the label. The effect of label size is significantly pronounced with microtubules, where antibody staining in STORM increases the apparent diameter of the cytoskeletal polymers to nearly double that of their true width of ~25 nm (83). The use of Fab fragments or direct immunofluorescence improves resolution. Additionally, biarsenical fluorophores derived from fluorescein, resorfin, and Cy3 have been developed that can bind specifically to short genetically encodable tetracysteine tags (84–86). Although switching with these fluorophores has not yet been reported, the potential exists for development of switchable analogues. Caged rhodamine derivatives featuring fluorescence-dampening protective groups, which are released in response to UV light, have recently been used in SRI-SML (74, 87). This chemistry can be applied to several other types of fluorophores, including biarsenicals, thus opening the door for the future direct labeling of cellular proteins with small, switchable, synthetic dyes.

Fluorescent Fusion Proteins

Switchable fluorescent proteins can be categorized into photoactivatable proteins that can be reversibly or irreversibly switched between a non-fluorescent off-state and a fluorescent on-state, and photoconvertible proteins whose emission changes color after activation with UV (405 nm) or blue (488 nm) light.

Two of the most popular switchable proteins are Dronpa and EosFP. Dronpa, engineered from the coral *Pectiniidae*, is a monomeric photoactivatable protein with a dark inactivated state and a green (518 nm) emission after activation with UV light. It is the one of the few reversibly switchable proteins and undergoes ~100 switching cycles. The switching mechanism is understood to involve a cis-trans isomerization of the chromophore that triggers further rearrangements of four proximal amino acids (88).

Dronpa has the highest extinction coefficient and contrast ratio of all switchable proteins, yet suffers from problematic switching behavior that ultimately affects its performance. In fact, PS-CFP2, a photoconvertible protein with a nearly two-fold lower extinction coefficient and higher background, could be localized more precisely than Dronpa in live cells (89). Several other variants have been engineered from Dronpa to improve or alter its switching properties (90, 91). EosFP, a photoconvertible protein isolated from the coral *Lobophyllia hemprichii*, exists as a tetramer in its native environment, but monomeric forms have been engineered (92). In response to UV irradiation, a peptide bond near the chromophore undergoes a cleavage reaction, causing the green (569 nm) emission to be bathochromatically shifted to orange (581 nm) (93). EosFP is currently the brightest switchable fluorescent protein, leading to an excellent contrast ratio, and is largely considered the best switchable fluorescent protein available for SRI.

Multi-Label Imaging

A powerful feature of fluorescence microscopy is the ability to visualize multiple targets in a single sample, wherein each target is labeled with a spectrally distinct probe. A broad color palette of fluorescent probes has now made conventional fluorescence imaging with up to four colors commonplace. When combined with objective lenses corrected for chromatic aberration, these images become unparalleled analytical tools for monitoring the relative spatial distributions of cellular macromolecules. Analysis of such images can vary from qualitative evaluations to quantitative measurements; either way most often the fundamental derived facet indicative of molecular interaction is spatial colocalization. Several methods for defining colocalization exist; yet regardless of the analytical method applied, the biological significance of colocalization ultimately depends on image resolution. Targets that appear to be colocalized or interacting may, in fact, be merely proximal to one another at an unresolvable distance. Multi-label SRI-SML thus has the potential to improve our current knowledge of molecular interactions and distributions in cells.

The combinatorial pairing of cyanine activator and reporter dyes has created a pallet of distinct synthetic switch-pairs for STORM imaging. Switches consisting of Cy3 activators and far-red reporter dyes, such as Cy5, Cy5.5, and Cy7, can all be activated and read out with green and red (657 nm) laser light, respectively, and their signals can be unmixed with dichroic mirrors and bandpass emission filters. Alternatively, spectrally distinct activators—Alexa405, Cy3, and Cy2—have been paired with a Cy5 reporter and selectively activated with minimal cross-talk using laser light corresponding to the absorbance of each activator dye. An advantage of this strategy is that fewer laser wavelengths are needed compared to multiple color readout. By alternatively activating Cy2-Alexa647 (the latter a structural analog of Cy5) and Cy3-Alexa647, Bates et al. (93) were able to reconstruct two-color images of microtubules and clathrin-coated pits in fixed mammalian cells. Huang et al. (94) imaged Alexa405-Cy5 labeled mitochondria and A555-Cy5 labeled microtubules and observed an inchworm-like interaction where only portions of the mitochondria were in direct contact with the microtubules. Conventional TIRF images obscured these details and instead the microtubules appeared to be in direct contact along the full length of the mitochondria.

Multi-label imaging with fluorescent proteins is currently more challenging. The complication is that, unlike the multiple pairings of synthetic dyes, photoactivatable fluorescent proteins largely lack diversity. That is, they are similarly activated with UV-light, and their activation transitions fall into either one of two categories: from dark (non-fluorescent) to green emission or from green to orange emission. Since the signal from the preactivated form of the latter overlaps with the activated form of the former, this precludes the possibility of simply using complementary probes from each category. Even though dark-to-red photoactivatable fluorescent proteins have been discovered, they are generally plagued by poor contrast ratio, although recent derivatives of mCherry are sufficient for use in live-cell PALM (94, 95). One strategy for overcoming this lack of spectral diversity is to take advantage of another unique attribute: reversible switching. Dronpa and EosFP can be concurrently activated yet imaged serially by first collecting the orange emission of EosFP and then, after irreversible inactivation or bleaching of all EosFP molecules, collecting the green emission of Dronpa without interference from EosFP's preactivated state (89). While a clever workaround, this technique is not suitable for live cell imaging since the target proteins will diffuse over time. Future development of unique fluorescent proteins will lead to new possibilities for multi-label imaging. As an example, Andresen et al. (97) engineered a new variant of Dronpa aptly named Padron (a reordering of the letters in Dronpa) that undergoes inverse or "positive switching" in which blue light activates and excites the green fluorescent protein and UV-light switches the protein off. Although Padron has not yet been used for SRI-SML, multi-label imaging with Padron and another Dronpa variant, rsFastlime (whose switching characteristics are similar to Dronpa), was demonstrated in living budding yeast cells by 3D confocal microscopy (96).

3D Imaging

At their inception, STORM and PALM were applied to a single focal plane of a sample and the improvement in lateral resolution was not extended to the z-dimension. To enable complete 3D reconstructive mapping of cells, several modified versions have been devised, many of which were inspired by the pioneering efforts in 3D SPT (discussed above). In 2008 Zhuang and colleagues published two methods for 3D STORM utilizing either a cylindrical lens or focal plane scanning (97, 98). Both methods produced similar results with resolutions as high as 20–30 nm lateral and 50–70 nm axial. By combining these methods with multi-label imaging, it was possible to visualize complex mitochondrial networks entangled within a 3D matrix of crisscrossing microtubules and the spherical cage-like shape of clathrin-coated pits containing structural details approaching those of electron microscopy (Figures 9.3b and 3c). However, one disadvantage of these techniques is their limited use in imaging thick samples. Cylindrical lens-based methods can only accurately localize molecules within < 600 nm of the focal plane and must be coupled with focal plane scanning to increase image depth. While biplane detection can achieve similar resolution with a slightly enhanced axial localization range independent of focal plane scanning, and with increased image depth (up to 4 μm) when coupled with scanning (99), there still exists an intrinsic limit in imaging depth caused by scattering of the emission light as it passes through a thick sample.

Multi-photon microscopy overcomes this obstacle by using short pulses of longer wavelength light to excite the fluorophore. Pioneered by Webb, two-photon excitation (TPE) overcomes many of the problems pertaining to conventional fluorescence microscopy—limited penetration depth of incident light, loss by scattering of emitted light, and background due to cellular autofluorescence. TPE is based on the principle that two photons can also excite a fluorophore in one quantum event when the sum of energies of both photons equals that of the absorption energy of the fluorescent probe. The same principle can also be extended to multiple photons in which case the scheme is referred to as multi-photon excitation (MPE). As excitation by two photons is less probable than excitation by a single photon, higher light intensity is required for excitation (more than twice that required for single-photon excitation). TPE has mostly been used for imaging tissues or even live animals (100) but has recently gained popularity in high-resolution intracellular SMFM and SPT (101, 102). By combining two-photon microscopy with temporal focusing, which reduces background by minimizing the excitation volume, Vaziri et al. achieved PALM images over an axial depth of ~10 μm in mammalian and *Drosophila* S2 cells (103).

Live-Cell Imaging

To gain a better understanding of the molecular dynamics involved in biological processes, the ultimate frontier of super-resolution imaging has become the living cell. Live-cell SRI-SML is currently limited to imaging relatively slow dynamic events in a select set of cell lines. Since STORM/PALM/F-PALM requires many frames to build a single image, data acquisition rates are slower than conventional wide-field fluorescence microscopy. This ultimately precludes SRI-SML from imaging biological processes that unfold at rates faster than the acquisition rate limit. The prolonged exposure time and high laser power also requires samples that are tolerant to photo-induced damage. Several mammalian cell lines, including Chinese hamster ovary (CHO), NIH 3T3 fibroblasts, and African green monkey kidney Cos-7 cells, remain healthy under the harsh imaging conditions and resistant to changes in morphology after exposure (82, 104). Using these photo-tolerant cells and PALM, Shroff et al. (103) visualized the movements and dynamic organization of adhesion complexes (ACs)—networks of nearly 100 proteins involved in cellular motility and anchoring, organized in a thin lamina at the cell-(extra-cellular) matrix interface. By capturing frames every 40 ms and grouping 750 frames into single images, the formation, elongation, and retraction of ACs were imaged with a lateral resolution of ~70 nm and a temporal resolution of ~30 s. Since ACs evolve slowly at a rate of 120 nm/min, the movement of ACs over the course of a single frame was small enough to prevent blurring. In addition to imaging in living eukaryotic cells, SRI-SML has also been applied to *Caulobacter crescentus* bacteria (82, 105, 106).

The photoswitching-based imaging methodology of SRI-SML can also be combined with SPT to investigate faster dynamic processes. In this way, the goal is not to build a super-resolution image, but rather to measure the displacement of particles over short time and length scales as discussed earlier in the chapter. The incorporation of photoswitching-based imaging enables tracking of a small subset of molecules at a given time duration within densely populated structures, such as cell membrane domains or lamellipodia, which would be difficult or almost impossible using conventional SPT (107–109). In this case, cellular over-expression of the target is of advantage as it allows one to collect many more tracks from a single cell, increasing the statistical power of the analysis while decreasing imaging time.

SUPER-RESOLUTION MICROSCOPY—HARDWARE (OPTICS)-BASED METHODS

Beginning with confocal microscopy, several imaging techniques were developed starting in the 1990s with the vision of increasing resolution by reducing the focal spot size. These methods use specialized optical setups and can be considered "hardware-based" solutions for super-resolution imaging, as opposed to "software-based" methods like STORM/PALM/F-PALM, which rely instead on mathematical processing of a series of acquired diffraction limited images. Spot-scanning 4Pi microscopy (4PiM) (110) and wide-field I^5M (111) are two hardware-based techniques that can achieve ~3–7-fold greater axial resolution than traditional confocal microscopy. 4PiM uses two opposing high-NA objective lenses to illuminate a single focal spot. The coherent addition of wavefronts from both lenses narrows the focal dimension across the optical axis, resulting in higher resolution. I^5M implements the same opposing lens aperture enhancement for the detection of fluorescence but uses plane-parallel standing waves to illuminate the entire FoV.

Reversible Saturable Optical Fluorescence Transitions (RESOLFT) Techniques

RESOLFT techniques utilize sophisticated excitation configurations to physically decrease the width of the excitation volume and thus decrease the size of the PSF below theoretical limits of diffraction (112). These techniques can be described as targeted illumination/readout methods that do not require detection of single molecules, whereas STORM/PALM/F-PALM are stochastic readout methods based on the very features of single molecule detection. Implementation of RESOLFT can be loosely divided into three categories—stimulated emission depletion (STED) (113), ground state depletion (GSD) (114, 115), and saturated pattern excitation microscopy (SPEM)/saturated structured illumination microscopy (SSIM) (116, 117). In STED, two laser beams are used (Figure 9.4a), an excitation beam and a comparatively red-shifted "depletion" beam. The excitation laser is first applied, followed immediately (within picoseconds) by a doughnut-shaped depletion beam that depletes the excited state of all fluorophores except those at the very center of the original excitation beam. Fluorescence from the remaining molecules in the center of the focal volume is then detected. The intensity of the depletion beam can be increased to "squeeze" the final detection volume to a smaller size. The two beams are raster-scanned across the entire sample to form the complete image, as is generally done in scanning laser confocal fluorescence microscopy (16). Using STED microscopy, GFP-labeled viruses and endoplasmatic reticulum were imaged in fixed PtK2 cells with a lateral resolution of ~70 nm (118). Recently, Hell and coworkers implemented STED in a 4Pi microscope to achieve 40–45 nm in all three spatial dimensions by generating a spherical focal spot of the same diameter (119). Axial scanning of fixed Vero cells with this configuration revealed that immunologically labeled Tom20 (a subunit of the TOM complex) formed distinct clusters on the mitochondrial membrane. The group also implemented a dual-color imaging scheme in which they used a combination of two excitation-one depletion events to image both Tom20 and mtHsp70 (a mitochondrial matrix protein) simultaneously with ~40-nm resolution in 3D.

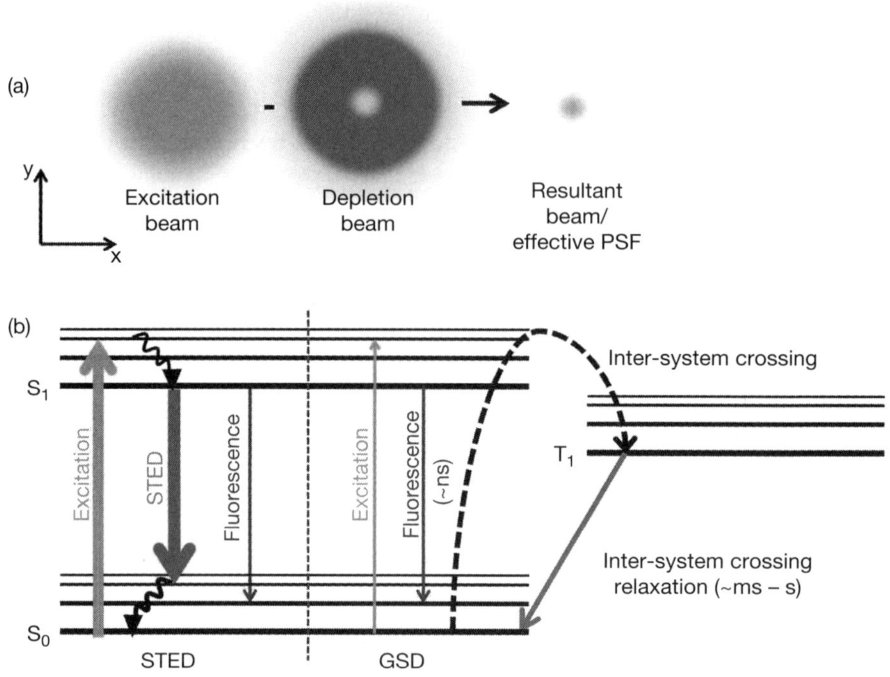

Figure 9.4 STED and GSD. (a) A 2D cross-section of the excitation, depletion, and resultant beams used in STED microscopy. The doughnut shape of the depletion beam can be achieved by placing a phase mask in the light path. (b) Jablonski diagram depicting STED and GSD. In STED, the depletion beam (labeled STED) de-populates the excited state (S_1). Fluorescence from very few fluorophores, which still remain in the excited state, is then recorded. In contrast, GSD works by "optical shelving." The electronic ground state (S_0) of fluorophores is depleted by pumping electrons into either the first triplet state (T_1) or other dark states. Electronic relaxations from these dark states occur at durations that are several orders of magnitude higher than that of fluorescence. At any given time, the ground state is thus sparsely populated, allowing only a few fluorophores to get excited.

As in STED, GSD requires reversible switching between on- and off-states of the fluorophore. However, one of the major differences between the two methods lies in the nature of the off-state. Rather than by stimulated depletion, the off-state in GSD is induced via ground state depletion through increasing the triplet state population (Figure 9.4b). Another difference between the two RESOLFT techniques is the use of a single continuous-wave laser for both excitation and depletion in GSD. A lateral resolution of ~50 nm has been demonstrated by applying GSD to the imaging of microtubules (114, 115) and to aggregates of SNAP-25 (114) or integrin-β proteins (115) in fixed cells.

SPEM/SSIM is a wide-field super-resolution technique that excites a fluorescent sample by structured or patterned illumination (116, 117). This way, unresolved information is first visible in the form of low-resolution moiré fringes, which are then computationally processed to obtain the final high-resolution image. Similar to other RESOLFT techniques, SSIM relies on the nonlinear relationship between the fluorescence emission rate and excitation intensity, and the practical resolving power is determined by the SNR. The improvement in resolution is, however, only twofold over that obtained by conventional fluorescence microscopy, which is almost an order of magnitude less than that obtained by STORM/PALM/F-PALM. Nevertheless, SIM is currently the only subdiffraction high-resolution imaging technique that can produce multicolor 3D images of whole cells with enhancement of resolution in both lateral and axial directions while using conventional fluorescent probes and low intensity light sources. Upon implementing 3D-SIM to visualize interphase chromatin architecture and nuclear pore complex structure in formaldehyde-fixed cells, never-before seen details of these macromolecular structures were unraveled in their native nuclear milieu (120). In another report, Kner et al. (121) used a combination of speckle microscopy and SIM to observe the evolution of microtubules in living cultured cells with ~100-nm lateral resolution and ~100-ms time resolution. Recently, impressive time-lapsed 3D images of zebrafish and *Drosophila* embryos undergoing development were recorded using SIM based on a tunable light sheet, taking the potential of SIM to new heights (122). By utilizing multiple fluorescent tags and imaging embryos at multiple angles the group was able to generate a "digital" fly embryo, highlighting the morphological changes of the entire embryo as it develops.

The theoretical resolution limit of RESOLFT techniques can be described by an extension of Abbe's law:

$$\Delta d = \frac{\lambda}{2n\sin\alpha\sqrt{1+\left(\frac{I_{max}}{I_s}\right)}} \tag{6}$$

where I_{max} is the intensity of light bordering the center and I_s is the saturation intensity required to outperform competing spectroscopic transitions (123). When $I_{max}/I_s \rightarrow \infty$, in theory "infinite" resolution can be achieved (i.e., down to a single molecule or even smaller), but this comes at the cost of using very high-intensity beams that might damage sensitive samples like living cells. Due to this risk, RESOLFT is still generally implemented as an ensemble imaging technique, although single molecule detection is definitely possible (124).

CONCLUSIONS

The gap in attainable limits of resolution between traditional microscopy and electron microscopy (EM) has recently been bridged by the high-resolution imaging techniques discussed in this chapter. Ultra-high-resolution microscopy with conventional optics, which for a long time was thought to be the realm of EM, near-field imaging or negative refractive index superlenses (125), has now become commonplace. Importantly, the ultimate contribution that these optical techniques will have to science depends on their accessibility to researchers. To this end, several "out of the box" microscopes that implement PALM, SSIM, and STED have recently been commercialized by manufacturers like Zeiss and Leica. Additionally, user-friendly software packages, both commercial and open-source (such as ImageJ), have made it easier to implement these powerful techniques without the need for technical proficiency or prior experience.

SMFM techniques are mainly limited by the number of observables accessible per experiment, imposing the need to perform multiple assays to study complex biological processes. The future of SMFM lies in the ability to combine multiple tools into one adaptive instrument, such that the wealth of information obtained from a single sample is broadened. Techniques like 3D correlated EM (126), which combines the technical prowess of EM with the noninvasive nature of fluorescence microscopy, and SPT/single molecule FRET (127), have emerged as powerful tools to study the spatiotemporal evolution of molecules during various stages of cellular processes at very high resolution. The latter method can particularly probe conformational changes of proteins as they diffuse inside living cells.

Realizing the true potential of any technique relies on understanding its scope and limitations. However, factors considered as a bane in certain methods can be a boon in others. For instance, reactive oxygen species produced by fluorescent probes, which are considered to be toxic to living cells, can be used to oxidize certain contrast-enhancing agents to improve image quality in EM (128). Even non-fluorescent chromophores like heme, which typically contribute to high background in intracellular fluorescence imaging, can be imaged by STED microscopy to provide unique details on subcellular super-structures (129). In this fashion, molecules can be visualized label-free in their native cellular environment without the need for fluorophore conjugation.

Chu and coworkers recently improved the precision of wide-field imaging of biological samples by a factor of two through correcting for microscope drift with active feedback control, reaching sub-nm accuracy and resolution (130, 131). At such resolutions, it should become possible to obtain structural information of single molecules, for instance the distance between two subunits of a protein, even at very low molecule abundance and without the need to cool the sample below ambient temperatures (as required for EM and X-ray diffraction). The near future of SMFM will involve the implementation of such image acquisition and processing algorithms to spatially resolve the complex biological processes that enable life. Spatial resolution is, however, just one side of the coin since sufficient temporal resolution is also necessary, and both types of resolution are typically inversely related. Reaching higher (sub-millisecond) temporal resolution during biological imaging is mainly impeded by hardware limitations (e.g., the maximal frame rate of the camera), limited photon collection at high frame rates, decreasing SNR, and fast diffusion of molecules inside the cell, which leads to image blurring. Stroboscopic imaging techniques, where laser excitation pulses are synchronized with the camera frame rate, have emerged as tools to address these limitations (132). Development of faster and more sensitive cameras may in the future enable higher resolution also during single-particle tracking based on simple optical and excitation schemes. Considering the large strides that SMFM has taken over the past two decades, one can easily envision that large-scale multiplexed imaging of biomolecules (e.g., proteins, DNA, RNA) at sub-nanometer and sub-millisecond spatiotemporal resolution within live cells or even animals will become possible within the next two decades.

REFERENCES

1. Lecuyer E, et al. (2007) Global analysis of mRNA localization reveals a prominent role in organizing cellular architecture and function. *Cell* 131:174–187.
2. Oleynikov Y, Singer RH (2003) Real-time visualization of ZBP1 association with beta-actin mRNA during transcription and localization. *Curr Biol* 13:199–207.
3. Montero Llopis P, et al. (2010) Spatial organization of the flow of genetic information in bacteria. *Nature* 466:77–81.
4. Fusco D, et al. (2003) Single mRNA molecules demonstrate probabilistic movement in living mammalian cells. *Curr Biol* 13:161–167.
5. Schuler B, Lipman EA, Eaton WA (2002) Probing the free-energy surface for protein folding with single-molecule fluorescence spectroscopy. *Nature* 419:743–747.
6. Merchant KA, Best RB, Louis JM, Gopich IV, Eaton WA (2007) Characterizing the unfolded states of proteins using single-molecule FRET spectroscopy and molecular simulations. *Proc Natl Acad Sci USA* 104:1528–1533.
7. Ditzler MA, Rueda D, Mo J, Hakansson K, Walter NG (2008) A rugged free energy landscape separates multiple functional RNA folds throughout denaturation. *Nucleic Acids Res* 36:7088–7099.
8. Solomatin SV, Greenfeld M, Chu S, Herschlag D (2010) Multiple native states reveal persistent ruggedness of an RNA folding landscape. *Nature* 463:681–684.
9. Zhuang X, et al. (2002) Correlating structural dynamics and function in single ribozyme molecules. *Science* 296:1473–1476.
10. Rueda D, et al. (2004) Single-molecule enzymology of RNA: essential functional groups impact catalysis from a distance. *Proc Natl Acad Sci USA* 101:10066–10071.
11. Lipman EA, Schuler B, Bakajin O, Eaton WA (2003) Single-molecule measurement of protein folding kinetics. *Science* 301:1233–1235.
12. Ha T, et al. (1999) Single-molecule fluorescence spectroscopy of enzyme conformational dynamics and cleavage mechanism. *Proc Natl Acad Sci USA* 96:893–898.
13. Kaufman RJ (2000) Overview of vector design for mammalian gene expression. *Mol Biotechnol* 16:151–160.
14. Toprak E, Kural C, Selvin PR (2010) Super-accuracy and super-resolution getting around the diffraction limit. *Methods Enzymol* 475:1–26.
15. Greenleaf WJ, Woodside MT, Block SM (2007) High-resolution, single-molecule measurements of biomolecular motion. *Annu Rev Biophys Biomol Struct* 36:171–190.
16. Walter NG, Huang CY, Manzo AJ, Sobhy MA (2008) Do-it-yourself guide: how to use the modern single-molecule toolkit. *Nat Methods* 5:475–489.
17. Joo C, Balci H, Ishitsuka Y, Buranachai C, Ha T (2008) Advances in single-molecule fluorescence methods for molecular biology. *Ann Rev Biochem* 77:51–76.
18. Lewis A, et al. (2003) Near-field optics: from subwavelength illumination to nanometric shadowing. *Nat Biotechnol* 21:1378–1386.
19. Edidin M (2001) Near-field scanning optical microscopy, a siren call to biology. *Traffic* 2:797–803.
20. Axelrod D (1981) Cell-substrate contacts illuminated by total internal reflection fluorescence. *J Cell Biol* 89:141–145.
21. Tokunaga M, Imamoto N, Sakata-Sogawa K (2008) Highly inclined thin illumination enables clear single-molecule imaging in cells. *Nat Methods* 5:159–161.
22. Konopka CA, Bednarek SY (2008) Variable-angle epifluorescence microscopy: a new way to look at protein dynamics in the plant cell cortex. *Plant J* 53:186–196.

23. Nakano A (2002) Spinning-disk confocal microscopy—a cutting-edge tool for imaging of membrane traffic. *Cell Struct Funct* 27:349–355.
24. Rehemtulla A, Hamilton CA, Chinnaiyan AM, Dixit VM (1997) Ultraviolet radiation-induced apoptosis is mediated by activation of CD-95 (Fas/APO-1). *J Biol Chem* 272:25783–25786.
25. Gossen M, Bujard H (1992) Tight control of gene expression in mammalian cells by tetracycline-responsive promoters. *Proc Natl Acad Sci USA* 89:5547–5551.
26. Prescher JA, Bertozzi CR (2005) Chemistry in living systems. *Nat Chem Biol* 1:13–21.
27. Lord SJ, et al. (2008) A photoactivatable push-pull fluorophore for single-molecule imaging in live cells. *J Am Chem Soc* 130:9204–9205.
28. Lord SJ, et al. (2009) DCDHF fluorophores for single-molecule imaging in cells. *Chemphyschem* 10:55–65.
29. Davidson MW, Campbell RE (2009) Engineered fluorescent proteins: innovations and applications. *Nat Methods* 6:713–717.
30. Heilemann M, et al. (2008) Subdiffraction-resolution fluorescence imaging with conventional fluorescent probes. *Angew Chem Int Ed* 47:6172–6176.
31. Kapanidis AN, Weiss S (2002) Fluorescent probes and bioconjugation chemistries for single-molecule fluorescence analysis of biomolecules. *J Chem Phys* 117:10953–10964.
32. Dave R, Terry DS, Munro JB, Blanchard SC (2009) Mitigating unwanted photophysical processes for improved single-molecule fluorescence imaging. *Biophys J* 96:2371–2381.
33. Aitken CE, Marshall RA, Puglisi JD (2008) An oxygen scavenging system for improvement of dye stability in single-molecule fluorescence experiments. *Biophys J* 94:1826–1835.
34. Olenych SG, Claxton NS, Ottenberg GK, Davidson MW (2007) The fluorescent protein color palette. *Curr Protoc Cell Biol* 36: 21.5.1-21.5.34
35. Bogdanov AM, et al. (2009) Cell culture medium affects GFP photostability: a solution. *Nat Methods* 6:859–860.
36. Billinton N, Knight AW (2001) Seeing the wood through the trees: a review of techniques for distinguishing green fluorescent protein from endogenous autofluorescence. *Anal Biochem* 291:175–197.
37. Kredel S, et al. (2008) Optimized and far-red-emitting variants of fluorescent protein eqFP611. *Chem Biol* 15:224–233.
38. Magde D, Elson E, Webb WW (1972) Thermodynamic Fluctuations in a Reacting System—Measurement by Fluorescence Correlation Spectroscopy. *Phys Rev Lett* 29:705–708.
39. Poo M, Cone RA (1974) Lateral diffusion of rhodopsin in the photoreceptor membrane. *Nature* 247:438–441.
40. Gelles J, Schnapp BJ, Sheetz MP (1988) Tracking kinesin-driven movements with nanometre-scale precision. *Nature* 331:450–453.
41. De Brabander M, Nuydens R, Geuens G, Moeremans M, De Mey J (1986) The use of submicroscopic gold particles combined with video contrast enhancement as a simple molecular probe for the living cell. *Cell Motil Cytoskeleton* 6:105–113.
42. Geerts H, et al. (1987) Nanovid tracking: a new automatic method for the study of mobility in living cells based on colloidal gold and video microscopy. *Biophys J* 52:775–782.
43. Moerner WE, Kador L (1989) Optical detection and spectroscopy of single molecules in a solid. *Phys Rev Lett* 62:2535–2538.

44. Orrit M, Bernard J (1990) Single pentacene molecules detected by fluorescence excitation in a p-terphenyl crystal. *Phys Rev Lett* 65:2716–2719.
45. Ambrose WP, Basche T, Moerner WE (1991) Detection and Spectroscopy of Single Pentacene Molecules in a p- Terphenyl Crystal by Means of Fluorescence Excitation. *J Chem Phys* 95:7150–7163.
46. Van Oijen AM, Köhlera J, Schmidta J, Müllerb M, Brakenhoffb GJ (1998) 3-Dimensional super-resolution by spectrally selective imaging. *Chem Phys Lett* 292:183–187.
47. Yildiz A, et al. (2003) Myosin V walks hand-over-hand: single fluorophore imaging with 1.5-nm localization. *Science* 300:2061–2065.
48. Bobroff N (1986) Position measurement with a resolution and noise-limited instrument. *Rev Sci Instrum* 57:1152–1157.
49. Thompson RE, Larson DR, Webb WW (2002) Precise nanometer localization analysis for individual fluorescent probes. *Biophys J* 82:2775–2783.
50. Hua W, Chung J, Gelles J (2002) Distinguishing inchworm and hand-over-hand processive kinesin movement by neck rotation measurements. *Science* 295:844–848.
51. Forkey JN, Quinlan ME, Shaw MA, Corrie JE, Goldman (2003) Three-dimensional structural dynamics of myosin V by single-molecule fluorescence polarization. *Nature* 422:399–404.
52. Yildiz A, Tomishige M, Vale RD, Selvin PR (2004) Kinesin walks hand-over-hand. *Science* 303:676–678.
53. Gordon MP, Ha T, Selvin PR (2004) Single-molecule high-resolution imaging with photobleaching. *Proc Natl Acad Sci USA* 101:6462–6465.
54. Qu X, Wu D, Mets L, Scherer NF (2004) Nanometer-localized multiple single-molecule fluorescence microscopy. *Proc Natl Acad Sci USA* 101:11298–11303.
55. Das SK, Darshi M, Cheley S, Wallace MI, Bayley H (2007) Membrane protein stoichiometry determined from the step-wise photobleaching of dye-labelled subunits. *Chembiochem* 8:994–999.
56. Kuszak AJ, et al. (2009) Purification and functional reconstitution of monomeric mu-opioid receptors: allosteric modulation of agonist binding by Gi2. *J Biol Chem* 284:26732–26741.
57. Churchman LS, Okten Z, Rock RS, Dawson JF, Spudich JA (2005) Single molecule high-resolution colocalization of Cy3 and Cy5 attached to macromolecules measures intramolecular distances through time. *Proc Natl Acad Sci USA* 102:1419–1423.
58. Goshtasby A (1988) Image registration by local approximation methods. *Image and Vision Computing* 6:255–261.
59. Grunwald D, Singer RH (2010) In vivo imaging of labelled endogenous beta-actin mRNA during nucleocytoplasmic transport. *Nature* 467:604–607.
60. Kao HP, Verkman AS (1994) Tracking of single fluorescent particles in three dimensions: use of cylindrical optics to encode particle position. *Biophys J* 67:1291–1300.
61. Speidel M, Jonas A, Florin EL (2003) Three-dimensional tracking of fluorescent nanoparticles with subnanometer precision by use of off-focus imaging. *Opt Lett* 28:69–71.
62. Schütz GJ, et al. (2000) 3D imaging of individual ion channels in live cells at 40 nm resolution. *Single Molecules* 1:25–31.
63. Toprak E, Balci H, Blehm BH, Selvin PR (2007) Three-dimensional particle tracking via bifocal imaging. *Nano Lett* 7:2043–2045.

64. Yajima J, Mizutani K, Nishizaka T (2008) A torque component present in mitotic kinesin Eg5 revealed by three-dimensional tracking. *Nat Struct Mol Biol* 15:1119–1121.
65. Sun Y, McKenna JD, Murray JM, Ostap EM, Goldman YE (2009) Parallax: high accuracy three-dimensional single molecule tracking using split images. *Nano Lett* 9:2676–2682.
66. Sun Y, et al. (2010) Single-molecule stepping and structural dynamics of myosin X. *Nat Struct Mol Biol* 17:485–491.
67. Pavani SR, et al. (2009) Three-dimensional, single-molecule fluorescence imaging beyond the diffraction limit by using a double-helix point spread function. *Proc Natl Acad Sci USA* 106:2995–2999.
68. Thompson MA, Lew MD, Badieirostami M, Moerner WE (2010) Localizing and tracking single nanoscale emitters in three dimensions with high spatiotemporal resolution using a double-helix point spread function. *Nano Lett* 10:211–218.
69. Shannon CE (1949) Communication in the Presence of Noise. *Proceedings of the IRE* 37:10–21.
70. Ando R, Hama H, Yamamoto-Hino M, Mizuno H, Miyawaki A (2002) An optical marker based on the UV-induced green-to-red photoconversion of a fluorescent protein. *Proc Natl Acad Sci USA* 99:12651–12656.
71. Bates M, Blosser TR, Zhuang X (2005) Short-range spectroscopic ruler based on a single-molecule optical switch. *Phys Rev Lett* 94:108101.
72. Lukyanov KA, Chudakov D, Lukyanov S, Verkhusha V (2005) Photoactivatable fluorescent proteins. *Nat Rev Mol Cell Biol* 6:885–891.
73. Rust MJ, Bates M, Zhuang X (2006) Sub-diffraction-limit imaging by stochastic optical reconstruction microscopy (STORM). *Nat Methods* 3:793–795.
74. Betzig E, et al. (2006) Imaging intracellular fluorescent proteins at nanometer resolution. *Science* 313:1642–1645.
75. Hess ST, Girirajan TPK, Mason MD (2006) Ultra-high resolution imaging by fluorescence photoactivation localization microscopy. *Biophys J* 91:4258–4272.
76. Gould TJ, Verkhusha VV, Hess ST (2009) Imaging biological structures with fluorescence photoactivation localization microscopy. *Nat Protocols* 4:291–308.
77. Geisler C, et al. (2007) Resolution of lambda/10 in fluorescence microscopy using fast single molecule photo-switching. *Applied Phys A Mat Sci Proc* 88:223–226.
78. Egner A, et al. (2007) Fluorescence nanoscopy in whole cells by asynchronous localization of photoswitching emitters. *Biophys J* 93:3285–3290.
79. van de Linde S, Sauer M, Heilemann M (2008) Subdiffraction-resolution fluorescence imaging of proteins in the mitochondrial inner membrane with photoswitchable fluorophores. *J Struct Biol* 164:250–254.
80. Dempsey GT, et al. (2009) Photoswitching mechanism of cyanine dyes. *J Am Chem Soc* 131:18192–18193.
81. Heilemann M, van de Linde S, Mukherjee A, Sauer M (2009) Super-resolution imaging with small organic fluorophores. *Angew Chem Int Ed* 48:6903–6908.
82. Conley NR, Biteen JS, Moerner WE (2008) Cy3-Cy5 covalent heterodimers for single-molecule photoswitching. *J Phys Chem B* 112:11878–11880.
83. Osborn M, Webster RE, Weber K (1978) Individual microtubules viewed by immunofluorescence and electron microscopy in the same PtK2 cell. *J Cell Biol* 77:R27–34.
84. Adams SR, et al. (2002) New biarsenical ligands and tetracysteine motifs for protein labeling in vitro and in vivo: synthesis and biological applications. *J Am Chem Soc* 124:6063–6076.

85. Cao H, et al. (2007) A red cy3-based biarsenical fluorescent probe targeted to a complementary binding peptide. *J Am Chem Soc* 129:8672–8673.
86. Griffin BA, Adams SR, Tsien RW (1998) Specific covalent labeling of recombinant protein molecules inside live cells. *Science* 281:269–272.
87. Folling J, et al. (2007) Photochromic rhodamines provide nanoscopy with optical sectioning. *Angew Chem Int Ed* 46:6266–6270.
88. Andresen M, et al. (2007) Structural basis for reversible photoswitching in Dronpa. *Proc Natl Acad Sci USA* 104:13005–13009.
89. Shroff H, et al. (2007) Dual-color superresolution imaging of genetically expressed probes within individual adhesion complexes. *Proc Natl Acad Sci USA* 104:20308–20313.
90. Ando R, Flors C, Mizuno H, Hofkens J, Miyawaki A (2007) Highlighted generation of fluorescence signals using simultaneous two-color irradiation on Dronpa mutants. *Biophys J* 92:L97–99.
91. Stiel AC, et al. (2007) 1.8 A bright-state structure of the reversibly switchable fluorescent protein Dronpa guides the generation of fast switching variants. *Biochem J* 402:35–42.
92. Wiedenmann J, et al. (2004) EosFP, a fluorescent marker protein with UV-inducible green-to-red fluorescence conversion. *Proc Natl Acad Sci USA* 101:15905–15910.
93. Nienhaus K, Nienhaus GU, Wiedenmann J, Nar H (2005) Structural basis for photo-induced protein cleavage and green-to-red conversion of fluorescent protein EosFP. *Proc Natl Acad Sci USA* 102:9156–9159.
94. Fernández-Suárez M, Ting AY (2008) Fluorescent probes for super-resolution imaging in living cells. *Nat Rev Mol Cell Biol* 9:929–943.
95. Stiel AC, et al. (2008) Generation of monomeric reversibly switchable red fluorescent proteins for far-field fluorescence nanoscopy. *Biophys J* 95:2989–2997.
96. Andresen M, et al. (2008) Photoswitchable fluorescent proteins enable monochromatic multilabel imaging and dual color fluorescence nanoscopy. *Nat Biotechol* 26:1035–1040.
97. Huang B, Jones SA, Brandenburg B, Zhuang X (2008) Whole-cell 3D STORM reveals interactions between cellular structures with nanometer-scale resolution. *Nat Methods* 5:1047–1052.
98. Huang B, Wang W, Bates M, Zhuang X (2008) Three-dimensional super-resolution imaging by stochastic optical reconstruction microscopy. *Science* 319:810–813.
99. Juette MF, et al. (2008) Three-dimensional sub-100 nm resolution fluorescence microscopy of thick samples. *Nat Methods* 5:527–529.
100. Helmchen F, Denk W (2005) Deep tissue two-photon microscopy. *Nat Methods* 2:932–940.
101. Levi V, Ruan Q, Gratton E (2005) 3-D particle tracking in a two-photon microscope: application to the study of molecular dynamics in cells. *Biophys J* 88:2919–2928.
102. Levi V, Gratton E (2008) Chromatin dynamics during interphase explored by single-particle tracking. *Chromosome Res* 16:439–449.
103. Vaziri A, Tang J, Shroff H, Shank CV (2008) Multilayer three-dimensional super resolution imaging of thick biological samples. *Proc Natl Acad Sci USA* 105:20221–20226.
104. Shroff H, Galbraith CG, Galbraith JA, Betzig E (2008) Live-cell photoactivated localization microscopy of nanoscale adhesion dynamics. *Nat Methods* 5:417–423.

105. Biteen JS, Moerner WE (2010) Single-molecule and superresolution imaging in live bacteria cells. *Cold Spring Harb Perspect Biol* 2:a000448.
106. Biteen JS, et al. (2008) Super-resolution imaging in live Caulobacter crescentus cells using photoswitchable EYFP. *Nat Methods* 5:947–949.
107. Hess ST, et al. (2007) Dynamic clustered distribution of hemagglutinin resolved at 40 nm in living cell membranes discriminates between raft theories. *Proc Natl Acad Sci USA* 104:17370–17375.
108. Manley S, et al. (2008) High-density mapping of single-molecule trajectories with photoactivated localization microscopy. *Nat Methods* 5:155–157.
109. Tatavarty V, Kim E-J, Rodionov V, Yu J (2009) Investigating sub-spine actin dynamics in rat hippocampal neurons with super-resolution optical imaging. *PLoS ONE* 4:e7724.
110. Lang MC, Engelhardt J, Hell SW (2007) 4Pi microscopy with linear fluorescence excitation. *Opt Lett* 32:259–261.
111. Gustafsson MG, Agard DA, Sedat JW (1999) I5M: 3D widefield light microscopy with better than 100 nm axial resolution. *J Microscopy* 195:10–16.
112. Hell SW, Wichmann J (1994) Breaking the diffraction resolution limit by stimulated emission: stimulated-emission-depletion fluorescence microscopy. *Opt Lett* 19:780–782.
113. Dyba M, Jakobs S, Hell SW (2003) Immunofluorescence stimulated emission depletion microscopy. *Nat Biotechnol* 21:1303–1304.
114. Bretschneider S, Eggeling C, Hell SW (2007) Breaking the diffraction barrier in fluorescence microscopy by optical shelving. *Phys Rev Lett* 98:218103.
115. Folling J, et al. (2008) Fluorescence nanoscopy by ground-state depletion and single-molecule return. *Nat Methods* 5:943–945.
116. Heintzmann R, Jovin TM, Cremer C (2002) Saturated patterned excitation microscopy—a concept for optical resolution improvement. *J Opt Soc Am A Opt Image Sci Vis* 19:1599–1609.
117. Gustafsson MG (2005) Nonlinear structured-illumination microscopy: wide-field fluorescence imaging with theoretically unlimited resolution. *Proc Natl Acad Sci USA* 102:13081–13086.
118. Willig KI, et al. (2006) Nanoscale resolution in GFP-based microscopy. *Nat Methods* 3:721–723.
119. Schmidt R, et al. (2008) Spherical nanosized focal spot unravels the interior of cells. *Nat Methods* 5:539–544.
120. Schermelleh L, et al. (2008) Subdiffraction multicolor imaging of the nuclear periphery with 3D structured illumination microscopy. *Science* 320:1332–1336.
121. Kner P, Chhun BB, Griffis ER, Winoto L, Gustafsson MG (2009) Super-resolution video microscopy of live cells by structured illumination. *Nat Methods* 6:339–342.
122. Keller PJ, et al. (2010) Fast, high-contrast imaging of animal development with scanned light sheet-based structured-illumination microscopy. *Nat Methods* 7:637–642.
123. Hell SW (2007) Far-field optical nanoscopy. *Science* 316:1153–1158.
124. Kasper R, et al. (2010) Single-molecule STED microscopy with photostable organic fluorophores. *Small* 6:1379–1384.
125. Smolyaninov, II, Hung YJ, Davis CC (2007) Magnifying superlens in the visible frequency range. *Science* 315:1699–1701.

126. Sun MG, et al. (2007) Correlated three-dimensional light and electron microscopy reveals transformation of mitochondria during apoptosis. *Nat Cell Biol* 9:1057–1065.
127. Sakon JJ, Weninger KR (2010) Detecting the conformation of individual proteins in live cells. *Nat Methods* 7:203–205.
128. Meisslitzer-Ruppitsch C, Rohrl C, Neumuller J, Pavelka M, Ellinger A (2009) Photooxidation technology for correlated light and electron microscopy. *J Microscopy* 235:322–335.
129. Min W, et al. (2009) Imaging chromophores with undetectable fluorescence by stimulated emission microscopy. *Nature* 461:1105–1109.
130. Pertsinidis A, Zhang Y, Chu S (2010) Subnanometre single-molecule localization, registration and distance measurements. *Nature* 466:647–651.
131. Mortensen KI, Churchman LS, Spudich JA, Flyvbjerg H (2010) Optimized localization analysis for single-molecule tracking and super-resolution microscopy. *Nat Methods* 7:377–381.
132. Elf J, Li GW, Xie XS (2007) Probing transcription factor dynamics at the single-molecule level in a living cell. *Science* 316:1191–1194.
133. Heuser J (1980) Three-dimensional visualization of coated vesicle formation in fibroblasts. *J Cell Biol* 84:560–583.

10

STED Microscopy

U. VALENTIN NÄGERL

INTRODUCTION

STED microscopy breaks the classic diffraction barrier of light microscopy (1), long believed to be an unsurpassable limit for the spatial resolution of a light microscope. STED microscopy can shine light on dynamic processes inside living cells and biological tissue that have been inaccessible for conventional light microscopy, whose spatial resolution is limited by the diffraction of light to more than 200 nanometers (nm). It shares many advantages with confocal microscopy regarding contrast and optical sectioning, scanning and speed, and labeling and sensitivity.

In a STED microscope, a spot of focused laser light is used to scan across the sample of interest point-by-point, reading out the fluorescence signal along the way, just like in confocal microscopy, which is also a laser-scanning technique. A second laser is used to cut down the size of the fluorescence spot and hence increase the spatial resolution of the microscope beyond the classic diffraction barrier.

As the acronym of the technique implies (STED, *st*imulated *e*mission *d*epletion), this feat relies on stimulated emission. Interestingly, stimulated emission was predicted in a 1917 paper by Albert Einstein, referring to a process whereby an excited molecule interacts with an electromagnetic wave by giving up its energy to that field.

The idea that stimulated emission could be used to break the diffraction resolution limit of optical microscopy was proposed by Stefan Hell in 1994 (2). His seminal paper provided a basic scheme for how a STED fluorescence scanning microscope might work and gave estimates for the gain in resolution over conventional fluorescence microscopes. The first successful implementation of this scheme to break the diffraction limit was published in 1999 using nanocrystals (3), followed a year later by a demonstration inside living cells (4). Since those early proof-of-principle days, STED microscopy has undergone tremendous development, making it a powerful tool to advance the frontier of biological research.

Fluorescence microscopy is one of the most powerful and widely used imaging techniques in the biosciences, allowing visualization of dynamic processes inside living cells with exquisite sensitivity, contrast, and specificity. Thousands of publications each year involve fluorescence microscopy, so there is a huge interest to make it even more useful and powerful.

Several major developments in different areas combined to give rise to modern fluorescence microscopy, including laser and detector technology, microscope design, fluorescent probes, molecular biology and data handling and analysis. As a result, it is possible to study large and complex biological systems, such as whole brains, embryos, or even animals at the single-cell level, under realistic conditions and in a wide range of physiological contexts.

The development of confocal microscopy in the 1980s and two-photon microscopy in the 1990s were major microscopy advances that vastly improved the ability to image with high contrast deep inside biological tissue. Subsequently, the green-fluorescent protein (GFP) revolution allowed labeling of specific proteins and organelles in living cells. Last but not least, better computers and software made it possible to acquire and analyze large sets of imaging data. The Nobel Prize was awarded for the work on GFP in 2008, and the inventors of two-photon microscopy are likely to be honored ultimately.

Despite tremendous progress in many areas, it seemed for a long time that fluorescence microscopy was facing a brick wall when it came to its spatial resolution. The received wisdom was that the spatial resolution of light microscopy was fundamentally limited by the diffraction of light, and for small things to be resolvable under the light microscope, they had to have a size of at least half the wavelength of the light (i.e., larger than 200 ~ 300 nanometers). This limit had been referred to as a physical law for more than one hundred years. In fact Ernst Abbe's famous resolution formula (Eq. 1) is chiseled in stone at his workplace at Carl Zeiss in Jena, Germany.

If the diffraction barrier was indeed a hard limit, the study of cell biological phenomena occurring on the "mesoscale" of 10 ~ 200 nm would forever remain out of reach for fluorescence microscopy, including macromolecular complexes, cellular organelles, and subcellular signaling domains.

Electron microscopy, which has nanometer spatial resolution, has been for decades the method of choice to get pictures of cells and organelles on this spatial scale. However, it requires tissue fixation, which causes artifacts and which makes it hard to understand dynamic events, and hence cannot be performed in parallel with other complementary functional assays.

In this review, I will explain the basic principles behind STED microscopy and how it is implemented in practice. In addition, I will illustrate its potential by reviewing recent proof-of-principle applications and by going over some promising new technical developments.

THE ADVENT OF SUPER-RESOLUTION MICROSCOPY

Because of its wave nature, light cannot be condensed into an infinitesimally small spot. Rather, the smallest spot size that can be achieved by focusing lenses is limited by diffraction, which refers to the phenomenon whereby a wave tends to spread out as it travels through small openings (5). The actual size of the focal spot can be accurately predicted by optical physics, varying linearly with the wavelength of the light and inversely with the numerical aperture (NA) of the lens. This relationship is at the origin of Abbe's spatial resolution formula:

$$\Delta r = \frac{\lambda}{2^*NA} \tag{1}$$

where Δr defines the minimal distance between two objects for them to be detected as separate, λ is the wavelength of light, and NA denotes the numerical aperture of the lens.

If an image is "diffraction-limited," it is considered an achievement because it means that all other problem sources that degrade the image resolution (e.g., aberrations, light scattering, or a poor signal-to-noise ratio) have been reduced to a point where their effects on image resolution are negligible compared with the effect of diffraction, which places the final—or so it was thought—limit on spatial resolution of the optical system.

Whereas STED microscopy was the first concrete concept to break the diffraction limit (2, 4), other powerful techniques, relying on different principles, have also been developed for nano-imaging of fluorescent samples. These include PALM (*p*hoto*a*ctivated *l*ocalization *m*icroscopy) (6, 7), STORM (*s*tochastic *o*ptical *r*econstruction *m*icroscopy) (8), and nonlinear SIM (*s*tructured *i*llumination *m*icroscopy) (9, 10).

These new super-resolution techniques fall into two main categories. One group is those based on stochastic switching and computational localization of single molecules (PALM/STORM), and another is based on imaging ensembles of molecules with structured or patterned illumination (STED/nonlinear SIM) where super-resolution is achieved optically. Based on different optical principles, the techniques have specific strengths and weaknesses in terms of temporal resolution, depth penetration, multi-color imaging, instrumentation requirements, practical handling, and other characteristics.

They all share a common feature: in theory, no hard resolution limit exists and a resolution as high as a few nanometers can be achieved under ideal conditions. However, in practice they are limited by signal noise (e.g., from drift inherent in samples, particularly in living biological samples, and from detector noise) to some tens of nanometers.

THE BASIC CONCEPT OF STED MICROSCOPY

In general, the spatial resolution of an optical system is defined by the minimal distance between two fluorescent objects at which they can still be detected as separate objects. For a laser-scanning microscope, this distance is determined by the size of the focal spot because fluorescent objects that fall within the focal spot volume are excited at the same time and thus cannot be distinguished from each other by the detector. Therefore, to increase the spatial resolution of a laser-scanning microscope, one must decrease the size of the spatial extent over which dye molecules are co-excited.

In conventional (confocal or two-photon) laser-scanning microscopy, the excitation light is focused by the microscope's objective to a small focal spot that is systematically moved across the specimen, typically in two or three spatial dimensions. Thus, images are constructed one pixel at a time by measuring the fluorescence signal at each position along the trajectory of the laser focal spot with a sensitive and fast photodetector. Because of the diffraction of light, any microscope objective will always produce an intensity distribution of finite size, no matter how jitter-free the scanner or how well aligned the laser beam is. The spatial extent of this blurry spot is called the point-spread function (PSF), which is typically > 250 nm for confocal microscopy and even wider for two-photon microscopy (> 350 nm) because of the use of longer wavelength light (see Eq. 1).

The basic idea of STED microscopy is to reduce the size of the focal spot, and hence improve spatial resolution, by actively inhibiting the fluorescence on the edge of the focal spot. This is achieved by routing in a second laser beam (called the STED laser) tuned to a longer wavelength, which has a donut-shaped or annular intensity distribution in the focal plane. The intense light of the doughnut inhibits the fluorescence on the outer edge of the PSF by the process of stimulated emission at a wavelength that is longer than the fluorescence, allowing the fluorescence signal to be separately detected by standard optical filters.

The on/off switching of the fluorophores that underlies the engineering of the PSF, and hence the gain in spatial resolution, can be understood in terms of the energy diagram and transitions between ground (S_0) and excited (S_1) states of the fluorescent molecules, each of which has vibrational sublevels (Figure 10.1A).

After absorption of high-energy photons by the fluorescent molecules sending them into the first excited state (S_1), they rapidly pass (within picoseconds) to the lowest vibrational sublevel of the S_1 state. From there, they normally return to a vibrational sublevel of the S_0 state via spontaneous emission of a photon (= fluorescence) over an average period defined by the fluorescence lifetime τ_{fl}, which is typically a few nanoseconds for organic molecules and fluorescent proteins (e.g., ~4.1 nanoseconds for the Alexa Fluor 488 dye and ~ 2.8 nanoseconds for GFP).

In a STED microscope, the STED beam prevents fluorescence by forcing the molecules from the excited state to the ground state S_0 before fluorescence emission can happen. This occurs via stimulated emission, where an incident photon coming from the STED laser forces a downward transition from the excited state, thereby in effect "harvesting" a photon from the excited molecule, which has the exact same energy, momentum, and polarization. Like fluorescence, this process is also radiative, but the wavelength of the STED laser (λ_{STED}) is at the far right end of the fluorescence emission spectrum where it can be effectively filtered out without discarding too much of the fluorescence signal.

To achieve effective de-excitation and to prevent re-excitation by the STED laser, the molecules must be pumped into a higher vibrational sublevel of the S_0 state from which they rapidly escape to the final ground state S_0. This ensures that de-excitation by the STED laser predominates, and re-excitation is negligible. In this case, the switching off can be approximated by the following exponential relationship:

$$N \approx e^{-I_{STED}/I_S} \tag{2}$$

Where N is the number of molecules that can still fluoresce, I_{STED} is the intensity of the STED laser, and I_S is a constant defined by the fluorophore (11). Hence, if I_{STED} is three times larger than I_S, the fluorophore is essentially switched off.

If the inhibition is sufficiently strong ($I_{STED} >> I_S$), fluorescence can occur only from the center of the doughnut beam, where the intensity of the STED light is zero due

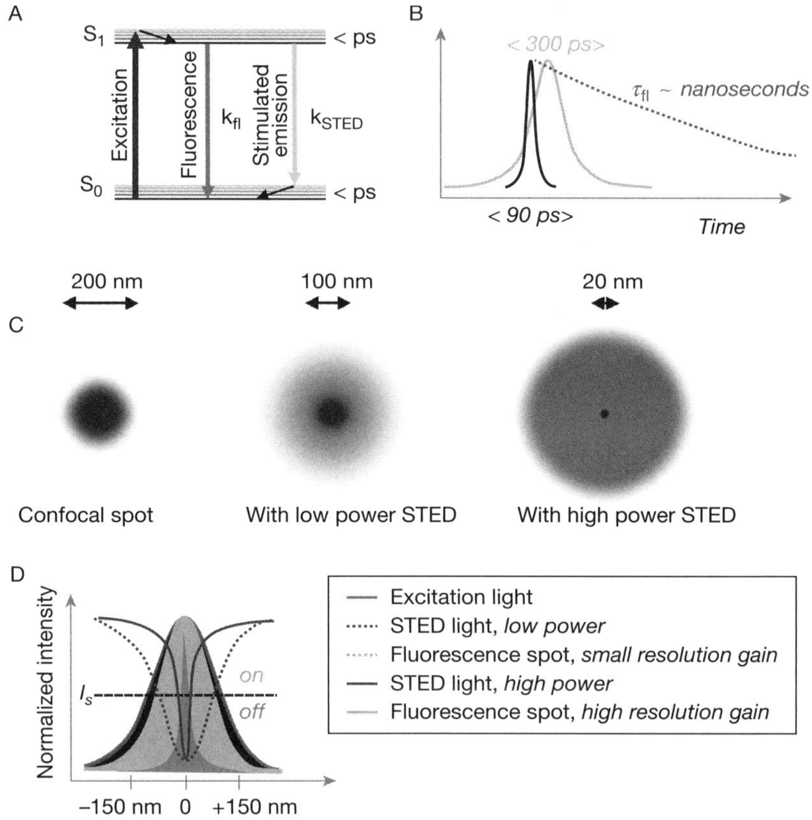

Figure 10.1 The principle behind STED microscopy. (A) Jablonksi diagram of the transitions of a fluorescent molecule, S_0 is the ground state, S_1 is the first excited singlet state, each with multiple vibrational sublevels. Absorption of a blue photon excites the dye molecule to S_1, from where it returns back to the ground state S_0 either via spontaneous emission of a green photon (fluorescence) or via stimulated emission by an orange photon. In fact, stimulated emission sends the molecules to an elevated (vibrational) sublevel of the ground state, from which they rapidly (in under a picosecond) decay to the final ground state. (B) The rationale behind using pulsed laser sources: After excitation by a brief pulse of blue laser light (90 picoseconds) the dye molecules normally relax back to the ground state via spontaneous emission of a green photon. The average time the molecules spend in the excited state is characterized by the lifetime (τfl), which is on the order of a few nanoseconds. If the STED pulse is timed so that it is delivered well within the lifetime τfl the fluorescence can be effectively inhibited. Provided the STED pulse is intense enough, most molecules will return to the ground state via stimulated emission instead of via spontaneous emission (fluorescence). Because of the short lifetime of the excited state, it makes sense to use synchronously pulsed lasers for STED microscopy, instead of continuous wave (cw) lasers, where the excitation and STED light is not synchronized. (C) Confining the fluorescence in space: the size of the green spot dictates the spatial resolution for a confocal microscope (*left*). The orange annulus (or "doughnut") is the focal intensity distribution of the STED laser. While it begins to suppress the fluorescence on the outer edge of the green spot for moderate STED laser intensities (*middle*), it drastically confines it at high STED laser power (*right*), leading in theory to a singularity of molecular size. (D) Normalized intensity profiles of the blue excitation laser, the STED laser at low (*dotted line*) and high (*solid line*) intensity, and the fluorescence distribution, before (*light green*) and after (*dark green*) strong STED suppression. See insert for color version. See insert for color version.

to destructive interference of the STED light by itself. The center area of the doughnut beam, which is called the null, can be made, in theory, arbitrarily small simply by increasing the power of the STED laser beam, I_{STED}. By saturating the quenching process on the rim of the doughnut, a very steep spatial gradient for molecules that are either "on" or "off" is created, constricting the area where fluorescence can still occur (Figure 10.1B).

In this way, the effective size of the fluorescent spot is reduced, and hence the spatial resolution of the optical system is improved. The current record in reducing the spot size is at 5.8 nm with diamond crystals. This is more than two orders of magnitude smaller than the wavelength of light that was used in the experiments (12).

Abbe's famous resolution formula picks up a second term in the denominator, which indicates that the spatial resolution barrier is not just shifted by a certain amount but is, in fact, fundamentally overcome (1):

$$\Delta r = \frac{\lambda}{2{*}NA{*}\sqrt{1 + I_{STED} / I_S}} \tag{3}$$

Because as the intensity of the STED beam becomes very large (I_{STED}, i.e., $I_{STED} >> I_S$), the smallest distance between two objects tends to zero (i.e., Δr). For most fluorophores tested so far, I_s is on the order of 100s of MW/cm^2 (peak power) when using STED pulses of 100–300 picoseconds, which corresponds to an average focal power of a few milliwatts. Whereas this is about three orders of magnitude greater than the average intensity of the excitation laser beam (which typically is just a few microwatts), the peak power of the STED beam is still substantially less than the one typically used in two-photon microscopy, which is on the order of 200 GW/cm^2.

Thus, the diffraction barrier of light microscopy can be effectively overcome by cleverly combining two laser beams, even though their intensity distributions in the focal plane are still dictated by diffraction. This may seem at first somewhat counterintuitive. However, one way to think about it is to look at I_S as a threshold value for the STED laser power, I_{STED}, above which the fluorescence is turned off. As I_{STED} can be arbitrarily increased on the rim of the doughnut while maintaining the null in the center, the threshold value I_S thus is reached closer to the center of the doughnut (Figure 10.1C). In this way, the distance to the threshold (i.e., where $I_S = I_{STED}$) can be made extremely short, much shorter than the spatial scale that diffraction normally imposes.

BUILDING A STED MICROSCOPE

Whereas the optical principle behind STED microscopy with its gain in spatial resolution is well established, the instrumentation and design for implementing it are still very much evolving, as the know-how and technology (e.g., lasers, scanners, detectors, and fluorophores) keep getting better and cheaper. Nevertheless, here we describe how a STED microscope based on pulsed excitation and pulsed STED lasers currently looks like. To begin with, a STED microscope resembles a regular laser-scanning confocal fluorescence microscope in terms of design and operation.

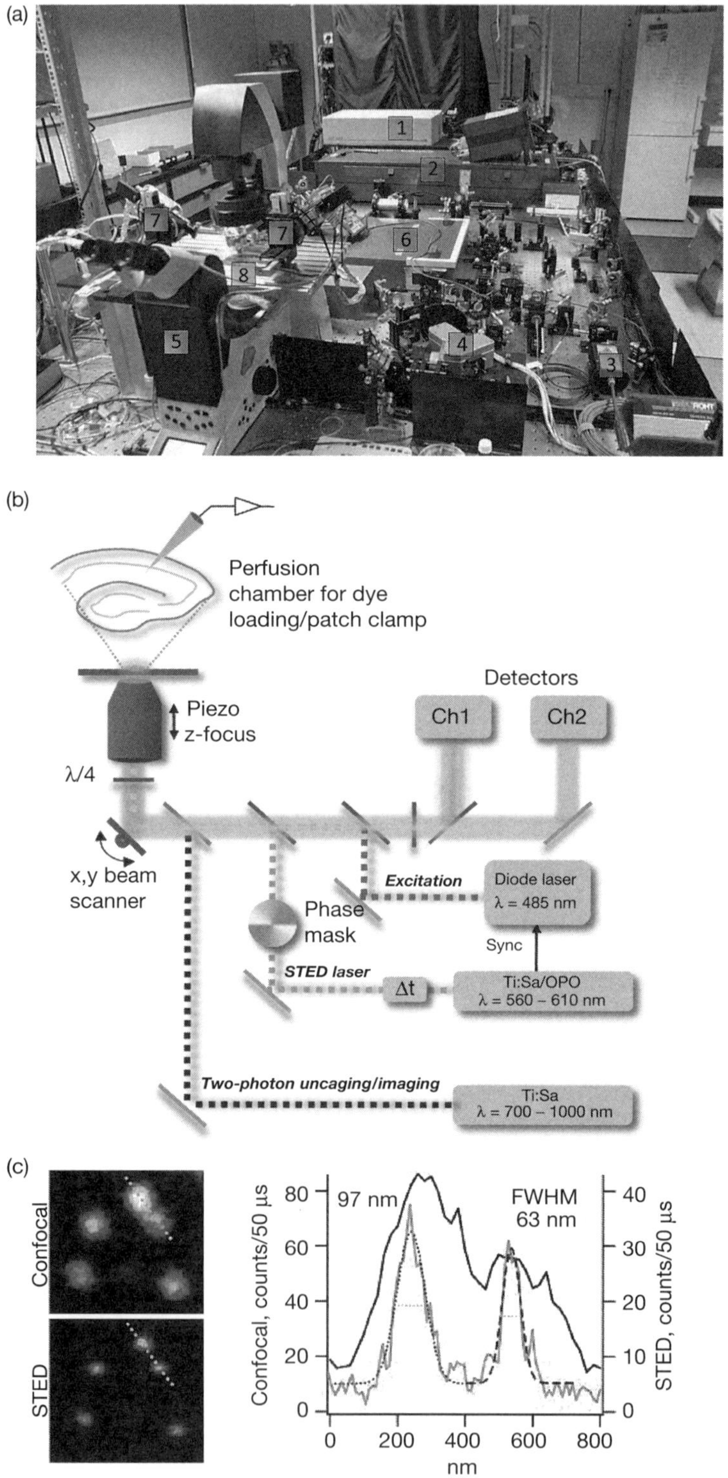

Figure 10.2 A homebuilt STED microscope. (A) Home-built STED microscope equipped for electrophysiological recordings and two-photon microscopy in living brain slices, including the following main components as indicated: 1: Femtosecond TiSa laser, 2: optical parametric oscillator (OPO), 3: Blue diode based excitation laser, 4: Beam

The Microscope Base

A home-built STED microscope for live-cell imaging can be built around an inverted microscope stand and equipped for electrophysiology and two-photon microscopy to carry out neurobiological experiments in brain slices (Figure 10.2A). Figure 10.2B presents a schematic of the system with the important optical components. Different kinds of standard research microscopes, upright or inverted, with or without motorized control of the wide-field epi-fluorescence and z-focus, can be used for this purpose.

Laser Sources

For excitation a fast pulsed laser diode (80 MHz pulses of 90 ps duration at a wavelength of 485 nm) can be used. For the STED light, a femtosecond Ti:Sapphire laser source is used in combination with an optical parametric oscillator (OPO), where the output is tuned to a wavelength of around 595 nm, which is suitable for quenching the fluorescence of GFP and yellow fluorescence protein (YFP).

The Doughnut

The light pulses of the STED beam must be stretched in time to reduce the chance of re-excitation of the fluorophore and two-photon absorption by the sample. This can be accomplished by sending the light from the OPO through a glass rod and a long optical fiber (100 meters). By pre-broadening the pulses from ~200 femtoseconds to a few picoseconds, the glass rod helps to protect its entry surface and avoid nonlinear effects inside the single-mode fiber. In the fiber, the pulses are stretched to about ~300 picoseconds. The STED beam then passes through a polymeric phase plate imprinting a helical phase delay on the wave front, which produces an annular intensity distribution in the focal plane of the lens, hence the term "doughnut."

Synchronization

The excitation and STED laser pulses need to be synchronized and overlaid in time. This can be achieved using external trigger pulses with variable delays. In the system shown here, the STED laser is the master and the excitation diode laser is the slave.

---

scanner, 5: Inverted confocal microscope (DMI 6000 from Leica), 6: Detector box with two avalanche photodiodes (APD) inside, 7: Micromanipulators for patch-clamp electrodes, 8: Microscope x-y stage with recording chamber. (B) Schematic representation of the system in (A). (C) Imaging fluorescent beads to determine the spatial resolution of the STED microscope, direct comparison with confocal case. Line profiles across beads to determine their FWHM with the STED laser switched off (black trace) and on (red trace). (reprinted from Tonnesen et al., *Biophys J*, 2011)

Objective Lenses

An objective lens with a high numerical aperture (e.g., HCX PL APO 100x/1.40 OIL CS or 63x/1.30 GLYC CORR CS) is crucial to focus the incident light into a diffraction-limited spot inside sample. The average focal power of the STED beam, which reduces the fluorescence by half, is on the order of 1 milliwatt for YFP at a repetition rate of 80 MHz. By increasing the STED power above that level, the resolution can be continuously adjusted. For excitation, an average focal power of 1–10 μW is sufficient; however, it can be adjusted depending on the brightness and photostability of the sample. The alignment of the excitation and STED lasers is done by looking at the back reflection of the excitation and doughnut beams using gold beads (diameter = 100 nm).

Scanning Schemes

Super-resolved STED images can be acquired by superimposing the excitation and STED laser beams in all spatial dimensions and in time and moving them together across the sample using one of the following schemes for scanning, each of which having specific strengths and weaknesses. 1) Beam-scanning with galvanometer-driven mirrors allows for versatile and fast scanning over large fields of view. 2) Stage-scanning, where it is the sample that moves using a piezo stage, is very precise, optically more simple, and stable; on the downside, it is relatively slow and makes it difficult, for instance, to include micromanipulators for electrophysiological recordings, which must not move relative to the sample during scanning. 3) In a hybrid of the two, the beam is rapidly moved along one axis by means of a resonant mirror, while the sample is moved more slowly along the perpendicular axis by the piezo stage (13). Although piezo stages usually do not come with large travel ranges, beam-scanning allows for relatively large fields of view (>200 micrometers).

Emission Detection

The epi-fluorescence signal is separated from the excitation light (Figure 10.3) by a dichronic mirror. After passing through a bandpass filter to discard the STED light and any remaining excitation light, the signal photons are collected by a photodetector, such as an avalanche photodiode in photon-counting mode, or a photomultiplier (PMT). APDs have higher quantum yields (50–60 percent at 535 nm) than PMTs and virtually no dark counts, but their signal response becomes nonlinear at count rates larger than 2 MHz. For convenience, a multimode optical fiber (MMF) can be used to pipe the photons to the APD, with the small fiber head serving as a confocal pinhole, which blocks out out-of-focus light. In principle, a pinhole is not needed, as the STED resolution gain comes from the interaction of the laser beams in the focal plane with the fluorophore.

Software

As scanning and image acquisition in STED microscopy is nearly identical to the one used for traditional scanning microscopy, any scanning software can be used for data

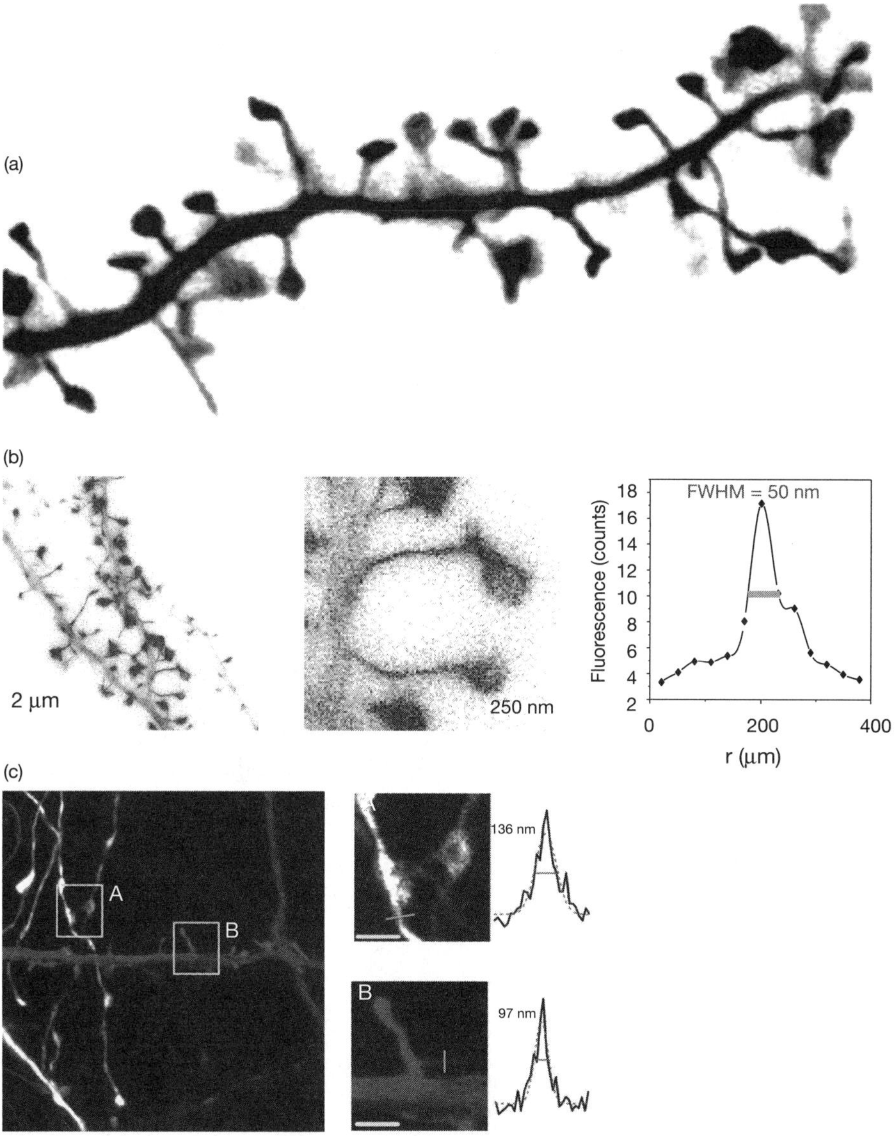

Figure 10.3 (A) Maximum intensity projection of a stretch of dendrite of a pyramidal neuron filled with YFP based on a 3D stack of STED images (scale bar, 1 micron). (B) Imaging the cytoskeleton of synapses at the nanoscale: Distribution of the cytoskeletal protein actin inside a dendritic spine imaged by STED using YFP-*lifeact* as label for actin. (reprinted from Urban et al, *Biophys J*, 2011). (C) Two-color STED microscopy of axons and dendrites using fluorescent proteins (GFP, YFP) and organic dyes (Alexa Fluor 488, Calcein-green) as volume markers (scale bar, 1 micron). (reprinted from Tonnesen et al., *Biophys J*, 2011). See insert for color version.

acquisition and instrument control of a STED setup. However, STED microscopy requires special attention to the adjustment of the excitation and the STED beam and conventional software packages can be somewhat clumsy for this purpose. The commercially available software package *ImSpector* (14) is a useful solution for this issue, as it facilitates rapid PSF measurements.

In contrast, with regard to the software to analyze images, there are no special requirements, and one can fully rely on standard commercial or freely available image analysis packages (e.g., ImageJ, Matlab, Imaris). Finally, STED images can be further enhanced by deconvolution algorithms, just like conventional confocal images (e.g. Huygens Professional from Scientific Volume Imaging, Hilversum, The Netherlands). However, unlike other nanoscopy techniques, image processing is, in principle, not needed to obtain super-resolution images.

Stability Considerations

Given the substantial improvement in spatial resolution in a STED microscope, the hardware requirements are accordingly more stringent. For instance, the electronic circuit boards that drive the galvanometers controlling the mirror position in a scan head are constructed such that the position noise of the mirror is substantially below the resolution limit of a two-photon or confocal microscope and, hence, this noise is not noticeable.

However, with an improvement in spatial resolution of about one order of magnitude these noise specs are not sufficiently good anymore: they would degrade the STED pictures substantially. Whereas in general the architecture of the hardware can stay the same, the requirements in terms of electronic noise, thermal and mechanical stability, are substantially more tight in a STED microscope, so that in many instances components have to be upgraded and improved.

Checking the Resolution

The performance of the optical system is checked by imaging sub-diffraction-sized fluorescent beads (e.g., diameter = 40 nm). After switching the STED laser on, the beads appear substantially smaller (Figure 10.2C), as confirmed by measuring their full-width-at-half-maximum (FWHM) using Gaussian or Lorentzian curve fits.

MAKING STED MICROSCOPY MORE USEFUL

While STED microscopy certainly is much more powerful than confocal/two-photon microscopy in terms of spatial resolution, it has some disadvantages regarding other important aspects, such as choice of suitable dyes, multi-color imaging, signal-to-noise ratio, temporal resolution, compatibility with live-cell imaging, imaging deep inside complex biological samples, and so on.

These drawbacks add on top of the relatively high cost and complexity to set up and maintain a STED microscope, either a commercial system or a home-built version. Not surprisingly, its use in biological research is still fairly limited, despite its ground-breaking potential and all the fanfare. Its adoption has been hampered by the view of many biologists that its design is prohibitively complex and its scope too limited. However, this view is bound to change as technical improvements and commercial developments make STED microscopy more affordable, easier, and versatile.

Labeling Strategies

An important challenge for STED microscopy lies in labeling and finding dyes with suitable photo-physical properties that can be targeted to specific organelles or proteins of interest inside cells. Several strategies exist, based on physical loading, transgenics, and/or chemistry, each having distinct weak and strong points (e.g., live-cell compatibility, specificity, choice of dyes, generalizability). The most common way to achieve strong and specific labeling is based on immune-histochemistry with particularly photostable and bright organic dyes (e.g., Atto647N and Chromeo494) conjugated to secondary antibodies. This approach, of course, requires tissue fixation and membrane permeabilization, and thus does not permit investigations in living cells.

More recently, other labeling strategies successfully used labels that are compatible with STED imaging in live preparations. Transgenic expression of fluorescent proteins (e.g., GFP, YFP, and variants) was used to image subcellular organelles (15) or neuronal morphology (16) with a resolution better than 100 nm in live cells. In addition, organic dyes can be introduced directly into single cells via patch pipettes, which was done for STED imaging of volume-labeled dendritic spines in brain slices with popular Alexa dyes (17, 18). Finally, SNAP-tag and related approaches were recently used to label genetically encoded fusion-proteins with exogenous organic dyes for STED imaging in live cells (19, 20).

Multi-Color Imaging

There is a clear need for multi-color solutions that integrate live-cell super-resolution laser scanning microscopy with existing fluorescent labeling strategies and markers. For example, synapse biology would benefit greatly from multi-color STED imaging, as synapses are composed of pre- and postsynaptic compartments ensheathed by glial processes, which are typically too small to be properly resolved by conventional microscopic techniques.

As STED microscopy uses two separate laser beams, one for fluorescence excitation and another for fluorescence quenching, it is more difficult to incorporate multi-color imaging than for conventional light microscopy.

However, several solutions exist for two-color imaging with STED microscopy, again each bearing particular strengths and weaknesses, in particular with regard to cost and complexity of the STED setup and the choice of fluorophores. The first viable scheme for two-color STED microscopy relied on separate excitation and STED lasers for combinations of green and red fluorophores (21, 22). The approach requires duplicating the lasers, worrying about their alignment, and chromatic aberrations. In addition, it runs the risk that the STED laser for the green dye ends up exciting the red dye, causing excessive bleaching and phototoxicity. Subsequently, a simpler approach was developed, which is also based on two excitation lasers, but only a single STED laser

(23). It uses a pair of dyes with a differential Stokes shift, where the green dye undergoes a much larger shift than the red dye, allowing both dyes to be quenched by a single STED laser beam. This approach also obviates the problem of excitation by the second STED laser. It was recently combined with SNAP-tag and CLIP-tag labeling to achieve two-color STED imaging in live cells (20). The idea to use two dyes with very different Stokes shifts for two-color nanoscopy is implemented in the commercial STED system from Leica with Chromeo494 and Atto647N as dye pair.

Additional two-color nanoscopy schemes have been developed, which use just a single pair of excitation and STED lasers. These approaches achieve spectral separation either by: 1) using dyes with differential fluorescence lifetimes (24); 2) using a pair of photo-switchable fluorescent proteins (padron and dronpa) (25) that are spectrally similar, but which can be specifically turned on or off by flashes of blue laser light (405 nm), allowing for sequential read-out of their signals; or 3) simultaneously imaging two spectrally similar fluorophores, followed by linear unmixing of the fluorescence intensity signal (18). The last approach works readily for pairs of popular green-yellow fluorescent labels, such as GFP, YFP, and Alexa Fluor 488, and is particularly easy to implement because it merely involves the addition of a second detector.

Importantly, since the same pair of lasers is used for different dyes, the measurements are effectively monochromatic, avoiding chromatic aberrations that could otherwise affect colocalization accuracy.

Temporal Resolution and Imaging Speed

The speed of STED imaging is defined by the number of pixels and the pixel dwell time, just like for any single point-scanning technique. Usually, low signal counts are rate-limiting. Only if the sample is extremely bright does the hardware (i.e., the speed of the scanner) become the limiting factor. Much like for confocal and two-photon microscopy, there is a trade-off between temporal and spatial resolution (e.g., acquisition speeds up to a few kilohertz can be achieved in line-scan mode, whereas larger images can take up to several seconds).

With the enhancement in spatial resolution, pixel sizes should be reduced to comply with the Nyquist sampling theorem, which says that images should be acquired with a pixel size of less than half the optical resolution, so typically 30 nm or less. This means that, for an increase in lateral spatial resolution of a factor of 5, the acquisition time for a field of view of same size should increase by a factor of $5 \times 5 = 25$. In practice, the field of view is reduced substantially to keep the frame rate reasonable.

The pixel dwell time is longer for STED than for confocal or two-photon imaging to account for the fact that signal intensity in the center of the doughnut is typically cut in half, which means the frame acquisition time is twice as long.

Using a stage-scanning system, moderate acquisition times of around 10 seconds per frame were used for imaging volume-labeled neuronal morphology over 15 micrometer fields of view (16). Using a faster scanning system based on a resonant mirror, STED imaging of synaptic vesicles at video-rate rate could be performed on small scan areas (2.5×1.8 micrometers) using a bright organic dye (13). More recently, multi-spot scanning with parallelized beams was used to increase the speed of STED microscopy by a factor of four (26).

Live-Cell Imaging

Much of the appeal of fluorescence microscopy stems from its ability to study dynamic cell biological processes inside living cells, setting it apart from electron microscopy. Naturally, there is a huge interest in getting the new super-resolution techniques to work in living cells.

A few years ago the idea of STED microscopy in living cells was met with quite a bit of skepticism, given difficulties with labeling at the time and concerns over photodynamic damage induced by the STED laser. Fortunately, these and other challenges and fears have largely been overcome.

Concerning the photodynamic damage that may be caused by the STED laser, it must be kept in mind that (a) the STED beam intensity ($I = 400$ MW/cm^2) is much lower than intensity typically used for two-photon microscopy (200 GW/cm^2), and (b) that the absorption by water is much lower at the wavelength of 598 nm than in the near infrared range, resulting in negligible temperature rises in the focal region (<0.2K) (27).

Whereas most early studies with STED were carried out on fixed tissue preparations, several papers over the last couple of years have used it with living samples, such as cell lines (15, 20), dissociated neurons (13), and brain slices (16–18, 28) (Figure 10.3A, B, C).

Depth Penetration and Spatial Resolution

Many important questions in biology require complex, thick tissue preparations, such as brain slices or even intact brains. Confocal microscopy provides a way to optically section three-dimensional fluorescent samples and acquire images with high signal-to-noise ratios several cell layers below tissue surface. Two-photon microscopy further improves on depth penetration and optical sectioning by the use of infrared light for two-photon fluorescence excitation.

However, with regard to the new nanoscopy techniques, the gain in spatial resolution has so far been demonstrated mostly for very thin or surface preparations. The achievable spatial resolution usually deteriorates quickly with imaging depth because of scattering, absorption, and aberrations induced along the optical path and by the sample. While this is true for any laser-scanning light microscopy technique, these effects are particularly damaging for super-resolution microscopy.

In general, spherical aberrations are caused by a mismatch of refractive indices between the immersion medium of the objective lens and the tissue sample (29). Several strategies exist to reduce both effects, and thus extend the ability to acquire super-resolved images deeper into living tissue.

The use of two-photon light for excitation reduces scattering of the excitation spot, and it was recently shown to be compatible with STED microscopy for imaging deep inside living brain slices using a water-immersion objective with a long working distance (34, 35).

Adaptive optics can be used to reduce aberrations, and thus to ensure a high-quality STED doughnut. Using a glycerine immersion objective with an aberration-reducing correction ring, it was recently demonstrated that it is possible to achieve a lateral spatial resolution of 60 nm at imaging depths of around 100 micrometers below the surface (28) of living biological tissue.

CONCLUDING REMARKS

Progress in expanding the scope and refining the use of STED microscopy for the biosciences is taking place at a rapid clip, which is going to accelerate as more labs worldwide join the effort.

RESOLFT: A More General Principle for Nanoscopy

De-excitation by stimulated emission is not the only process by which a fluorophore can be deactivated to achieve super-resolution. It was shown that other switching mechanisms than quenching, such as photo-switching between excitable and non-excitable states, can also achieve spatial resolution beyond the diffraction limit (36, 37). The more general concept behind STED microscopy is referred to as RESOLFT (reversible saturable optical fluorescence transitions) (38), which was recently implemented in a massively parallelized way, dramatically improving temporal resolution (39).

Key Areas of Development

Other promising technical developments (beyond the ones reviewed here) include "iso-STED," where the gain in resolution is extended into the z axis (23, 32), "gated-STED," which refers to time-resolved schemes to improve the spatial resolution of STED microscopy even further (33), "in-vivo STED" (i.e., intra-vital super-resolution imaging in intact brains of anesthetized animals,), which was recently demonstrated (40), and "integrated STED", where other modern biophysical approaches (2-photon photolysis, electrophysiology, FRAp experiments) are carried out in parallel to the STED imaging (41).

Thanks to the development of super-resolution microscopy, it is now possible to resolve details at the nanoscale in living biological specimens using fluorescence microscopy, which was dismissed as a pipe dream just a few years ago. Therefore, STED microscopy and other super-resolution techniques stand to significantly impact bioscience research in the foreseeable future.

REFERENCES

1. Hell SW (2007) Far-field optical nanoscopy. *Science* 316(5828):1153–1158.
2. Hell SW & Wichmann J (1994) Breaking the diffraction resolution limit by stimulated emission: stimulated-emission-depletion fluorescence microscopy. *Opt Lett* 19(11):780–782.
3. Klar TA & Hell SW (1999) Subdiffraction resolution in far-field fluorescence microscopy. *Opt Lett* 24(14):954–956.
4. Klar TA, Jakobs S, Dyba M, Egner A, & Hell SW (2000) Fluorescence microscopy with diffraction resolution barrier broken by stimulated emission. *Proc Natl Acad Sci U S A* 97(15):8206–8210.

5. Born M & Wolf E (1999) *Principles of optics: electromagnetic theory of propagation, interference and diffraction of light* (Cambridge University Press, Cambridge; New York) 7th expanded Ed pp xxxiii, 952 p.
6. Betzig E, et al. (2006) Imaging intracellular fluorescent proteins at nanometer resolution. *Science* 313(5793):1642–1645.
7. Hess ST, Girirajan TP, & Mason MD (2006) Ultra-high resolution imaging by fluorescence photoactivation localization microscopy. *Biophys J* 91(11):4258–4272.
8. Bates M, Huang B, Dempsey GT, & Zhuang X (2007) Multicolor super-resolution imaging with photo-switchable fluorescent probes. *Science* 317(5845):1749–1753.
9. Gustafsson MG (2005) Nonlinear structured-illumination microscopy: wide-field fluorescence imaging with theoretically unlimited resolution. *Proc Natl Acad Sci U S A* 102(37):13081–13086.
10. Heintzmann R, Jovin TM, & Cremer C (2002) Saturated patterned excitation microscopy—a concept for optical resolution improvement. *J Opt Soc Am A Opt Image Sci Vis* 19(8):1599–1609.
11. Dyba M & Hell SW (2002) Focal spots of size lambda/23 open up far-field fluorescence microscopy at 33 nm axial resolution. *Phys Rev Lett* 88(16):163901.
12. Rittweger E, Han KY, Irvine SE, Eggeling C, & Hell SW (2009) STED microscopy reveals crystal colour xcentres with nanometric resolution. *Nature Photonics* 3(March 2009).
13. Westphal V, et al. (2008) Video-rate far-field optical nanoscopy dissects synaptic vesicle movement. *Science* 320(5873):246–249.
14. Schonle A (2011) Imspector. www.max-planck-innovation.de/de/industrie/technologieangebote/software.
15. Hein B, Willig KI, & Hell SW (2008) Stimulated emission depletion (STED) nanoscopy of a fluorescent protein-labeled organelle inside a living cell. *Proc Natl Acad Sci U S A* 105(38):14271–14276.
16. Nägerl UV, Willig KI, Hein B, Hell SW, & Bonhoeffer T (2008) Live-cell imaging of dendritic spines by STED microscopy. *Proc Natl Acad Sci U S A* 105(48):18982–18987.
17. Ding JB, Takasaki KT, & Sabatini BL (2009) Supraresolution imaging in brain slices using stimulated-emission depletion two-photon laser scanning microscopy. *Neuron* 63(4):429–437.
18. Tonnesen J, Nadrigny F, Willig KI, Wedlich-Soldner R, & Nägerl UV (2011) Two-color STED microscopy of living synapses using a single laser-beam pair. *Biophys J* 101(10):2545–2552.
19. Hein B, et al. (2010) Stimulated emission depletion nanoscopy of living cells using SNAP-tag fusion proteins. *Biophys J* 98(1):158–163.
20. Pellett PA, et al. (2011) Two-color STED microscopy in living cells. *Biomed Opt Express* 2(8):2364–2371.
21. Donnert G, et al. (2007) Two-color far-field fluorescence nanoscopy. *Biophys J* 92(8):L67–69.
22. Meyer L, et al. (2008) Dual-color STED microscopy at 30-nm focal-plane resolution. *Small* 4(8):1095–1100.
23. Schmidt R, et al. (2008) Spherical nanosized focal spot unravels the interior of cells. *Nat Methods* 5(6):539–544.
24. Buckers J, Wildanger D, Vicidomini G, Kastrup L, & Hell SW (2011) Simultaneous multi-lifetime multi-color STED imaging for colocalization analyses. *Opt Express* 19(4):3130–3143.

25. Willig KI, Stiel AC, Brakemann T, Jakobs S, & Hell SW (2011) Dual-label STED nanoscopy of living cells using photochromism. *Nano Lett* 11(9):3970–3973.
26. Bingen P, Reuss M, Engelhardt J, & Hell SW (2011) Parallelized STED fluorescence nanoscopy. *Opt Express* 19(24):23716–23726.
27. Schonle A & Hell SW (1998) Heating by absorption in the focus of an objective lens. *Opt Lett* 23(5):325–327.
28. Urban NT, Willig KI, Hell SW, & Nägerl UV (2011) STED nanoscopy of actin dynamics in synapses deep inside living brain slices. *Biophys J* 101(5):1277–1284.
29. Hell SW, Reiner G, Cremer C, & Stelzer EH (1993) Aberrations in confocal fluorescence microscopy induced by mismatches in refractive index. *Journal of Microscopy* 169(3):64.
30. Li Q, Wu SS, & Chou KC (2009) Subdiffraction-limit two-photon fluorescence microscopy for GFP-tagged cell imaging. *Biophys J* 97(12):3224–3228.
31. Moneron G & Hell SW (2009) Two-photon excitation STED microscopy. *Opt Express* 17(17):14567–14573.
32. Wildanger D, Medda R, Kastrup L, & Hell SW (2009) A compact STED microscope providing 3D nanoscale resolution. *J Microsc* 236(1):35–43.
33. Vicidomini G, et al. (2011) Sharper low-power STED nanoscopy by time gating. *Nat Methods* 8(7):571–573.
34. Takasaki KT, Ding JB, Sabatini BL (2013) Live-cell superresolution imaging by pulsed STED two-photon excitation microscopy. Biophys J 104:770–777.
35. Bethge P, Chereau R, Avignone E, Marsicano G, Nägerl UV (2013) Two-photon excitation STED microscopy in two colors in acute brain slices. Biophys J 104:778–785.
36. Grotjohann T, Testa I, Leutenegger M, Bock H, Urban NT, Lavoie-Cardinal F, Willig KI, Eggeling C, Jakobs S, Hell SW (2011) Diffraction-unlimited all-optical imaging and writing with a photochromic GFP. Nature 478:204–208.
37. Schwentker MA, Bock H, Hofmann M, Jakobs S, Bewersdorf J, Eggeling C, Hell SW (2007) Wide-field subdiffraction RESOLFT microscopy using fluorescent protein photoswitching. Microsc Res Tech 70:269–280.
38. Hofmann M, Eggeling C, Jakobs S, Hell SW (2005) Breaking the diffraction barrier in fluorescence microscopy at low light intensities by using reversibly photoswitchable proteins. Proc Natl Acad Sci U S A 102:17565–17569.
39. Chmyrov A, Keller J, Grotjohann T, Ratz M, d'Este E, Jakobs S, Eggeling C, Hell SW (2013) Nanoscopy with more than 100,000 'doughnuts'. Nat Methods 10:737–740.
40. Berning S, Willig KI, Steffens H, Dibaj P, Hell SW (2012) Nanoscopy in a living mouse brain. Science 335:551.
41. Tonnesen J, Katona G, Rozsa B, Nagerl UV (2014) Spine neck plasticity regulates compartmentalization of synapses. Nat Neurosci 17:678–685.

11

Imaging the (Macro)Molecular Composition

Mass Spectrometry Imaging

BRENDAN PRIDEAUX

Coupling the selectivity of mass spectrometry with the spatial information offered by traditional molecular imaging techniques has opened up the potential for a whole new field of ex-vivo label-free biomedical imaging. The unique value of this approach, known as mass spectrometry imaging (MSI), is the ability to directly image molecules on a tissue surface, thereby providing vital information for the understanding of disease progression and potential pharmaceutical therapy.

Each pixel of an MS image consists of a mass spectrum, and that spectrum may cover a wide mass range encompassing many spectral peaks, diagnostic fragment ions (product) of a selected ion (precursor), or specific ion transitions (precursor ion to a single product ion). A selected spectral peak (or mass range) can be plotted to produce an ion map of the relative amount of that compound over the entire imaged tissue.

As gas phase ions generated from the surface of a biological tissue are directly analyzed based on their mass, there is no requirement for a tracer label. This enables the distribution of thousands of analytes to be imaged in a single experiment and eliminates the need for expensive and time-consuming labeling steps.

MASS SPECTROMETRY AS AN IMAGING TOOL

It would not be an exaggeration to say that mass spectrometry has revolutionized the field of bioanalytics since first applied. The advent of soft ionization sources, specifically matrix assisted laser desorption ionization (MALDI) by Karas, Bachman, and Hillenkamp in 1985 (1) (and for larger intact proteins by Tanaka et al. in 1988 (2)) and electrospray ionization (ESI) by Fenn et al. in 1989 (3), enabled the ionization of intact biomolecules covering a mass range from small lipids and metabolites up to large proteins. The ability to simultaneously detect thousands of analytes within one sample in a single experiment allows for a quick and efficient profiling of a biological sample as well as identification of anaylte ions based upon their specific mass-to-charge (*m/z*) ratio or the diagnostic fragments produced upon dissociation of intact ions in tandem MS (MS/MS) mode.

Mass spectrometry in its simplest form involves the formation of gaseous ions (positive or negative) from a sample, their subsequent separation based upon their *m/z* ratio, and detection.

A mass spectrometer consists of three main components:

1. Ion source—to produce ions in the gas phase from the sample
2. Mass analyzer—where ions are resolved according to their *m/z* ratio
3. Detector—to detect the ions and record their relative abundances

Multiple technologies are available for each of these key components, and each has specific analytical strengths and weaknesses. The ion source and mass analyzer are of the upmost importance in an MS imaging experiment, and these will be discussed in greater detail later in the chapter.

RESOLUTION

The term "resolution" is frequently referred to in MS imaging. There are two main types of resolution of particular significance for any imaging experiment: the spectral resolution (otherwise known as the mass resolution) and spatial resolution.

Mass resolution refers to the ability to distinguish two peaks of an *m/z* spectrum that are closely related in mass. It is of particular importance in both the resolution and identification of the many peaks with similar *m/z* values that exist in complex biological samples.

The achievable mass resolution in an imaging experiment depends upon the type of mass analyzer used. Fourier transform ion cyclotron resonance (FT-ICR) and orbital trapping (Orbitrap) analyzers can produce the highest mass resolution (within the early parts per million range), and both have been used for MALDI-MS imaging (4, 5). One striking application was provided by Cornett et al. who utilized the high mass resolution capabilities of FT-ICR to resolve the distribution of a primary metabolite of the drug olanzapine within rat kidney from three near-isobaric species within an *m/z* range of 0.2 daltons (6).

MS imaging datasets produced from such high mass resolution experiments will be invariably large (correlating to the total mass range measured) and pose significant problems in the subsequent data handling and processing steps. Current time-of-flight (TOF) analyzers have mass resolving powers as high as 60,000 with mass accuracy capabilities below 10 ppm (7) and can maintain high-speed acquisition over a wide mass range. They are the most common mass analyzers in MS imaging and have been used to image the distributions of biological species (lipids (8), peptides (9), and proteins (10, 11)), drugs, and metabolites (12). Quadrupole mass analyzers offer the lowest resolving power, and as such, they only have a limited use in the imaging and identification of analytes from a complex sample when operated in a full scanning mode. Triple quadrupole mass spectrometry imaging instruments have recently been introduced (13). The advantage offered by the quadrupole mass analyzer is the sensitivity, selectivity, and speed of the analysis when monitoring specific ion transitions either in selected reaction monitoring (SRM) or multiple reaction monitoring (MRM) imaging modes. Due to their suitability for targeted MS imaging, they have rapidly found acceptance for pharmaceutical (13, 14) and toxicology applications (15). Hybrid quadrupole time of flight instruments (QqTOF) combine attributes of TOF such as the wider *m/z* dynamic range and higher mass resolution capabilities with the collisional induced fragmentation and precursor ion isolation capabilities of quadrupoles. They have proved extremely popular for the imaging of low molecular weight analytes, such as drugs and metabolites (16, 17), lipids (18) peptides, and small proteins (19).

The term spatial resolution when applied to MS imaging refers to the ability to resolve spatial features within the image. Typically, this is discussed in terms of pixel size in the resulting MS image; however, it is important to realize that factors due to sample preparation and instrumental limitations can result in the actual spatial resolution being lower than the pixel size selected during the setup of the experiment (such as lateral diffusion of the analyte during wet sample preparation and the diameter of the ionizing beam profile). The spatial resolution capabilities of the different MS imaging ionization methods will be discussed in detail later in this chapter.

Increasing the spatial resolution of an imaging experiment will linearly increase the time required to complete the acquisition of the image and the size of the resulting data file. With this consideration in mind, a spatial resolution should be selected allowing localization of the analyte within the desired tissue structures, but within time and data constraints.

IONIZATION MODES FOR MS IMAGING

There are many ways of converting molecules from the tissue surface into gas phase ions, and many of these have been utilized for mass spectrometry imaging. This chapter discusses the techniques that are the most commonly used in current imaging experiments, namely MALDI, SIMS, and DESI (a diagrammatic overview is shown in Figure 11.1).

In addition to the aforementioned techniques, several other ionization methods have been utilized including laser desorption ionization (LDI) (21), matrix enhanced surface assisted laser desorption ionization (ME-SALDI) (22), nanostructure initiator

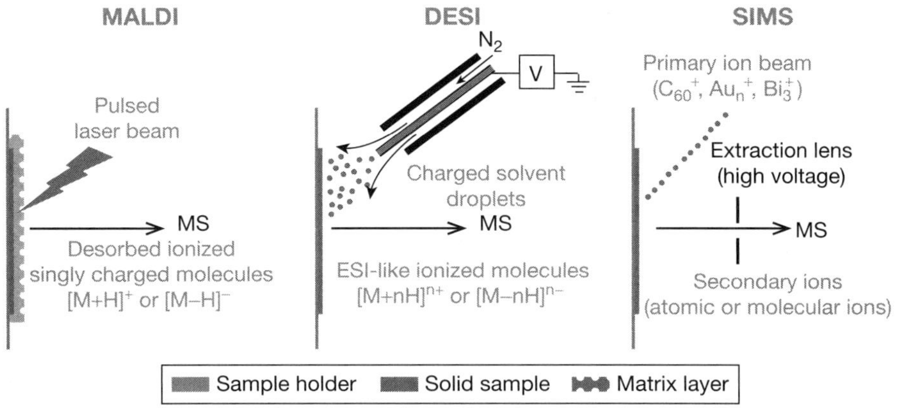

Figure 11.1 Diagrammatic overview of the three main MS imaging ionization techniques. Reproduced with permission (19)

mass spectrometry (NIMS) (23), and laser ablation electrospray ionization (LAESI) (24). These techniques, although more limited than the common approaches, do have advantages for certain niche applications and with further development may play a more prominent role in the future of MSI.

MATRIX ASSISTED LASER DESORPTION IONIZATION–MASS SPECTROMETRY IMAGING

First demonstrated for biological imaging in 1997 by Caprioli et al. (25), MALDI-MS imaging has since become the most frequently used imaging technology for the analysis of biological tissues. The MALDI ionization process is complex, and the exact mechanisms are still being debated (26–28). However, the UV absorbing matrix plays the crucial role, and both its selection and application will define the success of the imaging experiment. In most instances, the matrix is applied in solution to the tissue surface. The analytes within the tissue are extracted by the solution and become co-crystalized with the matrix layer at the surface of the tissue. A UV laser is directed at the tissue and ablates the matrix layer containing the analytes extracted from the tissue. During the ablation process, matrix and analyte molecules and clusters are ejected into the gas phase, causing the formation of an ion plume immediately above the surface of the tissue. The most frequently used matrices are organic acids, in particular α-cyano-4-hydroxycinnamic acid (for drugs, metabolites, and peptides), 2,5-dihydroxybenzoic acid (for drugs metabolites and lipids), and sinapinic acid (for larger peptides and proteins).

The MALDI ionization process is classified as a "soft" ionization technique because it enables the detection of intact biomolecules rather than the small fragments that result from harder ionization methods. As conclusive identification of biomolecules from hard ionization-derived fragments is difficult (often impossible) and as MALDI ionization is capable of delivering such a wide mass range of ions into the mass spectrometer, this has led to it becoming the most commonly utilized ionization method in biological mass spectrometry imaging experiments.

The ionizing laser is of great importance in the MALDI ion source. In the early days of MALDI-MS imaging, a N_2 laser at 337 nm wavelength was most frequently utilized. These were small, easy to incorporate into an MS instrument and relatively inexpensive. However, they have several important limitations for MS imaging applications, particularly the low repetition rate (typically 50 Hz in commercial instrumentation), which limits the speed of acquisition of the MS image. Additionally, the lifespan of a nitrogen laser is relatively short at 2×10^7–6×10^7 shots. Thus, imaging large tissue areas at high spatial resolution will limit the amount of tissues that can be analyzed before the laser needs to be replaced. Solid state lasers, in particular diode-pumped Nd:YAG (355 nm), have proved popular for imaging instrumentation in recent years due to their ability to achieve over 1-kHz repetition rates and lifespans of several 10^9 shots (29). The introduction of an Nd:YVO4 laser capable of operating at up to 20kHz has further increased the speed of MALDI-MS image acquisition. When incorporated into a standard MS instrument the Nd:YVO4 laser resulted in a greater than fourfold decrease in the time taken to image tumor tissue biopsies versus the Nd:YAG laser (30).

The laser spot size plays an important, but not always decisive, role in the achievable spatial resolution in a MS imaging experiment. Current commercial instruments have lasers that can be focused down to ~20 μm, and these have been utilized for high spatial resolution imaging of smaller tissues and regions where such spatial detail is necessary. Lagarrigue et al. demonstrated this for the imaging of protein changes occurring during germ cell development in rat testes (31). Lasers have been produced that can be focused down to less than 1 μm, but at such small spot sizes, the ability of the instrument to detect the much lower amounts of ions produced becomes limiting (32). Thus, only the very abundant species would be detectable. In most applications, images are acquired in the 100–200-μm range, enabling enough spatial detail to image features at the tissue level, while maintaining a reasonable acquisition time and protecting the laser life. For larger images, such as whole-body imaging of rat and mouse sections, lower spatial resolutions are utilized and are a good compromise. Such resolutions enable enough detail for localization of the analyte in all the organs of interest, but maintain a reasonable acquisition time. Another method of improving the spatial resolution without access to a MALDI source with such a highly focused laser is by oversampling, as first described by Jurchen et al. in 2005 (33). In the oversampling mode, the sample stage is moved at smaller increments than the laser spot size. The matrix layer (containing the analyte embedded in the matrix crystals) under the laser spot is completely ablated before the sample is moved preventing the crossover of signal between neighboring pixels. By this method, images of glycosphingolipids have been acquired at 15 μm from a laser with a ~150-μm laser spot in the spleen of a knockout mouse model of type 1 Gaucher disease (34). The greater the degree of oversampling, the lower the number of ions that will be produced from each "point." Thus, sensitivity of the mass spectrometer becomes paramount.

MALDI-MSI has been applied to the bioimaging of many different analyte classes in tissues. It is a natural choice for peptides and proteins as a larger mass range can be exploited in comparison to other ionization techniques (typically 5–100,000 Da). MALDI-MSI has also been readily accepted by the pharmaceutical research community for the analysis of low-molecular-mass analytes (<1000 Da) due to its high selectivity and sensitivity.

MALDI-MSI, like all MS imaging methods, can suffer from significant ionization suppression effects due to the presence of competing analytes within the tissue. The matrix crystallization itself will be directly affected by the surface composition of different tissues, such as salts, lipids, or blood. The MALDI ionization process does not result in equal ionization efficiencies for each species located on the tissue surface, and therefore, the anaylte with the highest signal intensity in an image is not necessarily the most abundant in the tissue (35). There is also the issue of extraction of analytes from the tissue and their incorporation into the matrix clusters. The extraction efficiencies differ from tissue to tissue and will depend upon the solvent systems used to apply the matrix.

SECONDARY ION MASS SPECTROMETRY

In SIMS-MS imaging, the ions are produced by directing a primary ion beam at the tissue surface, which results in a sputtering of neutral and charged secondary species from the tissue. The ejected species are composed of ionized atoms, atom clusters, and molecules and their fragments. SIMS was first applied to MS imaging in 1962, making it the first ionization method to be used for this purpose (36). However, it was primarily utilized for spatial inorganic and elemental analysis and did not find acceptance in the field of biological tissue imaging until much later (37, 38). The reasons why biologists have not embraced SIMS as enthusiastically as MALDI were discussed in detail by Heeren et al. (39).

SIMS can be either dynamic or static. In dynamic SIMS, a higher primary ion dose is used, limiting the ability to detect intact molecular ions (due to the induced molecular fragmentation during the ionization process), and therefore, this mode is primarily used for elemental and isotopic imaging of surfaces. One advantage of its more destructive nature is the ability to perform depth-profiling acquisitions below the surface of the tissue (typically up to depths of ~30 μm) as the surface is eroded away by the primary ion beam (40). As each ion impact results in damage to the surrounding molecules (within 5–10 nm), this prevents the acquisition of subsequent data from these affected surrounding molecules.

SIMS is predominantly used in biomedical imaging applications for the analysis of large molecules and molecular fragments, and therefore, static SIMS has traditionally been the approach utilized. In static SIMS only a very low primary ion dose is used ($<10^{13}$ ions cm^{-2}), which results in very minimal damage occurring to the tissue surface, as less than 1 percent of the surface layer of the atoms (or molecules) receives impact from a primary ion (41). This makes it statistically unlikely that any damaged surface area will be resampled and thus ensures that the resulting SIMS-MS image will be acquired from previously undisturbed tissue.

Traditionally, liquid metal ion sources, such as gallium, indium, or gold, have been used as the primary ion beam in SIMS imaging experiments. These have the advantage of producing small spot sizes (below 100 nm) but result in excessive fragmentation of the sputtered species (typically only producing ions > 200 Da). The recent development of ion sources producing cluster ions as their primary beam has led to an increase in the range of both biological tissues and analytes that can be studied. Commercially produced polyatomic ion sources, such as Au_n^+, Bi_n^+, SF_5^+, and C_{60}^+, are easy to operate and have long operational lifetimes (42, 43). They have the great advantage of producing a larger secondary ion yield and reducing sample damage and subsequent molecule fragmentation in comparison to the liquid metal ion sources (44).

Under static SIMS conditions, species can typically be detected in the region of 1–1000 Da, enabling the imaging of low-molecular-mass species of biological interest that include lipids, metabolites, toxins, and pharmaceutical compounds. The sensitivity of the analysis is generally much lower than that for MALDI-MSI particularly for larger molecules. However, as only the uppermost layers of the tissue are ablated, the same tissue sample can be coated with matrix and imaged using MALDI. This enables the acquisition of both high spatial resolution low mass images (SIMS) with higher mass imaging at lower spatial resolution (MALDI). Eijkel et al. used this approach for the imaging of lipids within a single rat brain tissue section (45). A particularly exciting recent area of SIMS instrumentation development is the potential for tandem MS analysis, facilitating the identification of TOF SIMS spectral peaks and increasing its usability for biological imaging applications (46).

DESORPTION ELECTROSPRAY IONIZATION MASS SPECTROMETRY IMAGING (DESI-MSI)

Desorption Electrospray Ionization (DESI) is the most recently introduced ionization method for MSI. It was developed in the Cook laboratory and first applied to image a range of anaytes, including peptides and proteins, from a range of nonbiological surfaces in 2004 (47). DESI has the primary advantage of being nondestructive and is performed under ambient laboratory conditions, therefore opening up new in situ MSI applications not possible under the traditional sealed high vacuum conditions. Applications to biological tissue sections soon followed (48, 49).

In the DESI ionization process, electrosprayed charged solvent droplets (most commonly water) are directed at the tissue surface. The impact of the droplet stream results in desorption of ions from the surface by electrostatic and pneumatic sources. The desorbed gas-phase ions are then transferred into the mass spectrometer by use of an ion transfer line.

In imaging mode, the sample is moved under the fixed spray nozzle until all the desired surface area has been analyzed with each fixed coordinate represented by a pixel in the resulting image. DESI requires no additional sample preparation steps after tissue sectioning, unlike MALDI or chemically enhanced SIMS methods. The method can also be used for direct analysis of intact living tissues, as has been demonstrated for bacteria cultures and in vivo skin analysis (50, 51). However, there are many impracticalities when applying this to the imaging of subjects that have the capability to move, due to the long acquisition time needed to acquire the MSI dataset and the requirement to maintain a consistent spray distance between droplet source and surface for a three-dimensional sample.

Several key factors limit the ultimate spatial resolution achievable in a DESI MSI experiment. The droplet spot size is primarily defined by the diameter of the originating capillary and by the distance from the capillary tip to the sample surface. However, spatial resolution is not limited by the droplet size alone as the achievable spatial resolution may be significantly poorer than the diameter of the impacting spot. DESI MS imaging experiments will typically produce a spatial resolution of 200–500 μm. However, the recent development of nano DESI demonstrated the potential for sub-20 μm spatial resolution imaging (52).

DESI MSI has been utilized to image a wide range of analytes in biological tissues, including pharmaceuticals (53), metabolites (54), and lipid species (55, 56). Many early DESI imaging studies have been conducted on lipids as these ionize extremely well by DESI and are located within the preferred mass range for DESI (<1500 Da). As with MALDI and SIMS, both positive and negative ion images may be acquired allowing the full range of lipid species to be localized.

DESI also has the potential to be used as a complimentary imaging method to MALDI when both lipid and protein information is desired from one tissue section. DESI analysis can be performed on the untreated surface for the imaging of lipid species before addition of a matrix coating and imaging by MALDI as demonstrated by Eberlin et al. (57).

SAMPLE PREPARATION

To create a high-quality 2D MS image, sample preparation is of upmost importance and can often be the limiting factor in the achievable spatial resolution. In the case of biological tissues, thin sections (10–50 μm) are cut using a cryomicrotome and placed onto a solid support (stainless steel plate, glass microscope slide) by thaw mounting or direct attachment to adhesive tape, which is then attached to the substrate. It is crucial that cross-contamination not occur during these steps as this may not be apparent by optical visualization and would only be discovered upon MS imaging analysis of the tissue section.

Tissues may be embedded in various media to facilitate the preparation of thin sections from fragile tissues by methods similar to traditional histology. Standard embedding media, such as OCT, can easily contaminate the tissue surface during the sectioning process, leading to suppression of analyte signals in the resulting MS imaging data (58). Therefore, several other "MS friendly" media, such as carboxymethylcellulose (16), gelatine (59), and ice (60), are frequently used.

Depending upon the analyte class to be measured, a washing step with water or common organic solvents may be incorporated to remove salts and endogenous species (such as blood or lipids) that may interfere with the MSI process. This is particularly advantageous for protein or peptide imaging as suppression of the protein ions is commonly observed by endogenous salts and abundant lipids (61). Washing may result in the delocalization or complete removal of lipids, drugs, or toxins from the tissue so these analytes are not usually subject to a washing step.

MATRIX APPLICATION

Arguably the most important step in producing a high-quality MALDI-MS image is the application of the matrix to the tissue surface. When applying the matrix dissolved in solvent, it is vital that the matrix spray is wet enough to extract the analyte from the tissue (particularly important for low abundance analytes) and into the surface matrix crystals, but not so wet that the analytes will delocalize from their positions leading to a loss of image spatial integrity. The ultimate size of the matrix crystals is also of primary importance, as they will become the limiting factor once they are larger in size than the laser beam profile.

In the years since MALDI-MSI imaging was first introduced, many different methods of matrix application have been utilized, and companies have been quick to produce commercial instrumentation for this purpose. Most initial work was performed manually using a pneumatic spray. This method is inexpensive and fast. However, there is potential for poor reproducibility especially between multiple users due to their individual spraying techniques (62). These concerns led to labs developing their own in-house matrix application devices (32, 62). Once there was a sufficient market for MSI sample preparation instrumentation, manufacturers soon developed novel instrumentation for matrix application purposes or adapted existing liquid deposition devices for this task (63, 64). Commercial instrumentation is fully automated and therefore less labor intensive and more reproducible. However, they can be significantly more expensive and not as customizable as the manual spray methods. Current commercial instrumentation is focused on either an automated spray deposition or droplet deposition, whereby many small liquid matrix droplets are deposited onto the tissue in an array. Spatial resolution is sacrificed in the droplet deposition method and is usually no better than 200 μm due to the size of the matrix droplets. However, sensitivity is high due to the high anaylte extraction efficiency of the droplets, and there is no risk of analyte delocalization outside of the matrix spot.

Droplet deposition is performed either by acoustic deposition (contactless) (65) or by a chemical inkjet printing method (62, 66).

Matrices have also been applied to biological tissues in an entirely solvent-free manner. Sublimation of the matrix at high temperature onto biological tissue surfaces has been performed for a range of different matrices (67) and results in a highly uniform coating with small crystal sizes. More recently, finely crushed or powdered matrices have been brushed or shaken directly onto the tissue surface to form homogeneous coatings (68). In these dry application methods, very little extraction of analytes occurs from the tissues, and as such, these techniques have been primarily used for species present in high abundance at the tissue surface, such as lipids.

SIMS ENHANCEMENT

Although SIMS allows for the analysis of tissues without the extensive sample preparation or tissue handling involved in MALDI, several enhancement steps may be added during the process. These steps involve the addition of a coating layer to the tissue before the image acquisition. This improves the secondary ion yield and therefore sensitivity of the analysis, in addition to increasing the ionization mass range of biological analytes within the tissue, thus enabling the imaging of larger lipids and proteins by SIMS.

Applying a layer of organic matrix to the tissue (in a method similar to MALDI-MSI) significantly increases the signal intensity of larger analytes directly from tissue (69). This enables the acquisition of peptides and small proteins in the mass range of 1–5000 Da. Larger peptides and proteins have been ionized from surfaces, but only when present in high abundance and after purification steps before analysis (70).

One of the main advantages that SIMS offers over MALDI for MS imaging is the potential for submicrometer and, thus, cellular level imaging. The addition of a matrix application step increases the potential for analyte delocalization during the matrix application (particularly if a solvent-based application is used), therefore limiting the ultimate spatial resolution possible. A second issue is the size of the matrix crystals once applied to the tissue. Crystal size will become the ultimate limiting factor in determining the achievable spatial resolution, as the crystal size (typically 2–10 μm) will be much larger than the focusing ability of the primary ion beam.

The second method by which the signal of intact molecular species can be enhanced is by the application of a thin metal layer (1–5 nm) to the surface of the tissue before analysis (71). A conducting metal, such as gold or silver, applied by sputter coating is typically used for this purpose. As the sputter coating is entirely solvent-free, there is no risk of analyte delocalization during the process. However, the formation of analyte adducts with the coating metal and the presence of metal clusters further complicate the mass spectrum (72).

APPLICATIONS OF MS IMAGING

Drug Discovery and Development

The high sensitivity, specificity, and rapid speed of analysis have led to MS becoming one of the most utilized technologies in pharmaceutical research. With many companies and academic research organizations already embracing MS, it was a natural progression that MS imaging soon caught much attention as a method of localizing both compound and metabolite distributions in tissue. Traditionally, in absorption, distribution, metabolism and excretion, pharmacokinetic (PK), and toxicology studies, compound and metabolite localization was determined by the extraction of drugs and metabolites from organ homogenates from animals dosed with the compound of interest and analyzed by LC-MS or LC-MS/MS. This technique offers full quantitative analysis of the compounds and metabolites in the tissues; however, any spatial information produced is limited to the size of the tissue before homogenation.

Spatial distribution is determined by whole-body autoradiography (WBA) used in a complimentary fashion to the LC-MS analysis. In the WBA method, a radiolabel is attached to the pharmaceutical compound before dosing. The radiolabel is imaged in the whole-body section of the animal. WBA can be quantitative and allows for spatial information in the micrometer range. However, there are two significant drawbacks to this approach. First, drug compounds must be synthesized with an incorporated radioactive isotope (a costly and time consuming procedure). Second and most important, in the WBA method, measurement is not specific. As only the radioactivity of the label is measured, and not the drug molecule itself, the parent drug cannot be distinguished from its metabolites (73).

MALDI-MSI enables the production of molecular images of both compounds and any detectable metabolites in one imaging experiment and is an effective method in terms of time and cost (after the initial outlay for the mass spectrometer). This ability to fuse the specificity offered by mass spectrometry with the spatial information provided by WBA has led to a surge in interest for MALDI-MS imaging within the pharmaceutical industry. There are also several disadvantages to using MALDI-MSI over WBA: the lower sensitivity offered, the requirement of the drug to ionize well by MALDI, and the difficulties with producing fully quantitative data (73). A direct comparison of compound imaging by MALDI-MSI and WBA on adjacent tissue sections from the same dosed rat is shown in Figure 11.2. MALDI-MS imaging shows significant promise in the field of biopharamceuticals as it can analyze species present within the large peptide and protein mass range (74).

Maximum sensitivity in MALDI-MS imaging is achieved by using a targeted MS imaging approach. This is a particularly appropriate approach for drug imaging as often the mass or fragmentation pattern of the compound and its major metabolites are already known. With the advent of high repetition and continuous rastering lasers, it is now possible to rapidly acquire MS images in selected and multiple reaction modes enabling a number of compounds and metabolites to be acquired in one targeted imaging experiment (75). This targeted approach has been used to produce semi-quantitative images of the anti-tuberculosis drug Moxifloxacin in lung biopsies and has been validated by LC-MS/MS (14).

An untargeted approach can be used when the metabolite masses are unknown before analysis. The parallel detection of 100s to 1000s of analytes in such an imaging approach opens up the ability to co-localize the distribution of compounds and metabolites with endogenous species, which may be of interest as biomarkers of disease, drug efficacy, or as structural tissue markers.

SIMS-MS imaging has not proved effective for pharmaceutical analysis within biological tissues even though it offers the highest spatial information of the three

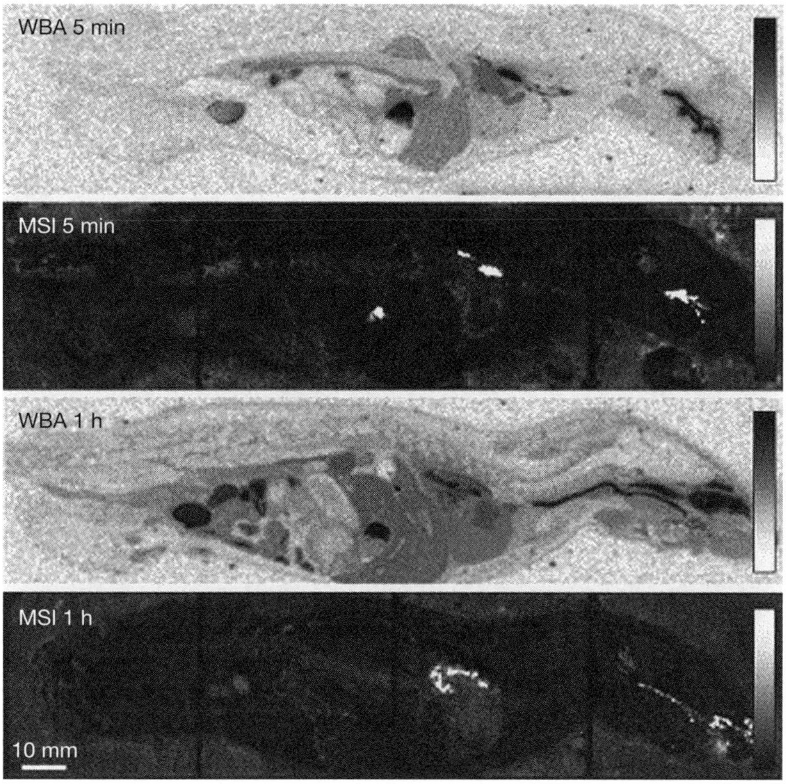

Figure 11.2 Comparison of MSI with WBAL using whole-body sections after intra-tracheal administration of a pharmaceutical compound (0.5 mg/kg) to rats. The highest levels are observed in the trachea, lung, and stomach by both methods. Reproduced with permission (33)

discussed ionization methods. While the effective mass range of SIMS is within that of most drugs and metabolites, the fragmentation that occurs during the ionization process can result in difficulties in identifying compounds. As commercial SIMS instruments do not typically have MS/MS capabilities, this further inhibits conclusive identification of the ionized species. The lower sensitivity of SIMS in comparison to MALDI is also a cause for concern, especially at the therapeutic dosing levels used in both animal studies and clinical trials.

DESI-MS imaging is the most recently developed technique applied to drug and metabolite imaging in biological tissues. While to date DESI-MS imaging has only been applied to a small number of pharmaceutical analyses, it is an effective method of analyzing compounds and metabolites in a range of biological samples, including skin and urine (76). Whole-body DESI-MS imaging in mice has been demonstrated for the drug propranolol and compared to WBA. Very similar distributions were observed using both techniques (Figure 11.3). When a surface scanning speed of 7 mm/sec was used, the whole mouse could be imaged in less than 80 minutes, making it a viable alternative to high-speed MALDI-MSI in this instance (54). However, it must still be noted that the sensitivity of the DESI analysis is lower and that the potential spatial information is much higher with MALDI. Therefore, these issues

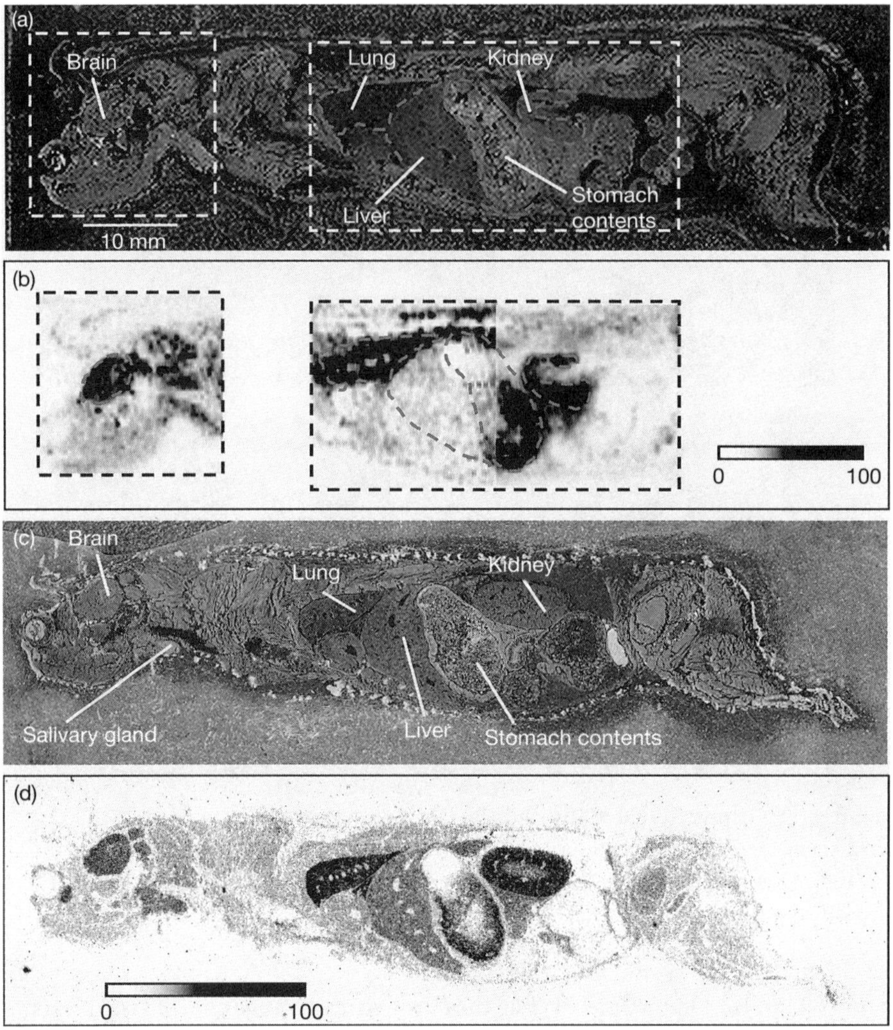

Figure 11.3 Comparison of WBAL with DESI-MSI. Propanolol distribution determined by DESI-MSI (b) is in agreement with the industry standard WBAL (d). Scanned optical reference sections (a) and (d) are used to show the compound is located at high levels in the kidney, lung, and stomach. Reproduced with permission (52)

must be addressed for it to emerge as a serious competitor to MALDI for use in drug discovery applications.

Biomedical Applications

The label-free nature of MS imaging offers the unique ability to acquire localization information for multiple unknown analytes within a biological tissue. These unknown species can subsequently be identified through tandem MS. Alternatively, high mass resolving MS instruments, such as FTICR and Orbitrap, allow for the confirmation of the identity of the analytes via exact mass imaging measurements and database searching.

Biological analytes of interest cover a wide mass range from low-molecular-mass metabolites and lipids (typically <1000 Da) through to large proteins over 100,000 Da. Identifying and localizing such biological species is of obvious importance in many areas of biological research. The pathology of many diseases remains poorly understood and the specificity of MS imaging allows for the biochemical changes occurring during disease progression to be spatially monitored. MSI has found particular application in disease areas, such as cancer, Parkinson's, and cardiovascular disease, where the localized tissue composition has a direct effect upon disease progression and the development, and delivery of, novel pharmaceutical therapies.

Lipidomics

Lipids are abundant in most of the biological tissues typically analyzed by MS imaging and can be ionized in both positive and negative modes. Sphingomyelins, phosphatidylcholines, and sterols ionize well in positive ion mode; phosphatidylinositols, phosphatidylserines, and sulfatides ionize in negative ion mode. Phoshatidylethanolamine can be measured using both (77).

The choice of matrix is of particular importance in MALDI-MS imaging of lipids due to the highly complex spectra that are typically produced. In positive ion mode, such spectra include the protonated molecular species as well as species cationized by sodium or potassium, which is present at high levels within the tissue. The addition of a lithium salt to the matrix results in a spectrum in which the lipid species are solely cationized by lithium, thus reducing the spectrum complexity (78). Lipid salt adducts fragment to provide more structural detail than protonated lipids, allowing for structural identification of lipids by tandem MS analysis.

Lipids may also fragment in source, resulting in extremely complex mass spectra from which they cannot be conclusively identified. This is especially troublesome in SIMS during which the "harder" ionization may result in a total loss of intact parent ions. As described earlier, the recent incorporation of "softer" ion cluster sources have lessened this effect. Lipid identification is predominantly based upon tandem MS or exact mass measurements and these are not commonly available on SIMS instruments. Where SIMS excels in comparison to MALDI and DESI is in the achievable spatial resolution. Indeed, SIMS has been applied to image the distribution of several common lipids in individual cells at spatial resolutions of 1 μm or higher (71, 79). Figure 11.4 shows one example of the localization of cholesterol within individual neuroblastoma cells. Currently, such spatial detail can only be obtained through SIMS-MS imaging.

DESI, SIMS, and MALDI-MS have been applied to image lipids in many different biological tissues. Many of these initial studies were conducted in brain tissue sections from rodent models due, in part, to their availability and distinct distributions of ionisable lipids. Figure 11.5 shows the distributions of several lipid species in mouse brain sections analyzed by the three technologies. Information from MS imaging of lipid species in tissues has been used to increase understanding of the pathology of many diseases, including macular degeneration (80), Fabry's disease (81), liver disease (82), various cancers (57, 83), and atherosclerosis (84). Other tissues in which lipids have been imaged are spinal cord (85), spleen (33), muscle, and whole mouse embryos (86, 87, 88).

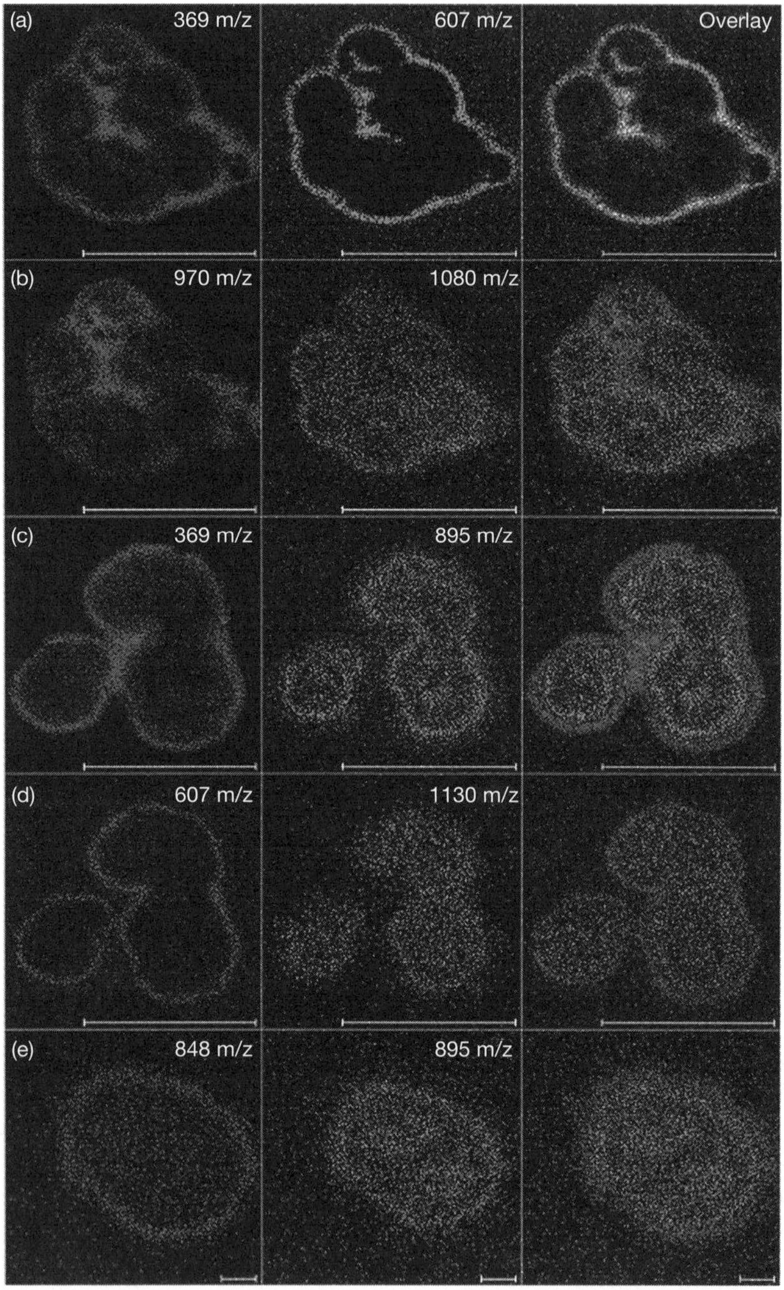

Figure 11.4 SIMS images of individual neuroblastoma cells after deposition of 1 nm gold. Cholesterol $[M\text{-}OH]^+$ *m/z* 369, Diacyglcerol *m/z* 607, cholesterol $[2M + Au]^+$ *m/z* 970, unidentified species *m/z* 1080. Scale bar 100 μm. Adapted with permission (70)

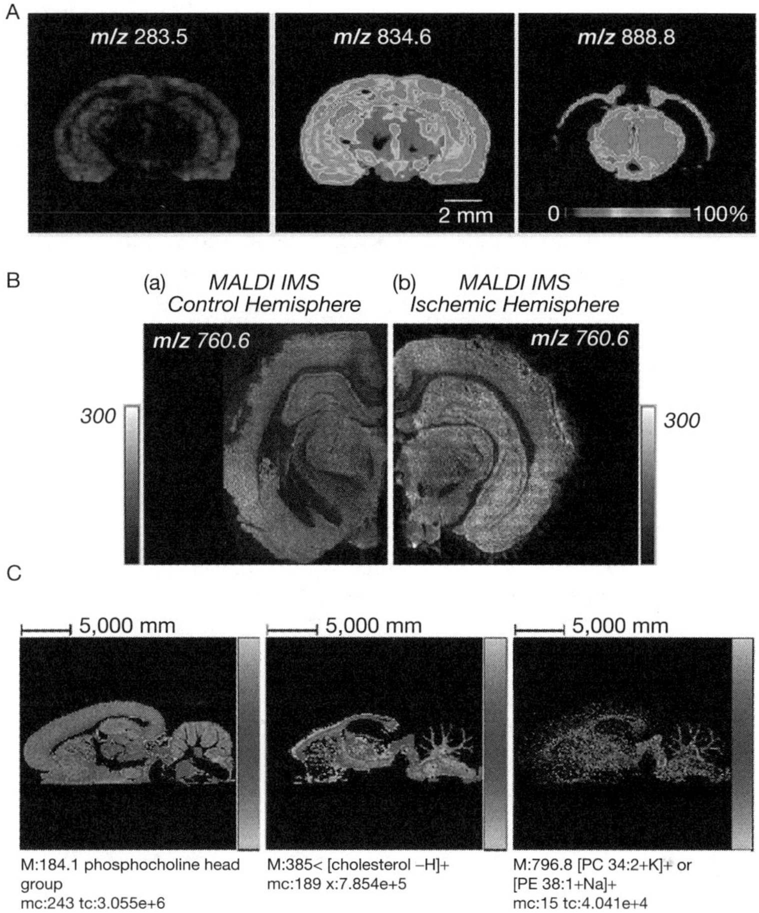

Figure 11.5 (A) Distribution of stearic acid, PS(18:0/22:6) and ST(24:1) in coronal mouse brain sections by DESI-MSI. (B) Distribution of palmitoyl-oleoyl phosphatidylcholine in control and ischemic rat brain by MALDI-MSI. (C) Distribution of phosphatidylcholine head group, cholesterol, and PC(32:0) in mouse brain by SIMS-MSI. Adapted with permission (56, 86, 87)

Metabolomics

Metabolomics is the study of low-molecular-mass species found in tissues, biological fluids, and cells participating in metabolic processes. Such species can be endogenous in origin or exogenous (for example from pharmaceutical therapy or toxin exposure). Combinations of NMR and MS approaches are typically performed, as it is not currently possible to analyze the entire metabolite content of cells by a single analytical method (89).

Confirmation of the identity of metabolites recorded in a mass spectrum may be obtained by searching the *m/z* values of spectral peaks against reference databases. Examples of these databases include METLIN (90) and the Human Metabolome Database (91). The reference databases enable identification based upon both the exact mass of the metabolites and diagnostic fragment ions produced through tandem MS analysis.

MS imaging approaches allow both the identification and localization of metabolites to be determined within tissues. Localization information is of great importance due to the ability to gain further understanding of the metabolites based upon their co-localization and potential interactions with other analytes of interest to the genome, such as proteins or transcripts. Although MS is still an emerging approach for metabolomics studies, several studies have been performed for both biomedical and systems biology research purposes. However, as MS imaging is a direct tissue analysis approach, it is susceptible to ion suppression and enhancement. This is a particularly important consideration when many metabolites are required to be analyzed simultaneously from a tissue. Sample preparation approaches may promote one analyte class (for example lipids) and further suppress the ionization of another. In an untargeted MS imaging approach when hundreds of species may need to be identified within one experiment, it may prove difficult to resolve unique metabolite ions from adducts and fragments formed in the mass spectrometer source.

Miura et al. used a combination of MALDI-MS imaging and metabolic pathway analysis of a rat transient middle cerebral artery occlusion model to visualize the spatiotemporal behavior of metabolites in the central metabolic pathway regulated by an ischemia reperfusion (92). Other metabolites imaged in brain tissue include AMD, ADP, and GTP (93).

Proteomics

MALDI-MS imaging is the preferred ionization method for the imaging of large peptides and proteins. This is due to the ability of the MALDI ionization process to ionize larger proteins and protein digest products than SIMS or DESI. Intact proteins larger than 1 kDa have been successfully ionized directly from biological tissue. However, sensitivity decreases with increasing mass. Identification of proteins in this "top down" approach is performed by fragmentation of the intact protein (94). An additional approach to image and identify proteins in tissue by MS has emerged in recent years. Proteolytic cleavage, achieved by spraying or spotting a digesting enzyme (such as trypsin) to the tissue before the application of the matrix, increases the number of detectable and identifiable proteins in tissues (95). This is known as the "bottom-up" approach to proteomic MS imaging, whereby the parent proteins are either identified by peptide mass finger printing or tandem MS. It must be noted that proteolytic cleavage of intact tissue produces extremely complex spectra, making identification of the parent proteins difficult, particularly for the less abundant proteins.

One important consideration with the bottom-up approach is the effect the enzymatic digestion process may have upon the achievable spatial resolution. Spotting the enzyme onto the tissue limits any delocalization to within the droplet size on the surface (typically 200–500 μm). Spraying the enzyme solution can increase the spatial resolution, but it can be difficult to monitor the extent of any delocalization of the digested proteins on the tissue that may occur.

The development of methods for dewaxing formalin-fixed paraffin embedded (FFPE) tissues has enabled the ability to perform MALDI-MS imaging and profiling archived tissue biopsies (96). This opens up access to huge historical tissue archives (stretching back over a hundred years) that may be used to investigate the pathology of many diseases. MALDI-MS has been applied to image many tissues of interest in biomedical research and has found particular acceptance for localization of tumorous tissue in cancer biopsies. An example of this approach is shown in Figure 11.6,

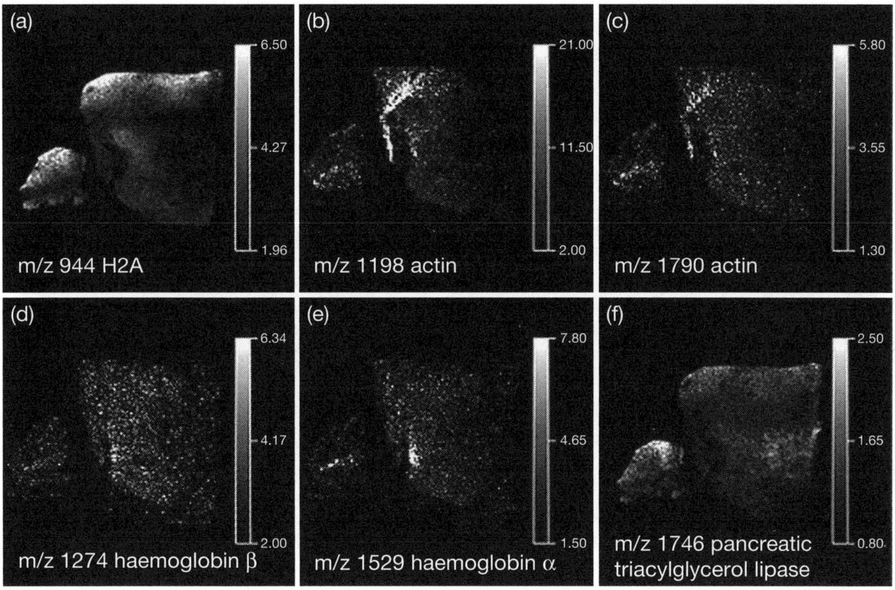

Figure 11.6 MALDI-MSI images of peptide distribution within FFPE pancreatic tumor tissue. The localizations of histone H2A, actin, hemoglobin, and pancreatic triacyglycerol lipase are demonstrated. Reproduced with permission (96)

for which Djidja et al. directly imaged the localization of several peptides of interest in cancer research within FFPE pancreatic tumor sections (97). It has been applied to identify protein markers of cancerous tissue to discriminate tumor grades and whose specific localizations are used to define tumor boundaries not often visible through traditional histological staining methods (98). Other diseases studied include Parkinson's disease (99), cataract development (100), cardiovascular and kidney disease (101).

CURRENT FRONTIERS AND FUTURE DIRECTIONS

MS imaging is a young and rapidly developing technique for bioimaging, and it is fast gaining acceptance as an essential tool in biological research. Recent technology advancements are opening up the technique to new tissues and analytes. Of particular importance is the increasing spatial resolution capabilities that have been presented over the last few years, advancing MS imaging from a tissue level tool to one that can be used for analyte localization within individual cells. The typical human cell is approximately 10 μm^2, and SIMS and MALDI-MS imaging are both capable of achieving maximum spatial resolutions higher than this (500 nm and 5 μm, respectively). Sensitivity is an obvious concern at pixel sizes below 10 μm. However, lipid species abundant within cell membranes, such as cholesterol and even drugs, have been imaged at resolutions of 10 μm and below (71, 102).

Data handling is a concern in MS imaging experiments as, depending upon the spatial and mass resolution, a single 2D imaging dataset may be several gigabytes in size. Hence, processing of such datasets requires powerful computing capabilities. Advanced data processing and statistical analysis tools have been used to extract meaningful data from the vastly complex spectra produced in many MS imaging experiments. Examples of these include hierarchical clustering and principal component analysis (103, 104). However, while several commercial and shareware packages exist, more user-friendly software and standard data storage formats must be developed for the technology to become accepted for routine analysis in clinical and biomedical research settings.

The advent of high-speed MS imaging instrumentation and robust and reproducible sample preparation methods have paved the way to make 3D MS imaging of biological tissues into a reality. Both MALDI and DESI MS have been used to create multiple sequential 2D ion images from sections cut from a biological tissue. These images can be stitched together to form a 3D stack as has been demonstrated in rodent brain (105). Data handling is an important issue to consider due to the large datasets generated and because the acquisition of the sequential 2D ion images is time consuming. With the increasing automation of sample preparation making high throughput MS imaging a possibility and the development of powerful data analysis software, there is potential for 3D MS imaging to become a more routine bioimaging method.

While mass spectrometry imaging has found widespread acceptance in qualitative imaging, until recently there has been little progression into the fully quantitative imaging of analytes in tissue. Quantitation has proved extremely difficult in MALDI-MS imaging analysis due to the non-homogeneous ion suppression and tissue extraction effects explained earlier in the chapter. For drug and metabolite imaging purposes, this has been addressed by the application of an internal standard to the tissue against which the drug or metabolite signal is normalized (14, 34). By further advancement of this approach, Bonnel et al. have demonstrated full drug quantitation in whole-body rodent sections (75).

For lipid and protein imaging, normalization is usually performed against a matrix ion or the total ion chromatogram (total recorded ions within the mass spectrum). While not leading to full quantitation, this significantly reduces the inhomogeneity of the analyte signal caused by ion suppression effects.

As SIMS and DESI do not require a matrix to be applied to the surface, they should, in theory at least, be more suited for quantitative imaging of unlabeled analytes. However, this has not proved to be the case, as ion suppression from competing species in the tissue is still present. SIMS has frequently been utilized for quantitation of elements from nonbiological surfaces and films, but quantitation from biological tissues has been predominantly performed by isotopic labeling of the analyte (106), thus negating many of the benefits offered by the label-free MS imaging approach. DESI has proved successful for the relative quantitative imaging of the drug clozapine in lung tissue biopsies after application of an internal standard (53). However, full quantitation has yet to be demonstrated.

A recent development with the potential to improve the sensitivity and selectivity of MS imaging is the incorporation of ion mobility spectrometry into the MS. This enables isobaric species to be resolved due to differences in their respective size and shape and has proved particularly useful for separating and identifying peptides following in situ enzyme digestion (97).

CONCLUSION

MS imaging has emerged as a valuable method of localizing molecules on tissues without the need for expensive and time-consuming labeling steps. This ability has led to it becoming a popular technology for systems biology, disease pathology, and pharmaceutical compound and metabolite distribution studies. While not currently capable of approaching the spatial resolution offered by microscopy methods, it offers valuable complementary information, and the MS ion images can be co-registered with those obtained by optical microscopy, enabling the determination of both molecular and structural information from a single tissue section.

With continued development of mass spectrometry and sample preparation methods and technologies, it has the potential to become a more routine tool in biomedical research for the imaging of molecules in tissues. As cellular resolution capabilities and advanced data processing technologies open up new avenues for MS imaging in biomedical research, the next few years promise to be a very exciting time.

REFERENCES

(1) Karas M, Bachman D, Hillenkamp F (1985) Influence of the Wavelength in High-Irradiance Ultraviolet Laser Desorption Mass Spectrometry of Organic Molecules. *Anal Chem* 57: 2935–2939

(2) Tanaka K, Waki, H, Ido Y, Akita S, Yoshida Y, Yoshida T, Matsuo T (1988) Protein and polymer analyses up to m/z 100 000 by laser ionization time-of-flight mass spectrometry. *Rapid Commun Mass Spectrom* 2:151–153

(3) Fenn JB, Mann M, Meng CK, Wong SF, Whitehouse CM (1989) Electrospray ionization for mass spectrometry of large biomolecules. *Science* 246:64–71

(4) Taban IM, Altelaar AF, van der Burgt YE, McDonnell LA, Heeren RM, Fuchser J, Baykut G. (2007) Imaging of peptides in the rat brain using MALDI-FTICR mass spectrometry. *J Am Soc Mass Spectrom* 18:145–151

(5) Landgraf RR, Prieto Conaway MC, Garrett TJ, Stacpoole PW, Yost RA (2009) Imaging of lipids in spinal cord using intermediate pressure MALDI-lit/orbitrap MS. *Anal Chem* 81:8488–8495

(6) Cornett DS, Frappier SL, Caprioli RM (2008) MALDI-FTICR imaging mass spectrometry of drugs and metabolites in tissue. *Anal Chem* 80:5648–5653

(7) Chughtai K, Heeren RMA (2010) Mass Spectrometric Imaging for biomedical tissue analysis. *Chem Rev* 110: 3237–3277

(8) Sjövall P, Lausmaa J, Johansson B (2004) Mass spectrometric imaging of lipids in brain tissue. *Anal Chem* 76: 4271–4278

(9) Stoeckli M, Staab D, Schweitzer A, Gardiner J, Seebach D Imaging of a beta-peptide distribution in whole-body mice sections by MALDI mass spectrometry. *J Am Soc Mass Spectrom* 18: 1921–1924

(10) Fournier I, Wisztorski M, Salzet M (2008) Tissue imaging using MALDI-MS: a new frontier of histopathology proteomics. *Expert Rev Proteomics* 5:413–424

(11) Grey AC, Chaurand P, Caprioli RM, Schey KL (2009) MALDI imaging mass spectrometry of integral membrane proteins from ocular lens and retinal tissue. *J Proteome Res* 8: 3278–3283

(12) Nilsson A, Fehniger TE, Gustavsson L, Andersson M, Kenne K, Marko-Varga G, Andrén PE (2010) Fine mapping the spatial distribution and concentration of unlabeled drugs within tissue micro-compartments using imaging mass spectrometry. *PLoS One* 5:e11411
(13) Hopfgartner G, Varesio E, Stoeckli M (2009) Matrix-assisted laser desorption/ionization mass spectrometric imaging of complete rat sections using a triple quadrupole linear ion trap. *Rapid Commun Mass Spectrom* 23:733–736
(14) Prideaux B, Dartois V, Staab D Weiner DM, Goh A, Via LE, Barry CE, Stoeckli M (2011) High-sensitivity MALDI/MRM/MS imaging of moxifloxacin distribution in tuberculosis-infected rabbit lungs and granulomatous lesions. *Anal Chem* 83:2112–2118
(15) Porta T, Grivet C, Kraemer T, Varesio E, Hopfgartner G (2011) Single hair cocaine consumption monitoring by mass spectrometric imaging. *Anal Chem* 83:4266–4272
(16) Prideaux B, Staab D, Stoeckli M (2010) Applications of MALDI-MSI to pharmaceutical research in *Methods in Molecular Biology* 656:405–413
(17) Trim PJ, Henson CM, Avery JL, McEwen A, Snel MF, Claude E, Marshall PS, West A, Princivalle AP, Clench MR (2008) Matrix-assisted laser desorption/ionization-ion mobility separation-mass spectrometry imaging of vinblastine in whole body tissue sections. *Anal Chem* 80: 8628–8634
(18) Hart PJ, Francese S, Claude E, Woodroofe MN, Clench MR (2011) MALDI-MS imaging of lipids in ex vivo human skin. *Anal Bioanal Chem* 401:115–125
(19) Cole LM, Djidja MC, Bluff J, Claude E, Carolan VA, Paley M, Tozer GM, Clench MR (2011) Investigation of protein induction in tumour vascular targeted strategies by MALDI MSI. *Methods* 54:442–453
(20) Porta T, Varesio E, Kraemer T, Hopfgartner G (2011) Molecular imaging by mass spectrometry: application to forensics. *Spectroscopy Europe* 23:6–12
(21) Spengler B, Hubert M (2002) Scanning microprobe matrix-assisted laser desorption ionization (SMALDI) mass spectrometry: instrumentation for sub-micrometer resolved LDI and MALDI surface analysis. *J Am Soc Mass Spectrom* 13:735–748
(22) Liu Q, Xiao Y, Pagan-Miranda C, Chiu YM, He L (2009) Metabolite imaging using matrix-enhanced surface-assisted laser desorption/ionization mass spectrometry (ME-SALDI-MS). *J Am Soc Mass Spectrom* 20:80–88
(23) Yanes O, Woo HK, Northen TR, Oppenheimer SR, Shriver L, Apon J, Estrada MN, Potchoiba MJ, Steenwyk R, Manchester M, Siuzdak G (2009) Nanostructure initiator mass spectrometry: tissue imaging and direct biofluid analysis. *Anal Chem* 81: 2969–2975
(24) Nemes P, Woods AS, Vertes A (2010) Simultaneous imaging of small metabolites and lipids in rat brain tissues at atmospheric pressure by laser ablation electrospray ionization mass spectrometry. *Anal Chem* 82: 982–988
(25) Caprioli RM, Farmer TB, Gile J (1997) Molecular imaging of biological samples: Localization of peptides and proteins using MALDI-TOF MS. *Anal Chem* 69:4751–4760
(26) Knochenmuss R, Zenobi R (2003) MALDI ionization: the role of in- plume processes. *Chem Rev* 103:441–452
(27) Karas M, Glückmann M, Schäfer J (2000) Ionization in matrix- assisted laser desorption/ionization: singly charged molecular ions are the lucky survivors. *J Mass Spectrom* 35:1–12
(28) Chang WC, Huang LC, Wang YS, Peng WP, Chang HC, Hsu NY, Yang WB, Chen CH (2007) Matrix-assisted laser desorption/ionization (MALDI) mechanism revisited. *Anal Chim Acta* 582:1–9

(29) Holle A, Haase A, Kayser M, Höhndorf J (2008) Optimizing UV laser focus profiles for improved MALDI performance. *J Mass Spectrom* 41:705–16

(30) Trim PJ, Djidja MC, Atkinson SJ, Oakes K, Cole LM, Anderson DM, Hart PJ, Francese S, Clench MR (2010) Introduction of a 20 kHz Nd:YVO4 laser into a hybrid quadrupole time-of-flight mass spectrometer for MALDI-MS imaging. *Anal Bioanal Chem* 397: 3409–3419

(31) Lagarrigue M, Becker M, Lavigne R, Deininger SO, Walch A, Aubry F, Suckau D, Pineau C Revisiting rat spermatogenesis with MALDI imaging at 20-microm resolution. *Mol Cell Proteomics* 10: M110.005991

(32) Guenther S, Koestler M, Schulz O, Spengler B (2010) Laser spot size and laser power dependence of ion formation in high resolution MALDI imaging. *Int J Mass Spectrom* 294:7–15

(33) Jurchen JC, Rubakhin SS, Sweedler JV MALDI-MS imaging of features smaller than the size of the laser beam. *J Am Soc Mass Spectrom* 16: 1654–1659

(34) Snel MF, Fuller M High-spatial resolution matrix-assisted laser desorption ionization imaging analysis of glucosylceramide in spleen sections from a mouse model of Gaucher disease. *Anal Chem* 82: 3664–3670

(35) Stoeckli M, Staab D, Schweitzer A (2007) Compound and metabolite distribution measured by MALDI mass spectrometric imaging in whole-body tissue sections. *Int J Mass Spectrom* 260:195–202

(36) Castaing R, Slodzian G (1962) Optique corpusculaire—premiers essais de microanalyse par emission ionique secondaire. *Microscopie* 1:395–399

(37) Chandra S, Smith DR, Morrison GH (2000) Subcellular imaging by dynamic SIMS ion microscopy. *Anal Chem* 72:104A–114A

(38) Mas S, Touboul D, Brunelle A, Aragoncillo P, Egido J, Laprevote O, Vivanco F (2007) Lipid cartography of atherosclerotic plaque by cluster-TOF-SIMS imaging. *Analyst* 132:24–26

(39) Heeren RMA, McDonnell LA, Amstalden E, Luxembourg SL, Altelaar AFM, Piersma SR (2006) Why don't biologists use SIMS? A critical evaluation of imaging MS. *Appl Surf Sci* 252:6827–6835

(40) Jones EA, Lockyer NP, Vickerman JC (2008) Depth profiling brain tissue sections with a 40 keV C60+ primary ion beam. *Anal Chem* 80:2125–2132

(41) Linton RW, Goldsmith JG (1992) The role of secondary ion mass spectrometry (SIMS) in biological microanalysis: technique comparisons and prospects. *Biol Cell* 74:147–160

(42) Weibel D, Wong S, Lockyer N, Blenkinsopp P, Hill R, Vickerman JC (2003) A C60 primary ion beam system for time of flight secondary ion mass spectrometry: its development and secondary ion yield characteristics. *Anal Chem* 75: 1754–1764

(43) Touboul D, Kollmer F, Niehuis E, Brunelle A, Laprévote O (2005) Improvement of biological time-of-flight-secondary ion mass spectrometry imaging with a bismuth cluster ion source. *J Am Soc Mass Spectrom* 16: 1608–1618

(44) Nagy G, Gelb LD, Walker AV (2005) An investigation of enhanced secondary ion emission under Au(n)+ (n = 1-7) bombardment. *J Am Soc Mass Spectrom* 16: 733–742

(45) Eijkel GB, Kükrer Kaletaş B, van der Wiel IM, Kros JM, Luider TM, Heeren RMA (2009) Correlating MALDI and SIMS imaging mass spectrometric datasets of biological tissue surfaces. *Surf Interface Anal* 8:675–685

(46) Piehowski PD, Carado AJ, Kurczy ME, Ostrowski SG, Heien ML, Winograd N, Ewing AG (2008) MS/MS methodology to improve subcellular mapping of cholesterol using TOF-SIMS. *Anal Chem* 80: 8662–8667

(47) Takats Z, Wiseman JM, Gologan B, Cooks RG (2004) Mass spectrometry sampling under ambient conditions with desorption electrospray ionization. *Science* 306:471–473
(48) Takats Z, Wiseman JM, Cooks RG (2005) Ambient mass spectrom- etry using desorption electrospray ionization (DESI): Instrumen- tation, mechanisms and applications in forensics, chemistry, and biology. *J Mass Spectrom* 40:1261–1275
(49) Ifa DR, Wiseman JM, Song Q, Cooks RG (2007) Development of capabilities for imaging mass spectrometry under ambient conditions with desorption electrospray ionization (DESI). *Int J Mass Spectrom* 259:8–15
(50) Song Y, Talaty M, Datsenko K, Wanner BL, Cooks G (2009) In vivo recognition of Bacillus subtilis by desorption electrospray ionization mass spectrometry (DESI-MS). *Analyst* 134: 838–841
(51) Justes DR, Talaty N, Cotte-Rodriguez I, Cooks RG (2007) Detection of explosives on skin using ambient ionization mass spectrometry. *Chem Commun (Camb)* 7: 2142–2144
(52) Roach PJ, Laskin J, Laskin A (2010) Nanospray desorption electrospray ionization: an ambient method for liquid-extraction surface sampling in mass spectrometry. *Analyst* 135: 2233–2236
(53) Wiseman JM, Ifa DR, Zhu Y, Kissinger CB, Manicke NE, Kissinger PT, Cooks RG (2008) Desorption electrospray ionization mass spectrometry: Imaging drugs and metabolites in tissues. *Proc Natl Acad Sci U S A* 105: 18120–18125
(54) Kertesz V, Van Berkel GJ, Vavrek M, Koeplinger KA, Schneider BB, Covey TR (2008) Comparison of drug distribution images from whole-body thin tissue sections obtained using desorption electrospray ionization tandem mass spectrometry and autoradiography. *Anal Chem* 80:5168–5177
(55) Pol J, Vidova V, Kruppa G, Kobliha V, Novak P, Lemr K, Kotiaho T, Kostiainen R, Havlicek V, Volny M (2009) Automated ambient desorption-ionization platform for surface imaging integrated with a commercial Fourier transform ion cyclotron resonance mass spectrometer. *Anal Chem* 81:8479–8487
(56) Dill AL, Ifa DR, Manicke NE, Ouyang Z, Cooks RG (2009) Mass spectrometric imaging of lipids using desorption electrospray ionization. *J Chromatogr B Anal Technol Biomed Life Sci* 877:2883–2889
(57) Eberlin LS, Liu X, Ferreira CR, Santagata S, Agar NY, Cooks RG (2011) Desorption electrospray ionization then MALDI mass spectrometry imaging of lipid and protein distributions in single tissue sections. *Anal Chem* 83: 8366–8371
(58) Schwartz SA, Reyzer ML, Caprioli RM (2003) Direct tissue analysis using matrix-assisted laser desorption/ionization mass spectrometry: practical aspects of sample preparation. *J Mass Spectrom* 38:699–708
(59) Meriaux C, Arafah K, Tasiemski A, Wisztorski M, Bruand J, Boidin-Wichlacz C, Desmons A, Debois D, Laprévote O, Brunelle A, Gaasterland T, Macagno E, Fournier I, Salzet M (2011) Multiple changes in peptide and lipid expression associated with regeneration in the nervous system of the medicinal leech. *PLoS One* 6:e18359
(60) Khatib-Shahidi S, Andersson M, Herman JL, Gillespie TA, Caprioli RM (2006) Direct molecular analysis of whole-body animal tissue sections by imaging MALDI mass spectrometry. *Anal Chem* 78:6448–6456
(61) Agar NY, Yang HW, Carroll RS, Black PM, Agar JN (2007) Matrix solution fixation: histology-compatible tissue preparation for MALDI mass spectrometry imaging. *Anal Chem* 79: 7416–7423

(62) Baluya DL, Garrett TJ, Yost RA (2007) Automated MALDI matrix deposition method with inkjet printing for imaging mass spectrometry. *Anal Chem* 79: 6862–6867
(63) Marshall P, Toteu-Djomte V, Bareille P, Perry H, Brown G, Baumert M, Biggadike K (2010) Correlation of skin blanching and percutaneous absorption for glucocorticoid receptor agonists by matrix-assisted laser desorption ionization mass spectrometry imaging and liquid extraction surface analysis with nanoelectrospray ionization mass spectrometry. *Anal Chem* 82:7787–7794
(64) Deutskens F, Yang J, Caprioli RM (2011) High Spatial Resolution Imaging Mass Spectrometry and Classical Histology on a Single Tissue Section. *J Mass Spectrom* 46:568–571
(65) Aerni HR, Cornett DS, Caprioli RM (2006) Automated acoustic matrix deposition for MALDI sample preparation. *Anal Chem* 78: 827–834
(66) Patel SA, Barnes A, Loftus N, Martin R, Sloan P, Thakker N, Goodacre R (2009) Imaging mass spectrometry using chemical inkjet printing reveals differential protein expression in human oral squamous cell carcinoma. *Analyst* 134: 301–307
(67) Hankin JA, Barkley RM, Murphy RC (2007) Sublimation as a Method of Matrix Application for Mass Spectrometric Imaging. *J Am Soc Mass Spectrom* 18: 1646–1652.
(68) Trimpin S, Herath TN, Inutan ED, Wager-Miller J, Kowalski P, Claude E, Walker JM, Mackie K (2010) Automated solvent-free matrix deposition for tissue imaging by mass spectrometry. *Anal Chem* 82:359–367
(69) Luxembourg SL, McDonnell LA, Duursma M, Guo X, Heeren RMA (2003) Effect of local matrix crystal variations in matrix-assisted ionization techniques for mass spectrometry. *Anal Chem* 75:2333–2341
(70) Wu KJ, Odom RW (1996) Matrix-enhanced secondary ion mass spectrometry: a method for molecular analysis of solid surfaces. *Anal Chem* 68:873–882
(71) Altelaar AFM, Klinkert I, Jalink K, de Lange RPJ, Adan RAH, Heeren RMA, Piersma SR (2006) Gold-enhanced biomolecular surface imaging of cells and tissue by SIMS and MALDI mass spectrometry. *Anal Chem* 78:734–742
(72) McDonnell LA, Heeren RMA (2007) Imaging mass spectrometry. *Mass Spectrom Rev* 26: 606–643
(73) Solon EG, Schweitzer A, Stoeckli M, Prideaux B (2010) Autoradiography, MALDI-MS, and SIMS-MS imaging in pharmaceutical discovery and development. *AAPS J* 12: 11–26
(74) Stoeckli M, Staab D, Schweitzer A, Gardiner J, Seebach D (2007) Imaging of a beta-peptide distribution in whole-body mice sections by MALDI mass spectrometry. *J Am Soc Mass Spectrom* 18:1921–1924
(75) Bonnel D, Legouffe R, Willand N, Baulard A, Hamm G, Deprez B, Stauber J (2011) MALDI imaging techniques dedicated to drug-distribution studies. *Bioanalysis* 3:1399–1406
(76) Williams JP, Patel VJ, Holland R, Scrivens JH (2006) The use of recently described ionisation techniques for the rapid analysis of some common drugs and samples of biological origin. *Rapid Commun Mass Spectrom* 20:1447–1456.
(77) Woods AS, Jackson SN (2006) Brain tissue lipidomics: Direct probing using matrix-assisted laser desorption/ionization mass spectrometry. *AAPS J* 8:E391–E395
(78) Cerruti CD, Touboul D, Guérineau V, Petit VW, Laprévote O, Brunelle A (2011) MALDI imaging mass spectrometry of lipids by adding lithium salts to the matrix solution. *Anal Bioanal Chem* 401:75–87

(79) Piehowski PD, Davey AM, Kurczy ME, Sheets ED, Winograd N, Ewing AG, Heien ML (2009) Time-of-flight secondary ion mass spectrometry imaging of subcellular lipid heterogeneity: Poisson counting and spatial resolution. *Anal Chem* 81:5593–5602
(80) Garrett TJ, Dawson WW (2009) Lipid geographical analysis of the primate macula by imaging mass spectrometry. *Methods Mol Biol* 579:247–260
(81) Touboul D, Roy S, Germain DP, Chaminade P, Brunelle A, Lapré vote O (2007) MALDI-TOF and cluster-TOF-SIMS imaging of Fabry disease biomarkers. *Int J Mass Spectrom* 260:158–165
(82) Debois D, Bralet MP, Le Naour F, Brunelle A, Laprévote O (2009) In situ lipidomic analysis of nonalcoholic fatty liver by cluster TOF-SIMS imaging. *Anal Chem* 81: 2823–2831
(83) Shimma S, Sugiura Y, Hayasaka T, Hoshikawa Y, Noda T, Setou M (2007) MALDI-based imaging mass spectrometry revealed abnormal distribution of phospholipids in colon cancer liver metastasis. *J Chromatogr B Analyt Technol Biomed Life Sci* 855: 98–103
(84) Malmberg P, Börner K, Chen Y, Friberg P, Hagenhoff B, Månsson JE, Nygren H (2007) Localization of lipids in the aortic wall with imaging TOF-SIMS. *Biochim Biophys Acta* 1771:185–95
(85) Landgraf RR, Prieto Conaway MC, Garrett TJ, Stacpoole PW, Yost RA (2009) Imaging of lipids in spinal cord using intermediate pressure matrix-assisted laser desorption-linear ion trap/Orbitrap MS. *Anal Chem* 81: 8488–8495
(86) Burnum KE, Cornett DS, Puolitaival SM, Milne SB, Myers DS, Tranguch S, Brown HA, Dey SK, Caprioli RM (2009) Spatial and temporal alterations of phospholipids determined by mass spectrometry during mouse embryo implantation. *J Lipid Res* 50: 2290–2298
(87) Hankin JA, Farias SE, Barkley RM, Heidenreich K, Frey LC, Hamazaki K, Kim HY, Murphy RC (2011) MALDI mass spectrometric imaging of lipids in rat brain injury models. *J Am Soc Mass Spectrom* 22:1014–1021
(88) Benabdellah F, Seyer A, Quinton L, Touboul D, Brunelle A, Laprévote O (2010) Mass spectrometry imaging of rat brain sections: nanomolar sensitivity with MALDI versus nanometer resolution by TOF-SIMS. *Anal Bioanal Chem* 396:151–162
(89) Miura D, Fujimura Y, Tachibana H, Wariishi H (2010) Highly sensitive matrix-assisted laser desorption ionization-mass spectrometry for high-throughput metabolic profiling. *Anal Chem* 82:498–504
(90) Smith CA, O'Maille G, Want EJ, Qin C, Trauger SA, Brandon TR, Custodio DE, Abagyan R, Siuzdak G (2005) METLIN: a metabolite mass spectral database. *Ther Drug Monit* 27:747–51
(91) Wishart DS, Tzur D, Knox C, Eisner R, Guo AC, Young N, Cheng D, Jewell K, Arndt D, Sawhney S, Fung C, Nikolai L, Lewis M, Coutouly MA, Forsythe I, Tang P, Shrivastava S, Jeroncic K, Stothard P, Amegbey G, Block D, Hau DD, Wagner J, Miniaci J, Clements M, Gebremedhin M, Guo N, Zhang Y, Duggan GE, Macinnis GD, Weljie AM, Dowlatabadi R, Bamforth F, Clive D, Greiner R, Li L, Marrie T, Sykes BD, Vogel HJ, Querengesser L (2007) HMDB: the Human Metabolome Database. *Nucleic Acids Res* 35: D521–526
(92) Miura D, Fujimura Y, Yamato M, Hyodo F, Utsumi H, Tachibana H, Wariishi H (2010) Ultrahighly sensitive in situ metabolomic imaging for visualizing spatiotemporal metabolic behaviors. *Anal Chem* 82:9789–9796

(93) Benabdellah F, Touboul D, Brunelle A, Laprévote O (2009) In situ primary metabolites localization on a rat brain section by chemical mass spectrometry imaging. *Anal Chem* 81: 5557–5560

(94) Rauser S, Deininger SO, Suckau D, Höfler H, Walch A (2010) Approaching MALDI molecular imaging for clinical proteomic research: current state and fields of application. *Expert Rev Proteomics* 7:927–941

(95) Groseclose MR, Andersson M, Hardesty WM, Caprioli RM (2007) Identification of proteins directly from tissue: in situ tryptic digestions coupled with imaging mass spectrometry. *J Mass Spectrom* 42:254–262

(96) Lemaire R, Desmons A, Tabet JC, Day R, Salzet M, Fournier I (2007) Direct analysis and MALDI imaging of formalin-fixed, paraffin-embedded tissue sections. *J Proteome Res* 6: 1295–1305

(97) Djidja MC, Claude E, Snel MF, Scriven P, Francese S, Carolan V, Clench MR (2009) MALDI-ion mobility separation-mass spectrometry imaging of glucose-regulated protein 78 kDa (Grp78) in human formalin-fixed, paraffin-embedded pancreatic adenocarcinoma tissue sections. *J Proteome Res* 8: 4876–4884

(98) Schwamborn K, Krieg RC, Reska M, Jakse G, Knuechel R, Wellmann A (2007) Identifying prostate carcinoma by MALDI-Imaging. *Int J Mol Med* 20:155–159

(99) Stauber J, Lemaire R, Franck J, Bonnel D, Croix D, Day R, Wisztorski M, Fournier I, Salzet M (2008) MALDI imaging of formalin-fixed paraffin-embedded tissues: application to model animals of Parkinson disease for biomarker hunting. *J Proteome Res* 7: 969–978

(100) Grey AC, Schey KL (2009) Age-related changes in the spatial distribution of human lens alpha-crystallin products by MALDI imaging mass spectrometry. *Invest Ophthalmol Vis Sci* 50: 4319–4329

(101) Herring KD, Oppenheimer SR, Caprioli RM (2007) Direct tissue analysis by matrix-assisted laser desorption ionization mass spectrometry: application to kidney biology. *Semin Nephrol* 27:597–608

(102) Römpp A, Guenther S, Takats Z, Spengler B (2011) Mass spectrometry imaging with high resolution in mass and space (HR(2) MSI) for reliable investigation of drug compound distributions on the cellular level. *Anal Bioanal Chem* 401:65–73

(103) Deininger SO, Ebert MP, Fütterer A, Gerhard M, Röcken C (2008) MALDI imaging combined with hierarchical clustering as a new tool for the interpretation of complex human cancers. *J Proteome Res* 7:5230–5236

(104) Djidja MC, Claude E, Snel MF, Francese S, Scriven P, Carolan V, Clench MR (2010) Novel molecular tumour classification using MALDI-mass spectrometry imaging of tissue micro-array. *Anal Bioanal Chem* 397: 587–601

(105) Andersson M, Groseclose MR, Deutch AY, Caprioli RM (2008) Imaging mass spectrometry of proteins and peptides: 3D volume reconstruction. *Nat Methods* 5:101–108

(106) Lechene C, Hillion F, McMahon G, Benson D, Kleinfeld AM, Kampf JP, Distel D, Luyten Y, Bonventre J, Hentschel D, Park KM, Ito S, Schwartz M, Benichou G, Slodzian G (2006) High-resolution quantitative imaging of mammalian and bacterial cells using stable isotope mass spectrometry. *J Biol* 5:20

12

Nondestructive Molecular Mapping and Imaging

Synchrotron FTIR Spectral Microscopy

HOI-YING N. HOLMAN AND LIANG CHEN

Synchrotron radiation–based Fourier transform infrared (SR-FTIR) spectromicroscopy is a label-free chemical imaging technology with diffraction-limited spatial resolution. All macromolecular components within biological samples can be obtained from a single data acquisition, at the single cell level, without the addition of chemical stains or reagents. SR-FTIR spectromicroscopy emerged from the established infrared spectroscopy in the late 1990s. It takes advantage of three technologies: (i) the well-known sensitivity of the mid-infrared spectroscopy to chemical functional groups in molecules and their conformations, (ii) the convenience of a light microscope to locate areas for molecular and composition analysis, and (iii) the high signal-to-noise ratio provided by a bright synchrotron radiation-based infrared light source. Using photons in the mid-infrared region (2.5–15.5 μm wavelength, or 4000–650 wave number in cm^{-1}), synchrotron radiation-based Fourier transform infrared (SR-FTIR) spectromicroscopy can detect subtle biochemical changes in single cells in tissues with a micrometer spatial resolution (1, 2). When combined with a microfludics platform, SR-FTIR can be used to measure and image chemical changes inside the same individual living cells as biological processes are occurring in situ (3, 4).

BACKGROUND

Development of Conventional Infrared Spectroscopy

In the early nineteenth century, during his investigations of the solar spectrum, William Herschel discovered infrared radiation by passing sunlight through a prism and measuring the temperature beyond the red end of the visible spectrum. In 1835 Macedonio Melloni invented the first thermopile infrared detector. The potential of using infrared light energy as a source for spectroscopy for molecular analysis was not realized until almost eighty years later when Abney and Festing successfully used a dispersive infrared approach to acquire infrared spectra of almost fifty-two organic compounds (5). Infrared radiation can excite molecular vibrations and rotations if the incoming radiation is of the same frequency as one of the fundamental modes of vibration of the molecule, and if the transition dipole moment of the vibrational group in the molecules changes during vibration (e.g., C–C or N=N bonds) (6). The larger the change in the transition dipole moment, the more intense the absorption band(s) will be. From a simplistic point of view, the bending or stretching of bonds involving atoms with different electronegativities (e.g., C=O) will lead to intense absorption bands. All vibrations symmetric to the molecular center of symmetry are infrared inactive because the dipole moment does not change (6).

Abney and Festing used a prism monochromator to separate the wavelengths of infrared light from a thermal emission light source, and reflective mirrors to direct each wavelength individually through a slit to the detector. For each compound, they made point-by-point measurements of infrared light intensity as a function of infrared wavelength. They correlated absorption bands with the presence of certain organic groups in the molecule. In 1903 Coblentz began to acquire the infrared spectra of hundreds of molecules, both inorganic and organic (including gelatin or chitin), many are still usable today. This large volume of data enabled Coblentz to generalize that different atomic and molecular groupings absorbed specific and characteristic wavelengths; even when they were in different molecules, they gave absorption bands that were found at approximately the same wavelength (7). Since each material is a unique combination of atoms, the presence, position, and intensity of vibrational bands create a specific absorption pattern representing a molecular fingerprint of the materials.

The point-by-point dispersive measurement approach was tedious, and the levels of signal of individual wavelength that reached the detector were low. In 1911 Rubens and Woods invented the Fourier transform infrared (FTIR) approach to measure all the infrared wavelengths simultaneously (8). They directed infrared light through the Michelson interferometer and the sample, and then used Fourier transformation to convert the raw infrared data into the sample's infrared spectrum. However, difficulties associated with computing Fourier transformations manually had limited the applications of the FTIR approach (9). Nevertheless, they lay the foundation for the modern FTIR spectroscopy approach.

Throughout the first half of the twenieth century, the dispersive infrared approach dominated the field of infrared spectroscopy. Obtaining a spectrum remained to be a time-consuming job requiring 3–4 hours or more, since each point in the spectrum had to be measured separately. Until 1940s its applications were limited mostly to researchers in physics and astronomy (10). A notable exception was that it had found its place in WWII as a useful diagnostic tool in determining the concentration and purity of butadiene in synthetic rubber(10) because of the shortage of natural rubber caused by the war.

A dispersive infrared spectrometer (also called grating or scanning infrared spectrometers) emerged in the late 1940s (11). Together with advances in electronics, it became possible to obtain a spectrum in 1–2 hours, and infrared spectroscopy became a common analytical technique for the characterization of not only organic compounds (12, 13), but also molecules in biological specimens.

Infrared Spectroscopy of Cells and Tissues

Cells are composed primarily of water, lipids, proteins, nucleic acids, and carbohydrates, which all give characteristic absorption bands in the mid-infrared region (**Figure 12.1A**). The infrared spectrum of a mammalian cell can be closely approximated with a linear combination of the spectra of the biochemical components (i.e., the spectra of DNA, RNA, phospholipids, glycogen, and protein) that constitute the cell (**Figure 12.1B**). As early as the 1950s, a number of research groups reported using bright field microscopes to visually examine the morphology and structure of the biological specimens before obtaining detailed molecular information in biological specimens with infrared spectroscopy (14–19). These researchers demonstrated in tissue and blood smear samples that even though an infrared spectrum was a sum of the contributions gathered from all biomolecules (i.e., proteins, amino acids, lipids, and nucleic acids) in the samples, distinct absorption bands exist that can be related to known chemical groups in biomolecules (**Table 12.1**). This idea was soon applied by a number of researchers to investigate the chemical composition of biological samples and cells in the absence of distinctive morphological features (20–23).

Rise of Modern Infrared Microscopy

The increasing needs in probing spatial distributions of molecules in specimens first led to the design of infrared spectroscopic microscopes in the 1950s (24). However, it did not become popular because the data acquisition process was very slow and the signals of individual wavelength that reached the pyroelectric detectors were often not sufficient. For over thirty years, infrared spectral microscopy had taken a backseat as fluorescence microscopy became the dominant method for studying cellular systems; this was also due in part to the advent of genetically encoded fluorescent proteins for imaging dynamic behavior in live cells. These difficulties were completely overcome by the 1990s as a result of three groundbreaking technology developments: (i) the innovative application of the Fast Fourier transform algorithm to the FTIR spectroscopy approach of Rubens and Woods; (ii) the availability of inexpensive fast digital computers which enabled the replacement of dispersive spectrometers with FTIR interferometers; and (iii) the introduction of fast-response, high-sensitivity photoconductive single-element mercury cadmium telluride (MCT) detector which is up to fifteen times more sensitive than the previous pyroelectric triglycine sulfate detector (25).

Infrared Microscopes

The microscope shown in **Figure** 12.2**A** is an example of the modern FTIR microscope designed to operate in either the transmission mode (Figure 12.2B, left panel), the standard reflection (transflectance) mode (**Figure 12.2B**, middle panel), or the more recent reflection beamsplitter mode (**Figure** 12.**2B**, right panel) for improving the spatial resolution (**Figure 12.2C**). The infrared sources used are thermal emission elements (or thermal globars) that produce a graybody spectrum from a filament physically moderate in size (at least several millimeters). They are heated to 1000–2000 K and typically radiate in all directions. Infrared light emitted from

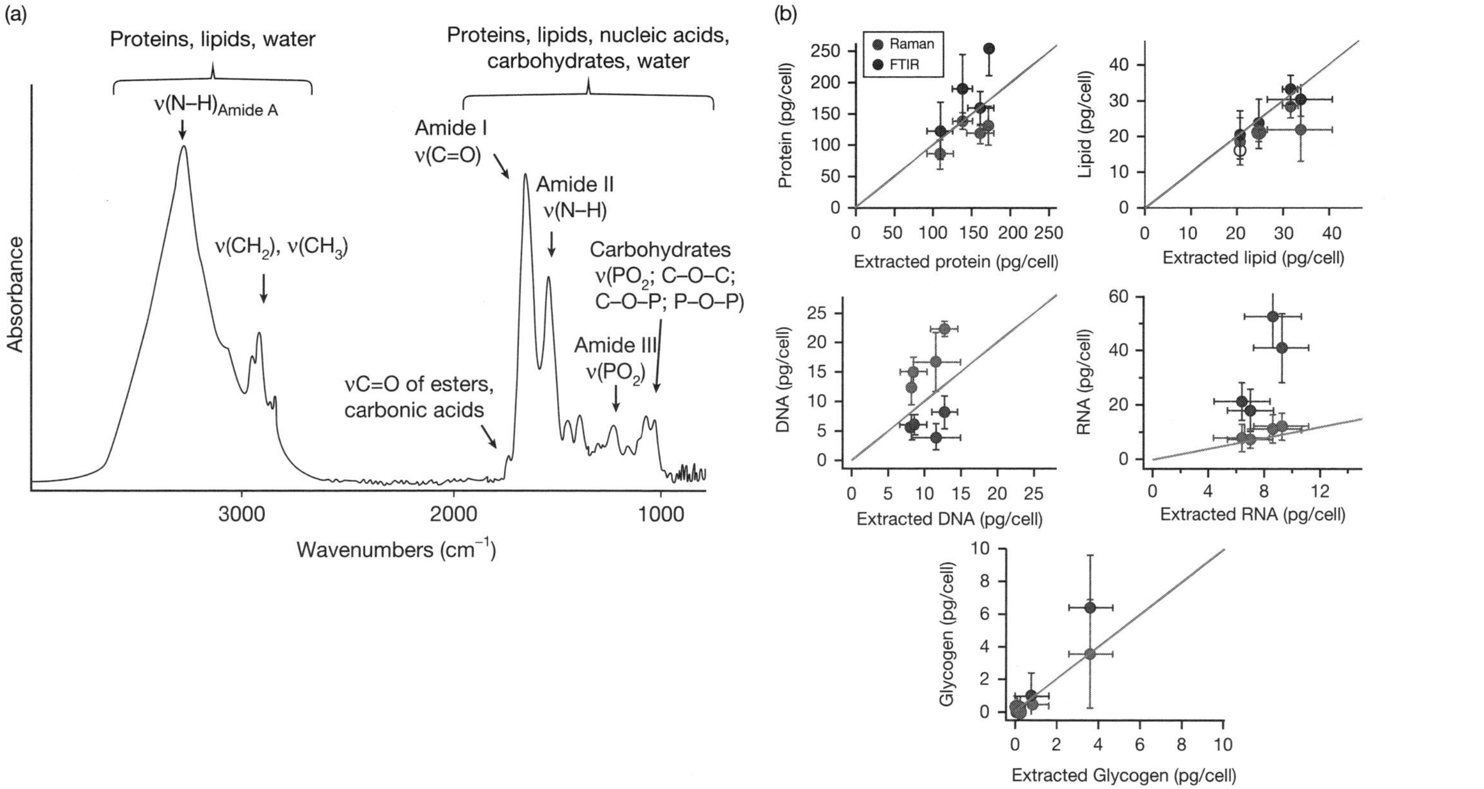

Figure 12.1 (A) Typical infrared micro-spectra recorded on an individual HeLa cells. The area probed is 20 20 mm^2. The spectrum was obtained with a laboratory-based infrared microscope and required a recording time of thirty seconds. It is important to note that the spectral quality (namely the signal-to-noise ratio) is essential in distinguishing the subtle changes that have to be observed among the overall line shape of the features (see text). In this spectrum, the high DNA content is clearly detected: the band centered around 1000 cm^{-1} is relatively high in intensity compared with the protein band at 1650/1550 cm^{-1}. (B) FTIR (in gray) and Raman (in black) spectroscopic measurements of biochemical components from Rat1-T1 fibroblast cells compared with results from extracting and other methods. The gray lines denote the position of perfect correlation. The graph of lipid content also shows NMR results versus extracted results in purple open circles. [From J.R. Mourant et al., *J. Biomedical Optics*, 11(6), 2006]

Table 12.1. Wavenumber (cm^{-1}) and assignment of the major bands in the infrared spectra of differentiated and non-differentiated PC12 cells (also see reference list in Chen et al., 2012)

Wavenumber (cm^{-1})	Vibrational mode assignment and main contribution
~3412	νOH of water, νNH of proteins
~2960	$\nu_{as}CH_3$ of lipids, proteins and nucleic acids
~2924	$\nu_{as}CH_2$ of lipids
~2852	$\nu_{s}CH_2$ of lipids
~1734	νC=O of lipids
~1643	Amide I of proteins
~1547	Amide II of proteins
~1462	δCH_2 of lipids
~1402	$\delta_s CH_3$ of lipids and proteins and $\nu_s COO^-$ of amino acid side
~1237	chains and fatty acids
~1151	$\nu_{as}PO_2^-$ of nucleic acids, lipids and phosphorylated proteins
~1080	νC-O of protein side chain groups and carbohydrate
~1053, ~1036	$\nu_s PO_2^-$ of nucleic acids, νPO_4^{2-} of phosphorylated proteins
990-970	νC-O-C and νC-O-P of polysaccharides
	νC-O coupled with δC-O of carbohydrates
	νPO_4^{2-} of phosphorylated proteins and nucleic acids
	νPO_4^{2-} of phosphorylated proteins and nucleic acids

a thermal emission element enters the Michelson interferometer that modulates the wavelength of the infrared light. The modulated light then passes into infrared microscopes that are very much like a visible light microscope, except all of the optics are reflective mirrors to enable the infrared light to reach the sample without being absorbed. The objective (the Schwarzchild objective) focuses the beam into a small spot where the sample is mounted. A mechanical aperture placed ahead of the objective defines the infrared illuminated area and its size, and thereby the area of interrogation. Closing the aperture ahead of the condenser improves spatial resolution (**Figure 12.2A**, inserts a versus b), but results in poor signal-to-noise ratios (SNRs). The motorized translation microscope stage raster scans the sample pixel-by-pixel to obtain an image with a spectrum at each pixel.

When the microscope is operated in the transmission mode (**Figure 12.2A; Figure 12.2B**, left panel), the infrared radiation that is transmitted through the sample is collected by the second objective (or condenser) and passes a second aperture before being sent to the detector. In the standard reflection mode (**Figure 12.2B**, middle panel), light that is reflected off the sample is collected by the same objective that focuses the IR light. The light is then directed to the detector by a "sluice" mirror, which is removed during transmission mode measurements. The mirror, however, blocks a portion of the incident light and effectively shadows half of the objective's secondary mirror. This half illumination degrades the spatial resolution of the microscope. In the reflection beamsplitter mode (**Figure 12.2C**), the reflective mirror is replaced with a 50 percent beamsplitter to allow full illumination of the objective and retain a way to direct the reflected light to the detector (26). The improvement in the resolving power is shown in **Figure 12.2C** (left panel versus right panel).

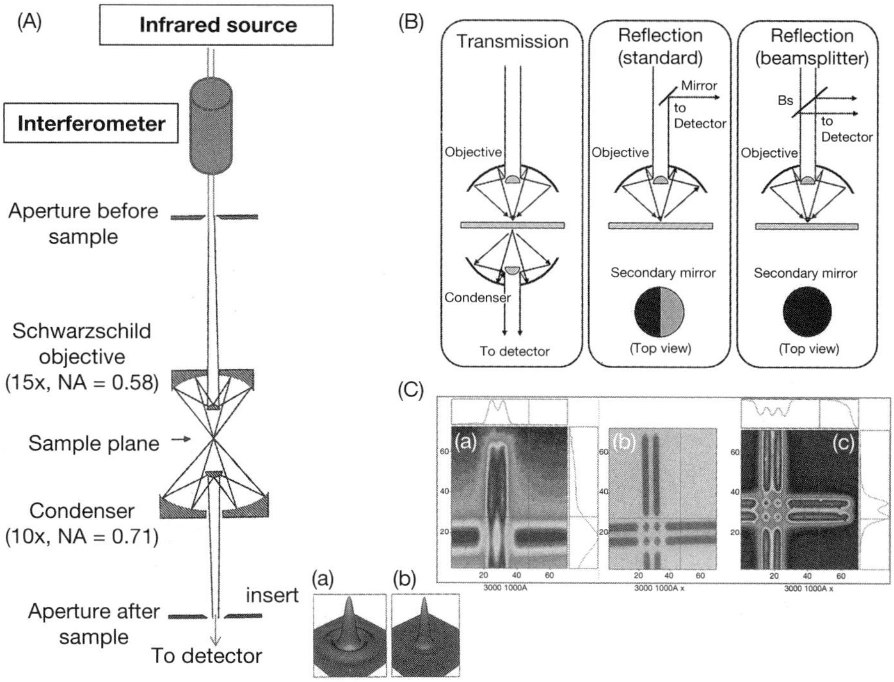

Figure 12.2 (A) Dual-aperture microscope similar to the optics of an infrared microscope. Insert: (a) Energy profile showing the diffracted radiation outside the area defined by a single aperture; (b) corresponding profile obtained with masks located before and after the sample. (Modified from Griffiths, 2009). (B) Schematic diagram of three different modes of IR microscopy: (left panel) transmission, (middle panel) reflection with the standard mirror design, and (right panel) reflection with the beamsplitter design. (C) infrared images of a patterned sample obtained with (left panel) standard reflection mode and (right panel) beamsplitter reflection mode using a globar source and a focal plane array. (Middle panel) Optical image of the sample (26).

Detectors: Single versus FPA

The pixel-by-pixel sequential collection of data, or mapping, can be slow when the measurement area is large. In 1995 Lewis and Levin introduced the technology of the focal plane array (FPA) detectors with a large number of small MCT detector elements (27, 28). The idea was that all the MCT detector elements of the FPA detector, also called "pixels" or "voxels," could be read simultaneously to provide a multidimensional data set from which chemical images can be constructed. Typical array sizes are 32×32, 64×64, and 128×128 detector elements with dimensions of 40 μm × 40 μm. By the 2000s most FTIR infrared microscopes have the capability of using either a single-element detector for mapping or a FPA for imaging.

Chemometrics for FTIR Image Analysis

Along with the advents of the FTIR microscopy came the challenge of generating molecular images from biological samples. FTIR mapping or imaging produce an array of massive spectra, or called hyperspectral data cube. It contains spectral information at each small area called "a pixel" of the samples. The data cube consists of four dimensions: the x- and y-coordinate of each pixel, the wavenumber, and intensity of the spectral vector. In order to extract the molecular information and display the image, the data cube must be reduced to three dimensions (3D), with the first two being the x- and y-coordinates.

There were various methods to generate the third dimension value. One common approach is univariate analysis, which in conjunction with the Beer-Lambert bilinear model can provide easily visualizable presentation, as illustrated in **Figure 12.3**. The model assumes that the measured absorbance intensity at a distinct wavenumber (or wavelength λ) is linearly related to both the concentration of the species at the pixel and the path length of the infrared light through the pixel.

However, the univariate approach does not take advantage of the full FTIR measurement, which is especially important when the molecular information is usually distributed over a wide spectral range, and the spectral variance between tissue types and pathological states are invariably small. The sheer size of the data cube, however, makes it exceedingly time-consuming to visually inspect each image over all the wavelengths of each spectrum. The multivariate image analysis, such as hierarchical cluster analysis (HCA) or principal component analysis (PCA), compresses the spectral data, allowing for an examination of the entire hyperspectral data cube. In HCA method, it examines the complete data cube using a minimum number of classes with the minimum within-class variation, and the classes are spatially mapped to result in a cluster map. In PCA method, it examines the image data structure using a minimum number of PCs. The projections of data onto individual PCs axes (in terms of scores) are spatially mapped, resulting in highly compressed score images.

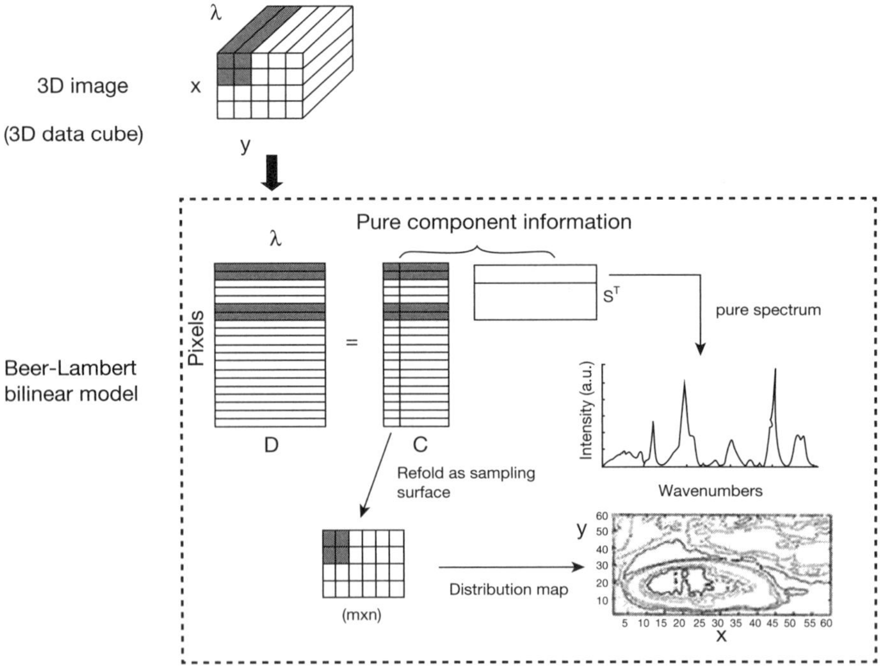

Figure 12.3 The underlying Beer-Lambert law bilinear model of hyperspectral images. (Adapted from 39.)

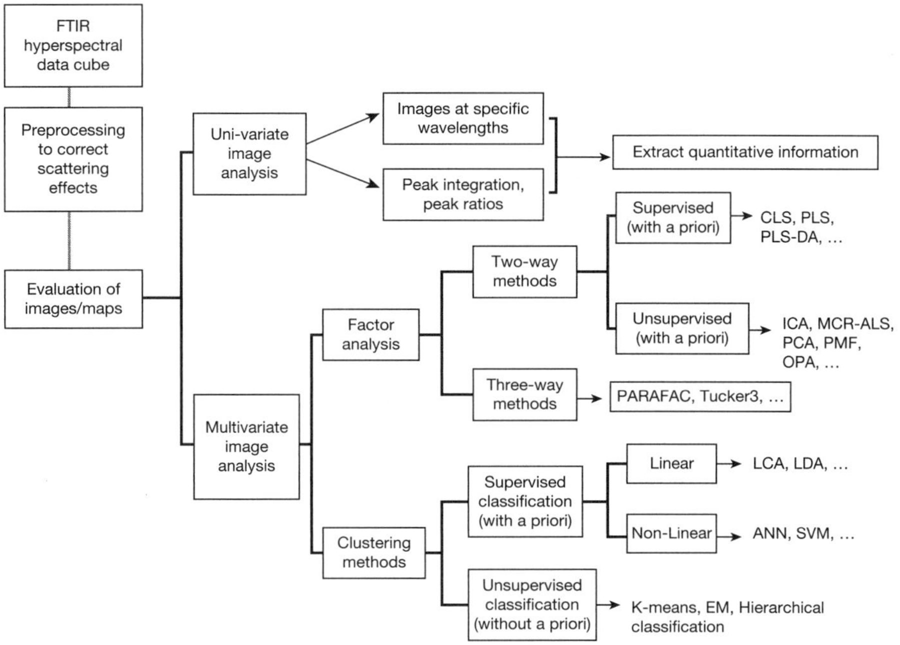

Figure 12.4 A typical chemometrics framework for the analysis of FTIR chemical mapping and imaging. ANN: Artificial Neural Network; BPSS: Bayesian Positive Source Separation; CLS: Classical Least Squares; ICA: Independent Component Analysis; KNN: K Nearest Neighbor; LCA: Linear Correlation Analysis; LDA: Linear Discriminant Analysis; MCR-ALS: Multivariate Curve Resolution-Alternating Least Squares; NMF: Non-negative Matrix Factorization; OPA: Orthogonal Projection Analysis; PARAFAC: Parallel Factor Analysis; PCA: Principal Component Analysis; PLS: Partial Least Squares; PLS-DA: Partial Least Squares–Discriminant Analysis; PMF: Positive Matrix Factorization; SVM: Support Vector Machine.

Independent of the approaches used, the quality of the images also depends on the quality of the raw data which are often affected by noise and instrumentation variations: the presence of intense and irregularly shaped baseline contributions (29), and the existence of anomalous pixel spectra. A wealth of user-friendly chemometric methods and algorithms is available for signal preprocessing, data pretreatments, image generation, analysis, and visualization (30). An appreciation of a typical infrared hyperstpectal imaging and analysis work flow and framework is shown in **Figure 12.4**.

At the Histopathology and Challenges

The merging of FTIR microscopy with chemometrics has opened wide perspectives in biomedical applications of FTIR imaging, especially to complement classical methods of diagnosis and to monitor a wide range of disease states in tissue (31–33). The classical methods include fixation, embedding in a wax block, the cutting of tissue sections, followed by their mounting on slides, staining, and inspection using optical microscopy. More recent methods include fluorescence microscopy and molecular biology. However, these methods require substantial preparation procedure, labeling of molecules, and qualitative interpretation by experts. In the mid-2000s Diem and his colleagues realized the potential significance of infrared spectral microscopy as a label-free diagnostic technique with minimum sample preparation. They presented the first organized strategies of applying FTIR absorption/reflection spectral microscopy together with unsupervised hierarchical clustering (34, 35).

An important step in the development of their applications is to demonstrate the sensitivity of the spectral regions in identifying chemical characteristics in soft tissues from patients. Diem and his colleagues prepared 4 μm-thick tissue sections from millimeter-size paraffin-embedded tissue blocks and mounted them on low-e slides. The low-e slides were made of glass and coated with a thin Ag/SnO_z layer, which are nearly transparent to visible light, but reflect mid-infrared radiation almost completely, therefore allow both visual and infrared images to be collected from the same samples, which are ideal for histopathology. Diem and his colleagues imaged the tissue sections in transflectance mode at 20 μm spatial resolution and with 4 cm^{-1} or 8 cm^{-1} spectral resolution. In the transflectance (also called double transmission) mode, the beam passed the tissue sample, the light was reflected by the silver layer of the low-e slide, and passed the sample again before being sent to detectors. For visual comparison and validation, they also stained the tissues for histology after FTIR imaging was performed.

The sensitivity of different mid-infrared spectral regions as a diagnostic tool is shown in **Figure 12.5**. In **Figure 12.5A** is a photomicrograph of a thin section of a cervical squamous epithelium from a patient diagnosed with low-grade squamous intracepithelial lesions (LSIL). The tissue was stained H&E after the FTIR data acquisition, and the FTIR images show the spatial variations in the tissue chemical properties as revealed by cluster analysis in the entire fingerprint region (1800–800 cm^{-1}) (**Figure 12.5B**), the protein region (1740–1470 cm^{-1}) (**Figure 12.5C**), and the Glycogen/nucleic acid region (1200–1000 cm^{-1}) (**Figure 12.5D**). The FTIR images can distinguish the basal and parabasal layers and the stroma (**Figures 12.5B–D**), which is an important result because cervical disease frequently originates in the basal layer of the squamous tissue or at the junction of squamous and columnar epithelium. Furthermore, several distinct regions in the FTIR image are consistent with the main tissue layers observed in the H&E-stained section. These include the outer layer (various blue hues), the intermediate layer (green), the parabasal and basal layers (orange and yellow), and the connective tissue (brown). In **Figure 12.5E**, the mid-blue and light blue traces show typical spectral features of a glycogen-rich superficial layer with strong 1151, 1078, and 1028 cm^{-1} bands due to coupled C–O and C–C stretching, and C–O–H deformations motions of glycogen. The red spectral trace is indicative of the intermediate layer, the green trace represents the parabasal layer, while the brown trace is characteristic of the connective tissue. The green trace shows a dramatic decrease in the glycogen band intensity and a reduction in the amide II/amide I intensity ratio compared to the other mean extracted spectra. The brown trace, correlating to connective tissue, exhibits a broad relative intense feature centered at 1240 cm^{-1} from strong collagen bands overlapping nucleic acid phosphate groups and virtually no contributions from glycogen in the 1200–1000 cm^{-1} region.

FTIR spectromicroscopy has also been applied to determine the chemical composition and quality of hard tissues including the adaptive response of bone and cartilage tissues. The application to bone has been especially fruitful in characterizing the

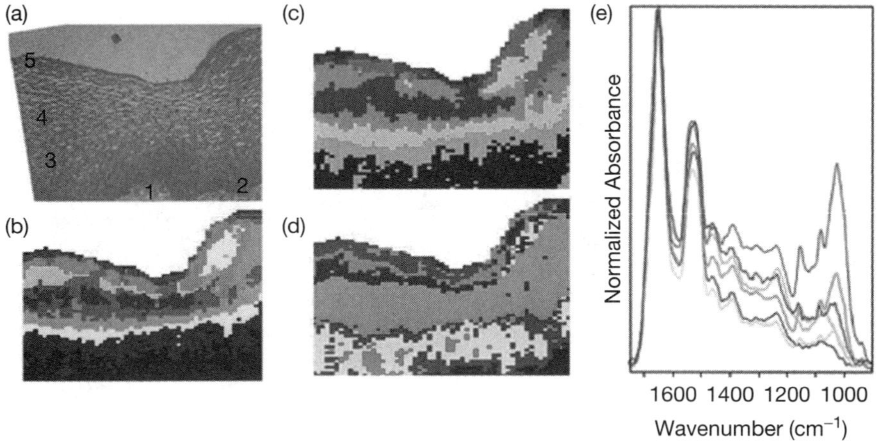

Figure 12.5 Sensitivity of different mid-infrared spectral regions as a diagnostic tool. (A) Photomicrograph of a section of cervical squamous epithelium from patient diagnosed with LSIL. The section was stained (H&E) after FTIR data acquisition. 1: Connective tissue; 2: basal layer; 3: parabasal layer; 4: intermediate layer; and 5: superficial layer. (B) 9-cluster maps using the entire mid-IR spectral range (1800–800 cm^{-1}). (C) 9-cluster maps using the "protein region" (1740–1470 cm^{-1}). (D) 9-cluster maps using the "glycogen/nucleic acid region" (1200– 1000 cm^{-1}). (E) Mean cluster spectra from six of the seven clusters color coded to enable correlation with 2C. Note: the individual colors in panels B–D are assigned by the algorithm for each data set and are not transferable between separate analyses (35). See insert for color version.

adaptation of bone in response to external stress such as microdamage and deformation/fracture (36) and to Osteoporosis treatments (37), and alternation of cartilage chemistry (38). In one bisphosphonate study, Durchschlag and coworkers (39) examined the effect of three- and five-year risedronate treatment on trabecular bone. The sensitivity of FTIR as a diagnostic technique to detect pathogenesis and response to treatment is shown in **Figure 12.6**, which reveals patients receiving placebo for three years show a significant increase in crystallinity-and-collagen-crosslink ratio when compared to baseline values. The patients who received risedronate for three or five years retained baseline values for both the crystallinity and crosslink ratio.

Infrared spectral microscopy applications in biology and biomedicine are making a resurgence with the emergence of user-friendly chemometrics, especially in diagnosis of cancers (40–42), analysis of bone and cartilage disease (43, 44), and prion-infected nervous tissue (45, 46). Many databanks of important cell signatures were retained from samples that were directly reassessed by pathologists and cytologists for comparison and validation. Analysis of the large volume of data reveals that improvements in two technical parameters could significantly improve the FTIR spectromicroscopy performance: (i) spatial resolution, which determines the measurement area within the biological sample and therefore the length scale of the heterogeneity that can be studied, and (ii) detection sensitivity, which determines the data collection time and therefore time resolution for investigating ongoing biological processes in real time. Both parameters can be dramatically readily improved

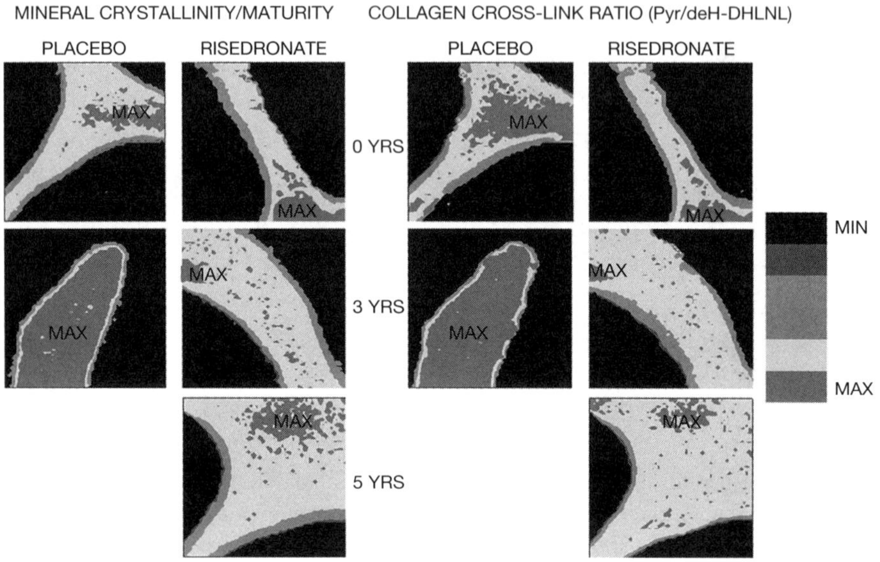

Figure 12.6 FTIR images of thin sections of trabecular bone from a placebo- and a risedronate-treated patient at baseline, and after three and five years of treatment. In placebo-treated patients, both mineral crystallinity and collagen cross-link ratio were significantly higher at three years compared with baseline. In contrast, in risedronate-treated patients, there were no significant differences (39).

by replacing the thermal emission source (e.g., a globar) in the conventional FTIR spectromicroscope with a bright synchrotron infrared source (1, 47, 48).

THE BRILLIANCE OF SYNCHROTRON RADIATION

A synchrotron is a high-energy electron storage ring optimized for the production and collection of light radiated by relativistic electrons as they traverse a curved path through a magnetic field (**Figure 12.7A**). Synchrotron radiation (SR) spans a large electromagnetic spectral range, beginning in the far-IR and extending through much of the X-ray region (**Figure 12.7B**). The primary advantage of the high brightness (i.e., photon flux, or power emitted per source area and per solid angle) of synchrotron infrared light was first demonstrated in 1983 by Duncan and Williams (49), who showed that a SR infrared source is 100–1000 times brighter than a conventional thermal IR source because the SR is concentrated in a very small opening angle such that the size of the SR infrared source can closely resemble an ideal point source.

When connecting the SR infrared beam to an infrared microscope and using a single MCT detector element for detection (**Figure 12.8A**), the near ideal point source means that the photons can be focused to a spot with a dimension approaching a diffraction-limited or the highest resolution. Levin and Martin demonstrated that the size of the area (and thereby the resolution) depends on the numerical aperture (NA) of the microscope objective and the wavelength (λ) of light as given by $0.61\times(\lambda/\mathrm{NA})$ (50, 51). The larger the NA of the objective, the finer the spatial resolution. Typical infrared microscopes use Schwarzschild objectives with NA in the range of

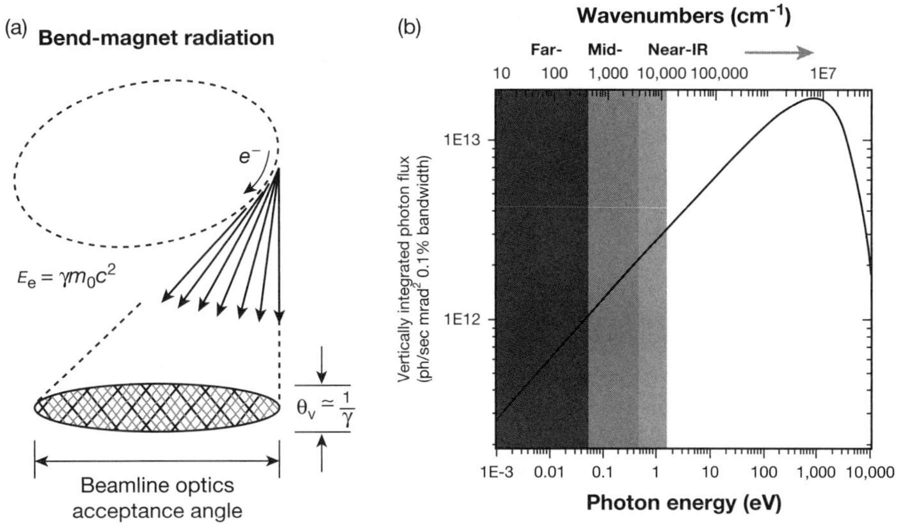

Figure 12.7 An overview of synchrotron radiation. (A) Guided by a series of bending magnets and straight sections, relativistic electrons inside a storage ring complete a loop. When the relativistic electrons encounter a magnetic field, they are deflected and they emit electro-magnetic radiation with energy photons up to hard X-rays. (B) This bending magnet spectrum extends from very low energies (far-infrared) continuously to a critical energy in the soft or hard X-ray, depending on the energy of the synchrotron. The radiation pattern from relativistic electrons is such that its effective source size can be considered as very close to an ideal point source (52).

0.4 to 0.7, and spot sizes (and hence spatial resolutions) of 1.2λ to 0.5λ are achievable. Although many more infrared photons are focused on the biological sample when using a synchrotron source, the low photon energy and low peak powers (relative to the alternative infrared laser sources) of < ~50 mW have shown to introduce no detectable effects on the quality of the samples (52). One important consideration when making SR infrared spectral measurements is the size of the targeted feature relative to the spot size of the infrared beam. In the mid-infrared region, these 0.5λ to 1.2λ (i.e., ~2 to ~10 μm) diffraction-limited spot sizes are smaller than eukaryotic cells, bigger than most of the prokaryotes or archaea, and are comparable to a small cluster of prokaryote or archaea cells, except for a few mega-bacteria such as the sulfur bacterium *Thiomargarita namibiensis* that can reach a diameter of 700 μm or the thermophilic archaea *Staphylothemus marinus* which can grow to 15 μm in diameter.

Unlike infrared microscopy with a conventional thermal source that requires using both apertures (and appropriate pixel size) to achieve micrometer spatial resolution, infrared microscopy with a synchrotron source has no requirement for apertures and thereby little loss in signal-to-noise ratio. However, it is important to match the mapping pixilation to the NA of the objective by using the appropriate pixel size. Too large a pixel size would lead to resolution loss; too small a pixel size would lead to long acquisition time with little improvement in the spatial resolution. For a microscope objective with a numerical aperture of 0.65, the diffraction-limited resolution over the entire mid-infrared spectrum can only be achieved with a pixel spacing less than ~ λ/4 (53).

Improved spatial resolution is of great benefit to molecular measurements involving heterogeneous samples such as cells within the tissues. The accuracy with which a given feature is measured is greatly increased because a synchrotron infrared beam allows the measurement to be taken only on the area of interest (1, 3). As illustrated in **Figure 12.8B**, the SR measurement is targeted to a small area, more accurately pinpointing a part of the sample and giving rise to a feature of interest. In **Figure 12.8C** the improvement in S/N and time resolution of SR-FTIR is demonstrated by comparing FTIR spectra from a single cell, as measured with a synchrotron and with a conventional thermal source using a 10-μm aperture. The collection time using the synchrotron source was only 16 seconds (32 scans), whereas the thermal source took 500 seconds (1000 scans). Even after extensive averaging, the quality of the thermal source measurement is not sufficient to reveal the fine molecular features within the vibrational spectrum.

These unique analytical capabilities of synchrotron IR spectromicroscopy could potentially meet the demands of modern biological research for studying molecular reactions in individual living cells. Presently, there are thirty-one synchrotron infrared spectromicroscopy facilities around the world with several more under construction or planned (see, for instance, http://infrared.als.lbl.gov/content/web-links/45-srir). Within the United States there are four active synchrotron infrared facilities with microscopy capabilities: the National Synchrotron Light Source (NSLS, Brookhaven National Laboratory), the Synchrotron Radiation Center (SRC, University of Wisconsin-Madison), the Center for Advanced Microstructures and Devices (CAMD, Louisiana State University), and the Advanced Light Source (ALS, Lawrence Berkeley National Laboratory); each have similar as well as unique capabilities. All four are user facilities open to qualified scientists, although the availability of beamtime can be highly competitive at times.

MOLECULAR MEASUREMENTS OF SINGLE CELLS IN TISSUES

The SR infrared brightness and high spatial resolution coupled with its noninvasive nature (52) have open new opportunities in the fields of biology and biomedicine. In 1998 synchrotron FTIR was first applied to map the distribution of macromolecules in a single human cell (54), and shortly afterwards to characterize human lung epithelial cells during the cell cycle and the early stage of apoptosis (55). This is followed by a growing trend of diagnosis of tissues for biomedical applications such as to identify the structures of misfolded protein aggregates in the brain tissue of Alzheimer's disease patients (56–59) to detect the evolution of Huntington's Disease (60), to characterize the infectious prions in scrapie-infected tissues (2, 61–63), to analyze the bone (64–69) and cartilage (66, 70) disease, and to detect advanced atherosclerotic plaques in aortas (71). More recently, new applications were carried out for the molecular characterization of single cells in tissue, including stem cells (72–77) and cancer cells (78), epithelial cancer cells (79, 80), brain tumors (81–83), breast cancer (84, 85), colon carcinoma cells (86), or prostate cancer (87–89).

One new practical approach involves integrating infrared focal plan array (FPA) imaging with synchrotron infrared mapping to detect subtle chemical and structural changes during early pathogenesis; which cannot be observed by

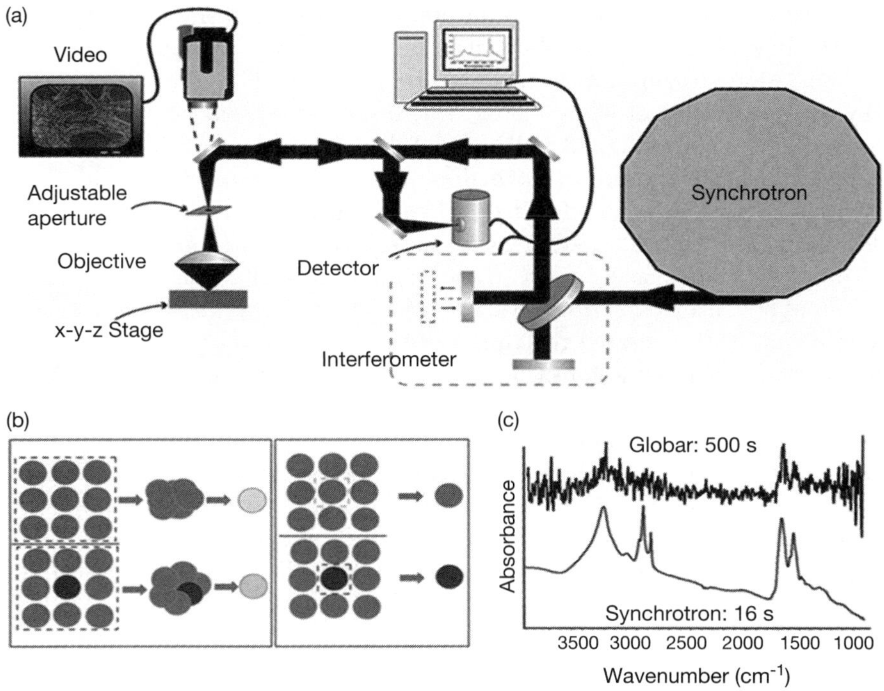

Figure 12.8 (A) Schematic of an SR-FTIR spectromicroscopy experiment, where the thermal source of a conventional FTIR microscope is replaced by a synchrotron source. Mid-IR radiation from a synchrotron is transported to an FTIR interferometer bench. After modulation by the interferometer, an IR microscope with reflective optics focuses the beam onto the sample. The stage is computer controlled and rasters the sample in the x-y plane with 0.1 μm precision to obtain spectral maps across the sample. (B) Demonstration of how improved spatial resolution can enhance detection sensitivity and spatial accuracy of heterogeneous samples within a larger population. The left panel demonstrates that with a larger spot size (from a poor brilliance source such as a conventional thermal emission source), the abnormal (black) circle is averaged with the normal (gray) circles and can be hidden. In the right panel, the smaller spot size (from a high brilliance source such as a synchrotron), the abnormal circle is clearly identified. (Adapted from Dumas et al., 2007.) (C) Comparison of FTIR spectra of a single cell using a 6×6 μm^2 aperture from a synchrotron source (grey trace) and a thermal (black trace) source. The collection time for the synchrotron source system was about 16 seconds (32 scans) compared to about 500 seconds (1000 scans) for the thermal source system (courtesy P. Dumas).

conventional histology or FPA alone. The application of this approach, in conjunction with bioinformatics to gain new insights into the early biochemical and protein conformational changes during neuropathology of multiple sclerosis (MS), are illustrated in **Figure 12.9**. Heraud and coworkers (90) analyzed formalin-fixed, paraffin-embedded sections of brain, cerebellum and spinal cord samples on the low-e reflective slides. They compared the early chemical changes during the induction and prevention of autoimmune-mediated demyelination in experimental autoimmune encephalomyelitis (EAE), an animal model for investigating the molecular mechanisms involved in the etiopathology of MS and to test therapies for MS. They examined the cerebellum of mice taken seven days after induction of EAE but before the appearance of clinical signs. Using a FPA microspectrometer, they imaged an area of white matter, where some clustering of nucleated cells were visualized following H&E staining of the adjacent section (**Figure 12.9A**); however, ambiguity existed as to whether these nucleated cells were indicative of lesional activity or was simply part of the structure of the underlying granular. FPA chemical images showed areas of some loss of lipid (**Figure 12.9B**) colocalizing with an increasing concentration of nucleic acid (**Figure 12.9C**) in this region. When using synchrotron SIR spectromicroscopy mapping, a detailed examination of the selected region showed distinctly lower lipid concentration equivalent to advanced lesions (**Figure 12.9D**) in areas approximately 20–30 μm in diameter, which co-localized with an increased nucleic acid concentration (**Figure 12.9F**). Furthermore, regions with high nucleic acid content were smaller than the regions of decreased lipid content (**Figures 12.9D** and **F**), which correspond with the positions of the nucleated cells identified by DAPI staining using confocal immunofluorescence microscopy on the same sections (**Figures** 12.**9G–I**). The lower lipid concentration in the synchrotron IR images together with the presence of some perivascular leukocyte infiltrations supported the view that the areas contain very early small EAE lesions. Additional corroboration for the pathological nature of the microlesions came from the classification of spectra within these regions (as lesion spectra based on an ANN trained with spectra taken from the known advanced stages of EAE) (**Figures 12.9G–I**, ANN). The ability of the ANN to discriminate unambiguously pathology on a fine spatial scale is shown in **Figure 12.9H** (bottom right corner). They confirm that the adjacent granular layer (green area) is clearly differentiated from microlesion (gray area) by ANN classification.

More recently, synchrotron FTIR has been combined with other label-free spectromicroscopy techniques to obtain more detailed chemical information on the same tissue. Petit and coworkers (91) integrated synchrotron FTIR and VU spectromicroscopy with time-of-flight secondary ion mass spectrometry (TOF-SIMS) to probe 6–10 μm thick tissue section from a cirrhotic liver which was deposited on a gold-coated glass slide. **Figure 12.10** illustrates the means by which synchrotron FTIR was used to visualize the distribution of fibrosis and cirrhotic modules before TOF-SIMS was used to identify certain molecular lipid species amid the fibrosis or cirrhotic nodules; lastly, UV spectromicroscopy was performed to confirm the presence of collagen by detecting the autofluorescence from fibrous septa. Gough and coworkers (92) integrated synchrotron FTIR and X-ray fluorescence spectromicroscopy to detect pigmented creatine deposits in amyotrophic lateral sclerosis central nervous system tissues (**Figure 12.11**).

In additional to biomedical applications, synchrotron FTIR was also used to derive characteristic cell wall architecture in single plant cells (93), plant tissues (94–97), and biofilms (98, 99). During the Deepwater Horizon blowout in the Gulf of Mexico in 2010, the worst oil spill in the US history, SR-FTIR spectromicroscopy played an integral role in demonstrating molecular measurements of in situ microbial processes by revealing how indigenous deep-sea microbes could immediately lead to significant molecular changes in oil hydrocarbon composition at 4°C in the deep water oil plume (100).

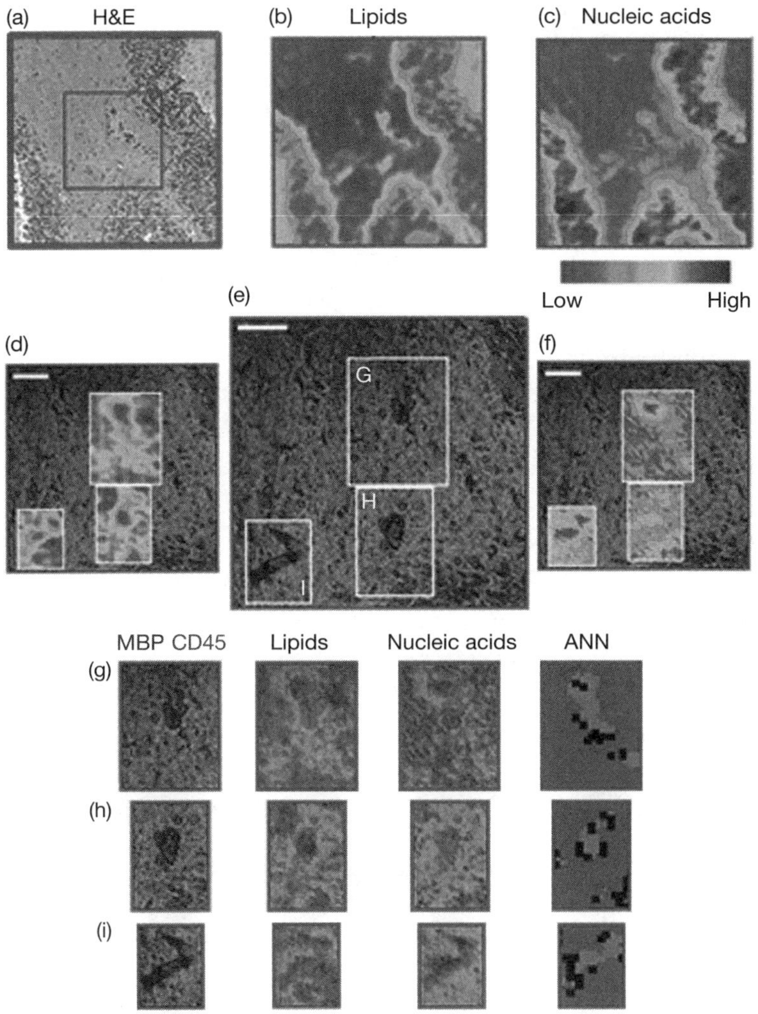

Figure 12.9 Synchrotron-based chemical mapping detects early experimental autoimmune encephalomyelitis (EAE). (A) Adjacent cerebellum section stained with H&E; blue square is the area studied by infrared FPA imaging; white square indicates area examined using the synchrotron FTIR. (B, C) FPA images of cerebellum showing the relative concentration of lipid and nucleic acid derived from FPA spectral data. Scale bar = 100 μm. (D–F) Areas interrogated with the synchrotron light source overlaid onto areas stained with immunofluorescence of 56 percent optical opacity. Note that the infrared analysis showing myelin organization and cellular infiltrate following staining with an anti-MBP antibody (green) and DAPI (blue). Corroboration of the pathological nature of the microlesion provided by ANN classification (G–I). These areas are classified in a similar manner to that seen in the advance stages of EAE; red classifies the cerebellum white matter, green classifies the granular layer, gray classifies the lesion spectra, and black pixels are unclassified spectra. Scale bar = 50 μm (88). See insert for color version.

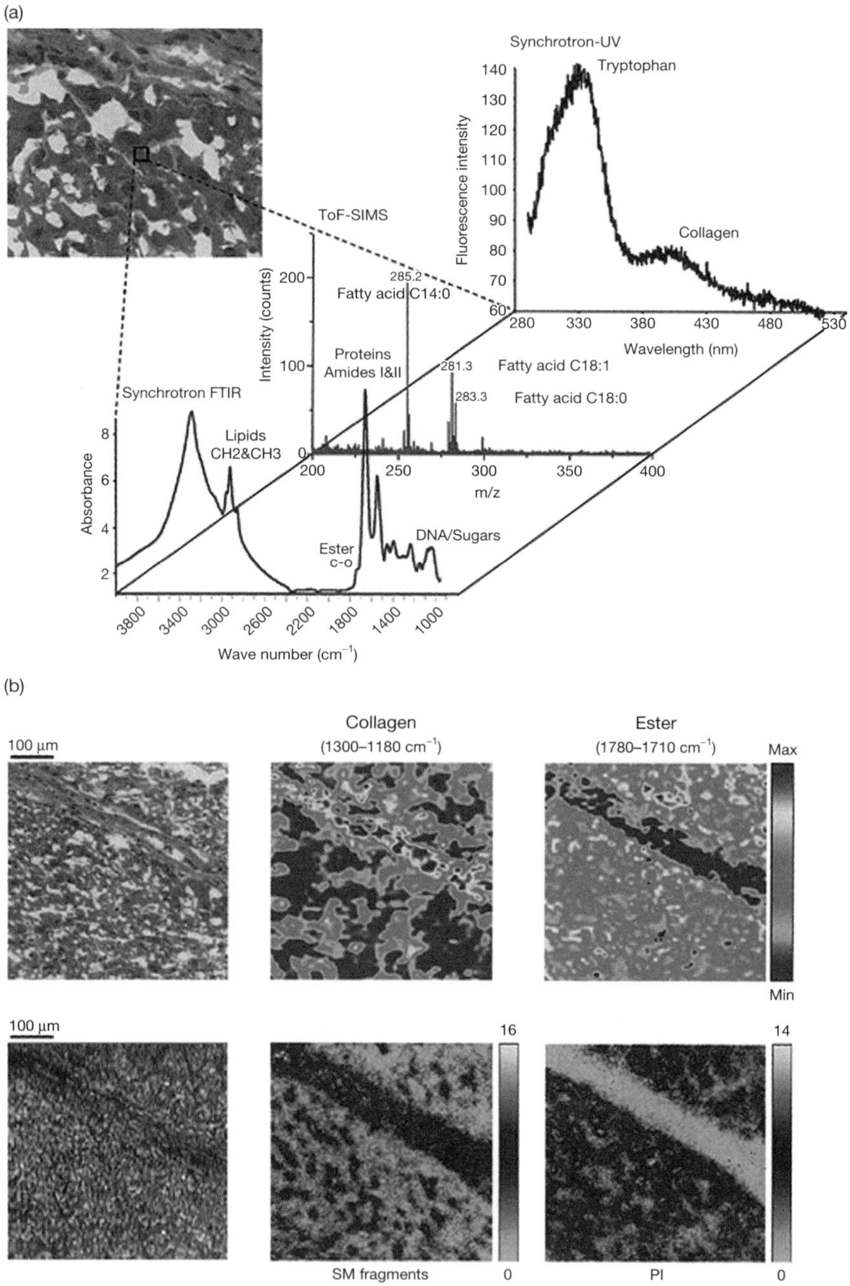

Figure 12.10 Combination of multiple high spatial resolution spectromicroscopy. (A) The histological tissue section of liver cirrhosis stained with HES (upper left). The square represents the $10 \times 10\ \mu m^2$ region analyzed. Sequential acquisitions were performed using synchrotron-FT-IR, TOF-SIMS, and synchrotron-UV spectromicroscopy. The corresponding spectra are shown. (B) The histological serial tissue section of liver cirrhosis stained with HES is shown (upper left). Synchrotron-FTIR spectromicroscopy images of infrared absorption bands corresponding to collagen or ester (upper panel). Mass spectrometry images using TOF-SIMS were recorded on the same tissue section (video image, lower left), and distributions of sphingomyelin fragments or phosphatidylinositol (PI) (lower panel) (89).

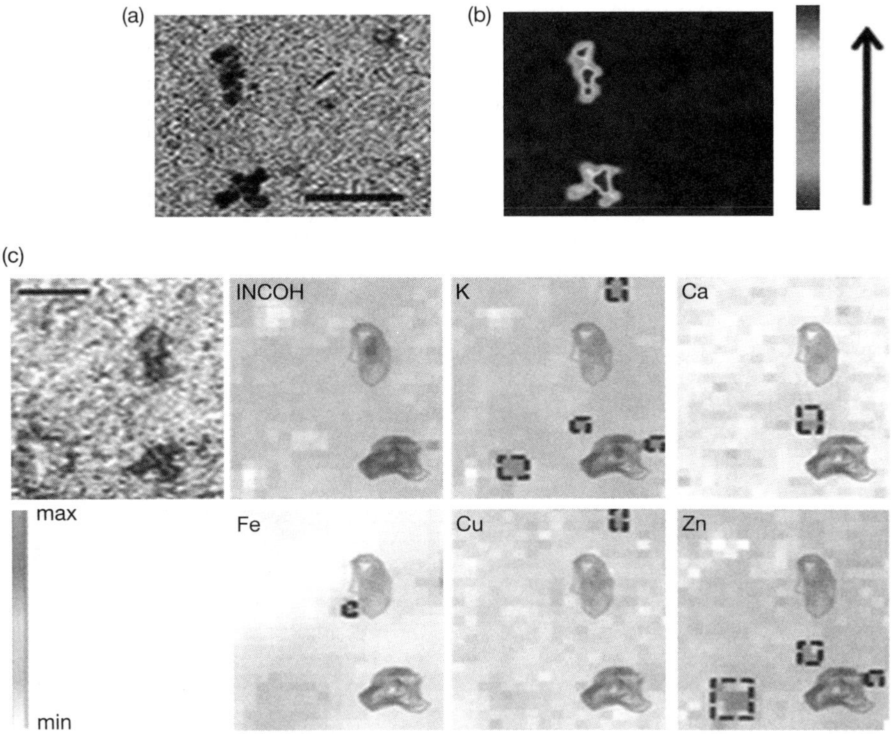

Figure 12.11 High spatial resolution SR-FTIR and X-ray fluorescence spectromicroscopy analysis of Amyotrophic Lateral Sclerosis (ALS). Central nervous system tissues reveal higher incidents of pigment deposit rich in Ca, Zn, Fe, and Cu. (A) Bright field image reveals two creatine deposits. Scale bar, 100 μm. (B) SR-FTIR heat map for creatine absorption peak at 2800 cm^{-1}. (C) XRF maps of selected elements obtained for ALS spinal cord sample region containing creatine deposits (white light image on the left top corner). Contour map created from incoherent scattering (INCOH) data indicate locations of dense creatine deposits (gray scales display range: 1900–3900 cm^{-1} for INCOH). There is no co-localization between creatine deposits and elements K, Ca, Fe, Cu, and Zn. Elemental hot spots are outlined with dashed black line. Scale bar, 100 μm (92).

IN SITU REAL-TIME MEASUREMENTS OF CHEMICAL CHANGES IN LIVING CELLS

At about the same time the synchrotron infrared community was focusing on the molecular imaging of pathogenesis in tissue, researchers at the Lawrence Berkeley National Laboratory began to develop approaches for the in situ and real-time molecular measurements of transient cellular processes. In an early study from the Berkeley group, time-lapse SR-FTIR transflectance measurements showed the mechanism by which a small subset of microbes on a basalt rock specimen survived exposure to a high concentration of toxic chromium(VI) while most of the surrounding microbial cells died (101). This demonstrates that although geological materials inherently have very rough surfaces, the small spot size and the high brightness of the SR infrared source made it possible to measure a small region with a handful of microorganisms, revealing that they were reducing chromium(VI) to chromium(III). This capability of also yielded new insights into how the soil bacteria *Mycobacterium* sp. JLS solubilize large recalcitrant organic pollutants like polycyclic aromatic hydrocarbons and metabolize them into biomass (102).

Today, much of the real-time SR-FTIR application is for the measurement and understanding of stress-adaptive-response in cells. The classical approaches, which often involve biochemistry experiments that are averaged over large populations of cells, have clarified many detailed mechanisms. However, they are not sufficient to reveal the heterogeneities that exist even within genetically homogeneous populations. Recent single-molecule imaging studies of living cells have revealed that the cellular reactions that occur in complex networks of regulatory and metabolic processes in living cells during adaptive responses may exhibit both spatial and temporal separation within a population or a community. This heterogeneity, which can be of ecological and medical significance, clearly cannot be evaluated in experiments that are averaged over large populations. The challenge is to identify those cells of ecological importance within a large population and track their biochemical reactions in situ in real time. By using the synchrotron light, and with the "just right" experimental setup, one can precisely target subpopulations with diffraction-limited spatial resolution and accuracy to track their chemical reactions with high molecular specificity without the use of labels—all at timescales that can be associated with important biological processes (3).

A "JUST RIGHT" MICROFLUIDIC PLATFORM FOR REAL-TIME SR-FTIR

Water presents one of the primary challenges of using infrared spectroscopy in studying living cells, even with a bright synchrotron source. Water strongly absorbs mid-infrared light, and even the absorption of a thin layer of water can completely dominate the spectrum. However, water is necessary for life and is the most common ingredient (>70 percent) in living cells. The trick is to get the optical thickness of the water just right: enough to support life but not so much that it masks the molecular signatures of interest. Typically less than ten micrometers of bulk water is preferred if one wants to exploit the molecular information across the full mid-infrared spectral range.

A traditional approach to minimizing water absorption during living-cell experiments is attenuated total reflectance (ATR) FTIR equipped with a flow chamber. In ATR-FTIR, the IR beam is internally reflected off a high-index crystal (e.g., diamond or germanium) and the attenuation of the beam is measured and analyzed (103). For living-cell experiments, cells are cultured on ATR crystals in a specially designed flow chamber that can carefully control the culturing/experimental conditions (pH, temperature, ionic strength, flow shear stress, and delivery of materials such as nutrients and drugs) for hours (104–106). The attenuated field formed at the cell/ATR crystal interface has a typical penetration depth of less than one micrometer, thereby reducing the optical path length in water. However, this small penetration depth, while important in reducing the path length through water, prevents the ATR-FTIR technique from examining chemical and biological processes in cells of biological systems with an extensive extracellular matrix. Another important consideration is the effect of growing the cells on the ATR crystal, which may affect extracellular matrix properties that are important in cellular phenotype.

Recently, the Berkeley team introduced an open-channel microfluidic approach instead (99). **Figure 12.12** shows an early version of the device. It was fabricated on a silicon chip using deep reactive ion etching to create hydrophilic microstructures 10–15 μm deep and at least 40 μm wide. A continuous thin-film (< 10 μm) laminar flow is maintained by balancing the hydrostatic pressure in a microliter-size feeder droplet at the microchannel inlet with the capillary forces present at the outlet. Mid-infrared photons emitted from the synchrotron are focused onto the targeted areas within the microstructures, and SR-FTIR measurements are made by the transflection or transmission modes. To image the biochemical properties and the distribution of bacterial activity across a biofilm, it is raster-scanned with equal micrometer-sized steps, collecting a full SR-FTIR spectrum at each position.

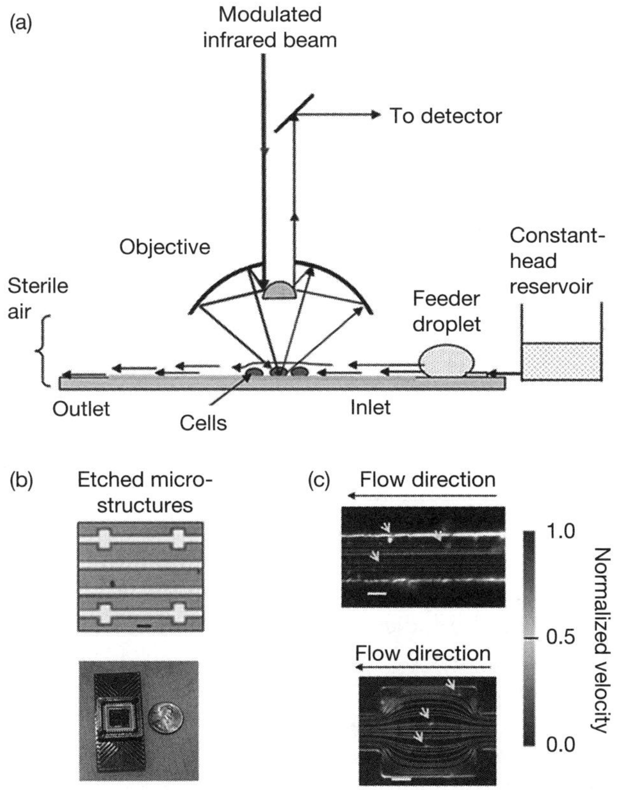

Figure 12.12 Microfluidic platform to minimize water absorption. (A) Microfluidic SR-FTIR microscopy platform design and setup. (B) Plane view depiction of an example chip with several parallel etched microstructures for multiple simultaneous experiments. (C) Flow maps of the microstructures, simulated from experimentally measured paths of near-neutral density polystyrene beads (arrows) and superimposed on a snapshot image: top, flow in a microchannel; bottom, flow in a microwell. Scale bars, 10 μm. Velocity = ~ 60 μm/s (99).

The ability of the platform to provide a sustainable environment for SR-FTIR measurements of living bacteria has been confirmed by using the reporter bacterium *E. coli* XL1-blue (99). Several other groups have since reported uncomplicated flow chamber techniques (107–109) that place living cells between two parallel infrared transparent windows (e.g., CaF_2) separated by a spacer of minimum thickness (< 12 μm). Although effective in its simplicity, the spectra from closed-chamber or closed-channel devices could be marred by interference fringes that probably arise from multiple reflections from the parallel surfaces. In this case, the sloping free interface between water and air has allowed the Berkeley team to completely eliminate the interference fringes during FTIR measurements.

IN SITU MOLECULAR MONITORING OF LIVING CELLS

Understanding Bacteria—Antibiotic Interactions in a Biofilm

A bacterial biofilm is a population of cells growing on a surface and enclosed in a self-produced polymeric matrix in aqueous environments. Bacteria within a biofilm can become up to 1000 times more resistant to antibiotics than those of free-floating bacteria of the same species (110), yet only about 1 percent of their genes show differential expression between these two types of population (111).

Figure 12.13 shows an example of using open-channel microfluidics with SR-FTIR spectromicroscopy to study why some *E. coli* in biofilms are not killed by the antimicrobial agent mitomycin-C (MMC) (99). MMC is activated only after it has entered a cell, where it is then reduced to a hydroquinone form that covalently cross-links to guanine residues in DNA to form DNA-MMC adducts (112). The presence of DNA-MMC adduct signals at ~986 cm^{-1} reflects cellular uptake and the action of MMC (99). **Figure 12.13B** shows chemical images before and after the introduction of MMC into the microfluidic system. Despite an uninterrupted supply of MMC-laden water, the SR-FTIR image plot shows that MMC uptake is highly localized. DNA-MMC adduct signals are especially strong in areas that were either closer to the MMC source (locations 1 and 2 in the figure) or in areas formerly rich in protein amide III (~1310 cm^{-1}). In areas with little DNA-MMC adduct, the protein amide III signal increased with MMC exposure. Shown in **Figure 12.13C** are analyses of vector-normalized SR-FTIR spectra (over the 900–1800 cm^{-1} region) at different locations; they also indicate localized small (~10 to ~20 μm) changes in biochemical contents consequent to MMC exposure. The Berkeley Group speculated that the localization of MMC may be caused by flow heterogeneities inside the biofilms, perhaps related to internal flow diversions around regions with higher protein amide III signal. The heterogeneous spectral features and behavior, however, may also reflect localized cellular diversification processes in response to MMC toxicity, such as metabolic modification and/or migration to more favorable living areas.

Revealing Biofilm Dynamics

The Berkeley group also used the open-channel SR-FTIR platform to improve understanding of the growth and development of bacteria on surfaces. Many microbial

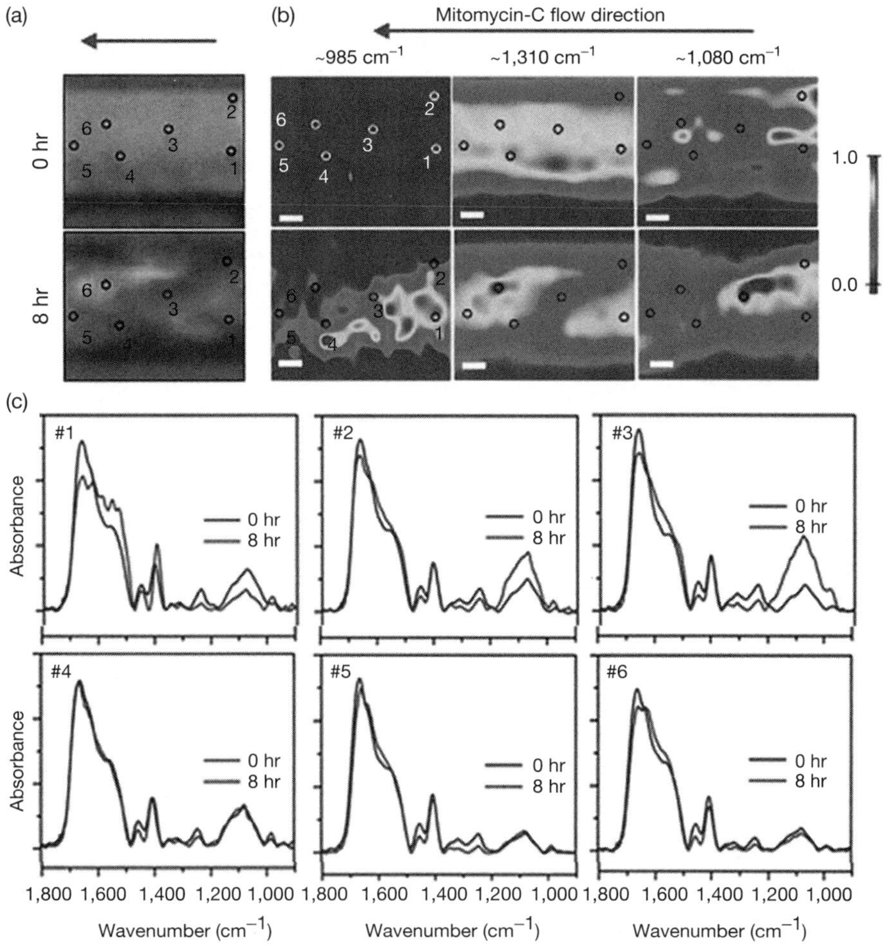

Figure 12.13 Interactions of antibiotics with *E. coli*. (A) Bright-field optical images of the one-day-old *E. coli* biofilm before (0 hr) and during (8 hr) 0.15 μg/mL concentration mitomycin-C (MMC) exposure. (B) SR-FTIR chemical image plots of the intensity, before (upper panels) and during (lower panels) MMC exposure, at ~985 cm^{-1} (DNA-MMC adducts), ~1310 cm^{-1} (protein amide III), and ~1080 cm^{-1} (polysaccharides). Three distinct biochemical regions formed during MMC treatment: a high MMC-uptake and a low MMC-uptake region separated by an area of high polysaccharide content. Arrows indicate MMC stream direction. "O" in panels mark selected locations. Scale bars, 10 μm. (C) Preprocessed SR-FTIR spectra at selected locations before (gray spectra) and during (black spectra) MMC exposure (99).

processes important in pathogenesis and ecology are initiated in confined microscopic spaces, and recent reports indicate that the colonizing bacterial cells actively seek out confined spaces where biofilm initiation, formation, and evolution can be influenced by fluid dynamics, nutrient supplies, or waste removal.

They compared biofilm formation in microchannels (with higher rates of nutrient supplies and waste product removal) to that in microwells (with lower rates of nutrient supplies and waste removal) (see **Figure 12.14**) (99). For microchannels, four different markers of biofilm formation increase asymptotically: polysaccharide (~1080 cm^{-1}), glycocalyx (~1130 cm^{-1}), DNA/RNA (~1240 cm^{-1}), and protein amide III (~1310 cm^{-1}) (see **Figure 12.14A**). This result provides chemical evidence that biofilm formation proceeds via multiple convergent genetic pathways. The abundance of glycocalyx carbohydrates in sites with large biofilm growth confirms the belief that glycocalyx synthesis is crucial to the formation of bacterial biofilms (113). In contrast, biofilm formation in microwells exhibits cyclic growth patterns

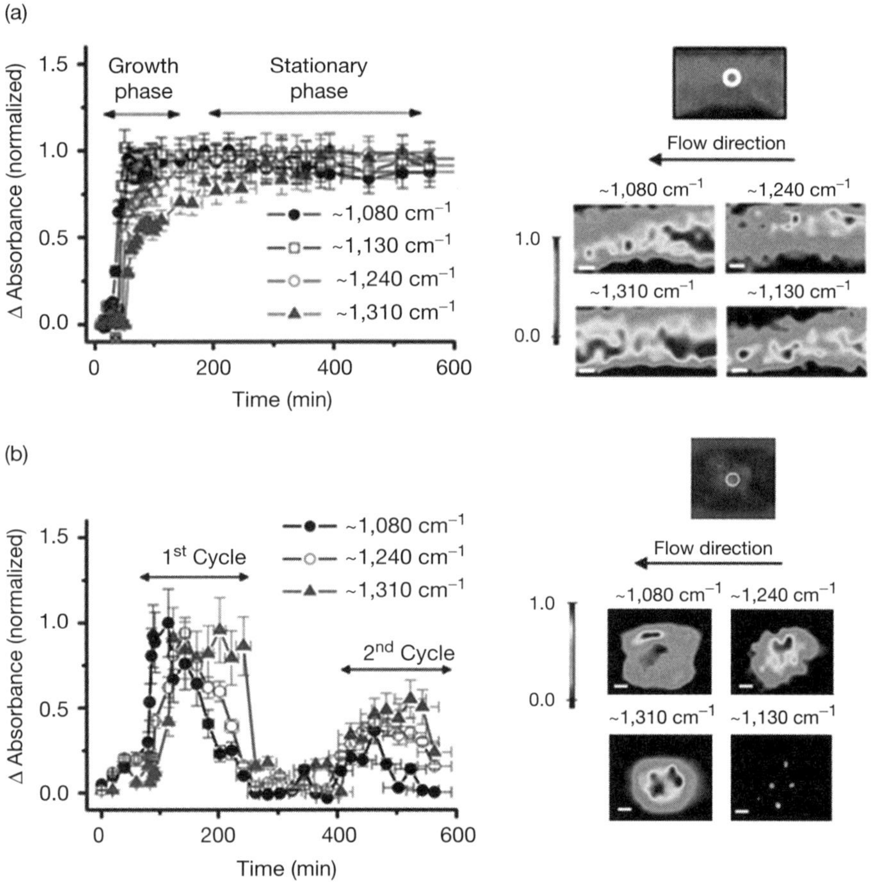

Figure 12.14 Biofilm dynamics. SR-FTIR time course analyses and chemical images of biofilms (A) in a microchannel and (B) in a microwell (as shown by four molecular markers at ~1,080 cm^{-1} (polysaccharides), ~1130 cm^{-1} (glycocalyx), ~1,240 cm^{-1} (DNA/RNA polysaccharides), and ~1,310 cm^{-1} (protein amide III). Unlike the microchannel data, where signal intensity of key biomolecules appeared to approach an asymptotic state, the microwell SR-FTIR data were cyclic (cell growth and release). The chemical image plots obtained after the second cycle show locally higher signal intensities of protein amide III (~1,310 cm^{-1}) and DNA/RNA polysaccharides (at ~1,240 cm^{-1}) near the microwell center, whereas the polysaccharide matrix accumulated near the microwell edge. There is little spectroscopic evidence of glycocalyx (~1,130 cm^{-1}) to facilitate strong adherence to the microwell substrate. Scale bars, 10 μm (99).

as indicated by the sequential rise and fall of the different biomolecule signals (see **Figure 12.14B**). The lack of spectroscopic evidence of glycocalyx suggests that biofilm attachment was weaker in the microwell. It was discovered most recently that some bacteria can produce a factor (mixtures of D-amino acids) that could prevent biofilm formation and cause the breakdown of existing biofilms (114).

Delineating Chemistry of Oxygen Stress and Adaptive Response

One of the earlier live-cell applications was to study how descendants of some the oldest bacteria can survive the oxygen stress of modern Earth. Microbes such as the ubiquitous sulfate-reducing bacteria (SRB) are the oldest (> 3.5 billion years) and smallest (about 1/8000th the volume of a human cell) living organisms on Earth. It is believed that their high surface-to-volume ratio and strong selection pressure had prepared the microbes to cope with environmental fluctuation. Genome sequencing shows that some SRB descendants have acquired the genetic "blueprints" to survive transiently an oxygenated atmosphere. This includes the potential capabilities to reduce (di)oxygen and cope with the subsequent toxic reactive oxygen species (ROS) in the modern atmosphere, such as superoxide anion radicals, peroxides, and hydroxyl radicals.

In the case of the model SRB, *Desulfovibrio vulgaris* Hildenborough, only a small number of cells in a population can survive transient exposure to atmospheric oxygen (see Figure 12.**15A-a**). These few survivors often exist individually or in small groups of several cells to form localized microscopic communities. They must be probed without spectroscopic influence from neighboring cells that are dying. The time scales of the oxygen adaptive response processes can potentially range from minutes to hours (115). To avoid long data acquisition times and thus capture the response dynamics at shorter time scales, a brilliant infrared source providing high SNR is required.

Previous observations showed that only cells entering the early stationary phase of the cell cycle and have accumulated an internal reserve of polyglucose and elemental sulfur are likely to survive transient air exposure.(115) The FTIR spectra of polyglucose-accumulated *D. vulgaris* (see **Figures 12.15A-b** and 12.**15A-c**) show the distinct spectral features of the non-glycosidic polyglucose vibration (νC–OH) band between 1055 and 1045 cm^{-1} and the glycosidic linkage vibration (νC–O–C) at ~1175 cm^{-1} (see **Figure 12.15A-b**). This spectral information is used to look for potential survivors in a thin layer of *D. vulgaris* clones that are maintained in an oxygen-free moist atmosphere inside the microscope-stage environmental chamber. By rastering the diffraction-limited synchrotron infrared beam across the thin film, spots where the spectrum exhibits features similar to the typical spectrum of polyglucose-bearing living *D. vulgaris* can be identified.

The graph in **Figure 12.15B-a** is an overview of the striking molecular changes in *D. vulgaris* survivors during their exposure to atmospheric oxygen. The trends seen in **Figures 12.15B-b** and 12.**15B-c**, which were derived from spectrally integrated absorption intensities of polyglucose and water bands (115), reveal an unanticipated multiphasic pattern. These two molecules are of immediate interest because the reduction of di-oxygen to water through aerobic respiration of internal polyglucose reserves was previously believed to be the primary step in air-tolerant SRB.

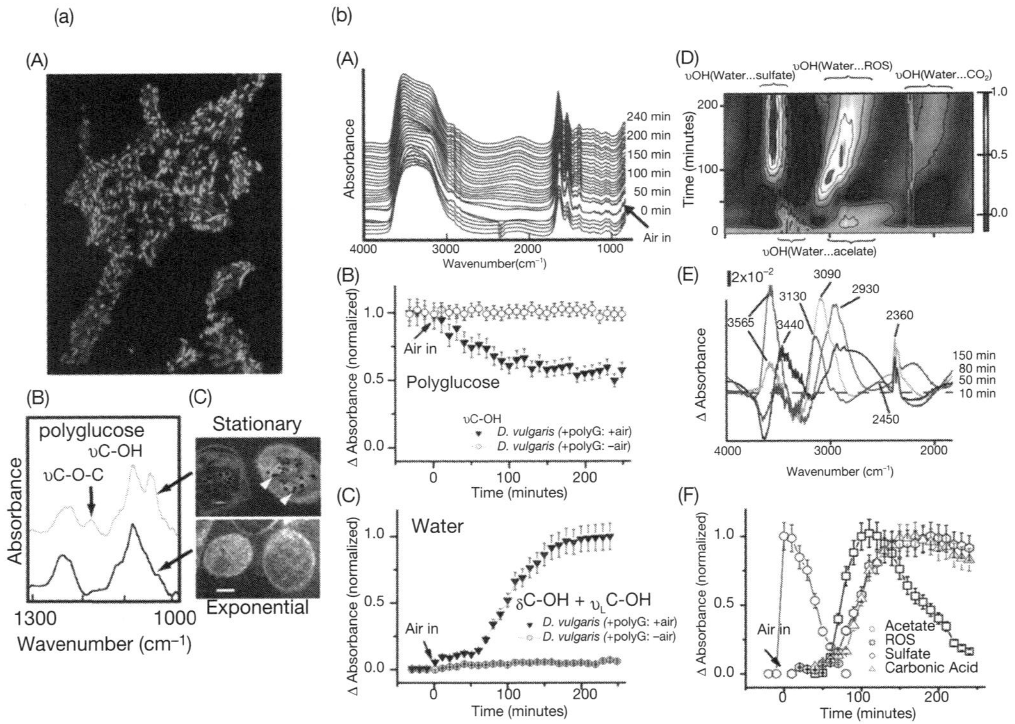

Figure 12.15 (A) Survivor sulfate reducing bacteria. (a) Viability of SRB after exposing to moist air for eight hours. Live (green) and dead (red) were assessed using reagents in LIVE/DEAD *Bac*Light Bacterial Viability kit. (b) Typical infrared absorption spectra of stationary-phase (red) and exponential-phase (blue) *D. vulgaris*. (c) Transmission electron microscopy (TEM) images of thin sections post-stained by the periodic acid thiosemicarbazide-osmium (PATO) method show intracellular polyglucose granules in stationary-phase but not exponential-phase *D. vulgaris*. (B) Oxidative Stress and adaptation. (a)Typical real-time SR-FTIR spectra of *D. vulgaris* during the transition from an anaerobic to aerobic environment. Since all spectra are derived using air as a reference, the abrupt change in the spectral feature at ~2348 cm^{-1} is associated with the presence of atmospheric CO_2. Typical time-course of infrared intensity (normalized by the maximum value) of (b) polyglucose content and (c) water. (d) Time versus frequency contour plot of SR-FTIR time-difference spectra in the hydride–OH dominated stretch region. (e) Snapshots of time-difference spectra for selected different time points. The dashed line marks zero difference absorbance. (F) Typical time-course of infrared intensity (normalized by the maximum value) of H-bonded species (115). See insert for color version.

Comparing trends in these two figures shows that from 0 to ~50 minutes there is a substantial decrease in the polyglucose band intensity but little change in the water band intensity. After ~50 minutes, the water band intensity increases abruptly while the rate of polyglucose disappearance continues unabated until slowing down distinctly later (after ~100 minutes). For an elucidation of the mechanism(s) underlying this behavior, the Berkeley Group analyzed the time-difference spectra in the hydride-OH region carefully. **Figure 12.15B-d** is a two-dimensional time-frequency contour plot of the time-difference spectra (negative values are shown in dark blue), with difference spectrum snapshots beneath (see **Figure 12.15B-e**). The contour map reveals the remarkable changing absorption features from νOH as a function of time, which reflects a cascade of chemical reactions (see **Figure 12.15B-f**) that sheds light on genetically controlled pathways (115).

The mechanisms by which SRB survive transient exposure to atmospheric oxygen have been controversial, partly because of the approach of averaging over a large population. Our SR-FTIR technique provides the first direct molecular observation of the reduction of di-oxygen to water by a small number of SRB through aerobic respiration of internal polyglucose reserves as one crucial step and identifies the transient chemical reactions that allow the survivor subpopulation to adapt to extreme changes in its environment.

Monitoring Protein Phosphorylation Events in Real-Time

The Berkeley Group has most recently demonstrated the suitability of using the label-free synchrotron FTIR for monitoring protein phosphorylation events in living mammalian cells instead of fluorescence-based assays (116). Protein phosphorylation is a post-translational modification that is essential for the regulation of many important cellular activities, including proliferation and differentiation. The Group studied the differentiation of rat pheochromocytoma PC12 cell line into neural-like phenotype induced by nerve growth factor (NGF), a process where phosphorylation pathways are massive and well characterized. An important step in achieving their goal is through a careful experimental design that complements SR-FTIR spectromicroscopy with fluorescence microscopy of transiently transfected PC12 cells, FACS flow cytometry, quantum mechanical calculations. They showed that a strong and statistically relevant correlation exists between the protein phosphorylation levels and the intensity of the IR-bands centered at 1080 cm^{-1} and 990–970 cm^{-1} (**Figure 12.16**), and that one could directly link these temporal spectral changes to the development of neuron-like phenotypes in the cells and can gain insights into the evolving molecular events. From these results, many advantages could potentially be taken by the bio-community in the future.

CURRENT FRONTIERS

The next step in the development of infrared spectromicroscopy is the improvement of imaging speed. Today synchrotron FTIR with a single-element MCT detector has provided the finest possible spatial resolution in biological images. However, even with the huge improvements in signal-to-noise from a synchrotron source,

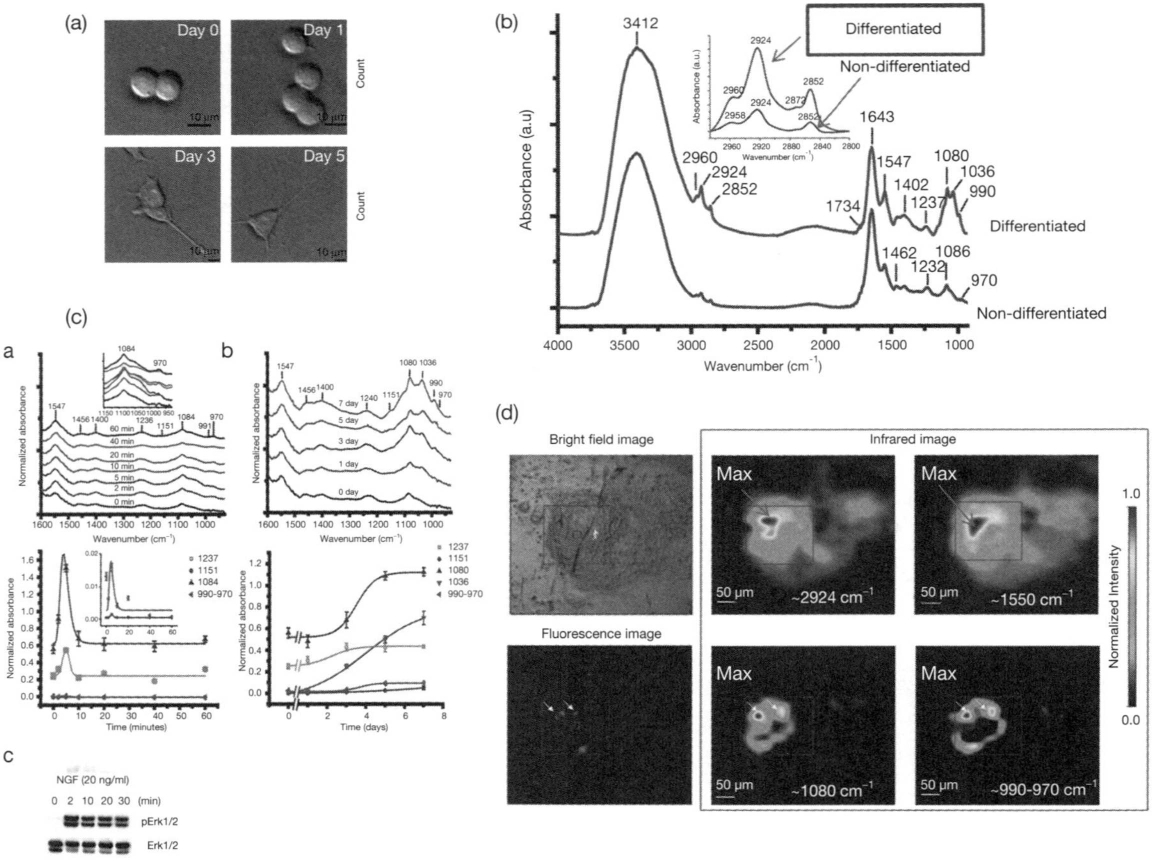
(a)
Day 0
Day 1
Day 3
Day 5
10 μm
Count
(b)
Differentiated
Non-differentiated
3412
2960
2924
2852
1643
1547
1402
1237
1080
1036
990
1734
1462
1232
1086
970
Absorbance (a.u)
Wavenumber (cm⁻¹)
(c)
a
b
c
Normalized absorbance
Wavenumber (cm⁻¹)
Time (minutes)
Time (days)
60 min
40 min
20 min
10 min
5 min
2 min
0 min
7 day
5 day
3 day
1 day
0 day
1237
1151
1084
1080
1036
990-970
NGF (20 ng/ml)
0 2 10 20 30 (min)
pErk1/2
Erk1/2
(d)
Bright field image
Fluorescence image
Infrared image
Max
50 μm
~2924 cm⁻¹
~1550 cm⁻¹
~1080 cm⁻¹
~990-970 cm⁻¹
Normalized Intensity
1.0
0.0

Figure 12.16 Monitor protein phosphorylation during neuron differentiation. (A) Bright field images of PC12 cells in one week treatment with NGF. Pictures were taken at the 0, 24, 72, and 120 hours respectively. (B) Comparison of typical SR-FTIR spectra of differentiated (red) and non-differentiated (blue) PC12 cells. The differentiated PC12 cells were treated with 25 ng/ml NGF for seven days. The inset shows the expanded spectral region between 2800 cm^{-1} and 3000 cm^{-1} after baseline correction. The spectra are normalized to the amide II band strength at 1550 cm^{-1} to normalize variations in sample biomasses[19]. (See Table 12.1 for assignment of the marked bands.) (C) Progressive spectral changes in the fingerprint region during PC12 cell differentiation. (a) Changes during the first hour of NGF treatment. (Top) FTIR spectra. Inset: enlarged spectrum in the 1150–900 cm^{-1} region. (Bottom) The intensity curves of prominent IR bands. The trend lines were fitted to the experimental data using nonlinear curve fit methods. $n = 150$; error bars represent the standard error of the mean (SEM). Inset: enlarged curves of the 1151 cm^{-1}, 990–970 cm^{-1} bands. (b) Changes during a 7-day NGF treatment. (Top) FTIR spectra. (Bottom) The intensity curves of prominent IR bands during 7 days of NGF treatment. $n = 150$; error bars represent SEM. The short term data points (within the first hour) are not re-plotted here. All band intensities were calculated by integrating the normalized spectra (i.e., normalized to the amide II band strength at 1550 cm^{-1}). (c) Western blot of pErk1/2 (phosphorylated Erk1/2), and total Erk1/2 protein levels in PC12 cells after treatment with 20 ng/ml NGF for the indicated time periods. The pErk1/2 level increases significantly at 2 min and gradually decreases from 10 to 30 min. The total Erk1/2 blot shows equal loading of Erk1/2 proteins. (D) In situ bimodal registration of phosphorylation in PC12 cell clusters transfected with TrkA-EGFP fusion plasmid. The bright field image (Top) and the fluorescent image (Bottom) of the same cluster of cells. The infrared images are intensity maps for the corresponding infrared bands as annotated in the figure. The ~2914 cm^{-1} band is associated with lipids and the ~1550 cm^{-1} band is associated with proteins. The bands between 990 and 970 cm^{-1} are integrated together. The large red squares mark the region of interest. The white arrows indicate the location of transfected cells (fluorescent spots) and the corresponding spots (black arrows) in the infrared map. The bright fluorescent spot at the lower edge does not register in any of the four infrared images. It may come from non-specific auto-fluorescence of necrotic cells. (116).

mapping large areas at high spatial resolution with the raster scanning technique can lead to exceedingly long acquisition time. For example, imaging a 30 μm × 30 μm as a diffraction-limited map at the most advanced synchrotron facility, such as the state-of-the-art third-generation synchrotron, would take 2–4 hours. Lengthy acquisition times lead many researchers to choose larger step sizes, and therefore compromise the deflection-limited spatial resolution. Nasse and Hirschmugl at the Synchrotron Radiation Center (SRC) at the University of Wisconsin-Madison made significant contributions to meet the challenge by combining multiple synchrotron beams and using the FPA (pixel size: 0.54 μm × 0.54 μm) as the wide-field detector. Their idea is to focus the image of the sample onto an array detector so that the intensity of the infrared radiation passing through each region of the sample can be measured at each pixel simultaneously. In 2011 Hirschmugl and Nasse published synchrotron infrared images of cancerous prostate tissue that display high quality infrared diffraction-limited resolution (53). In 2013, Carr and Miller at the Brookhaven National Laboratory elegantly simplified the approach by means of a beam splitting approach (125). This landmark improvement in the imaging resolution and speed could greatly augment the use of synchrotron FTIR in biomedical and environmental research. We have now entered an era of SR-FTIR imaging using the focal plane array (FPA) technology to increase the speed of image acquisition by multiplexing the acquisition of data, and with high signal-to-noise ratios (53, 125). The benefit has been demonstrated in imaging tissue samples. Most recently, researchers at the Lawrence Berkeley National Laboratory and at IRENI expanded the 2D spectral image acquisition capability to a full 3D spatial dimensions plus a full mid-IR spectral dimension capability (126). This method, Fourier transform infrared (FTIR) spectro-microtomography, provides rich spectral information and chemical fingerprinting, including 3D distributions of these signatures throughout the sample. It will be interesting to see in the coming years how these new technical advances will impact high-resolution real-time imaging of living cells.

Currently, there is still a need to improve the spatial resolution beyond the diffraction limit. Indeed, diffraction-limited spot sizes in the mid-infrared are still larger than most bacteria, and as a consequence it is impossible to localize subcellular components. A number of near-field infrared imaging techniques are being developed, including aperture-less scanning near-field infrared microscopy (117, 118), photothermally induced resonance imaging (119, 120), and photoexpansion imaging (121), but these techniques typically use narrow band lasers that have limited spectral range. The challenge is to couple these techniques with broadband radiation, such as synchrotron infrared radiation, to create a "nanoscope" with sub-micron resolution and the capability to acquire full infrared spectra. Many research groups are tackling the nanoscope challenge at synchrotrons worldwide (122, 123) and their progress will be monitored closely.

Finally, another direction is to integrate high-density microfluidic chips with advanced large-scale plumbing networks (124). This advance will enable the automatic and precise manipulation of fluids to provide accurate "on-demand" conditions to study the chemistry of living cells. This will permit investigations of many important cellular systems in aqueous environments over long time periods, including harmful processes like chronic bacterial infections or beneficial processes like energy production by microbes, and important biological processes such as protein phosphorylation.

ACKNOWLEDGMENTS

This work was supported by the US Department of Energy Office of Biological and Environmental Research's Structural Biology Program and the Berkeley Synchrotron Infrared Structural Biology (BSISB) program through contracts DE-AC02-05CH11231, and KP1501021 with Lawrence Berkeley National Laboratory. The Advanced Light Source is supported by the Director, Office of Science, Office of Basic Energy Sciences, of the US Department of Energy under contract DE-AC02-05CH11231.

REFERENCES

1. Dumas P, Sockalingum GD, & Sule-Suso J (2007) Adding synchrotron radiation to infrared microspectroscopy: what's new in biomedical applications? (Translated from English) *Trends Biotechnol* 25(1):40–44 (in English).
2. Miller LM & Dumas P (2006) Chemical imaging of biological tissue with synchrotron infrared light. (Translated from English) *Bba-Biomembranes* 1758(7):846–857 (in English).
3. Holman HYN, Bechtel HA, Hao Z, & Martin MC (2010) Synchrotron IR Spectromicroscopy: Chemistry of Living Cells. (Translated from English) *Analytical Chemistry* 82(21):8757–8765 (in English).
4. Quaroni L & Zlateva T (2011) Infrared spectromicroscopy of biochemistry in functional single cells. (Translated from English) *Analyst* 136(16):3219–3232 (in English).
5. Abney W & Festing ER (1882) On the Influence of the Atomic Grouping in the Molecules of Organic Bodies on Their Absorption in the Infrared Region of the Spectrum. *Phil. Trans. Roy. Soc. London* (172):887–918.
6. Coates JD (2000) Interpretation of infrared spectra, a practical approach. *Encyclopedia of Analytical Chemistry*, ed Meyer RA (John Wiley & Sons Ltd., Chichester), pp 10815–10837.
7. Coblentz WW (1905) Investigations of Infrared Spectra. *Physical Review* 16:35–50.
8. Rubens H & Wood RW (1911) Focal Isolation of Long Heat-Waves. *Phiolosophical Magazine* 21(122):249–261.
9. Strong J & Vanasse GA (1959) Interferometric Spectroscopy in the Far Infrared. (Translated from English) *J Opt Soc Am* 49(9):844–850 (in English).
10. Davies JK (1992) Infrared Astronomy from the Uk Infrared Telescope. (Translated from English) *Contemp Phys* 33(6):357–368 (in English).
11. Bullock BW & Silverman S (1949) A Rapid-Scanning Spectrometer for the near Infrared. (Translated from English) *Analytical Chemistry* 21(4):535–535 (in English).
12. Bent HA & Crawford B (1959) Infrared Studies of Propellant Flames. (Translated from English) *J Phys Chem-Us* 63(6):941–952 (in English).
13. Fateley WG, Bent HA, & Crawford B (1959) Infrared Spectra of the Frozen Oxides of Nitrogen. (Translated from English) *J Chem Phys* 31(1):204–217 (in English).
14. Barer GR (1954) Quantitative Measurements on Spreading Phenomena in Skin. *British Journal of Pharmacology and Chemotherapy* 9(3):346–351.
15. Barer R (1949) The Reflecting Microscope in Anatomical Research. *Journal of Anatomy* 83(1):73–74.
16. Gore RC (1949) Infrared Spectrometry of Small Samples with the Reflecting Microscope. *Science* 110(2870):710–711.

17. Bird GR & Blout ER (1952) Infrared Microspectroscopy of Biologic Materials. *Laboratory Investigation* 1(2):266–272.
18. Blout ER (1953) Infrared Microspectroscopy—Instrumentation and Some Biological Applications. *Transactions of the New York Academy of Sciences* 15(8):280–281.
19. Blout ER & Mellors RC (1949) Infrared Spectra of Tissues. (Translated from eng) *Science* 110(2849):137–138 (in eng).
20. Wood DL (1951) Infrared microspectrum of living muscle cells. (Translated from eng) *Science* 114(2950):36–38 (in eng).
21. Woernley DL (1952) Infrared absorption curves for normal and neoplastic tissues and related biological substances. (Translated from eng) *Cancer Res* 12(7):516–523 (in eng).
22. Stevenson HJ & Bolduan OE (1952) Infrared spectrophotometry as a means for identification of bacteria. (Translated from eng) *Science* 116(3005):111–113 (in eng).
23. Stevenson HJ & Levine S (1952) Infrared spectra of pneumococcal polysaccharides. (Translated from eng) *Science* 116(3026):705–706 (in eng).
24. Coates VJ, Offner A, & Siegler EH (1953) Design and Performance of an Infrared Microscope Attachment. *J Opt Soc Am* 43(11):984–989.
25. Schmidt DA (2011) A matter of scale: from far-field microscopy to near-fiedl nanoscopy. *Laser & Photonics Reviews* (10):1–37.
26. Bechtel HA, Martin MC, May TE, & Lerch P (2009) Improved spatial resolution for reflection mode infrared microscopy. (Translated from English) *Rev Sci Instrum* 80(12):126-106 (in English).
27. Lewis EN, et al. (1995) Fourier-Transform Spectroscopic Imaging Using an Infrared Focal-Plane Array Detector. (Translated from English) *Analytical Chemistry* 67(19):3377–3381 (in English).
28. Lewis EN & Levin IW (1995) Real-Time, Midinfrared Spectroscopic Imaging Microscopy Using Indium-Antimonide Focal-Plane Array Detection. (Translated from English) *Appl Spectrosc* 49(5):672–678 (in English).
29. Mohlenhoff B, Romeo M, Diem M, & Woody BR (2005) Mie-type scattering and non-Beer-Lambert absorption behavior of human cells in infrared microspectroscopy. (Translated from English) *Biophysical Journal* 88(5):3635–3640 (in English).
30. de Juan A, Maeder M, Hancewicz T, Duponchel L, & Tauler R (2009) Chemometric Toos for Image Analysis. Infrared and Raman Spectroscopic Imaging, eds Salzer R & Siesler HW (Wiley-VCH Verlag GmbH & Co. KGaA, Weinheim), p 510.
31. Yano K, et al. (2000) Direct measurement of human lung cancerous and noncancerous tissues by Fourier transform infrared microscopy: Can an infrared microscope be used as a clinical tool? (Translated from English) *Anal Biochem* 287(2):218–225 (in English).
32. Sahu RK, et al. (2005) Can Fourier transform infrared spectroscopy at higher wavenumbers (mid IR) shed light on biomarkers for carcinogenesis in tissues? (Translated from English) *J Biomed Opt* 10(5):054017.0-054017.10 (in English).
33. Krafft C, Steiner G, Beleites C, & Salzer R (2009) Disease recognition by infrared and Raman spectroscopy. (Translated from English) *Journal of Biophotonics* 2(1–2):13–28 (in English).
34. Romeo MJ & Diem M (2005) Infrared spectral imaging of lymph nodes: Strategies for analysis and artifact reduction. *Vib Spectrosc* 38(1–2):115–119.
35. Wood BR, et al. (2004) Fourier transform infrared (FTIR) spectral mapping of the cervical transformation zone, and dysplastic squamous epithelium. (Translated from English) *Gynecol Oncol* 93(1):59–68 (in English).

36. Carden A, Rajachar RM, Morris MD, & Kohn DH (2003) Ultrastructural changes accompanying the mechanical deformation of bone tissue: A Raman imaging study. (Translated from English) *Calcified Tissue Int* 72(2):166–175 (in English).
37. Boskey AL, DiCarlo E, Paschalis E, West P, & Mendelsohn R (2005) Comparison of mineral quality and quantity in iliac crest biopsies from high- and low-turnover osteoporosis: an FT-IR microspectroscopic investigation. (Translated from English) *Osteoporosis Int* 16(12):2031–2038 (in English).
38. Crombie DE, et al. (2005) Destructive effects of murine arthritogenic antibodies to type II collagen on cartilage explants in vitro. (Translated from English) *Arthritis Res Ther* 7(5):R927–R937 (in English).
39. Durchschlag E, et al. (2006) Bone material properties in trabecular bone from human iliac crest biopsies after 3-and 5-year treatment with risedronate. (Translated from English) *Journal of Bone and Mineral Research* 21(10):1581–1590 (in English).
40. Argov S, et al. (2002) Diagnostic potential of Fourier-transform infrared microspectroscopy and advanced computational methods in colon cancer patients. (Translated from English) *J Biomed Opt* 7(2):248–254 (in English).
41. Gazi E, et al. (2003) Applications of Fourier transform infrared microspectroscopy in studies of benign prostate and prostate cancer. A pilot study. (Translated from English) *J Pathol* 201(1):99–108 (in English).
42. Walsh MJ, et al. (2007) IR microspectroscopy: potential applications in cervical cancer screening. (Translated from English) *Cancer Lett* 246(1–2):1–11 (in English).
43. Carden A & Morris MD (2000) Application of vibrational spectroscopy to the study of mineralized tissues (review). (Translated from English) *J Biomed Opt* 5(3):259–268 (in English).
44. Boskey A & Camacho NP (2007) FT-IR imaging of native and tissue-engineered bone and cartilage. (Translated from English) *Biomaterials* 28(15):2465–2478 (in English).
45. Beekes M, Lasch P, & Naumann D (2007) Analytical applications of Fourier transform-infrared (FT-IR) spectroscopy in microbiology and prion research. (Translated from English) *Vet Microbiol* 123(4):305–319 (in English).
46. Kneipp J, Beekes M, Lasch P, & Naumann D (2002) Molecular changes of preclinical scrapie can be detected by infrared spectroscopy. (Translated from English) *J Neurosci* 22(8):2989–2997 (in English).
47. Carr GL, Hanfland M, & Williams GP (1995) Midinfrared Beamline at the National Synchrotron Light-Source Port U2b. (Translated from English) *Rev Sci Instrum* 66(2):1643–1645 (in English).
48. Carr GL, Reffner JA, & Williams GP (1995) Performance of an Infrared Microspectrometer at the Nsls. *Rev Sci Instrum* 66(2):1490–1492.
49. Duncan WD & Williams GP (1983) Infrared Synchrotron Radiation from Electron Storage-Rings. (Translated from English) *Appl Optics* 22(18):2914–2923 (in English).
50. Levenson E, Lerch P, & Martin MC (2006) Infrared imaging: Synchrotrons vs. arrays, resolution vs. speed. *Infrared Phys Techn* 49(1–2):45–52.
51. Levenson E, Lerch P, & Martin MC (2008) Spatial resolution limits for synchrotron-based spectromicroscopy in the mid- and near-infrared. (Translated from English) *Journal of Synchrotron Radiation* 15:323–328 (in English).
52. Holman HYN, et al. (2002) Synchrotron infrared spectromicroscopy as a novel bioanalytical microprobe for individual living cells: cytotoxicity considerations. (Translated from English) *J Biomed Opt* 7(3):417–424 (in English).

53. Nasse MJ, et al. (2011) High-resolution Fourier-transform infrared chemical imaging with multiple synchrotron beams. (Translated from English) *Nat Methods* 8(5):413–U458 (in English).
54. Jamin N, et al. (1998) Highly resolved chemical imaging of living cells by using synchrotron infrared microspectrometry. *P Natl Acad Sci USA* 95(9):4837–4840.
55. Holman HN, Martin MC, Blakely EA, Bjornstad K, & McKinney WR (2000) IR spectroscopic characteristics of cell cycle and cell death probed by synchrotron radiation based fourier transform IR spectromicroscopy. *Biopolymers* 57(6):329–335.
56. Leskovjan AC, Kretlow A, & Miller LM (2010) Fourier Transform Infrared Imaging Showing Reduced Unsaturated Lipid Content in the Hippocampus of a Mouse Model of Alzheimer's Disease. (Translated from English) *Analytical Chemistry* 82(7):2711–2716 (in English).
57. Dumas P, Miller LM, & Tobin MJ (2009) Challenges in Biology and Medicine with Synchrotron Infrared Light. (Translated from English) *Acta Phys Pol A* 115(2):446–454 (in English).
58. Miller LM, et al. (2007) A new sample substrate for imaging and correlating organic and trace metal composition in biological cells and tissues. (Translated from English) *Anal Bioanal Chem* 387(5):1705–1715 (in English).
59. Miller LM, et al. (2006) Synchrotron-based infrared and X-ray imaging shows focalized accumulation of Cu and Zn co-localized with beta-amyloid deposits in Alzheimer's disease. (Translated from English) *J Struct Biol* 155(1):30–37 (in English).
60. Bonda M, et al. (2011) Synchrotron Infrared Microspectroscopy Detecting the Evolution of Huntington's Disease Neuropathology and Suggesting Unique Correlates of Dysfunction in White versus Gray Brain Matter. (Translated from English) *Analytical Chemistry* 83(20):7712–7720 (in English).
61. Kretlow A, Wang Q, Beekes M, Naumann D, & Miller LM (2008) Changes in protein structure and distribution observed at pre-clinical stages of scrapie pathogenesis. (Translated from English) *Bba-Mol Basis Dis* 1782(10):559–565 (in English).
62. Kneipp J, et al. (2003) In situ identification of protein structural changes in prion-infected tissue. (Translated from English) *Bba-Mol Basis Dis* 1639(3):152–158 (in English).
63. Didonna A, Vaccari L, Bek A, & Legname G (2011) Infrared Microspectrosco py: A Multiple-Screening Platform for Investigating Single-Cell Biochemical Perturbations upon Prion Infection. (Translated from English) *Acs Chem Neurosci* 2(3):160–174 (in English).
64. Huang RY, Miller LM, Carlson CS, & Chance MR (2002) Characterization of bone mineral composition in the proximal tibia of cynomolgus monkeys: Effect of ovariectomy and nandrolone decanoate treatment. (Translated from English) *Bone* 30(3):492–497 (in English).
65. Huang RY, Miller LM, Carlson CS, & Chance MR (2003) In situ chemistry of osteoporosis revealed by synchrotron infrared microspectroscopy. (Translated from English) *Bone* 33(4):514–521 (in English).
66. Miller LM, Novatt JT, Hamerman D, & Carlson CS (2004) Alterations in mineral composition observed in osteoarthritic joints of cynomolgus monkeys. (Translated from English) *Bone* 35(2):498–506 (in English).
67. Ruppel ME, Burr DB, & Miller LM (2006) Chemical makeup of microdamaged bone differs from undamaged bone. (Translated from English) *Bone* 39(2):318–324 (in English).

68. Miller LM, Feldman TC, Schirmer A, Smith RJ, & Judex S (2007) Microscopic imaging of bone composition En block with synchrotron infrared microspectroscopy. (Translated from English) *Journal of Bone and Mineral Research* 22:S258-S258 (in English).
69. Fuchs RK, et al. (2008) In situ examination of the time-course for secondary mineralization of Haversian bone using synchrotron Fourier transform infrared micro spectroscopy. (Translated from English) *Matrix Biol* 27(1):34–41 (in English).
70. Wetzel DL, Post GR, & Lodder RA (2005) Synchrotron infrared micro spectroscopic analysis of collagens I, III, and elastin on the shoulders of human thin-cap fibroatheromas. (Translated from English) *Vib Spectrosc* 38(1–2):53–59 (in English).
71. Holman HYN, et al. (2008) Mid-infrared reflectivity of experimental atheromas. *J Biomed Opt* 13(3):030503.1-030303.4.
72. Heraud P & Tobin MJ (2009) The emergence of biospectroscopy in stem cell research. (Translated from English) *Stem Cell Res* 3(1):12–14 (in English).
73. Chonanant C, et al. (2011) Characterisation of chondrogenic differentiation of human mesenchymal stem cells using synchrotron FTIR microspectroscopy. (Translated from English) *Analyst* 136(12):2542–2551 (in English).
74. Tanthanucha W, Thumanu K, Lorthongpanich C, Parnpai R, & Heraud P (2010) Neural differentiation of mouse embryonic stem cells studied by FTIR spectroscopy. (Translated from English) *J Mol Struct* 967(1–3):189–195 (in English).
75. German MJ, et al. (2006) Characterization of putative stem cell populations in the cornea using synchrotron infrared microspectroscopy. (Translated from English) *Invest Ophth Vis Sci* 47(6):2417–2421 (in English).
76. Kelly JG, Nakamura T, Kinoshita S, Fullwood NJ, & Martin FL (2010) Evidence for a stem-cell lineage in corneal squamous cell carcinoma using synchrotron-based Fourier-transform infrared microspectroscopy and multivariate analysis. (Translated from English) *Analyst* 135(12):3120–3125 (in English).
77. Nakamura T, et al. (2010) Microspectroscopy of spectral biomarkers associated with human corneal stem cells. (Translated from English) *Mol Vis* 16(42):359–368 (in English).
78. Walsh MJ, et al. (2007) FTIR microspectroscopy coupled with two-class discrimination segregates markers responsible for inter- and intra-category variance in exfoliative cervical cytology. (Translated from English) *Mutagenesis* 22(6):453–453 (in English).
79. Tobin MJ, et al. (2004) Infrared microscopy of epithelial cancer cells in whole tissues and in tissue culture, using synchrotron radiation. (Translated from English) *Faraday Discuss* 126:27–39 (in English).
80. Acerbo AS & Miller LM (2009) Assessment of the chemical changes induced in human melanoma cells by boric acid treatment using infrared imaging. (Translated from English) *Analyst* 134(8):1669–1674 (in English).
81. Ali K, et al. (2010) A new approach to concordance in mid-infrared spectromicroscopy mapping of malignant tumors. (Translated from English) *Oncol Rep* 24(4):857–860 (in English).
82. Ali K, et al. (2010) Biomolecular diagnosis of human glioblastoma multiforme using Synchrotron mid-infrared spectromicroscopy. (Translated from English) *Int J Mol Med* 26(1):11–16 (in English).
83. Ali K, et al. (2008) Fourier transform infrared spectromicroscopy and hierarchical cluster analysis of human meningiomas. (Translated from English) *Int J Mol Med* 21(3):297–301 (in English).

84. Kelly JG, et al. (2009) Derivation of a subtype-specific biochemical signature of endometrial carcinoma using synchrotron-based Fourier-transform infrared micro-spectroscopy. (Translated from English) *Cancer Lett* 274(2):208–217 (in English).
85. Liu CL, et al. (2006) Infrared absorption of human breast tissues in vitro. (Translated from English) *J Lumin* 119:132–136 (in English).
86. Yousef I, et al. (2011) Infrared spectral signatures of CDCP1-induced effects in colon carcinoma cells. (Translated from English) *Analyst* 136(24):5162–5168 (in English).
87. Kwiatek WM, et al. (2009) Application of mu-FTIR-SR Spectroscopy to Prostate Tissue Analysis. (Translated from English) *Acta Phys Pol A* 115(2):602–605 (in English).
88. Gazi E, et al. (2005) A study of cytokinetic and motile prostate cancer cells using synchrotron-based FTIR micro spectroscopic imaging. (Translated from English) *Vib Spectrosc* 38(1–2):193–201 (in English).
89. Gazi E, et al. (2004) Imaging ToF-SIMS and synchrotron-based FT-IR micro spectroscopic studies of prostate cancer cell lines. (Translated from English) *Appl Surf Sci* 231:452–456 (in English).
90. Heraud P, et al. (2010) Early detection of the chemical changes occurring during the induction and prevention of autoimmune-mediated demyelination detected by FT-IR imaging. (Translated from English) *Neuroimage* 49(2):1180–1189 (in English).
91. Petit VW, et al. (2010) Multimodal Spectroscopy Combining Time-of-Flight-Secondary Ion Mass Spectrometry, Synchrotron-FT-IR, and Synchrotron-UV Microspectroscopies on the Same Tissue Section. (Translated from English) *Analytical Chemistry* 82(9):3963–3968 (in English).
92. Kastyak MZ, et al. (2010) Pigmented Creatine Deposits in Amyotrophic Lateral Sclerosis Central Nervous System Tissues Identified by Synchrotron Fourier Transform Infrared Microspectroscopy and X-Ray Fluorescence Spectromicroscopy. (Translated from English) *Neuroscience* 166(4):1119–1128 (in English).
93. Lacayo CI, et al. (2010) Imaging Cell Wall Architecture in Single Zinnia elegans Tracheary Elements. (Translated from English) *Plant Physiol* 154(1):121–133 (in English).
94. Dokken KM & Davis LC (2007) Infrared imaging of sunflower and maize root anatomy. (Translated from English) *J Agr Food Chem* 55(26):10517–10530 (in English).
95. Yu P (2007) Ultra-spatial synchrotron radiation for imaging molecular chemical structure: Applications in plant and animal studies. (Translated from English) *Spectrosc-Int J* 21(4):183–192 (in English).
96. Yu P (2005) Molecular chemistry imaging to reveal structural features of various plant feed tissues. (Translated from English) *J Struct Biol* 150(1):81–89 (in English).
97. Yu PQ, et al. (2003) Chemical imaging of microstructures of plant tissues within cellular dimension using synchrotron infrared microspectroscopy. (Translated from English) *J Agr Food Chem* 51(20):6062–6067 (in English).
98. Murdock JN, Dodds WK, Reffner JA, & Wetzel DL (2010) Measuring Cellular-Scale Nutrient Distribution in Algal Biofilms with Synchrotron Confocal Infrared Microspectroscopy. (Translated from English) *Spectroscopy* 25(10):32–41 (in English).
99. Holman HYN, et al. (2009) Real-Time Chemical Imaging of Bacterial Activity in Biofilms Using Open-Channel Microfluidics and Synchrotron FTIR Spectromicroscopy. (Translated from English) *Analytical Chemistry* 81(20):8564–8570 (in English).

100. Hazen TC, et al. (2010) Deep-Sea Oil Plume Enriches Indigenous Oil-Degrading Bacteria. (Translated from English) *Science* 330(6001):204–208 (in English).
101. Holman HYN, et al. (1999) Real-time characterization of biogeochemical reduction of Cr(VI) on basalt surfaces by SR-FTIR imaging. (Translated from English) *Geomicrobiol J* 16(4):307–324 (in English).
102. Holman HYN, et al. (2002) Catalysis of PAH biodegradation by humic acid shown in synchrotron infrared studies. (Translated from English) *Environ Sci Technol* 36(6):1276–1280 (in English).
103. Iwamoto R & Ohta K (1984) Quantitative Surface-Analysis by Fourier-Transform Attenuated Total Reflection Infrared-Spectroscopy. (Translated from English) *Appl Spectrosc* 38(3):359–365 (in English).
104. Hutson TB, Mitchell ML, Keller JT, Long DJ, & Chang MJW (1988) A Technique for Monitoring Mammalian-Cell Growth and Inhibition Insitu Via Fourier-Transform Infrared-Spectroscopy. (Translated from English) *Anal Biochem* 174(2):415–422 (in English).
105. Suci PA, Vrany JD, & Mittelman MW (1998) Investigation of interactions between antimicrobial agents and bacterial biofilms using attenuated total reflection Fourier transform infrared spectroscopy. *Biomaterials* 19(4–5):327–339.
106. Kuimova MK, Chan KLA, & Kazarian SG (2009) Chemical Imaging of Live Cancer Cells in the Natural Aqueous Environment. *Appl Spectrosc* 63(2):164–171.
107. Nasse MJ, Ratti S, Giordano M, & Hirschmugl CJ (2009) Demountable liquid/flow cell for in vivo infrared microspectroscopy of biological specimens. (Translated from eng) *Appl Spectrosc* 63(10):1181–1186 (in eng).
108. Tobin MJ, et al. (2010) FTIR spectroscopy of single live cells in aqueous media by synchrotron IR microscopy using microfabricated sample holders. *Vib Spectrosc* 53:34–38.
109. Birarda G, et al. (2010) Fabrication of a microfluidic platform for investigating dynamic biochemical processes in living samples by FTIR microspectroscopy. (Translated from English) *Microelectronic Engineering* 87(5–8):806–809 (in English).
110. Stewart PS & Franklin MJ (2008) Physiological heterogeneity in biofilms. *Nat Rev Microbiol* 6(3):199–210.
111. Whiteley M, et al. (2001) Gene expression in Pseudomonas aeruginosa biofilms. *Nature* 413(6858):860–864.
112. Portugal J & Sanchezbaeza FJ (1995) Mitomycin-C Binding to Poly[D(G-M(5)C)]. *Biochemical Journal* 306:185–190.
113. Costerton JW, Irvin RT, & Cheng KJ (1981) The Bacterial Glycocalyx in Nature and Disease. *Annual Review of Microbiology* 35:299–324.
114. Kolodkin-Gal I, et al. (2010) D-amino acids trigger biofilm disassembly. (Translated from eng) *Science* 328(5978):627–629 (in eng).
115. Holman HYN, et al. (2009) Real-time molecular monitoring of chemical environment in obligate anaerobes during oxygen adaptive response. (Translated from English) *P Natl Acad Sci USA* 106(31):12599–12604 (in English).
116. Chen L, et al. (2012) Synchrotron infrared measurements of protein phosphorylation in living single PC12 cells during neuronal differentiation. *Analytical Chemistry* 84(9): 4118-4125
117. Gomez L, et al. (2006) Apertureless scanning near-field optical microscopy: a comparison between homodyne and heterodyne approaches. (Translated from English) *J Opt Soc Am B* 23(5):823–833 (in English).

118. Aizpurua J, Taubner T, de Abajo FJ, Brehm M, & Hillenbrand R (2008) Substrate-enhanced infrared near-field spectroscopy. (Translated from English) *Opt Express* 16(3):1529–1545 (in English).
119. Policar C, et al. (2011) Subcellular IR Imaging of a Metal-Carbonyl Moiety Using Photothermally Induced Resonance. (Translated from English) *Angew Chem Int Edit* 50(4):860–864 (in English).
120. Dazzi A (2009) PhotoThermal Induced Resonance. Application to Infrared Spectromicroscopy. (Translated from English) *Top Appl Phys* 118:469–503 (in English).
121. Lu F & Belkin MA (2011) Infrared absorption nano-spectroscopy using sample photoexpansion induced by tunable quantum cascade lasers. (Translated from English) *Opt Express* 19(21):19942–19947 (in English).
122. Clède S, et al. (2013) Synchrotron radiation FTIR detection of a metal-carbonyl tamoxifen analog. Correlation with luminescence microscopy to study its subcellular distribution,. *Biotechnology Advances* 31(3)393- 395.
123. Ikemoto U, et al. (2012) Development of scattering near-field optical microspectroscopy apparatus using an infrared synchrotron radiation source. *Optics Communications* 285(8):2212–2217.
124. Goff KL, Quaroni L, & Wilson KE (2009) Measurement of metabolite formation in single living cells of Chlamydomonas reinhardtii using synchrotron Fourier-Transform Infrared spectromicroscopy. (Translated from English) *Analyst* 134(11):2216–2219 (in English).
125. Stavitski E, Smith RJ, Bourassa MW, Acerbo AS, Carr GL, Miller LM (2013) Dynamic Full-Field Infrared Imaging with Multiple Synchrotron Beams. *Analytical Chemistry*, 85 (7), 3599–3605
126. Martin MC, Dabat-Blondeau C, Unger M, Sedlmair J, Parkinson DY, Bechtel HA, Illman B, Castro JM, Keiluweit M, Buschke D, Ogle B, Nasse MJ, Hirschmugl CJ (2013) 3D spectral imaging with synchrotron Fourier transform infrared Spectromicrotomography. *Nature Methods*, 10(9): 861-868

13

Raman Spectroscopic Imaging of Biological Systems

MARTIN SCHMIDT, PRADEEP N. PERERA, ALEXANDER WEBER-BARGIONI, PAUL D. ADAMS, AND P. JAMES SCHUCK ■

Previously predicted by A. Smekal (1), C. V. Raman, and K. S. Krishnan (2), as well as G. Landsberg and L. Mandelstam (3), the inelastic scattering of light was first reported in 1928. This effect, eponymous with its principal discoverer, forms the basis of Raman spectroscopy. Its use as a practical microscopic method, however, was impeded for many decades due to instrumental shortcomings. Only in the early 1990s did Raman microscopy begin to come into its own after significant advances in instrumentation and technology. Since then, the dormant potentials and underused capabilities for chemical and structural microanalyses have been increasingly tapped, as evidenced by an impressive body of literature. The field has made great strides in recent years, establishing applications in a multitude of disciplines and advancing and improving the technique with exciting new developments. Therefore, it seems expedient to review the pertinence of Raman microscopy and imaging to the study of biological questions. In this chapter, we will give a brief description of the fundamental principles and typical instrumentation, identifying unique characteristics and also capabilities that are complementary to other imaging techniques. By design, this review cannot be exhaustive and we rather focus on presenting selected examples from a few areas of biological research to illustrate trends, general capabilities, and current challenges. We especially put an emphasis on highlighting emerging technological innovation and invention and want to accentuate how advances in technology and methodology are useful and even transformative in biologically related applications. For further reading and in-depth presentations we would like to refer to a number of excellent textbooks (4–9).

BASIC CONCEPTS

Ideally, analytical tools should permit the specific identification and accurate rapid characterization of biomolecules and biomaterials without the need for expensive, tedious, and destructive sample preparation. In principle, Raman spectroscopy has such potential. The technique relies on inelastic scattering of monochromatic light, like that from a laser, where the shift in energy of the scattered laser photons is related to molecular vibrations and presents an intrinsic label-free molecular "fingerprint" of the sample. The information contained within these molecular fingerprints is specific in the sense that spectral frequencies are directly related to certain functional groups or chemical bonds as well as sensitive to conformation, and that spectral intensities correlate with molecular concentrations. Raman spectra are thus useful to investigate the chemical composition and the biochemical architecture in a sample. Biological materials are often hierarchical and heterogeneous with regard to both structure and composition. This means that, in addition to chemical sensitivity, spatial resolution and the ability to study samples in the native context are desirable instrumental attributes. Coupling Raman spectroscopy with microscopy yields an analytic methodology with microspectroscopic capabilities. Raman spectroscopic imaging affords relatively nondestructive and comparatively inexpensive measurements of microscopic or heterogeneous samples with minimal sample preparation, giving insights into chemical composition and molecular structure in a close-to-native state. The Raman spectra simultaneously register contributions of active vibrations from all components in the sample volume. The technique is very well suited to noninvasively acquire molecular-level information on complex multicomponent and multiphase systems in aqueous environments or at interfaces. Importantly, native molecules are studied *without* exogenous labeling and possible interference of molecular tags, which are bulky appendages and can interfere with normal function. Applying this label-free spectroscopic imaging method under in situ conditions enables one to explore spatiotemporal behaviors and gradients in heterogeneous systems and to dynamically acquire compositional and structural information.

In the following, typical instrumentation and methodology of Raman microspectroscopy and Raman imaging are presented. Strictly speaking, there is a differentiation between spectroscopic mapping, where point-by-point spectra are measured sequentially, and imaging, where multiple spectra are recorded simultaneously at different points (10). The resulting maps or images, though, are both hyper-spectral data sets that are in essence the same. Therefore, we decided not to adhere to the distinction rigorously and use the term imaging somewhat more generically.

RAMAN PRINCIPLES AND INSTRUMENTATION

Fundamentally, Raman spectra arise from transitions between quantized vibrational energy states of molecules. In its most common form, a sample is illuminated with a beam of monochromatic radiation from a laser and some of the incident photons raise some of the molecules from the ground state to a virtual state of higher potential energy. While most molecules relax through elastic or Rayleigh scattering, a small subset of molecules drops back to the first excited vibrational state of a given vibrational mode. This inelastic scattering is called Stokes Raman scattering where the energy of the scattered photon is red-shifted. Raman spectra record this *Raman shift* in energy. Traditionally, scanning monochromator spectrometers using double or triple monochromators and photomultiplier tube detectors were used. Through the development of optical filters that have very low transmittance for the Rayleigh or laser line and high transmittance for the Raman photons, more sensitive spectroscopic setups could be implemented that measure the whole wavelength range simultaneously. In principle, there are two types of spectrometers commonly in use. The first one is the Fourier transform Raman spectrometer, that usually uses long wavelength 1064 nm near-infrared laser excitation, which greatly minimizes fluorescence, and liquid-nitrogen-cooled InGaAs or germanium detectors. The second type uses low-pass optical filters to eliminate Rayleigh-scattered light, a single monochromator, and cooled CCD array detectors. This type of spectrometer, using shorter wavelength visible excitation, is the instrument of choice for high-resolution Raman imaging. The coupling of a spectrometer with a microscope yields a Raman microspectrometer or simply a Raman microscope. Because array detectors are used in Raman microspectroscopy, there are different possible modalities to acquire Raman images. In the most commonly used mode the sample is raster scanned through the focal point of the probing beam. This is called point mapping and a Raman spectral map is acquired one pixel at a time collecting Raman spectra of adjacent pixels in a grid. Alternatively, global illumination techniques can be used for so-called widefield or direct imaging. In this case the entire region of interest or field of view is illuminated, simultaneously recording an image at discrete energy increments. Widefield Raman imaging can be achieved implementing dielectric filters in combination with tunable lasers, rotating dielectric filters, acousto-optic tunable filters, or liquid crystal tunable filters. Line mapping represents an intermediate case where the excitation laser beam is spread into a line by scanning a laser spot with a scanning mirror or using cylindrical optics collecting line-scanned Raman maps. In many applications, attaining high spatial resolution is desirable. In conventional far-field microscopy, spatial resolution is fundamentally limited by the diffraction of light and ultimately by the wavelength according to the Rayleigh criterion. For highest resolution, confocal Raman microscopes are used where a pinhole blocks out-of-focus light. This confocal design provides depth resolution and capabilities for optical sectioning and three-dimensional mapping and can help to suppress spurious fluorescence originating from out-of-focus sample regions. For Raman imaging there inevitably exists a trade-off between sensitivity and temporal resolution, and fast acquisition times may be an economical necessity. In the context of many biological systems, possible photodamage and photodegradation also bring about an upper limit in terms of laser exposure time and laser power. There are different ways to compute or reconstruct Raman images but essentially the chemical contrast is based on intrinsic Raman vibrational frequencies instead of labeling or staining.

APPLICATIONS USING CONVENTIONAL RAMAN IMAGING

Raman spectroscopic imaging has been established as a method that can chemically distinguish compartments and subcellular architecture within individual cells without the use of exogenous labels. Thus, it has unique capabilities for the noninvasive and label-free detection and monitoring of the distribution of biochemical components and intracellular processes (11). Applications in cellular biology are prominent and therefore subject to several recent reviews. Here we present a few select examples that are illustrative of the technique's advantages and abilities. Matthäus et al. recorded Raman microspectroscopic images of human epithelial cancer cells at different stages of mitosis (12). Utilizing inherent protein and DNA vibrational spectral markers, the distribution of condensed nuclear chromatin and other biochemical components could be monitored and the complete cell division cycle was mapped. It was suggested that the recognition of mitotic cell division without the use of stains combined with the possibility of doing so on living cells may be useful for an automatic spectroscopic determination of the proliferation rates of cells and tissues. This work can also be seen as part of a broader scheme to characterize the Raman spectral (and hence chemical) patterns of cells in response to cell division, cell differentiation, cell death, and onset of disease or drug treatment. Another example pertains to the imaging of the intracellular delivery and degradation of polymeric nanoparticle systems (13). Specifically, the subcellular degradation patterns of two biodegradable nanocarrier systems, poly-(ε-caprolactone) and poly-(lactic-*co*-glycolic acid), were monitored by observing changes in Raman peaks attributable to these molecules. Conventionally, the intracellular behavior of such nanocarriers would be studied by encapsulating fluorophores or a labeled drug derivative and imaging with fluorescence microscopy. However, there is the potential for leakage of the fluorophores into the system and for the labels to possibly interfere with biochemical behavior. Using the Raman approach, the authors instead noninvasively mapped variations in characteristic vibrational spectroscopic peak intensities and positions in order to follow the uptake, subcellular specificity, metabolism, and fate of the carrier systems. In a related application, Meister et al. imaged the uptake and intracellular distribution of a bioactive metal complex, a manganese-based CO-releasing compound which has photoinducible cytotoxic activity against cancer cells (14). Their investigation benefitted from the fact that metal-carbonyl complexes have strong C≡O stretching vibrations between 1800 and 2200 cm^{-1}, a spectral region where vibrational signals from cellular constituents are negligible. It was shown that these characteristic and spectroscopically isolated vibrational frequencies can be used as ideal markers for the label-free identification and localization of such complexes in living cancer cells. In a similar vein, nitrile and azide groups also have characteristic vibrations in this isolated spectroscopic window, which can be useful as intrinsic bioorthogonal markers. As a side note, it is important to remember that measurement details depend critically on the goals and needs of a specific experiment as well as its boundary conditions, such as whether live or preserved cells will be sampled, and at what times. When cell sample preservation methods such as fixation and drying are used, care must be taken and the spectroscopic implications considered. For example, optimized fixation methods are necessary to minimize auto-fluorescence and Raman spectral contamination from fixatives. Such considerations are discussed in detail in the literature (15).

A study presenting the Raman imaging of hemozoin within the food vacuole of *Plasmodium falciparum* trophozoite-infected functional erythrocytes underscore the utility of the method to address disease processes, in this case related to malaria (16). For hemozoin, an insoluble aggregate—formed by reactive heme moieties released from hemoglobin—also called "malaria pigment," an extraordinary enhancement of a Raman band at 1374 cm^{-1} is observed when applying laser excitation at 780 nm. It was suggested that this effect was due to excitonic coupling between linked porphyrin moieties in the extended aggregate, leading to a resonant Raman condition. It is this unusual enhancement for hemozoin with the concomitant poor enhancement of hemoglobin that enables Raman imaging of the malaria pigment within the trophozoite, which could have implications for the intracellular detection and monitoring of therapeutic agents that are thought to target hemozoin formation.

In the cases of both the metal-carbonyl complex and hemozoin, specific advantageous spectroscopic effects or circumstances were exploited, rendering the imaging rather facile and straightforward. Yet for label-free Raman data of cells, it is generally found that the spectral differences between various organelles and other subcellular structures are typically small because of the similarity in chemical composition sampled at discrete pixel points. Therefore, sensitive methods that capture small spectral differences are required in order to reconstruct meaningful images. Instead of constructing "chemical images" of the sampled areas by a univariate approach—e.g., by color-coding the intensity of a particular Raman band, the position of this band, or band intensity ratios—multivariate methods are needed to extract more information and to unlock details from the available spectral contrast. Superior sensitivity is obtained because the entire spectrum, or at least large parts of it, is used for image reconstruction. Miljkovic et al. recently presented and compared in detail a number of multivariate image reconstruction methods (17). They showed that such methods can detect minor recurrent and correlated spectral variations that are due to small biochemical compositional changes over the spatial dimension of the sample and that would be lost in univariate analyses.

Vibrational spectroscopic studies of microorganisms form a related and active area of study. As an example, Raman microspectroscopy and imaging can be useful in detecting the heterogeneity of microbial populations on the single-cell level, which may be lost in ensemble-averaged bulk analyses. Hermelink et al. introduced the method for investigating and illustrating the spatial heterogeneity of microbial cell populations (18). They were able to identify and visualize single-component distributions within heterogeneous multicomponent systems of genetically homogeneous microbiological samples (*Legionella bozemanii* L2165, *Bacillus cereus* ATCC 10987, and *Bacillus thuringiensis* DSM 5815) without the need for any dye or contrasting agents. For *B. thuringiensis*, the authors imaged a binary mixture using diagnostic marker Raman bands for *B. thuringiensis* toxin and Ca^{2+}-dipicolinic acid respectively. For *L. bozemanii*, intra- and intercellular heterogeneity in terms of the poly-β-hydroxybutyric acid content was visualized using the integrated intensity of the C=O ester stretching vibration at 1726 cm^{-1}. In the case of *B. cereus*, three different components were detected and distinguished; namely 1) locally enriched poly-β-hydroxybutyric acid, 2) vegetative cells visualized by mapping the intensity of the C-H stretching, and 3) spores using the diagnostic Raman band for Ca^{2+}-dipicolinic acid. This is a powerful example of the ability to identify and visualize in situ chemical and structural heterogeneity of microscopic composite biological systems and materials. As a side note, axial chromatic aberrations have been observed in measurements of single microbial spores and a method for correcting this artifact in four-dimensional (three spatial and the spectral dimension) data sets is described in reference (19).

Raman imaging is also proving a valuable tool for research on bone and biomineralization. Bone is a fibrous composite material with a hierarchical structure consisting of carbonated apatite mineral and an organic collagen matrix, and bone strength is dependent on structural and material properties. Importantly, the Raman technique can grant in situ insights into the compositional and structural organization of heterogeneous bone tissue at the ultrastructural level. The spatial resolution makes possible the analysis of biologically important localities, such as individual lamellae and cement lines, soft-hard tissue interfaces (enthuses), and boundaries around microcracks or human dentin tubules. Moreover, Raman spectra capture information on both the mineral component and the organic matrix, thus providing simultaneous access to the major bone constituents in the sampled area. Raman spectroscopic imaging is therefore very well suited to characterize the heterogeneous chemical microstructure of bone. Kazanci et al. made additional use of the polarization dependence of the Raman signal on molecular bond alignment, reporting on the orientation of mineral and collagen components in osteonal lamellae (20). Spatial variations of the mineral and organic matrix were imaged as a function of the linear polarization of incident light. It was shown that the prominent phosphate ν_1 band at approximately 960 cm^{-1} and the amide I band typically used for the determination of mineral and organic compositions respectively are indeed sensitive to orientation and the polarization direction of the incident light. This means that these bands are useful for exploring bone anisotropy and orientational features. But it also underlines the point that polarization must be considered when analyzing and interpreting Raman bone data. Also, other bands were identified that are less susceptible to orientational effects.

Another instructive example of the polarization and orientation dependence is an investigation imaging two orthogonal planes within cortical bone, where identical bone lamellae in both longitudinal and transverse directions were analyzed and information about lamellar bone orientation and variation in bone composition could be gleaned (21). A Raman study on single diatom (*Stephanopyxis turris*) cell walls determined the distribution of both silica and the embedded organic material, underscoring again the multiphase and multicomponent capabilities (22). The ability to distinguish and localize amorphous and crystalline mineral phases is a key characteristic that will further promote a multitude of applications addressing biomineralization, such as time- and space-dependent phase transitions and polymorphism. A related example pertains to the study of ancient individual microscopic fossils, including those regarded to be the oldest known fossils (23). Combining optically discernible morphology with chemical composition, it becomes possible to discriminate between true microbial fossils and microscopic pseudofossils, that is, nonbiological mineral microstructures. In particular, geochemically highly altered carbonaceous prokaryotic filamentous microbial fossils in petrographic thin sections of Precambrian cherts were investigated. The noninvasive, nondestructive Raman imaging approach preserved the samples and, having high sensitivity for distinctive graphitic carbon signals of organic matter, it provided evidence of the biogenicity of these poorly preserved microstructures.

Another interesting application involves research related to marine biofouling and its many fundamental and commercial implications. Schmidt et al. used confocal Raman microscopy for the in situ imaging of barnacle (*Balanus amphitrite*) cyprid cement (24). Barnacles are a model for research on permanent underwater adhesion and the wider process of marine biofouling. The permanent adhesive secreted by the cypris larva for permanent settlement, the so-called cyprid cement, is the key agent in the colonizing stage. Detailed insights into the chemistry and the adhesion mechanism have the potential to lead to novel antifouling solutions. Along with visualizing the morphology, information on the chemical composition of the cement plaque of attached individual live juvenile barnacles were gleaned *without the need* for cumbersome sample isolation or sample invasion. Chemical images recorded in the native state in artificial seawater showed that the cement is hydrated, which is supportive of the theory that water might play an important role in cement curing and function. Recent advances in technology with regard to sensitivity and temporal resolution (e.g., electron multiplying CCD detectors) should enable the application of this technique to dynamic chemical imaging during the settlement of cypris larvae in order to monitor cement curing. Carotenoids, which are components of crustacean cuticle with suggested roles as antioxidants and as membrane reinforcers, were identified at the periphery of the antennular attachment disc, suggesting that this region of the disc makes contact with the substratum and perhaps contributes to adhesion. Carotenoids have very characteristic Raman signatures, including overtones and combination bands, where spectral differences arise from the variations between different carotenoid species, structural conformation, and molecular association in biological matrices. Carotenoid signals are intense when the laser used for excitation of the Raman scattering is in resonance or preresonance with electronic transitions in the carotenoid pigments. Raman microscopy and imaging are hence very useful for the extraction-free chemical and structural characterization of carotenoids, with improved sensitivity and structural selectivity compared to extractive procedures. This is also exemplified by studies on individual pollen grains (25, 26) and on the tissue-specific accumulation of carotenoids in carrot roots (27). Subsequent to the study on barnacles, Harrington et al. recently reported on the chemical investigation of the byssus cuticle and the presence and distribution of metal-protein complexes by resonance Raman microscopy and imaging (28). The byssus is a fibrous holdfast made by marine mussels. Raman imaging was performed on thin sections of *Mytilus californianus* and *M. galloprovincialis* threads. Byssal threads are shielded from abrasion by an outer cuticle that is harder than the thread core and largely proteinaceous, containing a protein rich in the catecholic amino acid 3,4-dihydroxyphenylalanine (dopa). Intense resonance-enhanced peaks due to interactions of iron with the catecholic moiety of dopa were found to be localized exclusively within the outer cuticle of the thread. It was further demonstrated that the cuticle is a polymeric scaffold stabilized by catecholato-iron chelate complexes having an unusual clustered or granular distribution, where the granules contain a higher dopa-metal cross-linking density than the surrounding matrix. The cuticles exhibit a remarkable combination of high hardness and high extensibility, and it was proposed that the dense cross-linking in the granules provides hardness, whereas the less cross-linked matrix provides extensibility.

Mutant Plant Cell Wall Imaging

Plant biomass constitutes a renewable energy source, which harnesses solar energy through photosynthesis, and is pivotal in providing solutions to pressing challenges facing humanity, such as feeding a growing global human population, mitigating global climate change, and the need of many major national economies to resort to renewable resources within the next decades. Scientific and technological advances in plant biology, genetics, and biotechnology are crucial for delivering abundant renewable resources for the increased and sustainable production of food, biofuels, and chemicals. The potential of Raman microscopy and Raman imaging in plant research was discussed in detail in a review by Gierlinger and Schwanninger (29). Plant cell walls are the most abundant renewable resource of photosynthetically fixed carbon. This lignocellulosic biomass is poised to become the primary source for the production of liquid biofuels. The recalcitrance of plant cell wall materials to cost-effective and efficient degradation, however, poses a major impediment in the production of transportation fuels and chemicals. Plant cell walls have a heterogeneous and complex hierarchical structure, which is primarily made up of cellulose, hemicelluloses, and lignin (30). The detailed structure of plant cell walls, especially at the ultrastructural level, however, remains to be fully elucidated. Yet such a detailed knowledge is required to understand the specific chemical and physical obstacles to breakdown, enabling the redesign of specific chemical (pre)treatments as well as the reengineering of plant cell walls for the purpose of commercially viable production of lignocellulosic biofuels (31). Analytical tools that permit the accurate rapid characterization of plant cell walls become increasingly important for evaluating specific traits associated with a large number of breeding populations. Accurate investigation of specific chemical characteristics of lignocellulosic materials can be difficult since extractive procedures for the isolation of native components are inevitably cumbersome and destructive, bringing about significant chemical and structural modifications. Analytical chemical in situ methods are, therefore, essential for the compositional and structural characterization of lignocellulosic materials.

Confocal Raman imaging has proven a valuable emerging tool in plant cell wall research. It facilitated the chemical imaging of poplar (32) and spruce wood cell walls (33) and enabled the in situ visualization of the spatial distribution of cell wall polymers, in particular cellulose and lignin. A recent study made use of the polarization dependence to determine the cellulose microfibril orientation of *Picea abies* and its variability at the micrometer-level (34). Based on the earlier work of Gierlinger et al. and Agarwal, we have recently adopted and expanded the method to compare lignification in wild type and lignin-deficient transgenic *Populus trichocarpa* (black cottonwood) stem wood (35, 36). Lignin is a complex and irregular poly-phenylpropanoid heteropolymer present in the cell walls of vascular plants and it is problematic to the postharvest deconstruction of lignocellulosic biomass. For example, in biomass conversion for biofuels, lignin inhibits saccharification in processes aimed at producing simple sugars for fermentation to ethanol. In general, the effective use of plants for industrial purposes is largely dependent on the extent to which the plant cell wall is lignified and the removal of lignin is a costly and limiting factor. Therefore, lignin has become a key plant breeding and genetic engineering target in order to improve cell wall conversion. Wood is one of the possible feedstocks for the commercial production of liquid biofuels and genetic engineering of trees has generated substantial interest aimed at the development of genotypes with improved wood traits. Low-lignin wood offers the potential for improved biomass conversion into fermentable sugars. In our study, we compared chemical images of wood sections of wild type poplar and lignin-deficient poplar grown with transgenic suppression of the monolignol pathway gene encoding 4-coumarate-CoA ligase. Analyzing the lignin Raman bands in the spectral region between 1600 and 1700 cm^{-1}, lignin signal intensity and localization were mapped in situ without staining or labeling of the tissues. Our approach visualized differences in lignin content, localization, and chemical composition (Figure 13.1). Lignin was found to be significantly reduced in the secondary wall layers of the transgenic poplar fibers. Given that most wood biomass is in the secondary layers of fibers, this finding may have important implications toward addressing biomass recalcitrance to degradation.

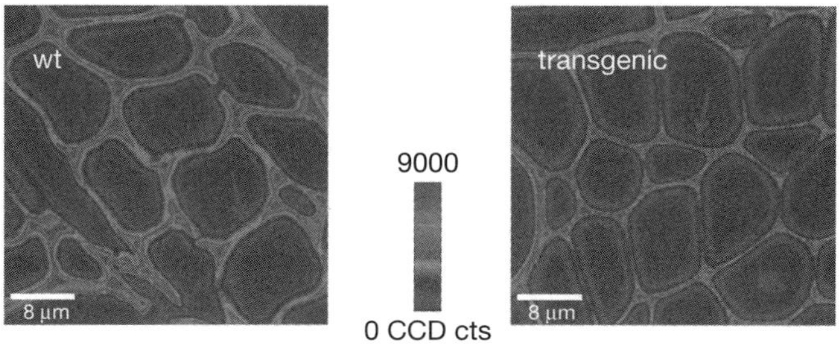

Figure 13.1 Label-free in situ imaging of lignification in the cell walls of wild type (wt) and low-lignin transgenic *Populus trichocarpa*. Lignin distributions are visualized by integrating the Raman intensity from 1550 to 1700 cm^{-1}. The red arrows point to secondary walls in which lignin was found to be significantly reduced upon transgenic suppression of 4-coumarate-CoA ligase.

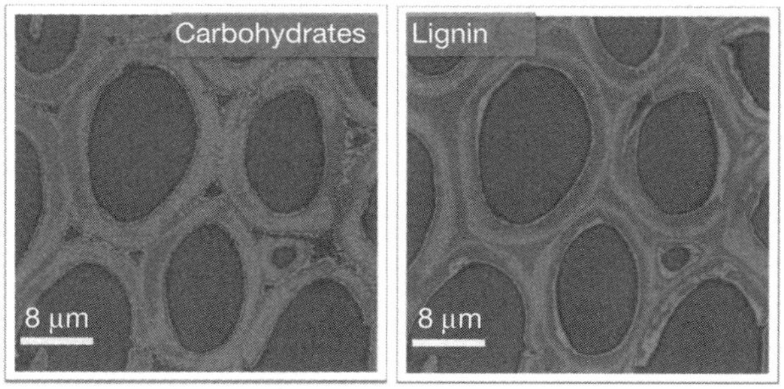

Figure 13.2 Chemical images of interfascicular cells of senescent *Arabidopsis thaliana* based on the C–H and CH_2 stretching of carbohydrates (2820–2935 cm^{-1}) and lignin Raman bands (1550–1700 cm^{-1}).

The measurements were carried out in D_2O in a fluid cell. This is advantageous in several respects. Laser-induced fluorescence is limited, overlap between the O–H and C–H stretching bands is minimized, and additional chemical contrast and insights into accessibility and reactivity (H–D exchange) of different structural parts is provided (37). As an extrinsic marker, D_2O also presents a convenient internal reference for intensity normalization based on the O–D stretching band around 2500 cm^{-1} in the average lumen spectra. This normalization is critical and means our method can afford quantitative comparisons between different measurements, tissues, samples, or species.

Recently, we reported on the Raman imaging of cell wall polymers in *Arabidopsis thaliana* (38). We demonstrated high-quality Raman imaging of *A. thaliana* with visible (532 nm) wavelength excitation and high spatial resolution (~ 300 nm lateral resolution), overcoming autofluorescence by using senescent samples (Figure 13.2). The plant *Arabidopsis* is a model system for studying a variety of biological processes in plant biology on the cellular and molecular level and provides an excellent model for probing the plant cell wall. Its small size and fast generation time allow rapid characterization of treatments and mutations. Its small genome was the first plant genome sequenced and *Arabidopsis* research also benefits from a wide array of tools, including collections of gene knockouts that cover much of the genome and collections of gene expression, gene function, and proteomics data. Our results open up new avenues for the noninvasive, relatively fast and inexpensive label-free in situ characterization and screening of different tissues and mutants of this important model plant and for the correlation of mutations with variations in cell wall composition and structure on the sub-μm scale.

We have subsequently worked to establish a method for mutant microanalysis in *A. thaliana* using advanced chemometrics (39). As a starting point, we chose two well-characterized mutants of *A. thaliana* that show significantly altered cell wall structure compared to wild type in order to test and validate our findings. The first mutant, *irx5*, is a null mutation to the cellulose synthase 4 gene, an essential component of the complex that synthesizes cellulose in the secondary cell wall, and *irx5* mutant plants have reduced cellulose content by 70–90 percent in the secondary cell wall with no apparent change in both lignin and hemicelluloses. The second mutant, *irx6*, has a down-regulated monolignol-pathway-specific cinnamoyl-CoA reductase enzyme, and *irx6* mutant plants have diminished overall lignin content and alterated lignin structure in the secondary cell wall. The changes that were induced in the cell wall by these mutants lent themselves well to a comparative analysis and were leveraged to dissect the chemical and structural makeup. By analyzing and comparing spectroscopic data from wild type *Arabidopsis* and the two mutants with impaired cellulose and lignin biosynthesis, we could spectrally and spatially resolve in situ lignin-carbohydrate complexes (LCCs), which are implicated in plant biomass recalcitrance, and the hemicellulose glucuronoxylan. We also detected variabilities in carbohydrate distribution and concentration, cellulose microfibril angle, lignin distribution and concentration, and lignin chemistry as a function of mutation. The capability to semi-quantitatively detect and spatially resolve LCCs in the native cell wall has interesting implications for the reengineering of the cell wall. LCCs provide a strong association between the lignin matrix and the carbohydrate network. They thus have a strong contribution to the cell wall recalcitrance towards saccharification. Importantly, we do not detect LCC-associated Raman peaks in *irx6*, suggesting a possible candidate mutation to target cell wall recalcitrance. We anticipate that our relatively fast and inexpensive method can overcome a bottleneck in the characterization and screening of a multitude of mutants, on the cellular and subcellular level. It should be noted here that there is a severe challenge in that native lignin in lignocellulosics has high spectral overlap with contributions from cellulose and hemicelluloses. To fully exploit the multicomponent and multiphase sensitivity of Raman imaging spectroscopy, it is mandatory to apply a chemometric approach to dissect the spatial distributions of the cell wall polymers, especially for the different carbohydrates, which bear strong structural and spectral similarities. We therefore implemented and applied chemometric tools, which are discussed in detail in the section on *Advanced Chemometric Analysis*.

CHALLENGES, INNOVATIONS, AND FRONTIERS

As is common with any technique, Raman spectroscopy intrinsically has both advantages and disadvantages. We want to mention the most severe challenges and describe in the following sections how these problems are addressed and overcome. Spontaneous Raman scattering inherently is a very weak process. This means that

Raman spectroscopy often lacks sensitivity, the implication being that long acquisition times may be required to obtain a sufficient signal-to-noise ratio, which is impractical for many imaging applications—and renders investigations of dynamics nearly impossible at all but the longest time scales. Surface-enhanced Raman scattering (SERS) as well as nonlinear optical techniques can be used to increase signal intensity by many orders of magnitude and remedy this problem. For confocal Raman imaging, there also exists a fundamental limit to the insights this technique offers in terms of lateral spatial resolution (~ 250 nm) due to diffraction. This diffraction limit is a problem since crucial (and often poorly understood) structural features and supramolecular arrangements of most biological systems have single- and double-digit nanometric length scales. Tip-enhanced Raman imaging, which combines the nanoscale spatial resolution of scanning probe microscopy techniques with the chemical information inherent in Raman spectroscopy, is emerging as a nano-Raman imaging tool. It provides access to these critical scales with the potential to grant unique chemical and structural insights into native biological assemblies. Besides limits to sensitivity and both temporal and spatial resolution, a very challenging impediment encountered in some samples is prohibitively strong fluorescence. Sample fluorescence is an unwanted background signal that can often swamp the much weaker Raman scattering. This is especially true when lasers with short wavelengths are used; while shorter wavelength photons yield stronger spontaneous Raman scattering efficiencies and better spatial resolution, their energies more often overlap with molecular electronic absorptions that lead directly to fluorescence. Time-gating techniques, fluorescence quenching by SERS substrates, and nonlinear optical approaches can mitigate this problem. As a multivariate technique, Raman spectroscopy gives complex multiplexed data that are rich in chemical and structural information, but these data can be difficult to interpret. In order to maximize the utility of the technique and the amount of extractable information, advanced chemometric methods for the data analysis are required. Hyphenation with other analytical and imaging methods can also be very useful in this respect. In the following, we highlight some of these technologies and methodologies and illustrate how they advance applicability and applications of Raman imaging in biological research.

Applications Using SERS Imaging

Advances in nanotechnology and the concomitant ability to control and to tune the properties of nanomaterials and nanoparticles have greatly reinvigorated the research on and with surface-enhanced Raman spectroscopy (SERS), which is reflected in a number of excellent reviews in recent years, such as (40).

Substrates made up of Au or Ag nanoparticles and/or nanofeatures exhibit highly localized and enhanced Raman signal at specific spots (rough edges and/or small gaps), sometimes with single molecule sensitivity (41–44), by combining sub-diffraction localization of the optical fields with substantial field enhancement (45, 46). While the creation of these Raman hot spots is often random, the basic underlying mechanisms (both electromagnetic and chemical) leading to the localized enhancement are mostly understood (47–53). The enhanced fields in SERS are created by localized Surface Plasmon-Polariton resonances (SPPs—a collective, coherent oscillation of charge) that can be excited when an incoming photon's electromagnetic field interacts with a metallic nanostructure or nanofeature. For Au and Ag nanostructures, these plasmon resonances are typically excited by light in the near-visible regime of the spectrum. The plasmon resonance of each metallic nanostructure on a SERS substrate depends specifically on its size, shape, and local environment. The localized SPPs efficiently interact with incoming photons of the proper energy, and therefore these types of metallic nanostructures are frequently thought of as optical antennas. Importantly, the plasmonic fields are localized on the nanostructure (manifested as evanescent electromagnetic fields that are also referred to as optical near-fields), and therefore concentrated in an area significantly below the diffraction limit. Due to the buildup of plasmons and the extreme concentration of fields, the localized SPP near fields are often orders of magnitude more intense than the incident electromagnetic field. These last two points are important for Raman applications because they mean that Raman signal in SERS is (a) orders of magnitude stronger than that originating from the same number of molecules in a standard Raman measurement, and (b) originate only from molecules within a couple nanometers of the SERS substrate's surface since the SPP fields are strongly localized near the metallic nanoparticles' edges and gaps.

As SERS has become more understood and better controlled, its use for bio sensing and imaging has been gaining traction (11, 54). For example, cell walls of living single yeast cells were imaged using SERS spectroscopy by adsorbing silver nanoparticles that can form nanoaggregates on the surface of the yeast cells (55). The spectral response was shown to originate mainly from mannoproteins which cover the outermost part of the cell wall. The spectral analysis may contribute towards an understanding of the surface chemistry and dynamics.

Very recently the applicability of SERS imaging for the label-free in situ analysis of the chemical and structural makeup of multispecies heterotrophic biofilms has been demonstrated (56). Low water background and multicomponent sensitivity in the Raman measurements are beneficial for the rapid chemical characterization of the highly hydrated biofilms, which consist of communities of microorganisms enclosed in a complex matrix of extracellular polymeric substances. The SERS effect provided a significant signal enhancement along with a larger number of sharp peaks, and hence larger chemical information content, as well as a suppression of the fluorescence background. Various components and their heterogeneous spatial distribution were mapped, detecting differences for samples cultivated for different times and also between different axial sections within samples. This approach can contribute towards an understanding of the influence of environmental factors on biofilm development and the variable physico-chemical properties of the complex biopolymer matrix since it enables spatially resolved chemical online monitoring of biofilm formation and aging under different environmental conditions. In an application towards addressing biofouling and related problems in cooling water systems, micro-SERS spectroscopy and imaging experiments were performed on algal and bacterial biofilms grown on titanium surfaces using bimetallic silver/copper colloids (57). Spatial distributions of polysaccharides and proteins in the biofilms were characterized aiming to improve the control measures of biofouling.

In the field of biomedical imaging, efforts are underway to improve sensitivity and specificity using a multimarker approach for molecular imaging. Nanoparticles have potential as multiplexed imaging probes for diagnostic and therapeutic applications, and preclinical imaging modalities based on these probes have emerged. Using several SERS nanotags, each being loaded with different predetermined "reporter" molecules and thus having unique Raman spectra, in vivo multiplex Raman imaging in skin and liver tissue of living nude mice was demonstrated (58). The nanotag design gave robust and inert nanoparticles consisting of a gold core, a Raman active molecular layer, and a silica coating. In general, spectral deconvolution of mutliple SERS nanoparticles is facilitated because the linewidth of the Raman peaks is narrow and it is made easier when the spectral fingerprints are known a priori. The ability to detect, spectrally separate, and hence colocalize, a combination of SERS tags gives the prospect of interrogating multiple molecular targets simultaneously using this approach. Similarly, Wang et al. reported on the first multiplex in vivo detection of SERS nanoprobes by Raman mapping in live zebrafish embryos, potentially enabling molecular imaging aimed at the spatial and temporal tracking of cells during the development of organs and tissues (59).

Applications Using Nonlinear Raman Imaging

As mentioned above, a primary limiting factor of Raman spectroscopic imaging is the inherent weakness of the spontaneous Raman scattering process, even when exploiting resonant Raman signal enhancement. This means relatively large incident powers are required to obtain a detectable amount of Raman signal, often exacerbating sensitivity issues related to direct laser absorption within the sample, which include the (power-dependent) excitation of background autofluorescence and sample damage. Ultimately, these issues limit the types of samples that can be studied as well as the imaging speeds used. Aiming to improve sensitivity while retaining the important chemical information, a number of efforts have been dedicated to the development of nonlinear coherent Raman imaging techniques. Included in these methods are, most notably, coherent anti-Stokes Raman scattering (CARS) imaging and, more recently, stimulated Raman scattering (SRS) imaging. The increased sensitivity afforded by CARS and SRS has enabled new bio-imaging applications and led to significantly increased imaging speeds, often reaching video rates. Here we discuss the primary pros and cons of these microscopies and describe a few examples highlighting recent breakthroughs.

Coherent Anti-Stokes Raman Scattering (CARS)

As pointed out by Evans and Xie (60), coherent anti-Stokes Raman scattering was first reported in 1965 by Maker and Terhune at the Ford Motor Company, though, ironically, the behavior was not named CARS until nearly a decade later. Particularly since 1999, there have been significant advances in understanding and developing CARS microscopy techniques. Excellent discussions of CARS and CARS imaging can be found, for example, in (60–62) and references therein. In short, the CARS process is driven by a pump laser beam at frequency ω_p and Stokes laser beam at frequency ω_s. These two beams interact within the sample via a wave-mixing process, beating against each other at a frequency $\omega_p - \omega_s$. When this beat frequency matches the frequency of a Raman active molecular vibration, the resonant vibrations are driven coherently by the excitation fields, thereby generating a strong anti-Stokes signal at frequency $\omega_{CARS} = 2\omega_p - \omega_s$ (60, 63).

Like conventional micro-Raman, chemically specific CARS-based imaging is relatively noninvasive and label-free. Because it is a nonlinear process proportional to the cube of the incident intensity, pulsed laser sources with high peak intensities are required for CARS investigations. Sources with picosecond pulse durations are typically used since they offer a good compromise between the desired narrow spectral bandwidth of the pulses and large peak intensities. Similar to two-photon fluorescence (TPF) imaging, the nonlinear contrast means that signal originates only from the focal volume where the incident beams are most intense, making it possible to create three-dimensional (3D) images serially by scanning the focus throughout the sample. Laser wavelengths in the near-infrared (nIR) region of the spectrum tend to be most advantageous. At these wavelengths, multiphoton absorption, which results in background fluorescence and sample damage, is minimized due to the low photon energy relative to visible light. In addition, penetration depth in bio/tissue samples is maximized because unwanted scattering of incident light by heterogeneous tissue is reduced in the nIR, enabling 3D imaging of thicker samples.

Of course, there are a number of distinct advantages over standard Raman imaging. For one, CARS signal can be orders of magnitude stronger than spontaneous Raman scattering (11, 64, 65). Though it is a nonlinear process and therefore much weaker when probing only a small number of resonant molecules, the coherence of CARS means that its signal intensity scales as the square of the number of scattering bonds contained within the microscope's focal volume ($\sim n^2$) and can eclipse the signal strength of conventional Raman for dense samples (spontaneous Raman scales linearly with number of molecules n). Secondly, the scattered light at ω_{CARS} is at higher frequency/energy than both of the incident frequencies ω_p and ω_s; that is, it is anti-Stokes-shifted relative to the incident beams. Thus, there is almost no spectral overlap with linear autofluorescence from the sample, eliminating a major source of background present in conventional Raman spectroscopy. And as mentioned above, spectrally overlapping background signal from multiphoton fluorescence can be minimized by proper choice of (usually nIR) incident wavelengths.

Besides the need for more sophisticated pulsed laser systems (which are becoming more user-friendly and less expensive every day), a primary disadvantage of CARS relative to conventional Raman is that CARS spectral information can be difficult to interpret. This is because the shape and energetic position of CARS spectral features depend sensitively on the relative contributions of resonant and non-resonant wave-mixing signals, making it difficult to apply the wealth of information from the Raman literature to the assignment of CARS spectra (66). Also, spectral interference from two or more resonances energetically near each other can make CARS spectra uninterpretable (60). Using a pulsed broadband light source (e.g., a white light supercontinuum source) for the Stokes beam enables the resonant excitation of multiple vibrations within the sample at once, in principle allowing for the collection of an entire CARS spectrum at each point in an image (CARS hyperspectral imaging). However, the relatively low intensities per unit wavelength of these sources (low power spectral density) makes long integration times necessary for collecting full spectra, which increases the chance of sample damage. This, along with the difficult interpretation of CARS spectra, has resulted in many groups supporting the use of narrow band CARS imaging, where the incident lasers are tuned to only one vibrational resonance (i.e., not hyperspectral), over broadband CARS hyperspectral imaging.

In fact, it is the use of narrow band CARS combined with the enhanced sensitivity due to much larger signals and reduced background that has resulted in the primary advantage of CARS: high-speed imaging. In 1999 the first 3D CARS image took nearly thirty minutes to complete (67). But by 2008, researchers were acquiring data at thirty frames per second. These fast rates make it possible to study dynamics and tracking within biological samples. For example, by tuning into the CH2 stretching frequency, Nan et al. (68), used CARS microscopy to monitor the transport of lipid droplets in steroidogenic Y-1 mouse adrenal cortical cells. The label-free visualization of lipids in this study is important because selective fluorescent lipid labels may be perturbative. Another noteworthy CARS application involves the investigation of metabolytes and their function within cells (60). Metabolytes are small molecules that, due to their small size and subsequent sensitivity to labeling, are difficult to image with conventional microscopy techniques. In one study, CARS microscopy has been been used to image lipid metabolism by following lipid storage in different strains of *Caenorhabditis elegans* (69). In other work, CARS imaging enabled the investigation of changes in lipid metabolism caused by the hepatitis C virus (70, 71). Murugkar et al. reported on the label-free chemical imaging of waterborne pathogens using CARS microscopy imaging, demonstrating the capability for the rapid detection of *Cryptosporidium parvum*, a protozoan parasite, at the single oocyst level, which is applicable to the development of a near real-time automated water testing system (72).

The 3D imaging capabilities and use of nIR light has made CARS microcsopy particularly useful for in vivo investigations. The first CARS in vivo studies were performed on mouse ear skin, where video-rate imaging was used to visualize the entire depth of the tissue (73). Evans, et al. resolved the abundant lipid structures throughout the 120 μm-thick skin layer and tracked in real time chemical diffusion into the skin. Researchers have now used CARS microscopy to study a number of biomedical samples, including lung (74, 75), nerve (76), and brain tissue (77); the interested reader is encouraged to consult the subject reviews and book chapters mentioned above and in the literature. It is important to note that, due to its coherence, CARS signal from thick bulk samples is quantitatively different than that originating from thin layers or nanoscale structures that are much smaller than the wavelength of light. Specifically, CARS in bulk samples propagates only in the forward direction (the same direction as the incident light's k-vector) due to phase-matching considerations (the backwards propagating signal destructively interferes within the sample), while CARS from non-bulk structures scatters in all directions. This complementary information makes it possible to differentiate between nanoscale structures (including thin films and interfaces) and bulk material. The two types of signal can be separately detected by collecting the forward CARS with an optic placed opposite the focusing objective on the other side of the sample (assuming the sample is not opaque) while using a backwards, or epi-CARS, collection geometry for the other (collecting the backwards CARS with the focusing objective). Schie, et al. presented a clever alternative method allowing both signals to be collected in an epi geometry using the same detector: By placing a mirror above the sample, forward CARS is reflected back into the microscope objective, spatially overlapping it with backwards CARS signal. The two signals could be easily resolved in the time domain, however, due to the known extra distance traveled by the forward signal (78). Using this novel method, they imaged living adipocyte cells derived from human mesenchymal stem cells, differentiating between lipid membrane interfaces, nanoscale droplets, and bulk volumes within the cells.

Although CARS imaging has experienced significant improvements in recent years, many experiments are still limited by nonresonant CARS background that can obscure weaker resonant CARS signals. This nonresonant background signal, so called because it is generated even when $\omega_p - \omega_s$ is not resonant with any vibrational frequency in the sample, arises from other nonlinear optical interactions and has a photon frequency equal to ω_{CARS}. A number of methods have been suggested for reducing the relative strength of this unwanted background. For example, polarization-sensitive detection (79) can completely suppress the nonresonant contribution, but at the cost of a significantly attenuated resonant signal. Ganikhanov et al. have developed a frequency-modulated technique, fm-CARS, that uses a lock-in amplifier to pull out the modulated resonant CARS signal while essentially ignoring the nonresonant background. This has resulted in a sensitivity nearly three orders of magnitude greater than conventional CARS imaging. The sensitivity and biomedical imaging capabilities of fm-CARS were then demonstrated by imaging lipids on human lung cancer cells (64). An alternative approach for improving signal strength over the nonresonant background involves the exploitation of multiple chemically specific resonances existing within the molecule of interest. In conventional CARS, two photons with frequency ω_p combine with a single ω_s photon, leading to signal at wCARS. If, however, the second ω_p photon is replaced by one with a third frequency ω'_p, such that $\omega_p - \omega_s$ and $\omega'_p - \omega_s$ are each resonant with different vibrational frequencies of the molecule, the resulting doubly-resonant signal can be significantly increased without any increase in the nonresonant background (80). Weeks, et al. have used this doubly-resonant CARS (DR-CARS) (also called doubly-resonant four-wave-mixing, or DR-4WM) technique to image lipids and glucose within a *C. elegans* worm body (81). Along similar lines, Min et al. have proposed and demonstrated a triply-resonant CARS scheme utilizing shaped femtosecond pulses for excitation (82).

Stimulated Raman Scattering Microscopy

SRS microscopy is an emerging coherent Raman bioimaging technique, which has the enhanced sensitivity of CARS but with added benefits. First observed in 1968 (83), it has since been used in many chemical spectroscopy studies (84–87). As in CARS, a pump beam at frequency ω_p and a Stokes beam at frequency ω_s are used for excitation. When $\omega_p - \omega_s$ corresponds to a molecular vibration frequency in the sample, the Raman signal can be amplified due to stimulated emission. When this occurs, the Stokes beam intensity I_s experiences a gain. Meanwhile, the pump beam intensity I_p experiences a loss since more pump photons are being scattered. Unlike CARS, there is no nonresonant background signal when $\omega_p - \omega_s$ does not match a vibrational frequency in the sample. Either the increase in I_s or the decrease in I_p can be exploited as the contrast mechanism in imaging applications, helped by the fact that SRS spectral features agree well with spontaneous Raman spectra, another advantage over CARS. SRS microscopy was first reported by Ploetz et al., where a femtosecond laser system was used for excitation (88). Since the high peak intensities present in amplified fs laser pulses can damage bio samples, Freudiger et al. developed an imaging technique based on lower power, but higher repetition rate, laser oscillators that is designed to be more bio-friendly (89).

To demonstrate the 3D label-free bioimaging capabilities of SRS, Freudiger et al. monitored the uptake of omega-3 fatty acids by living human lung cancer cells. The Xie group has gone on to demonstrate in vivo imaging capabilities in live mice and humans at video rates (90), and to use SRS microscopy to monitor biomass processing in real time (91). Using similar concepts for achieving chemical specificity, Yakovlev et al. used an SRS-type excitation scheme coupled with ultrasonic detection for deep-tissue photoacousting imaging with chemical resolution (92). In another study, SRS imaging was used to screen fat-regulating genes, yielding eight new genetic regulators for fat storage (93). More recent advances to SRS imaging include the use of a spectrally modulated broadband pump beam for distinguishing between chemical species with overlapping Raman spectra. When combined with its inherent sensitivity and straightforward interpretation, new advances in SRS microscopy should make it an important bioimaging method in the near future.

Towards Raman Imaging with Nanoscale Resolution: Advances in Tip-Enhanced Raman Spectroscopy and Imaging Methods

Much of the discussion in previous sections elucidated the potential of Raman spectroscopy and hyperspectral Raman imaging to determine the chemical structure of biological specimens with *diffraction-limited resolution* (>250 nm). However, the majority of biological systems are highly complex and functional, operating with incredible precision at many length scales, from meters down to the scale of individual molecules. To study nature's chemical functionality and complexity at the smallest length scales via Raman spectroscopy, a technique is necessary that pushes spatial resolution beyond the diffraction limit. In this section we will introduce and show the progress of tip-enhanced Raman spectroscopy that enables Raman spectroscopy and imaging down to single-digit-nm (sub-10 nm) resolution (94–100).

Before discussing the details, it is worth comparing the general properties of nano-Raman imaging and spectroscopy, which combine optics with scanning probe microscopy, with other high-resolution imaging methods. For example, in biology, super-resolution optical techniques based on fluorescence (e.g., PALM, F-PALM, STORM, SMACM, STED) have become important tools for subcellular investigations by offering nm-scale resolution (typically ~10 nm)—see (101–104) and references therein. Many of the advantages of nano-Raman imaging over these fluorescence-based super-resolution methods are the same as those offered by conventional micro-Raman: it is label-free, and unlike fluorescence it yields a large amount of direct chemical information. Regarding the former point, note that the super-resolution fluorescence measurements rely on dense labeling with exogenous luminescent reporters, which can interfere with normal biological function. Nano-Raman imaging also has other distinct benefits since it relies on the presence of evanescent optical fields located within a few nm of the end of a scanning tip. For example, because the evanescent field strength falls off quickly from the end of the tip, it allows one to specifically probe molecules only at a surface without generating significant background signal from molecules below the surface in the bulk sample (though this also makes it difficult to investigate sub-surface structures).

One can also compare nano-Raman imaging to the high-spatial-resolution imaging techniques based on electron microscopies, as well as scan-probe methods such as atomic force microscopy. While offering spatial resolution down to the atomic level, these methods generally yield only topographical/morphological, and sometimes atomic, information, but not the molecular chemical information afforded by Raman imaging. In addition, these techniques can be more invasive and/or destructive to bio samples than are optical methods. Ultimately, however, it should be remembered that each of these classes of high-resolution imaging yield specific, often unique, types of information, and should therefore be considered complementary tools.

Near-Field Imaging and Spectroscopy

Beating the optical diffraction limit using visible light was first demonstrated in 1984 (105). In this case, light was squeezed through a small metallic aperture at the end of a metal-coated tapered glass rod. Later efforts used pulled glass pipets and tapered optical fibers to deliver photons to the small metal apertures (106–108). The sub-diffraction-limit-sized aperture at the apex of the tip enables a small amount of light to leak out (compare Figure 13.3). In the direct vicinity of this nano-aperture, evanescent electromagnetic fields (also referred to as optical near-fields) are created. The spatial extent of these fields very near the aperture is determined by the size of the hole and not by the wavelength λ of incident light. Therefore, the nano-aperture at the end of a tip can be scanned over the substrate using scanning probe microscopy methods, leading to a sub-diffraction-limited optical image with a spatial

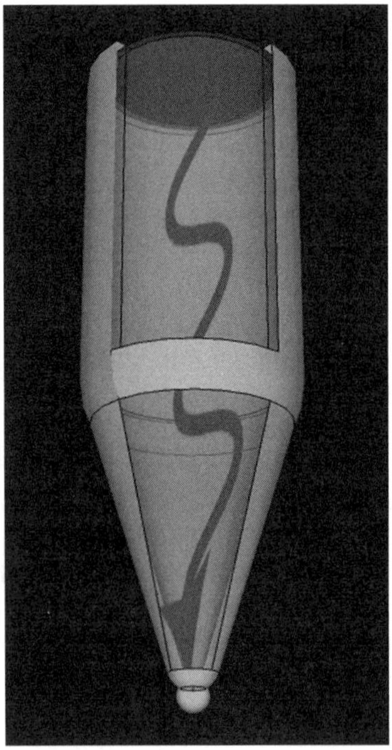

Figure 13.3 Basic concept of aperture-based NSOM. Light is coupled into a tapered, metal-coated optical fiber and squeezed through a significantly sub-wavelength-sized aperture at the end, creating a weak but localized optical near field. Here, the tapered probe is drawn without part of its metallized sidewall, showing the optical fiber.

resolution defined by the aperture size—typically on the order of λ/20 (105). This technique is referred to as Near-field Optical Microscopy (NSOM) or Scanning Near-field Optical Microscopy (SNOM). While conventional aperture-based NSOM beats the diffraction limit, the intensity of light squeezed through these apertures is so low that NSOM-based spectroscopy experiments relying on relatively weak scattering processes (e.g., Raman Spectroscopy) are often not feasible.

Along these lines, it was realized that the physical mechanisms responsible for Surface Enhanced Raman Spectroscopy (SERS) (45, 46)—i.e., the excitement of localized SPPs, could provide an alternative approach for beating the diffraction limit (see SERS section above). For example, a sharp Au or Ag tip can be employed as a crude optical antenna, with the tip apex acting as a sharp Au or Ag feature, that when illuminated by a laser, excites SPPs and creates a highly localized optical near field at the tip apex (109, 110), (111, 112). The SPP field confinement, and hence the resolution, is determined by the radius of curvature of the tip apex.

The technique of using these sharp metal tips as a local light source (or as a local light scatterer) is often referred to as apertureless NSOM (a-NSOM). In general, the fabrication of Au and Ag tips that produce a meaningful near-field enhancement is an art and not very reproducible. However, the near-field intensity enhancement of a working tip can be up to two orders of magnitude relative to the incident intensity, enabling sufficient signal to perform nanoscale fluorescence and Raman spectroscopy on some types of samples, including biological structures (94, 95, 113, 114), (115), (116–122). When a-NSOM is used for sub-diffraction-limited Raman spectroscopy, it is often referred to as Tip-Enhanced Raman Spectroscopy (TERS).

In TERS, the sharp metal (or metal-coated) tips are first typically scanned over the sample in Atomic Force Microscopy mode, creating a high-resolution topographic image. The tip is then placed over a specific feature on the substrate, excited by an incoming laser, and the tip-enhanced Raman Spectrum is recorded. TERS has been employed on a variety of biological samples (123–125), where it enabled label-free high-resolution measurements of the chemical information within, for example, DNA (126), the tobacco mosaic virus (127), and other cellular environments (128, 129).

Besides being used to take spectra at select points on the sample, this technique can also be adapted for tip-enhanced spectral imaging. In the most common tip-enhanced spectral imaging method, TERS signal from a spectral region correlated to a specific vibrational mode is binned together and collected at each point in an image. An optical filter that allows only photons to pass in the chosen spectral window is placed in the optical collection path and the transmitted signal is detected by a broadband point detector, such as an Avalanche Photo Diode. In this case, an entire spectrum is not collected, and so less integration time is necessary (94). This enables scanning over the sample in a reasonably fast time while still collecting the chemical information associated with the specific spectral window (94), (130).

In an effort to improve the reproducibility of TERS-capable tips, well-defined metallic nanoparticles that produce localized and enhanced plasmonic near-fields have been synthesized and placed on the end of AFM tips (131, 132). The nanoparticle acts as a plasmonic cavity (133) and represents a simple but well-defined optical (or plasmonic) antenna. In principle, nanoparticles can be synthesized such that their plasmon resonance matches a specific laser wavelength. A particle can then be picked up with an AFM tip, thereby creating a reproducible TERS tip. Tips of this type have been used to probe a number of interesting nanoscale objects, including individual plasma-membrane-bound Ca2+ pumps (PMCA4) immersed in aqueous environments via nano-fluorescence measurements (134, 135).

Clearly, however, an ultimate goal is to perform hyperspectral imaging (a full Raman spectrum obtained at each point in an image) with nanoscale resolution—that is, with the resolution of TERS. So far, the majority of tips used in TERS experiments have not demonstrated a near-field enhancement sufficiently large for collecting a full Raman spectrum in a period compatible with reasonable image acquisition times. Generally speaking, a spectral integration time of less than 100 ms per pixel is needed for this purpose. The relatively low enhancement values arise from the fact that many TERS tips are based on rather primitive optical antenna designs: *single* sharp metal tips (either tapered tips etched from metal wire or metal-coated sharp dielectric tips) or *single* metallic nanoparticles (e.g., Au nanospheres) attached to the apex of a scan probe, both in combination with *insulating* substrates.

Though many possible optical antenna designs exist (136–153), by far the most promising for near-field imaging purposes use *plasmonically coupled* structures (136, 154–157). These types of antennas take advantage of the ultra-enhanced and localized fields created around the nanogaps or crevices between constituent substructures, and are responsible for the largest reported scattering and fluorescence signal enhancements to date (e.g., single-molecule surface-enhanced Raman scattering) (158, 159). This has led to near-field imaging spectroscopy methods based on the so-called "gap mode" geometry, where a sharp metal tip is held ~ 1nm above a metallic substrate (with the sample located in the gap), effectively forming a vertically oriented coupled dipole—or bowtie-like plasmonic antenna (160–165). Recently, researchers have used the large field enhancements inherent in this geometry for full hyperspectral imaging of molecular and polymer layers, collecting a Raman spectrum at each image pixel with sub-second integration times (164). The technique has also been used to investigate single molecules (118–120), graphene surfaces (164), Hydrogen Bonding between Adenine and Thymine Adsorbed on Au (111) (166), as well as DNA absorbed on a Au substrate via TERS (126).

This tip-substrate gap geometry, however, is rather limiting. First, it requires a metallic substrate (it is often preferable to have biological samples on a dielectric substrate) and a very small tip-substrate gap, meaning only very thin samples (e.g., molecules) can be investigated, precluding many classes of nanoscale structures and devices. Secondly, highest enhancements result only when the light is polarized in the "z" direction normal to the sample surface, restricting one of the most useful variable parameters in Raman studies—the probed polarization.

To avoid these constraints, we and others have fabricated dipole-like optical antennas on scan probes, effectively creating the nano-gap directly on the tip (Figure 13.4), allowing for ultra-high enhancement without the need for metallic substrates and tip-substrate "gap-mode" experimental geometries (167–170). Unfortunately, experiments based on this type of scan probe had themselves been limited by a lack of reproducible tip fabrication techniques. But this issue has been recently overcome (168), and we have used these novel fabrication methods to create reproducible bowtie tips that yield repeatable tip-enhanced Raman scattering (TERS) signals similar to those obtained with the tip-substrate gap-mode geometry (168). Bowtie-based scan probes, however, also have a downside: namely, their relatively flat structure results in comparatively poor topographic imaging quality and large tip-sample interaction forces.

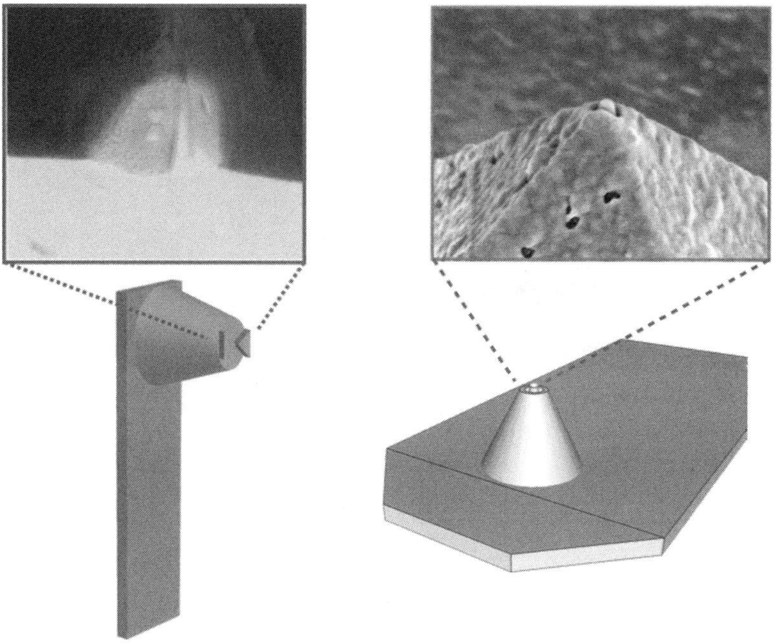

Figure 13.4 Two examples of optical antennas fabricated on scanning probe tips (bowtie antenna, left, and coaxial antenna, right). These enable the reproducible enhancement and localization of optical near fields for performing optical spectroscopy techniques with resolutions near, or even below, 10 nm.

This has led to the creation of a new generation of probes based on the coaxial (coax) plasmonic antenna concept (171). A coax aperture can be thought of as a metal-insulator-metal slit or antenna aperture wrapped back onto itself, yielding low-loss propagation of optical-frequency modes (172–175). Recent simulations have suggested that the optical coax geometry can be relatively broadband and efficient at coupling to far-field radiation (172, 173, 176). By fabricating the center conductor with a tapered end, a coax-based probe combines the advantages of plasmonically coupled antennas (polarization flexibility; larger fields) with sharp tips, and with the additional potential benefit of back-illumination (transmission mode) of the coax tip for virtually "background-free" imaging (177, 178).

This geometry enabled full hyperspectral tip-enhanced Raman imaging of carbon nanotubes (CNTs) with a resolution below 20nm. But before describing the coax-tip-based hyperspectral nanoimaging and its requirements in more detail, a couple additional geometries must be mentioned that also have the potential to perform hyperspectral TERS imaging. De Angelis et al. (178) combined a photonic crystal nanofabricated on the cantilever with a sharp tip in the center of the photonic crystal defect in order to efficiently couple highly coherent modes into a sharp adiabatically tapered Au tip (179–181). Raschke's group has efficiently coupled far-field light to plasmonic modes on an adiabatically tapered metalized tip via a nanofabricated grating coupler located on the tip but many microns from the apex (182, 183). These geometries have produced large, background-free TERS spectra and have the potential to enable hyperspectral TERS imaging.

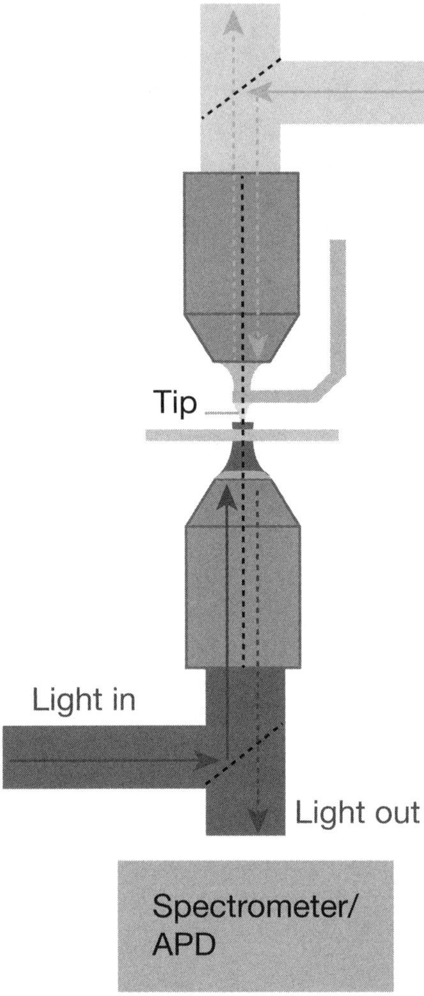

Figure 13.5 Example of a near-field optical microscopy setup. An optical antenna tip can be excited from the bottom (red path) or the top (green path), then the sample can be scanned. The signal can be collected by microscope objectives on the top or bottom depending on the antenna tip geometry, then directed onto a spectrometer or a point detector (e.g., avalanche photo diode).

A typical optical setup used for performing tip-enhanced Raman spectroscopy and imaging is shown in Figure 13.5. The optical axes of two opposing microscope objectives can be aligned and between them is placed the sample as well as the AFM attachment with the tip. This enables the excitation of the optical antenna either from below or from the backside of the cantilever, and the signal collection can be done again through either objective. For the work described below, the coaxial tips were excited from the bottom through a fused silica substrate using a linearly polarized HeNe laser focused by a 100x 1.4NA oil objective. The scattered Raman light was collected through the same objective.

We demonstrated the hyperspectral imaging capabilities of coaxial tips (171) by imaging carbon nanotubes that were grown directly on fused silica cover slips. CNTs

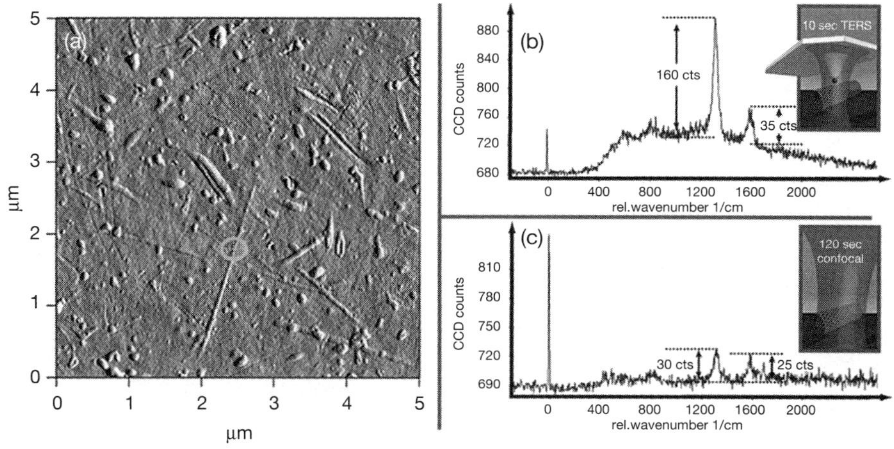

Figure 13.6 Tip-enhanced Raman spectroscopy on CNTs. a) Topographic error-signal image of CNTs grown on a fused silica cover slip acquired with the coaxial antenna scanning probe. b) Tip-enhanced Raman spectrum of the CNT taken at the marked position in a), integrated over 10 sec. The 10-second-long spectrum clearly shows the D and G modes at 1308 and 1586 wavenumbers along with a PL background originating from the Au antenna. c) Confocal Raman spectrum at the equivalent position. An integration time of 120 sec was necessary to achieve a similar G-mode signal.

have been investigated extensively via near-field optical spectroscopy and are therefore an ideal test sample (184–187). Figure 13.6a shows a topographic AFM scan acquired with a coax antenna probe. The lateral resolution is on the order of 20nm, consistent with the radius of curvature of the central coax pin.

The sample was then positioned so that the tip was placed on a CNT (marker in Figure 13.6a) and the Raman spectrum was recorded (Figure 13.6b). The ten-second-long spectrum clearly shows the D and G modes at 1308 and 1586 wavenumbers along with a PL background originating from the Au antenna. The AFM tip was then retracted to take a far-field confocal spectrum (Figure 13.6c) of the same area. The data shown here give a G-mode signal enhancement of 17 (D-mode enhancement value is 64). Note that there are no area or volume correction factors used in obtaining this number because of uncertainties in active area estimates and the corresponding controversial estimates of "true" field enhancement—i.e., this is the most conservative calculation of enhancement (and the one that ultimately matters most). The nanofabricated coaxial antenna tip yield approached 100 percent—all of the tips that look good in post-FIB SEM images demonstrated significant Raman enhancement. The coax tips were also sufficiently durable to support many hours of contact-mode imaging.

Increasing the laser power to 150μW led to satisfactory signal-to-noise ratios in the Raman spectra for integration times as low as 50 ms, enabling practical Raman mapping on dielectric substrates. Based on this, it was possible to acquire a full Raman Spectrum at each pixel of a near-field imaging scan. While performing a slow 1.5 μm by 1.5 μm AFM scan over our sample (see Figure 13.7a topography image), spectra were acquired at each pixel, allowing the detection of small variations in the sample chemistry and composition reflected by small changes in peak positions and intensities. A typical 0.1 s spectrum over the CNTs is shown in Figure 13.7b.

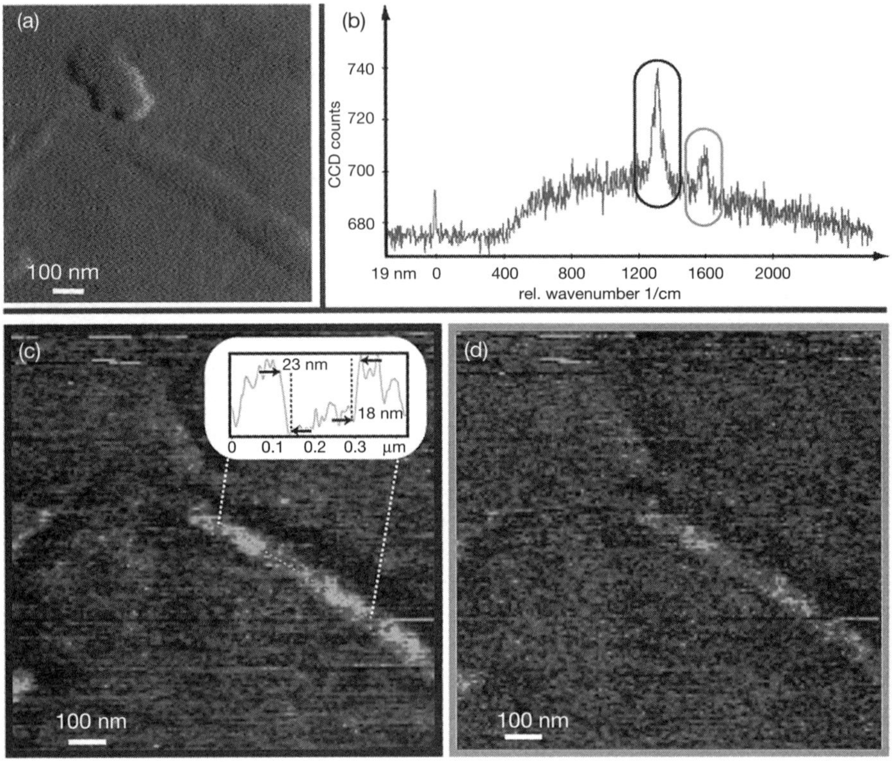

Figure 13.7 Raman spectral mapping of CNTs. a) Topography of mapped region (error signal). b) Example of 0.1 sec Raman spectrum. D and G peaks are shown with a blue and green circle, respectively. c) and d) are the corresponding intensity maps of the D and G peaks. The inset in c) is a line-scan along the CNT structure demonstrating an optical resolution of ~ 20 nm.

Once an entire hyperspectral image is collected, using software to map specific spectral regions by integrating signal intensity over various spectral windows yields similar information to what is collected via the spectral imaging technique described previously. In that case, however, a physical optical filter is used to select the equivalent spectral window and all photons that are transmitted through the filter are collected and summed by a broadband point detector. The advantage of obtaining this information via hyperspectral imaging is that: a) fluctuations in background signal can be filtered out via background subtraction, and therefore intensity fluctuation in the maps correspond to actual intensity fluctuation of the Raman peak (and not fluctuations in background); b) one does not need to know in advance what spectral regions will be of most interest (i.e., what chemical information will be contained in the sample); c) a large amount of chemical information is obtained in one scan; and d) collecting a full spectrum allows one to be very sensitive to small variations in spectral peak positions, which correspond to chemical modifications of a certain bond. This information can be extracted with the software and its spatial dependence/distribution mapped out, which is impossible with spectral imaging.

In another experiment, a CNT sample was heterogeneously "functionalized" with random absorbates (171, 188, 189). Since this technique makes it possible to

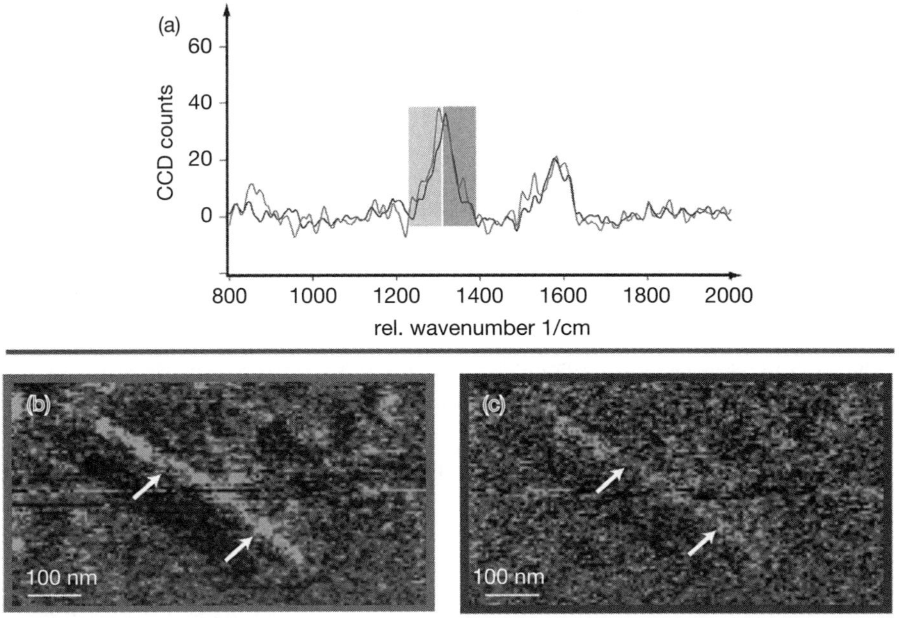

Figure 13.8 Fine structure mapping of CNTs. a) represents two Raman spectra taken at two different positions over a CNT. The D mode's peak position varies over a range of ~20 wave numbers along the structure. The lower energy D-peak, marked with the orange square, was mapped out in b) whereas the higher energy peak (blue square) is shown in c). The white arrows point out regions where the lower energy map seems to have low signal while the higher energy peak demonstrates stronger contrast in these regions.

distinguish between shifts in peak position and in peak intensity, small but statistically significant and spatially correlated position-dependent variations in the D-mode peak positions were observed after leaving the CNT samples in ambient conditions for four days, as shown in Figure 13.8. It should also be noted that in the freshly prepared sample we did not observe systematic, spatially correlated shifts of the D-peak position.

Nano-Raman and Biology: Looking toward the Future

Through the use of specifically engineered optical antennas—and the ability to reproducibly fabricate them at the end of scan probes—the ability to perform optical spectroscopy techniques with spatial resolution approaching the single digit nm scale is now being achieved. However, hyperspectral nano-Raman measurements have only been demonstrated on samples consisting of carbon nanostructures and laser dye molecules—i.e., samples with unusually large Raman scattering cross sections. A majority of samples will yield weaker Raman signal, and thus, to date, nanoscale hyperspectral Raman imaging on biological samples has not been realized.

A number of large challenges remain before the technique becomes a reliable tool in biology, but significant research efforts are underway and recent advances are quite promising. For example, theoretical models of novel nano-Raman tips that efficiently marry the concepts of ideal photonic waveguides, plasmonic waveguides, and optical antennas show they are quite capable of achieving enhancement values large enough to enable bio sample investigations. Additionally, very recent progress in modern scanning probe techniques (190) promise to allow for fast scanning of soft materials in liquid environments, such as living cells. The combination of these ultrasensitive AFM techniques with optical probes will enable in-depth investigation of cell membrane compositions as well as biochemical processes through cell membranes, which are crucial processes in the life cycle. Ultimately, nano-Raman hyperspectral imaging will have a large impact since it provides a label-free method for investigating previously inaccessible parameters in biological samples.

Advanced Chemometric Analysis

As is true for any multivariate technique, for maximum utility Raman spectroscopy needs to be paired with advanced data analysis methods extracting information from the entire spectral range. In the case of Raman spectroscopic imaging, the main goal usually will be twofold: understanding and unlocking details from the spectral data and at the same time providing a means to reconstruct meaningful images. Typically huge data sets are acquired in the measurements that are arranged in so-called hyperspectral cubes with two or three spatial dimensions, the spectral dimension, as well as a possible temporal dimension when time-dependent or dynamic studies are conducted. So another goal can be seen in reducing or compressing the amount of data while retaining and accentuating meaningful information—i.e., spectral variance or contrast. Broadly speaking, two classes of multivariate methods are distinguished: supervised methods based on a priori knowledge that require calibration or training sets of data to build a reliable model of the system, and blind or non-supervised data analysis methods that need little or no a priori information about the system and extract all relevant information from the system itself. It is beyond the scope of this review to discuss the principle and merits of the various methods used and we would like to refer the interested reader to the literature for commonly used techniques such as partial least squares (191), discriminant analysis (192), principal component analysis (193), k-means cluster analysis (194), (agglomerative) hierarchical cluster analysis (195, 196), divisive correlation cluster (197), vertex component analysis (198, 199), and artificial neural networks (200). Instead, we give a specific example and introduce our multivariate method for dissecting imaging data of mutant plant cell walls that we already discussed earlier. In fact, a vigorous multivariate approach is required to untangle the chemically heterogeneous and structurally complex plant cell wall. Therefore, we implemented a new non-supervised image analysis strategy for a detailed characterization that can elucidate chemical, structural, and spatial information of plant cell walls. Spectral images of plant cell walls present a stern challenge to multivariate data mining methods for the following reasons: a multicomponent system is present at any given image pixel/spectrum and often the exact number of components is unknown; pure components spectra are unknown and the spectrum of an isolated component (e.g., lignin) would be different from that in the native environment due to loss of hydrogen bonds, change of three-dimensional configuration and possible chemical alteration during extraction; owing to similarities of monomers, cellulose and hemicelluloses have similar spectral features and therefore high spectral overlap; spectral weights of pure components are also unknown and difficult to estimate due to the presence of a large number of components and severe spectral overlap; pure components could be spatially collinear at the submicron to micron level and identifying them as independent entities is difficult; auto-fluorescence from impurities as well as components such as lignin contributes to the noise and shape of the background significantly; and Raman spectra have poor signal-to-noise ratio due to practical constraints such as short integration time that is required to avoid thermal and photo damage to the sample. As a consequence, supervised methods are unsuitable for multivariate analysis of cell wall images because the prediction accuracy of the model depends on the extent of the similarity of the training set to the actual system and it would be difficult to build a reasonable training set that captures the essence of the chemical and structural complexity of the system. Our non-supervised method combines k-means clustering and spectral entropy minimization with Raman microspectroscopy to elucidate compositional and structural details of plant cell walls. The image analysis procedure consists of three main steps: spectral preprocessing, stepwise clustering, and estimation of spectral profiles of pure components and their weights. Initially, the spectral images are conditioned for the subsequent steps. Generally poor (i.e., low) signal-to-noise ratio Raman spectra are first subjected to wavelet transformation to mitigate the noise contribution. Wavelet analysis has been shown to remove noise contribution with minimal deterioration to the signal by decomposing the original spectrum into a high frequency part called "details" and a low frequency part called "approximations" with a preselected wavelet function. The high-frequency component contains the noise contribution to the signal while the low-frequency component carries the information content. The low-frequency component is subsequently further decomposed until an appropriate level of separation between noise and signal has been achieved. The essentially noise-free signal is constructed by mixing appropriate amounts of high- and low-frequency coefficients at each level. Denoised spectra are then converted into their second derivatives with the Savitzky-Golay second derivative method to remove fluorescence contributions. The second derivative of a spectrum diminishes the slow changing fluorescence contributions while preserving the signatures of Raman peaks. Although the Savitzky-Golay procedure has been shown to remove noise while calculating second derivatives, the effective removal of noise requires applying larger filter windows which in turn would broaden the Raman features and result in the loss of finer spectral information content. By contrast, noise filtering with wavelet prior to the Savitzky-Golay second derivative method gives second derivative spectra with low noise while retaining finer Raman features. Next, the second derivative spectral matrix is decomposed with principal component analysis to compress the data matrix into a much smaller size. Principal component analysis further removes noise and other undeterministic contributions while conserving the systematic variance in the system. The reduced matrix is then introduced to the second step, which is the stepwise clustering with k-means clustering. K-means cluster analysis provides an unsupervised approach that clusters objects into different groups, provided that the user defines the desired number of clusters. At first, the reduced matrix is partitioned into two clusters to separate cell wall spectra from the redundant spectra that arise from the lumen and the second derivative spectra of cell walls are separated from those of lumen and the latter set is discarded.

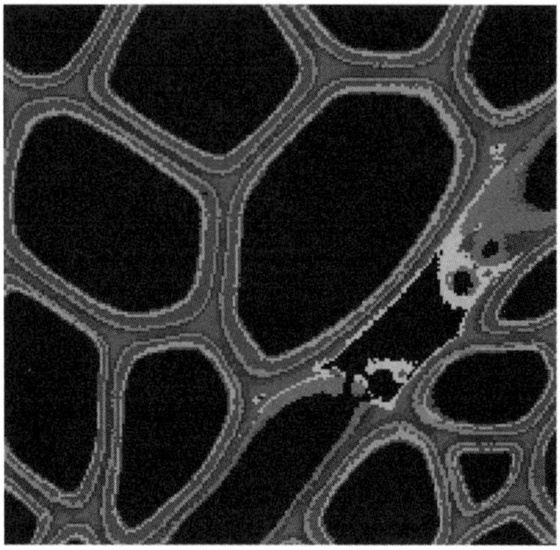

Figure 13.9 Reconstructed image of a xylem cross-section of poplar wood obtained by preprocessing and k-means cluster analysis. See insert for color version.

The cell wall set is subsequently subjected first to principal component analysis decomposition and then k-means clustering with the desired number of classes. The cell wall image is reconstructed at this stage with cluster identities instead of (univariate) peak intensity or peak area to identify and visualize the natural spatial segregation of the image. This clustering enables the identification of different sub-layers in the cell walls of the images. See Figure 13.9 for an example of a reconstructed clustered image for a xylem cross-section of poplar wood. The final step aims at identifying and quantifying the factors that contribute to the partitioning of the image. Spectral identities of the clusters are calculated by averaging all the spectra of each cluster. These average spectra are then subjected to pure spectral estimation. Spectral entropy minimization methodology can be used to recover pure components spectra without assuming their functional forms. The main advantage of the entropy minimization method is that no knowledge of the exact number of pure components and/or estimation of their spectral weights is required. Both these parameters are difficult to obtain for cell wall images. The entropy minimization method relies on the fact that pure spectra have lower entropies than their mixtures. Therefore, the entropy minimization method aims at finding the spectral profile that has the smallest entropy in space that is defined by the mixture spectra. A detailed description of the method is given in the literature (201, 202). In our case, pure components spectra are estimated by spectral entropy minimization criteria with simulated annealing optimization (203). This way we are able to estimate and recover full spectra of native lignin in situ without ambiguity. This is not possible with traditional data analysis methods because of significant spectral overlap with carbohydrate signals. We have demonstrated the ability of our method to determine and distinguish native lignin composition of different plant species in situ without physical extraction (204). Specifically, we determined the composition of native lignin of poplar, Arabidopsis, and Miscanthus, as well as a change in lignin composition upon transgenic suppression of 4-coumarate-CoA ligase in

P. trichocarpa. Secondly, we retrieve a pure spectral estimate of carbohydrates—i.e., both cellulose and hemicelluloses. Spatial segregation or spatially distinct distribution patterns of carbohydrates and lignin across the cell walls makes possible the recovery of pure lignin spectra without spectral contamination from the carbohydrate Raman signals, rendering the structural and quantitative analysis of lignin relatively straightforward. Unfortunately, this is not the case for the carbohydrate analysis. The spatial correlation of cellulose with hemicelluloses makes the spectral recovery of either component in pure form practically unfeasible without additional extrinsic information. The carbohydrate analysis is further complicated due to the fact that Raman spectra of cellulose and hemicelluloses bear significant similarities. However, the recovery of the carbohydrate spectrum free of lignin contribution allows us to quantify and analyze cellulose and hemicelluloses contribution more accurately with extrinsic information on those polymers as we demonstrated in our analysis on Arabidopsis mutants (205). In conclusion, it should be mentioned that the presented approach is more general such that it can be combined with any multivariate microscopic method (e.g., IR, mass spectrometry, fluorescence, etc.) to analyze natural or artificial images.

OUTLOOK

Despite the fact that this review could not deliver a complete cross-sectional overview of all the activities and directions in the field, we hope it has become clear that Raman spectroscopic imaging has matured into a powerful and versatile microanalytical tool for biological research. Innovative developments are emerging that will translate into an increasing number of applications and will further consolidate Raman imaging in its own right. We believe, however, that this modality will unfold its full potential when used in conjunction or hyphenation with other complementary techniques and methods, such as fluorescence microscopy, electron and X-ray microscopy and tomography, atomic force microscopy, optical tweezers, and microfluidics. In fact, some of these combinations have already been explored and implemented and we anticipate this trend will intensify in the future. In addition, ongoing efforts into establishing automated or hands-off data analysis routines will make Raman imaging more user-friendly and broadly accessible also to non-experts and contribute towards overcoming possible reluctance or latent antagonism in using this spectroscopic method. Overall, this should lead to Raman imaging being routinely added to the biological imaging methods toolkit.

ACKNOWLEGMENTS

The authors would like to acknowledge the support and scientific discussions from our colleagues at the Molecular Foundry, Lawrence Berkeley National Lab, and from our colleagues at the Energy Biosciences Institute (EBI), as well as from our other collaborators. We specifically thank Vincent L. Chiang, Manfred Auer, Purbasha Sarkar, Andrew Carroll, Adisorn Chaibang, and Adam Schwartzberg.

REFERENCES

1. Smekal, A., *Zur Quantentheorie der Dispersion*. Die Naturwissenschaften, 1923. **43**: p. 873–875.
2. Raman, C.V. and K.S. Krishnan, *A New Type of Secondary Radiation*. Nature, 1928. **121**: p. 501–502.
3. Landsberg, G. and L. Mandelstam, *Eine neue Erscheinung bei der Lichtzerstreuung in Krystallen*. Die Naturwissenschaften, 1928. **28**: p. 557–558.
4. Chalmers, J.M. and P.R. Griffiths, eds. *Handbook of Vibrational Spectroscopy*. 2002, John Wiley & Sons Ltd: Chichester.
5. Sasic, S., ed. *Pharmaceutical Applications of Raman Spectroscopy*. Wiley Series on Technologies for the Pharmaceutical Industry, ed. S. Ekins. 2007, Wiley.
6. Siebert, F. and P. Hildebrandt, eds. *Vibrational Spectroscopy in Life Science*. 2007, Wiley-VCH Verlag GmbH & Co. KGaA: Weinheim.
7. Diem, M., P.R. Griffiths, and J.M. Chalmers, eds. *Vibrational Spectroscopy for Medical Diagnosis*. 2008, J.Wiley-Interscience.
8. Lasch, P. and J. Kneipp, eds. *Biomedical Vibrational Spectroscopy*. 2008, Wiley-VCH Verlag GmbH & Co. KGaA: Weinheim.
9. Salzer, R. and H.W. Siesler, eds. *Infrared and Raman Spectroscopic Imaging*. 2009, WILEY-VCH Verlag GmbH & Co. KGaA: Weinheim.
10. Carter, E.A., et al., *Vibrational spectroscopic mapping and imaging of tissues and cells*. Biophys. Rev., 2009. **1**: p. 95–103.
11. Chan, J., et al., *Raman spectroscopy and microscopy of individual cells and cellular components*. Laser & Photonics Reviews, 2008. **2**(5): p. 325–349.
12. Matthaus, C., et al., *Raman and Infrared Microspectral Imaging of Mitotic Cells*. Appl. Spectrosc., 2006. **60**: p. 1–8.
13. Chernenko, T., et al., *Label-Free Raman Spectral Imaging of Intracellular Delivery and Degradation of Polymeric Nanoparticle Systems*. ACS Nano, 2009. **3**(11): p. 3552–3559.
14. Meister, K., et al., *Label-Free Imaging of Metal–Carbonyl Complexes in Live Cells by Raman Microspectroscopy*. Angew. Chem. Int. Ed., 2010. **49**(19): p. 3310–3312.
15. Mariani, M.M., et al., *Impact of fixation on* in vitro *cell culture lines monitored with Raman spectroscopy*. The Analyst, 2009. **134**(6): p. 1154–1161.
16. Wood, B.R., et al., *Raman imaging of hemozoin within the food vacuole of Plasmodium falciparum trophozoites*. FEBS Letters, 2003. **554**(3): p. 247–252.
17. Miljkovic, M., et al., *Label-free imaging of human cells: algorithms for image reconstruction of Raman hyperspectral data sets*. Analyst, 2010. **135**(8): p. 2002–2013.
18. Hermelink, A., et al., *Phenotypic heterogeneity within microbial populations at the single-cell level investigated by confocal Raman microspectroscopy*. Analyst, 2009. **134**(6): p. 1149–1153.
19. Lasch, P., A. Hermelink, and D. Naumann, *Correction of axial chromatic aberrations in confocal Raman microspectroscopic measurements of a single microbial spore*. The Analyst, 2009. **134**(6): p. 1162–1170.
20. Kazanci, M., et al., *Bone osteonal tissues by Raman spectral mapping: Orientation-composition*. J. Struct. Biol., 2006. **156**(3): p. 489–496.
21. Kazanci, M., et al., *Raman imaging of two orthogonal planes within cortical bone*. Bone, 2007. **41**(3): p. 456–461.
22. Kammer, M., et al., *Spatially resolved determination of the structure and composition of diatom cell walls by Raman and FTIR imaging*. Analytical and Bioanalytical Chemistry, 2010. **398**(1): p. 509–517.

23. Schopf, J.W., et al., *Laser-Raman imagery of Earth's earliest fossils*. Nature, 2002. **416**(6876): p. 73–76.
24. Schmidt, M., et al., In Situ *Imaging of Barnacle (Balanus amphitrite) Cyprid Cement Using Confocal Raman Microscopy*. J. Adhes., 2009. **85**(2): p. 139–151.
25. Schulte, F., et al., *Chemical Characterization and Classification of Pollen*. Anal. Chem., 2008. **80**(24): p. 9551–9556.
26. Schulte, F., et al., *Characterization of Pollen Carotenoids with* in situ *and High-Performance Thin-Layer Chromatography Supported Resonant Raman Spectroscopy*. Anal. Chem., 2009. **81**(20): p. 8426–8433.
27. Baranska, M., et al., *Tissue-specific accumulation of carotenoids in carrot roots*. Planta, 2006. **224**(5): p. 1028–1037.
28. Harrington, M.J., et al., *Iron-Clad Fibers: A Metal-Based Biological Strategy for Hard Flexible Coatings*. Science, 2010. **328**(5975): p. 216–220.
29. Gierlinger, N. and M. Schwanninger, *The potential of Raman microscopy and Raman imaging in plant research*. Spectrosc. Int. J., 2007. **21**: p. 69–89.
30. Burton, R.A., M.J. Gidley, and G.B. Fincher, *Heterogeneity in the chemistry, structure and function of plant cell walls*. Nat. Chem. Biol., 2010. **6**(10): p. 724–732.
31. Sarkar, P., E. Bosneaga, and M. Auer, *Plant cell walls throughout evolution: towards a molecular understanding of their design principles*. J. Exp. Bot., 2009. **60**(13): p. 3615–3635.
32. Gierlinger, N. and M. Schwanninger, *Chemical Imaging of Poplar Wood Cell Walls by Confocal Raman Microscopy*. Plant Physiol., 2006. **140**(4): p. 1246–1254.
33. Agarwal, U.P., *Raman imaging to investigate ultrastructure and composition of plant cell walls: distribution of lignin and cellulose in black spruce wood (Picea mariana)*. Planta, 2006. **224**(5): p. 1141–1153.
34. Gierlinger, N., et al., *Cellulose microfibril orientation of Picea abies and its variability at the micron-level determined by Raman imaging*. J. Exp. Bot., 2010. **61**(2): p. 587–595.
35. Schmidt, M., et al., *Label-free* in situ *imaging of lignification in the cell wall of low lignin transgenic Populus trichocarpa*. Planta, 2009. **230**(3): p. 589–597.
36. Schmidt, M., et al., *Label-free* in situ *Imaging of Lignification in Plant Cell Walls*. J. Vis. Exp., 2010(45): p. e2064.
37. Schmidt, M., et al., *Polarized infrared microspectroscopy of single spruce fibers: Hydrogen bonding in wood polymers*. Biopolymers, 2006. **83**(5): p. 546–555.
38. Schmidt, M., et al., *Raman imaging of cell wall polymers in Arabidopsis thaliana*. Biochem. Biophys. Res. Comm., 2010. **395**(4): p. 521–523.
39. Perera, P.N., et al., *Unraveling cell wall structure: In situ compositional analysis of Arabidopsis thaliana with Raman microspectroscopy and chemometrics*. submitted.
40. Stiles, P.L., et al., *Surface-Enhanced Raman Spectroscopy*. Annu. Rev. Anal. Chem., 2008. **1**(1): p. 601–626.
41. Kneipp, K., et al., *Detection and identification of a single DNA base molecule using surface-enhanced Raman scattering (SERS)*. Physical Review E, 1998. **57**(6): p. R6281–R6284.
42. Kneipp, K., et al., *Single-molecule detection of a cyanine dye in silver colloidal solution using near-infrared surface-enhanced Raman scattering*. Applied Spectroscopy, 1998. **52**(2): p. 175–178.
43. Kneipp, K., et al., *Single molecule detection using surface-enhanced Raman scattering (SERS)*. Physical Review Letters, 1997. **78**(9): p. 1667–1670.
44. Nie, S.M. and S.R. Emery, *Probing single molecules and single nanoparticles by surface-enhanced Raman scattering*. Science, 1997. **275**(5303): p. 1102–1106.

45. Fleischm.M, P.J. Hendra, and Mcquilla.Aj, *Raman-Spectra of Pyridine Adsorbed at a Silver Electrode*. Chemical Physics Letters, 1974. **26**(2): p. 163–166.
46. Jeanmaire, D.L. and R.P. Vanduyne, *Surface Raman Spectroelectrochemistry.1. Heterocyclic, Aromatic, and Aliphatic-Amines Adsorbed on Anodized Silver Electrode*. Journal of Electroanalytical Chemistry, 1977. **84**(1): p. 1–20.
47. Xu, H.X., et al., *Electromagnetic contributions to single-molecule sensitivity in surface-enhanced Raman scattering*. Physical Review E, 2000. **62**(3): p. 4318–4324.
48. Stiles, P.L., et al., *Surface-Enhanced Raman Spectroscopy*. Annual Review of Analytical Chemistry, 2008. **1**: p. 601–626.
49. Willets, K.A. and R.P. Van Duyne, *Localized surface plasmon resonance spectroscopy and sensing*, in Annual Review of Physical Chemistry. 2007, Annual Reviews: Palo Alto. p. 267–297.
50. Moskovits, M., *Surface-enhanced Raman spectroscopy: a brief retrospective*. Journal of Raman Spectroscopy, 2005. **36**(6–7): p. 485–496.
51. Zayak, A.T., et al., *Chemical Raman Enhancement of Organic Adsorbates on Metal Surfaces*. Physical Review Letters, 2011. **106**(8): p. 083003.
52. Jensen, L., C.M. Aikens, and G.C. Schatz, *Electronic structure methods for studying surface-enhanced Raman scattering*. Chemical Society Reviews, 2008. **37**(5): p. 1061–1073.
53. Morton, S.M. and L. Jensen, *Understanding the Molecule-Surface Chemical Coupling in SERS*. Journal of the American Chemical Society, 2009. **131**(11): p. 4090–4098.
54. Lee, S.E. and L.P. Lee, *Biomolecular plasmonics for quantitative biology and nanomedicine*. Current Opinion in Biotechnology, 2010. **21**(4): p. 489–497.
55. Sujith, A., et al., *Imaging the cell wall of living single yeast cells using surface-enhanced Raman spectroscopy*. Anal. Bioanal. Chem., 2009. **394**(7): p. 1803–1809.
56. Ivleva, N.P., et al., *Label-Free* in Situ *SERS Imaging of Biofilms*. J. Phys. Chem. B, 2010. **114**(31): p. 10184–10194.
57. Ramya, S., et al., *Detection of algae and bacterial biofilms formed on titanium surfaces using micro-Raman analysis*. Appl. Surf. Sci., 2010. **256**(16): p. 5108–5115.
58. Zavaleta, C.L., et al., *Multiplexed imaging of surface enhanced Raman scattering nanotags in living mice using noninvasive Raman spectroscopy*. Proc. Natl. Acad. Sci. USA, 2009.
59. Wang, Y., et al., *Biocompatibility and Biodistribution of Surface-Enhanced Raman Scattering Nanoprobes in Zebrafish Embryos:* In vivo *and Multiplex Imaging*. ACS Nano, 2010. **4**(7): p. 4039–4053.
60. Evans, C.L. and X.S. Xie, *Coherent Anti-Stokes Raman Scattering Microscopy: Chemical Imaging for Biology and Medicine*, in Annual Review of Analytical Chemistry. 2008, Annual Reviews: Palo Alto. p. 883–909.
61. Cheng, J.X., et al., *Laser-scanning coherent anti-stokes Raman scattering microscopy and applications to cell biology*. Biophysical Journal, 2002. **83**(1): p. 502–509.
62. Potma, E.O. and X.S. Xie, *Theory of spontaneous and coherent Raman Scattering*, in *Handbook of Biomedical Nonlinear Optical Microscopy*, B.R. Masters and P.T.C. So, Editors. 2008, Oxford University Press: New York. p. 164–186.
63. Duncan, M.D., J. Reintjes, and T.J. Manuccia, *SCANNING COHERENT ANTI-STOKES RAMAN MICROSCOPE*. Optics Letters, 1982. **7**(8): p. 350–352.
64. Ganikhanov, F., et al., *High-sensitivity vibrational imaging with frequency modulation coherent anti-Stokes Raman scattering (FM CARS) microscopy*. Optics Letters, 2006. **31**(12): p. 1872–1874.

65. Potma, E.O., C.L. Evans, and X.S. Xie, *Heterodyne coherent anti-Stokes Raman scattering (CARS) imaging*. Optics Letters, 2006. **31**(2): p. 241–243.
66. Lin-Vien, D., et al., *The handbook of infrared and Raman characteristic frequencies of organic molecules*. 1991, San Diego: Academic.
67. Zumbusch, A., G.R. Holtom, and X.S. Xie, *Three-dimensional vibrational imaging by coherent anti-Stokes Raman scattering*. Physical Review Letters, 1999. **82**(20): p. 4142–4145.
68. Nan, X.L., E.O. Potma, and X.S. Xie, *Nonperturbative chemical imaging of organelle transport in living cells with coherent anti-stokes Raman scattering microscopy*. Biophysical Journal, 2006. **91**(2): p. 728–735.
69. Hellerer, T., et al., *Monitoring of lipid storage in Caenorhabditis elegans using coherent anti-Stokes Raman scattering (CARS) microscopy*. Proceedings of the National Academy of Sciences of the United States of America, 2007. **104**(37): p. 14658–14663.
70. Rakic, B., et al., *Peroxisome proliferator-activated receptor alpha antagonism inhibits hepatitis C virus replication*. Chemistry & Biology, 2006. **13**(1): p. 23–30.
71. Nan, X.L., et al., *Intracellular imaging of HCV RNA and cellular lipids by using simultaneous two-photon fluorescence and coherent anti-Stokes Raman scattering microscopies*. Chembiochem, 2006. **7**(12): p. 1895–1897.
72. Murugkar, S., et al., *Chemically specific imaging of cryptosporidium oocysts using coherent anti-Stokes Raman scattering (CARS) microscopy*. J. Microsc., 2009. **233**(2): p. 244–250.
73. Evans, C.L., et al., *Chemical imaging of tissue* in vivo *with video-rate coherent anti-Stokes Raman scattering microscopy*. Proceedings of the National Academy of Sciences of the United States of America, 2005. **102**(46): p. 16807–16812.
74. Young, B. and J.W. Heath, *Wheater's Functional Histology: A Text and Colour Atlas*. 2000, London: Churchill Livingstone.
75. Toytman, I., et al., *Wide-field coherent anti-Stokes Raman scattering microscopy with non-phase-matching illumination*. Optics Letters, 2007. **32**(13): p. 1941–1943.
76. Fu, Y., et al., *Coherent anti-stokes Raman scattering imaging of myelin degradation reveals a calcium-dependent pathway in lyso-PtdCho-induced demyelination*. Journal of Neuroscience Research, 2007. **85**(13): p. 2870–2881.
77. Evans, C.L., et al., *Chemically-selective imaging of brain structures with CARS microscopy*. Optics Express, 2007. **15**(19): p. 12076–12087.
78. Schie, I.W., et al., *Simultaneous forward and epi-CARS microscopy with a single detector by time-correlated single photon counting*. Optics Express, 2008. **16**(3): p. 2168–2175.
79. Cheng, J.X., L.D. Book, and X.S. Xie, *Polarization coherent anti-Stokes Raman scattering microscopy*. Optics Letters, 2001. **26**(17): p. 1341–1343.
80. Weeks, T., S. Wachsmann-Hogiu, and T. Huser, *Raman Microscopy based on Doubly-Resonant Four-Wave Mixing (DR-FWM)*. Optics Express, 2009. **17**(19): p. 17044–17051.
81. Weeks, T., et al., *Signal generation and Raman-resonant imaging by non-degenerate four-wave mixing under tight focusing conditions*. Journal of Biophotonics, 2010. **3**(3): p. 169–175.
82. Min, W., et al., *Triple-Resonance Coherent Anti-Stokes Raman Scattering Microspectroscopy*. ChemPhysChem, 2009. **10**(2): p. 344–347.
83. Woodbury, E.J. and W.K. Ng, *RUBY LASER OPERATION IN NEAR IR*. Proceedings of the Institute of Radio Engineers, 1962. **50**(11): p. 2367.
84. Owyoung, A., *SENSITIVITY LIMITATIONS FOR CW STIMULATED RAMAN-SPECTROSCOPY*. Optics Communications, 1977. **22**(3): p. 323–328.

85. Levine, B.F., C.V. Shank, and J.P. Heritage, *SURFACE VIBRATIONAL SPECTROSCOPY USING STIMULATED RAMAN-SCATTERING*. IEEE Journal of Quantum Electronics, 1979. **15**(12): p. 1418–1432.
86. Levenson, M.D. and S.S. Kano, *Introduction to Nonlinear Laser Spectroscopy*. 1988, San Diego: Academic Press.
87. Kukura, P., D.W. McCamant, and R.A. Mathies, *Femtosecond stimulated Raman spectroscopy*. Annual Review of Physical Chemistry, 2007. **58**: p. 461–488.
88. Ploetz, E., et al., *Femtosecond stimulated Raman microscopy*. Applied Physics B-Lasers and Optics, 2007. **87**(3): p. 389–393.
89. Freudiger, C.W., et al., *Label-Free Biomedical Imaging with High Sensitivity by Stimulated Raman Scattering Microscopy*. Science, 2008. **322**(5909): p. 1857–1861.
90. Saar, B.G., et al., *Video-Rate Molecular Imaging* in Vivo *with Stimulated Raman Scattering*. Science, 2010. **330**(6009): p. 1368–1370.
91. Saar, B.G., et al., *Label-Free, Real-Time Monitoring of Biomass Processing with Stimulated Raman Scattering Microscopy*. Angewandte Chemie-International Edition, 2010. **49**(32): p. 5476–5479.
92. Yakovlev, V.V., et al., *Stimulated Raman photoacoustic imaging*. Proceedings of the National Academy of Sciences of the United States of America, 2010. **107**(47): p. 20335–20339.
93. Wang, M.C., et al., *RNAi screening for fat regulatory genes with SRS microscopy*. Nature Methods, 2011. **8**(2): p. 135–138.
94. Hartschuh, A., et al., *Nanoscale optical imaging of excitons in single-walled carbon nanotubes*. Nano Letters, 2005. **5**(11): p. 2310–2313.
95. Hartschuh, A., et al., *High-resolution near-field Raman microscopy of single-walled carbon nanotubes*. Physical Review Letters, 2003. **90**(9): 095503
96. Hayazawa, N., et al., *Towards atomic site-selective sensitivity in tip-enhanced Raman spectroscopy*. Journal of Chemical Physics, 2006. **125**(24): 244706
97. Deckert-Gaudig, T., E. Bailo, and V. Deckert, *Perspectives for spatially resolved molecular spectroscopy—Raman on the nanometer scale*. Journal of Biophotonics, 2008. **1**(5): p. 377–389.
98. Berweger, S. and M.B. Raschke, *Signal limitations in tip-enhanced Raman scattering: the challenge to become a routine analytical technique*. Analytical and Bioanalytical Chemistry, 2010. **396**(1): p. 115–123.
99. Raschke, M.B., et al., *Apertureless near-field vibrational imaging of block-copolymer nanostructures with ultrahigh spatial resolution*. Chemphyschem, 2005. **6**(10): p. 2197–2203.
100. Huber, A., et al., *Nanoscale resolved infrared probing of crystal structure and of plasmon-phonon coupling*. Nano Letters, 2006. **6**(4): p. 774–778.
101. Betzig, E., et al., *Imaging intracellular fluorescent proteins at nanometer resolution*. Science, 2006. **313**(5793): p. 1642–1645.
102. Hess, S.T., T.P.K. Girirajan, and M.D. Mason, *Ultra-high resolution imaging by fluorescence photoactivation localization microscopy*. Biophysical Journal, 2006. **91**(11): p. 4258–4272.
103. Rust, M.J., M. Bates, and X.W. Zhuang, *Sub-diffraction-limit imaging by stochastic optical reconstruction microscopy (STORM)*. Nature Methods, 2006. **3**(10): p. 793–795.
104. Moerner, W.E., *New directions in single-molecule imaging and analysis*. Proceedings of the National Academy of Sciences, 2007. **104**(31): p. 12596–12602.

105. Pohl, D.W., W. Denk, and M. Lanz, *Optical Stethoscopy—Image Recording with Resolution Lambda/20*. Applied Physics Letters, 1984. **44**(7): p. 651–653.
106. Betzig, E. and J.K. Trautman, *Near-Field Optics: Microscopy, Spectroscopy, and Surface Modification Beyond the Diffraction Limit*. Science, 1992. **257**(5067): p. 189–195.
107. Hecht, B., et al., *Scanning near-field optical microscopy with aperture probes: Fundamentals and applications*. Journal of Chemical Physics, 2000. **112**(18): p. 7761–7774.
108. Paesler, M.A. and P.J. Moyer, *Near-Field Optics: Theory, Instrumentation and Applications*. 1996, New York: Wiley.
109. Zenobi, R. and V. Deckert, *Scanning near-field optical microscopy and spectroscopy as a tool for chemical analysis*. Angewandte Chemie-International Edition, 2000. **39**(10): p. 1746–1756.
110. Suh, Y.D., et al., *Tip-enhanced Raman spectroscopy as a new analytical tool for nanomaterials*. Abstracts of Papers of the American Chemical Society, 2001. **221**: p. U91-U91.
111. Hartschuh, A., N. Anderson, and L. Novotny, *Near-field Raman spectroscopy using a sharp metal tip*. Journal of Microscopy-Oxford, 2003. **210**: p. 234–240.
112. Bouhelier, A., et al., *Near-field second-harmonic generation induced by local field enhancement*. Physical Review Letters, 2003. **90**(1): 013903
113. Ikeda, K., et al., *Resonant hyper-Raman scattering from carbon nanotubes*. Chemical Physics Letters, 2007. **438**(1–3): p. 109–112.
114. Saito, Y., et al., *Polarization measurements in tip-enhanced Raman spectroscopy applied to single-walled carbon nanotubes*. Chemical Physics Letters, 2005. **410**(1–3): p. 136–141.
115. Bharadwaj, P., P. Anger, and L. Novotny, *Nanoplasmonic enhancement of single-molecule fluorescence*. Nanotechnology, 2007. **18**(4): 044017
116. Hayazawa, N., et al., *Near-field Raman imaging of organic molecules by an apertureless metallic probe scanning optical microscope*. Journal of Chemical Physics, 2002. **117**(3): p. 1296–1301.
117. Neacsu, C.C., et al., *Scanning-probe Raman spectroscopy with single-molecule sensitivity*. Physical Review B, 2006. **73**(19): 193406.
118. Zhang, W.H., et al., *Single molecule tip-enhanced Raman spectroscopy with silver tips*. Journal of Physical Chemistry C, 2007. **111**(4): p. 1733–1738.
119. Steidtner, J. and B. Pettinger, *Tip-enhanced Raman spectroscopy and microscopy on single dye molecules with 15 nm resolution*. Physical Review Letters, 2008. **100**(23): 236101
120. Domke, K.F. and B. Pettinger, In Situ *Discrimination between Axially Complexed and Ligand-Free Co Porphyrin on Au(111) with Tip-Enhanced Raman Spectroscopy*. Chemphyschem, 2009. **10**(11): p. 1794–1798.
121. Watanabe, H., et al., *Tip-enhanced near-field Raman analysis of tip-pressurized adenine molecule*. Physical Review B, 2004. **69**(15): 155418
122. Hagen, A., et al., *Exponential decay lifetimes of excitons in individual single-walled carbon nanotubes*. Physical Review Letters, 2005. **95**(19): 197401
123. Hartschuh, A., et al., *Tip-enhanced optical spectroscopy for surface analysis in biosciences*. Surface and Interface Analysis, 2006. **38**(11): p. 1472–1480.
124. Schmid, T., et al., *Towards chemical analysis of nanostructures in biofilms II: tip-enhanced Raman spectroscopy of alginates*. Analytical and Bioanalytical Chemistry, 2008. **391**(5): p. 1907–1916.

125. Deckert-Gaudig, T. and V. Deckert, *Tip-enhanced Raman scattering (TERS) and high-resolution bio nano-analysis-a comparison*. Physical Chemistry Chemical Physics, 2010. **12**(38): p. 12040–12049.
126. Hennemann, L.E., A.J. Meixner, and D. Zhang, *Surface—and tip-enhanced Raman spectroscopy of DNA*. Spectroscopy-an International Journal, 2010. **24**(1–2): p. 119–124.
127. Cialla, D., et al., *Raman to the limit: tip-enhanced Raman spectroscopic investigations of a single tobacco mosaic virus*. Journal of Raman Spectroscopy, 2009. **40**(3): p. 240–243.
128. Budich, C., et al., *Cell wall investigations utilizing tip-enhanced Raman scattering*. Journal of Microscopy-Oxford, 2008. **229**(3): p. 533–539.
129. Richter, M., et al., *Laterally Resolved and Direct Spectroscopic Evidence of Nanometer-Sized Lipid and Protein Domains on a Single Cell*. Small, 2011. **7**(2): p. 209–214.
130. Qian, H., et al., *Visualizing the local optical response of semiconducting carbon nanotubes to DNA-wrapping*. Nano Letters, 2008. **8**(9): p. 2706–2711.
131. Kuhn, S., et al., *Enhancement of single-molecule fluorescence using a gold nanoparticle as an optical nanoantenna*. Physical Review Letters, 2006. **97**(1): 017402
132. Kim, Z.H. and S.R. Leone, *High-resolution apertureless near-field optical imaging using gold nanosphere probes*. Journal of Physical Chemistry B, 2006. **110**(40): p. 19804–19809.
133. Mock, J.J., et al., *Shape effects in plasmon resonance of individual colloidal silver nanoparticles*. Journal of Chemical Physics, 2002. **116**(15): p. 6755–6759.
134. Hoppener, C. and L. Novotny, *Antenna-based optical imaging of single Ca2+ transmembrane proteins in liquids*. Nano Letters, 2008. **8**(2): p. 642–646.
135. Hoppener, C. and L. Novotny, *Imaging of membrane proteins using antenna-based optical microscopy*. Nanotechnology, 2008. **19**(38): 384012
136. Schuck, P.J., et al., *Improving the Mismatch between Light and Nanoscale Objects with Gold Bowtie Nanoantennas*. Physical Review Letters, 2005. **94**(1): p. 017402–4.
137. Muhlschlegel, P., et al., *Resonant Optical Antennas*. Science, 2005. **308**(5728): p. 1607–1609.
138. Wang, H., et al., *Symmetry breaking in individual plasmonic nanoparticles*. PNAS, 2006. **103**(29): p. 10856–10860.
139. Vedantam, S., et al., *A Plasmonic Dimple Lens for Nanoscale Focusing of Light*. Nano Letters, 2009. **9**(10): p. 3447–3452.
140. Kosako, T., Y. Kadoya, and H.F. Hofmann, *Directional control of light by a nano-optical Yagi-Uda antenna*. Nature Photonics, 2010. **4**(5): p. 312–315.
141. Biagioni, P., et al., *Cross Resonant Optical Antenna*. Physical Review Letters, 2009. **102**(25): p. 256801.
142. Zhang, Z., et al., *Manipulating Nanoscale Light Fields with the Asymmetric Bowtie Nano-Colorsorter*. Nano Letters, 2009. **9**(12): p. 4505–4509.
143. Olmon, R.L., et al., *Determination of Electric-Field, Magnetic-Field, and Electric-Current Distributions of Infrared Optical Antennas: A Near-Field Optical Vector Network Analyzer*. Physical Review Letters. **105**(16).
144. Novotny, L., *Effective wavelength scaling for optical antennas*. Physical Review Letters, 2007. **98**(26): p. 266802.
145. Hrelescu, C., et al., *Single gold nanostars enhance Raman scattering*. Applied Physics Letters, 2009. **94**(15): p. 153113.

146. Ghenuche, P., S. Cherukulappurath, and R. Quidant, *Mode mapping of plasmonic stars using TPL microscopy*. New Journal of Physics, 2008. **10**: p. 105013.
147. Hicks, E.M., et al., *Controlling plasmon line shapes through diffractive coupling in linear arrays of cylindrical nanoparticles fabricated by electron beam lithography*. Nano Letters, 2005. **5**(6): p. 1065–1070.
148. Rogobete, L., et al., *Design of plasmonic nanoantennae for enhancing spontaneous emission*. Optics Letters, 2007. **32**(12): p. 1623–1625.
149. Murphy-DuBay, N., et al., *Nanopatterning using NSOM probes integrated with high transmission nanoscale bowtie aperture*. Optics Express, 2008. **16**(4): p. 2584–2589.
150. Matteo, J.A., et al., *Spectral analysis of strongly enhanced visible light transmission through single C-shaped nanoapertures*. Applied Physics Letters, 2004. **85**(4): p. 648–650.
151. Wang, D.X., T. Yang, and K.B. Crozier, *Charge and current reservoirs for electric and magnetic field enhancement*. Optics Express, 2010. **18**(10): p. 10388–10394.
152. Fischer, H. and O.J.F. Martin, *Engineering the optical response of plasmonic nanoantennas*. Optics Express, 2008. **16**(12): p. 9144–9154.
153. Fleischer, M., et al., *Three-dimensional optical antennas: Nanocones in an apertureless scanning near-field microscope*. Applied Physics Letters, 2008. **93**(11): p. 111114.
154. Chang, S.W. and S.L. Chuang, *Normal modes for plasmonic nanolasers with dispersive and inhomogeneous media*. Optics Letters, 2009. **34**(1): p. 91–93.
155. Kang, J.H., D.S. Kim, and Q.H. Park, *Local Capacitor Model for Plasmonic Electric Field Enhancement*. Physical Review Letters, 2009. **102**(9): p. 093906.
156. Li, K.R., M.I. Stockman, and D.J. Bergman, *Self-similar chain of metal nanospheres as an efficient nanolens*. Physical Review Letters, 2003. **91**(22).
157. Lim, D.K., et al., *Nanogap-engineerable Raman-active nanodumbbells for single-molecule detection*. Nature Materials, 2010. **9**(1): p. 60–67.
158. Kinkhabwala, A., et al., *Large single-molecule fluorescence enhancements produced by a bowtie nanoantenna*. Nature Photonics, 2009. **3**(11): p. 654–657.
159. Camden, J.P., et al., *Probing the structure of single-molecule surface-enhanced Raman scattering hot spots*. Journal of the American Chemical Society, 2008. **130**(38): p. 12616.
160. Pettinger, B., et al., *Nanoscale probing of adsorbed species by tip-enhanced Raman spectroscopy*. Physical Review Letters, 2004. **92**(9): p. 096101.
161. Neacsu, C.C., et al., *Scanning-probe Raman spectroscopy with single-molecule sensitivity*. Physical Review B (Condensed Matter and Materials Physics), 2006. **73**(19): p. 193406-4.
162. Sackrow, M., et al., *Imaging nanometre-sized hot spots on smooth Au films with high-resolution tip-enhanced luminescence and Raman near-field optical microscopy*. Chemphyschem, 2008. **9**(2): p. 316–320.
163. Deckert-Gaudig, T. and V. Deckert, *Ultraflat Transparent Gold Nanoplates—Ideal Substrates for Tip-enhanced Raman Scattering Experiments*. Small, 2009. **5**(4): p. 432–436.
164. Stadler, J., T. Schmid, and R. Zenobi, *Nanoscale Chemical Imaging Using Top-Illumination Tip-Enhanced Raman Spectroscopy*. Nano Letters, 2010. **10**(11): p. 4514–4520.
165. Taminiau, T.H., et al., *lambda/4 Resonance of an optical monopole antenna probed by single molecule fluorescence*. Nano Letters, 2007. **7**(1): p. 28–33.

166. Zhang, D., K.F. Domke, and B. Pettinger, *Tip-Enhanced Raman Spectroscopic Studies of the Hydrogen Bonding between Adenine and Thymine Adsorbed on Au (111).* Chemphyschem, 2010. **11**(8): p. 1662–1665.
167. Farahani, J.N., et al., *Single quantum dot coupled to a scanning optical antenna: A tunable superemitter.* Physical Review Letters, 2005. **95**(1): p. 017402.
168. Weber-Bargioni, A., et al., *Functional plasmonic antenna scanning probes fabricated by induced deposition mask lithography.* Nanotechnology, 2010. **21**: p. 065306.
169. Farahani, J.N., et al., *Bow-tie optical antenna probes for single-emitter scanning near-field optical microscopy.* Nanotechnology, 2007. **18**(12): p. 125506.
170. Iwami, K., T. Ono, and M. Esashi, *Optical Near-Field Probe Integrated With Self-Aligned Bow-Tie Antenna and Electrostatic Actuator for Local Field Enhancement.* Journal of Microelectromechanical Systems, 2006. **15**(5): p. 1201–1208.
171. Weber-Bargioni, A., et al., *Hyperspectral Nanoscale Imaging on Dielectric Substrates with Coaxial Optical Antenna Scan Probes.* Nano letters, 2011. **11**(3): p. 1201–1207.
172. Li, D. and R. Gordon, *Electromagnetic transmission resonances for a single annular aperture in a metal plate.* Physical Review A, 2010. **82**(4): p. 041801(R).
173. Catrysse, P.B. and S.H. Fan, *Understanding the dispersion of coaxial plasmonic structures through a connection with the planar metal-insulator-metal geometry.* Applied Physics Letters, 2009. **94**(23): p. 231111.
174. Lockyear, M.J., et al., *Microwave transmission through a single subwavelength annular aperture in a metal plate.* Physical Review Letters, 2005. **94**(19): p. 193902.
175. Poujet, Y., J. Salvi, and F.I. Baida, *9% Extraordinary optical transmission in the visible range through annular aperture metallic arrays.* Optics Letters, 2007. **32**(20): p. 2942–2944.
176. de Waele, R., et al., *Plasmon Dispersion in Coaxial Waveguides from Single-Cavity Optical Transmission Measurements.* Nano letters, 2009. **9**(8): p. 2832–2837.
177. Neacsu, C.C., et al., *Near-Field Localization in Plasmonic Superfocusing: A Nanoemitter on a Tip.* Nano Letters, 2010. **10**(2): p. 592–596.
178. De Angelis, F., et al., *Nanoscale chemical mapping using three-dimensional adiabatic compression of surface plasmon polaritons.* Nature Nanotechnology, 2010. **5**(1): p. 67–72.
179. Nerkararyan, K.V., *Superfocusing of a surface polariton in a wedge-like structure.* Physics Letters A, 1997. **237**(1–2): p. 103–105.
180. Babadjanyan, A.J., N.L. Margaryan, and K.V. Nerkararyan, *Superfocusing of surface polaritons in the conical structure.* Journal of Applied Physics, 2000. **87**(8): p. 3785–3788.
181. Stockman, M.I., *Nanofocusing of Optical Energy in Tapered Plasmonic Waveguides.* Physical Review Letters, 2004. **93**(13): p. 137404.
182. Ropers, C., et al., *Grating-coupling of surface plasmons onto metallic tips: A nanoconfined light source.* Nano Letters, 2007. **7**(9): p. 2784–2788.
183. Berweger, S., et al., *Adiabatic Tip-Plasmon Focusing for Nano-Raman Spectroscopy.* Journal of Physical Chemistry Letters, 2010. **1**(24): p. 3427–3432.
184. Anderson, N., A. Hartschuh, and L. Novotny, *Chirality changes in carbon nanotubes studied with near-field Raman spectroscopy.* Nano letters, 2007. **7**(3): p. 577–582.
185. Hartschuh, A., *Tip-Enhanced Near-Field Optical Microscopy.* Angewandte Chemie-International Edition, 2008. **47**(43): p. 8178–8191.
186. Hartschuh, A., et al., *Simultaneous fluorescence and Raman scattering from single carbon nanotubes.* Science, 2003. **301**(5638): p. 1354–1356.

187. Yano, T., et al., *Pressure-assisted tip-enhanced Raman imaging at a resolution of a few nanometres*. Nature Photonics, 2009. **3**(8): p. 473–477.
188. Strano, M.S., et al., *Electronic structure control of single-walled carbon nanotube functionalization*. Science, 2003. **301**(5639): p. 1519–1522.
189. Qu, L.W., et al., *Interactions of functionalized carbon nanotubes with tethered pyrenes in solution*. Journal of Chemical Physics, 2002. **117**(17): p. 8089–8094.
190. Sanii, B. and P.D. Ashby, *High Sensitivity Deflection Detection of Nanowires*. Physical Review Letters, 2010. **104**(14): 147203
191. Geladi, P. and B.R. Kowalski, *PARTIAL LEAST-SQUARES REGRESSION—A TUTORIAL*. Analytica Chimica Acta, 1986. **185**: p. 1–17.
192. Mattson, J.S., et al., *CLASSIFICATION OF PETROLEUM POLLUTANTS BY LINEAR DISCRIMINANT FUNCTION ANALYSIS OF INFRARED SPECTRAL PATTERNS*. Analytical Chemistry, 1977. **49**(3): p. 500–502.
193. Wold, S., K. Esbensen, and P. Geladi, *PRINCIPAL COMPONENT ANALYSIS*. Chemometrics and Intelligent Laboratory Systems, 1987. **2**(1–3): p. 37–52.
194. Hartigan, J.A., *Clustering algorithms*. Wiley Series in Probability and Mathematical Statistics. Vol. xiii. 1975, New York-London-Sydney: John Wiley \& Sons.
195. Almeida, J.A.S., et al., *Improving hierarchical cluster analysis: A new method with outlier detection and automatic clustering*. Chemometrics and Intelligent Laboratory Systems, 2007. **87**(2): p. 208–217.
196. Bondarenko, I., et al., *HIERARCHICAL CLUSTER-ANALYSIS WITH STOPPING RULES BUILT ON AKAIKES INFORMATION CRITERION FOR AEROSOL-PARTICLE CLASSIFICATION BASED ON ELECTRON-PROBE X-RAY-MICROANALYSIS*. Chemometrics and Intelligent Laboratory Systems, 1994. **22**(1): p. 87–95.
197. Sun, L.X., et al., *CLUSTER-ANALYSIS BY THE K-MEANS ALGORITHM AND SIMULATED ANNEALING*. Chemometrics and Intelligent Laboratory Systems, 1994. **25**(1): p. 51–60.
198. Nascimento, J.M.P. and J.M.B. Dias, *Vertex component analysis: A fast algorithm to extract endmembers spectra from hyperspectral data*, in *Pattern Recognition and Image Analysis, Proceedings*, F.J. Perales, Editor. 2003. p. 626–635.
199. Nascimento, J.M.P. and J.M.B. Dias, *Vertex component analysis: A fast algorithm to unmix hyperspectral data*. Ieee Transactions on Geoscience and Remote Sensing, 2005. **43**(4): p. 898–910.
200. Vandeginste, B.G.M., Massart, D.L, Buydens, LMC, De JONG, S, Lewi, P.J, Smeyers-Verbeke, J, *Hnadbook of Chemometrics and Qualimetrics; Part B*. Data Handing in Science and Technology. Vol. 20B. 2003: Elsevier.
201. Widjaja, E., et al., *Band-target entropy minimization (BTEM) applied to hyperspectral Raman image data*. Applied Spectroscopy, 2003. **57**(11): p. 1353–1362.
202. Tan, S.T., H.H. Zhu, and W. Chew, *Self-modeling curve resolution of multi-component vibrational spectroscopic data using automatic band-target entropy minimization (AutoBTEM)*. Analytica Chimica Acta, 2009. **639**(1–2): p. 29–41.
203. Perera, P.N., et al., *Chemical compositional analysis of plant cell wall Raman microspectroscopy and Chemometrics*. Submitted.
204. Perera, P.N., et al., *Raman spectroscopy-based non-invasive microanalysis of native lignin structure*. Submitted.
205. Perera, P.N., et al., In situ *compositional analysis of Arabidopsis thaliana with Raman microspectroscopy and chemometrics*. (Submitted).

14

Automated Microscopic Imaging and Survival Statistics

STEVEN FINKBEINER ■

HOW DO WE KNOW WHAT EVENTS MEAN?

Big Problems in Biology: A Brief History

The complete understanding of a physiological or pathophysiological process requires a definition of the molecules or environmental factors that are involved and the dynamical interactions between them. Though much remains to be done, advances in molecular biology and genomics in the late twentieth and early twenty-first century have gone a long way toward the identification of the molecules that make up our cells. To know how these biological systems work in health and how they might be manipulated effectively to treat disease, a principal challenge in the twenty-first century will be to understand how these myriad molecules work together dynamically.

The Special Problem of Pathobiology

Efforts to understand and treat the causes of disease highlight the critical need to elucidate these dynamic relationships. In most cases, our understanding of disease reflects decades of careful pathological analyses. In the case of neurodegenerative diseases, a focus of research in our laboratory, these studies provided detailed descriptions of the subpopulations of affected neurons for each disorder, the gross and microscopic pathological changes that characterize a disease, and a time course over which the changes typically occur. The use of specific antibodies, in situ labeling approaches, and ultrastructural techniques led to descriptions of pathological states with increasing spatial resolution and molecular specificity.

As valuable as pathological studies of human disease have been, the information they provide is static and therefore prone to misinterpretation. These limitations are even more acute in diseases that lead to cell death. Without a direct method to establish the future fate of a particular cell, it is impossible to ascertain the prognostic

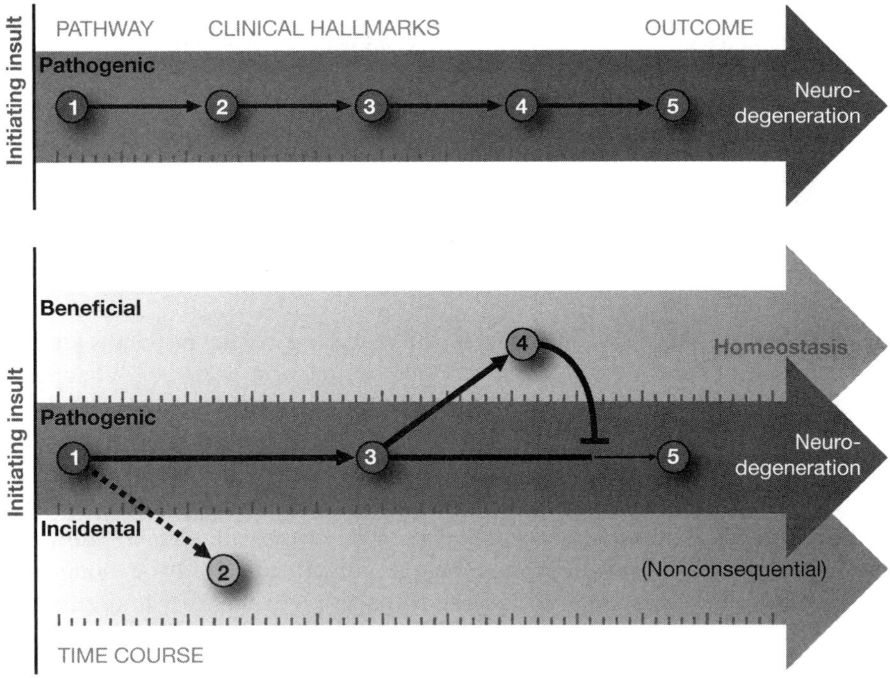

Figure 14.1 Neurodegenerative diseases display phenotypes that can be beneficial, pathogenic, or incidental. Whether an observed change is pathogenic or beneficial is a common question that has enormous therapeutic importance. Some "pathological" changes may actually be beneficial. Activation of homeostatic pathways may mitigate pathogenic perturbations, and promoting these beneficial coping responses could be effective therapies. The timing of the events can further confound the true disease pathways. If pathogenic processes elicit nearly simultaneous coping responses, they will be highly correlated with each other and with the course of disease. New strategies will enable us to identify disease-modifying pathways and may identify new therapeutic targets for these devastating diseases.

and pathogenic significance of an abnormality that it exhibits. In such cases, conventional approaches cannot answer even the simplest question, "Do cells with a particular abnormality experience a better or worse fate than those without the abnormality?" Though it is commonly assumed that any deviation from normality is pathogenic, the possibility that a particular change is incidental or even a beneficial coping responses is real and would completely change the interpretation of its significance for disease and its potential as a therapeutic target (Figure 14.1).

For example, it is tempting to infer stepwise disease mechanisms from the time course of pathological changes, as one might infer mechanisms of development by studying different stages of embryogenesis. However, such an approach could be misleading. With embryogenesis, it is reasonable to assume that intermediate stages are steps along a well-established path that lead to a properly developed organism. In contrast, pathological changes could be part of a single pathogenic pathway, or they could well represent components of parallel but independent pathways that emanate from the initiating insult. Moreover, some "pathological" changes may actually reflect beneficial responses. Indeed, synaptic plasticity, which is so critical for development and learning and memory, is induced in the central nervous system in response to injury (1).

Whether a change is pathogenic or beneficial is a fundamental question that has enormous therapeutic importance. Most therapies focus on inhibiting pathogenic processes (2, 3). But promoting beneficial coping responses could be equally effective (4). Unfortunately, the nature of most neurodegenerative diseases makes this determination extremely difficult with conventional approaches, which depend on inferring underlying relationships from "snapshots" in time. Though these diseases are devastating, many progress slowly over years or decades, possibly indicating that the initial trigger is only moderately toxic, that effective coping responses are triggered early and in parallel with disease, or both. If the most pathogenic abnormalities are also the most toxic, then the length of time that a cell lives and exhibits the most informative changes could be short, making them difficult to detect. By contrast, the pathological changes that are easiest to detect could be the ones that help cells live longer by enabling them to cope with disease. Correlations of particular changes with disease progression won't help unravel pathogenic changes from coping responses because, by definition, coping responses should only be mounted in response to and in proportion to inciting pathogenic changes. If pathogenic processes elicit nearly simultaneous coping responses, they will be highly correlated with each other and with the course of disease.

Imaging and "Systems Biology" as Solutions

Identifying effective therapies for adult-onset neurodegenerative diseases has been a major challenge. For example, although the gene that causes Huntington's disease (HD) has been identified, its protein product appears to lack receptor or enzymatic activity that could be disrupted with therapeutic small molecules (5). For more common diseases, such as Alzheimer's and Parkinson's, the underlying causes remain mostly unknown. (The incidence of ALS is about the same as HD.) In a minority of cases, these disorders have been linked to genetic mutations; however, the affected proteins also appear to be poor pharmacological targets. Because of these difficulties, indirect treatment strategies are appealing. Putative disease-modifying genes have been identified for some disorders (6, 7). These genes, especially those in which loss-of-function suppresses disease, could encode proteins that are more attractive therapeutic targets. For example, in HD, variation in the age of onset can be substantial. If the pathways responsible for the variation could be pharmacologically targeted, symptoms might be delayed for decades.

The discipline of "systems biology" has emerged as an approach to tackle biological problems in their full complexity. Though the field of systems biology encompasses a diverse array of techniques and approaches, the basic idea is to wed new technologies that are capable of collecting vast amounts of data from a biological system with powerful computational strategies that can organize those data in useful ways and uncover global relationships that might have been missed or misunderstood with conventional approaches. For example, the comprehensive state descriptions provided by systems biology approaches can reveal expected changes in biological pathways that result from genetic manipulations as well as any associated and unexpected compensatory changes in other biological pathways. Undoubtedly, the elucidation of complex biological networks with systems biology approaches will eventually help us to understand why some genetic manipulations produce no or unexpected phenotypic changes and why some putative therapies unexpectedly fail to work or produce unwanted side effects. This sort of understanding could accelerate and lower the cost of the development of effective therapeutics (8).

THE PARTICULAR PROBLEMS IN HD

Survival Analysis: A New Way to Look at Neurons

To overcome the limitations of conventional approaches to disentangle pathogenic, incidental, and coping responses, we developed a new strategy for the study of neurodegenerative disease. We reasoned that if we could observe the same neuron repeatedly, as neurodegeneration unfolded, we could link the appearance of particular changes within that neuron to its ultimate fate. Moreover, if we could quickly collect similar information for millions of neurons, we could identify natural variation in the levels or timing of the changes and any associated changes in future fate and establish predictive relationships. In turn, this information would help us categorize these changes qualitatively as pathogenic, incidental, or beneficial, and quantitatively as major or minor.

The practical embodiment of this strategy has two components. First, we developed an automated microscope and software that can acquire highly focused images of cultured primary neurons and return to precisely the same neuron or field of neurons, even after the culture plate has been removed from and returned to the microscope stage (9). To enable the microscope to return to the same cell, proprietary software algorithms were developed that store an image of a fiduciary mark on each plate. At the beginning of an imaging run, the algorithm instructs the motorized stage attached to the microscope to move to the approximate location of the fiduciary mark. The program then compares the newly obtained image of the fiduciary mark with the stored copy and, using a matrix multiplication algorithm, determines the stage movements required to align the plate perfectly with its location during any previous imaging epoch. Once aligned, the program instructs the microscope to collect images from a user-defined number of microscope fields. At each location, the microscope is automatically focused using image-based or interface-based strategies. Pairs of excitation and emission filters are rotated into and out of the light path as each image is collected. Information can be collected from up to 1 million neurons in as little as 10 minutes, and individual cells can be tracked as often and for as long as we want. For our HD studies, we use primary striatal neurons expressing mutant huntingtin and typically observe them once daily for about two weeks.

Second, we applied survival analysis to quantify relationships between particular changes within neurons and some future fate (9). Although it is a well-established statistical tool for quantifying relationships between two events separated by time, survival analysis is not limited to the study of survival per se. It is particularly well-suited to quantifying neurodegeneration, in which the precise timing of meaningful events (e.g., cell death) appears to be stochastic and affects a relatively small subpopulation of cells (10). The cumulative nature of the analysis makes it much more sensitive (~10^2–10^3-fold) and less variable than approaches that depend on average responses measured at particular time points (9).

Our approach follows the fate of every cell from the beginning of the monitoring period until the end. As a result, it is essentially insensitive to a common problem with conventional approaches—observer bias: that the duration of an event affects the likelihood of observing it. In neurodegeneration, short-lived changes may be the most critical steps and the most difficult to observe. Cells exhibiting these changes die more quickly and are less available to study. Conversely, homeostatic changes tend to prolong the lifetimes of affected cells, making these changes easier to observe. These beneficial changes could themselves become so closely associated with disease progression that they are mistaken for pathogenic factors.

Survival Analysis Applied to Mechanisms of HD

As one example, the abnormal deposition of mutant huntingtin (htt) protein in intracellular structures called inclusion bodies is a pathological hallmark of HD (11). The pathogenic role of inclusion bodies has been controversial, in part because conventional time-course data from animal and cell models of the disease were insufficient to determine if those structures were pathological, beneficial, or incidental (12, 13). To address this controversy, we applied automated microscopy and survival analysis to the study of striatal neurons expressing normal or disease-associated versions of an amino-terminal htt fragment. By fusing a fluorescent protein to the carboxy-terminal of htt and introducing it into live neurons, we were able to observe inclusion body formation and measure htt levels. By simultaneously introducing a different colored fluorescent protein, it was possible to observe the morphology of each neuron and determine how long it survived. Surprisingly, neurons that formed inclusion bodies lived longer than those that did not, and we were able to show that the level of more diffuse forms of htt predicted whether and when neurodegeneration would occur (14). Indeed, the ability to simultaneously observe levels of more diffuse forms of htt, inclusion body formation, and survival made it possible to construct a dynamical model that explained cell fate in which inclusion body formation is part of a cellular coping response that reduces levels of diffuse mutant htt.

The fact that inclusion bodies containing mutant htt form in HD suggests that the disease involves a defect in protein turnover. However, it has been difficult to establish directly that a defect in specific protein turnover pathways exists. Studies have conflicted over whether the proteasome, a major organelle for degrading intracellular proteins, is unaffected or impaired by mutant htt in vitro or in vivo (15). To try to resolve this controversy, we applied automated microscopy and longitudinal analysis to relate real-time measures of proteasome function to inclusion body formation and survival in our neuron model of HD. To do so, we introduced a third fluorescent protein that was engineered to report proteasome function in live neurons. We found that proteasome function became inhibited before inclusion bodies formed but their function improved after the inclusion body appeared (16). The result suggests that inclusion body formation disinhibits proteasome function, possibly by sequestering forms of mutant htt away from the proteasome pore. Importantly, the approach revealed the complex dynamic relationship that can exist between disease-causing proteins and the biological responses that they elicit. Given the asynchronous nature of inclusion body formation among different neurons and the correlated biphasic change in proteasome function, the discovery would have been impossible without such a longitudinal single-cell approach.

General Insights into Neurodegeneration Offered by Survival Analysis

The ability of survival analysis to uncover predictive relationships that are not susceptible to observer bias is one of the major reasons why this approach, unlike conventional correlations, can address causation per se. Indeed, gross correlations made with conventional approaches can be resolved further with survival analysis into changes that are associated with an increase, a decrease, or no change in the likelihood of reaching an end point within a defined interval. Factors associated with accelerated neurodegeneration are likely to be pathogenic, those associated with decelerated neurodegeneration are likely to be coping responses, and those associated with no significant relationship to the rate of neurodegeneration are likely to be incidental.

In addition to resolving qualitative relationships, survival analysis can quantify the magnitude of the change in risk associated with a factor. This is especially helpful because many biological processes may go awry during neurodegeneration, but only some may be relevant to patient's symptoms and therefore important therapeutic targets. Quantitative measures of the risk of neurodegeneration associated with identifiable factors provide a way to rank them in order of importance.

Finally, survival analysis and related statistical approaches make it possible to build quantitative models that incorporate each factor shown to bear a predictive relationship (including factors that promote or inhibit a fate) to the outcome of interest. This makes it more feasible to satisfy a critical requirement for establishing cause-and-effect relationships—to simultaneously measure (or exclude) the contributions of other potential mechanisms to a putative cause-and-effect relationship. This type of model building can reveal the extent to which neurodegeneration can be predicted, and therefore fully understood, or whether important stochastic or deterministic contributions remain to be elucidated (17).

HOW WILL THIS DEVELOP?

Phenotypic Assays

The application of automated microscopy and survival analysis to elucidate the relationship of proteasome function to the process of neurodegeneration in HD nicely illustrates one of the directions along which this technology will develop. For that application, a genetically encoded fluorescent protein targeted to the proteasome was used to make time-lapse measurements of relative changes in steady-state proteasome function. Over the past decade, the use of fluorescent protein technology to create faithful reporters of diverse cell biological functions has exploded and there is no end in sight (18). It is the full expectation of this author that the number of fluorescent proteins and the diversity, sophistication, and precision of reporter assays will only increase in the coming years.

Combined with automated microscopy and longitudinal analysis, the information that these reporters provide can be used in two ways. The time-dependent changes in a parameter of a specific pathway can be used to determine if changes in that pathway predict a fate of interest, as we did in the case of inclusion body formation and survival. Secondly, because survival analysis can be used to measure any fate, not just survival, changes in the pathway itself can become the fate that we seek to explain, as we did in the case of proteasome function. As long as a parameter can be measured from an image and a criterion of interest for that parameter defined, survival analysis can be applied to determine the probabilistic relationship between other factors and that outcome.

Statistics

The ability afforded by automated microscopy and longitudinal analysis to make repeated measures on individual cells permits the application of powerful statistical techniques such as survival analysis and related but more sophisticated methods such as Cox proportional hazards (CPH) analysis. CPH in particular makes it possible to create complex statistical models that integrate a user-defined number of parameters to approach solutions that explain fate to a greater and greater extent. There are many commercially available software packages that have the capability of performing standard survival analysis and CPH. However, it is important for biologists to keep in mind that data sets must meet certain criteria in order to validly apply Cox proportional hazards analysis (19–21).

There are other limitations to most commercially available implementations of survival analysis and Cox proportional hazards analysis. Typical implementations are based on frequentist analysis. A full discussion of frequentist analysis is beyond the scope of this chapter; however, one limitation of frequentist approaches is that they require specific a priori assumptions that can influence the final result in unwanted ways that can produce inaccurate results. In part to overcome this potential limitation, we are adapting Bayesian analysis to Cox proportional hazards analysis. Bayesian analysis is a statistical technique that allows one to infer or update one's "degree of belief" in a hypothesis or parameter value in light of new experimental information. For our purposes, Bayesian analysis starts with an assumed/given "prior" probability distribution for a parameter of interest. We may be as specific or nonspecific as we want with the assumed/given "prior" distribution. Evidence from experiments is then used to update the prior distributions with new probability distributions for the parameter. After a sufficient number of iterative updates, Bayesian analysis produces a "posterior" distribution, which gives the probability distribution for the parameter of interest taking into account all experimental results.

Spectral Imaging

That the aforementioned statistical techniques can accommodate a virtually unlimited number of parameters is exciting because it offers hope for achieving what may be the principal goal of twenty-first century biology—developing dynamical systems models that fully explain biology and pathobiology. Unfortunately, there remain significant technical challenges. One of the most important is the amount of information that can be collected from a cell at one point in time. Epifluorescence microscopy has been developed, in part, to overcome this limitation. Conventional implementation of epifluorescence microscopy relies on optical filters and polychroic mirrors to selectively excite and detect specific fluorescent molecules. With conventional approaches, therefore, the amount of information that can be reliable collected is largely limited by the power of the optical system to collect information from one fluorophore independent of another. The spectra of small molecule fluorescent dyes tend to have fairly narrow absorption and excitation peaks. This feature makes it easier to use several simultaneously and yet collect information from each fairly independently. However, small molecule fluorescent dyes tend to be metabolized by cells over intervals of hours to days making them unsuitable for long-term longitudinal analysis. On the other hand, genetically encoded fluorescent proteins can be produced continually by cells for their entire lifetime. Unfortunately, currently available fluorescent proteins have relatively broad absorption and emission spectra, making it difficult to measure more than three or four different fluorescent proteins at once.

The advent of commercially available spectral detectors provides a potential solution to expanding bandwidth. This approach relies on sophisticated technology to simultaneously excite multiple different fluorophores in a microscope field. In one common implementation, the complex fluorescence emissions are then collected and passed through an optical element that spatially disperses the fluorescence emissions according to their wavelength. Multiple detectors are positioned to collect emitted fluorescence from specific segments of the spectrum. If the technique is used to first determine the spectra for each fluorophore in the sample, complex emissions from multiple fluorophores can be deconvolved into their component parts with numerical curve-fitting approaches.

The prospect of incorporating spectral detection into high-content imaging seems so promising that it could affect the future development of fluorescent proteins in ways that could increase bandwidth further. Similar to small molecule fluorescent dyes, fluorescent proteins from different organisms have been obtained and optimized to create versions with spectra dominated by single major excitation and emission peaks. However, the existence of spectral detection creates the opportunity to develop fluorescent proteins with multiple emission peaks, akin to bar codes, that might be exploited to increase further the number of fluorescent proteins that can be measured simultaneously.

High-Throughput Screens

The method of automated microscopy and longitudinal analysis was developed to facilitate hypothesis-driven research. However, the speed and the sensitivity of the approach suggest that it might be a powerful new tool for conducting high throughput unbiased screens for genetic or small molecule modifiers of biological or pathobiological pathways. Indeed, we have recently developed a second-generation system with several modifications for that purpose. One necessary adaptation involved a change in our autofocusing algorithm. We discovered that the original image-based algorithms worked poorly with high density (>96) multi-well plates. The meniscus formed by the media distorted the transmitted light in ways that made it impossible to form sharply focused images. A commercially available image-independent method of autofocusing (PerfectFocus™) substantially solved the problem. The method reflects an infrared laser light off the bottom of a multi-well plate and through the optical system and uses the position of the reflection to determine the distance between the tip of the objective and the bottom of the plate. With a simple error-detection circuit and feedback control of the z-axis stepper motor, the technology is able to maintain the relative position of the objective while fluorescence images are collected. Another adaptation involved the integration of a robotic arm and the construction of a compatible stage for the automatic delivery and removal of multi-well plates. Finally, because screening large numbers of plates takes hours, even with high-throughput approaches, an enclosure was built around the system to control the temperature and humidity of the environment. Using a common industry measure of robustness, application of this method to a primary neuron model of HD (Q_{17} v Q_{72}) yielded a Z' of 0.96, which is considered outstanding for a cell-based assay.

CONCLUSION

Automated microscopy provides a method to avoid the seriously confounding problems that often occur in using more traditional imaging methods that rely on a "snapshot" of events at a particular moment. By following individual cells, we can use our statistical methods to understand how any chosen perturbation affects the cells over time. While we have used it primarily to examine neurons and neurodegenerative disease, the approach can be used for any cell type. Moreover, the ability to adapt automated microscopy for use in high-throughput screens further enhances its capabilities for drug discovery.

REFERENCES

1. Emery DL, Royo NC, Fishcher I, Saatman KE, McIntosh TK. (2003) Plasticity following injury to the adult central nervous system: Is recapitulation of a developmental state worth promoting. *J. Neurotrauma* 20:1271–1292.
2. Zhang X, Smith DL, Meriin AB, Engemann S, Russel DE, Roark M, Washington SL, Maxwell MM, Marsh JL, Thompson LM, Wanker EE, Young AB, Housman DE, Bates GP, Sherman MY, Kazantsev AG. (2005) A potent small molecule inhibits polyglutamine aggregation in Huntington's disease neurons and suppresses neurodegeneration *in vivo*. *Proc. Natl. Acad. Sci. U.S.A.* 102:892–897.
3. Thakur AK, Yang W, Wetzel R. (2004) Inhibition of polyglutamine aggregate cytotoxicity by a structure-based elongation inhibitor. *FASEB J.* 18:923–925.
4. Bodner RA, Outeiro TF, Altmann S, Maxwell MM, Cho SH, Hyman BT, McLean PJ, Young AB, Housman DE, Kazantsev AG. (2006) Pharmacological promotion of inclusion formation: A therapeutic approach for Huntington's and Parkinson's disease. *Proc. Natl. Acad. Sci. U.S.A.* 103:4246–4251.

5. Beal MF, Ferrante RJ. (2004) Experimental therapeutics in transgenic mouse models of Huntington's disease. *Nat. Rev. Neurosci.* 5:373–384.
6. Wexler NS, Lorimer J, Porter J, Gomez F, Moskowitz C, Shackell E, Marder K, Penchaszadeh G, Roberts SA, Gayan J, Brocklebank D, Cherny SS, Cardon LR, Gray J, Dlouhy SR, Wiktorski S, Hodes ME, Conneally PM, Penney JB, Gusella J, Cha JH, Irizarry M, Rosas D, Hersch S, Hollingsworth Z, MacDonald M, Young AB, Andresen JM, Housman DE, De Young MM, Bonilla E, Stillings T, Negrette A, Snodgrass SR, Martinez-Jaurrieta MD, Ramos-Arroyo MA, Bickham J, Ramos JS, Marshall F, Shoulson I, Rey GJ, Feigin A, Arnheim N, Acevedo-Cruz A, Acosta L, Alvir J, Fischbeck K, Thompson LM, Young A, Dure L, O'Brien CJ, Paulsen J, Brickman A, Krch D, Peery S, Hogarth P, Higgins DS, Jr., Landwehrmeyer B. (2004) Venezuelan kindreds reveal that genetic and environmental factors modulate Huntington's disease age of onset. *Proc. Natl. Acad. Sci. U.S.A.* 101:3498–3503.
7. Djoussé L, Knowlton B, Hayden MR, Almqvist EW, Brinkman RR, Ross CA, Margolis RL, Rosenblatt A, Durr A, Dode C, Morrison PJ, Novelletto A, Frontali M, Trent RJA, McCusker E, Gómez-Tortosa E, Mayo Cabrero D, Jones R, Zanko A, Nance M, Abramson RK, Suchowersky O, Paulsen JS, Harrison MB, Yang Q, Cupples LA, Mysore J, Gusella JF, MacDonald ME, Myers RH. (2004) Evidence for a modifier of onset age in Huntington disease linked to the HD gene in 4p16. *Neurogenetics* 5:109–114.
8. Butcher EC, Berg EL, Kunkel EJ. (2004) Systems biology in drug discovery. *Nat. Biotechnol.* 22:1253–1259.
9. Arrasate M, Finkbeiner S. (2005) Automated microscope system for determining factors that predict neuronal fate. *Proc. Natl. Acad. Sci. U.S.A.* 102:3840–3845.
10. Clarke G, Collins RA, Leavitt BR, Andrews DF, Hayden MR, Lumsden CJ, McInnes RR. (2000) A one-hit model of cell death in inherited neuronal degenerations. *Nature* 406:195–199.
11. Davies SW, Turmaine M, Cozens BA, DiFiglia M, Sharp AH, Ross CA, Scherzinger E, Wanker EE, Mangiarini L, Bates GP. (1997) Formation of neuronal intranuclear inclusions underlies the neurological dysfunction in mice transgenic for the HD mutation. *Cell* 90:537–548.
12. Sisodia SS. (1998) Nuclear inclusions in glutamine repeat disorders: Are they pernicious, coincidental or beneficial? *Cell* 95:1–4.
13. Orr HT. (2004) Neuron protection agency. *Nature* 431:747–748.
14. Arrasate M, Mitra S, Schweitzer ES, Segal MR, Finkbeiner S. (2004) Inclusion body formation reduces levels of mutant huntingtin and the risk of neuronal death. *Nature* 431:805–810.
15. Mitra S, Finkbeiner S. (2008) The ubiquitin-proteasome pathway in Huntington's disease. *ScientificWorldJournal* 8:421–433.
16. Mitra S, Tsvetkov AS, Finkbeiner S. (2009) Single neuron ubiquitin-proteasome dynamics accompanying inclusion body formation in Huntington disease. *J. Biol. Chem.* 284:4398–4403.
17. Herndon LA, Schmeissner PJ, Dudaronek JM, Brown PA, Listner KM, Sakano Y, Paupard MC, Hall DH, Driscoll M. (2002) Stochastic and genetic factors influence tissue-specific decline in ageing C. *elegans*. *Nature* 419:808–814.
18. Giepmans BNG, Adams SR, Ellisman MH, Tsien RY. (2006) The fluorescent toolbox for assessing protein location and function. *Science* 312:217–224.
19. Codori AM. (1997) Psychological opportunities and hazards in predictive genetic testing for cancer risk. *Gastroenterol. Clin. North Am.* 26:19–39.

20. Roodnat JI, Mulder PG, Tielens ET, van Riemsdijk IC, van Gelder T, Weimar W. (2004) The Cox proportional hazards analysis in words: Examples in the renal transplantation field. *Transplantation* 77:483–488.
21. Hastie T, Sleeper L, Tibshirani R. (1992) Flexible covariate effects in the proportional hazards model. *Breast Cancer Res. Treat.* 22:241–250.

PART III

Modeling of Complex Biological Functions

Introduction to Part III

MANFRED AUER, NATALIA PINZON, AND GARY C. HOWARD ■

> The sciences do not try to explain, they hardly even try to interpret, they mainly make models. By a model is meant a mathematical construct which, with the addition of certain verbal interpretations, describes observed phenomena. The justification of such a mathematical construct is solely and precisely that it is expected to work—that is, correctly to describe phenomena from a reasonably wide area.
>
> —JOHN VON NEUMANN

All of the techniques described in this book require the extensive use of computers and data processing to make sense of the data they produce. Efficient display of the raw and refined data, Fourier analysis of diffraction data, 3D reconstruction, Molecular Dynamics calculations, signal-to-noise improvement (e.g., through averaging or noise filtering) all involve heavy computational lifting. Undoubtedly, computational science will continue to improve image processing.

Here we want to explore another side of mathematics and computational science that is becoming increasingly important: bioimage analysis. By this, we mean quantitative volumetric data analysis, data abstraction, and representation, including modeling, simulation, and statistical analysis.

This post-data-reconstruction analysis step is integral to, yet almost buried in, the techniques featured in the first section of this book. Most of the techniques implicitly or explicitly average over a large number of molecules, to improve signal-to-noise and overcome experimental limitations that degrade data quality. No wonder these approaches lead to higher resolution.

What is often overlooked is that most of the information from these techniques is not simply the result of a reconstruction of the raw data. It is our interpretation of the data. When we say we have a structure to atomic resolution, we actually mean that we have a built an atomic model that is consistent with the observed data. Just looking at the density map by itself, even with high-quality high-resolution data, is not very helpful in itself. What makes the density map intelligible is its representation by models with various levels of detail. For example, having a simple ball and stick model of the primary sequence threaded through the 3D space yields exquisite insight into the molecular interactions of the amino acid side chains and is therefore crucial to our chemical understanding of, for example, an enzymatic reaction site or the molecular recognition patterns of an antibody-epitope or protein-protein complex. Yet, such ball-and-sticks models can be far too complex to reveal the overall organization of the protein, where certain secondary structure elements, such as alpha helices and beta sheets, are organized into folds. Proteins similar in function often can be different in sequence, yet adopt a similar secondary structure architecture (a.k.a. as fold), since the function of a protein is best accomplished by a certain set of organizational principles. Hence, the abstraction of alpha helices as cylinders and beta-sheet strands as flat arrows reveals such architecture much better than would be detectable at the raw density level or even at the stick-and-ball model. So while neither the alpha helix nor the beta strand strictly speaking looks like a cylinder or a flat arrow, this level of abstraction is very useful. Even less detailed depictions of protein shape and the changes it undergoes in its functional cycle can be more revealing about the essence of domain movement, than would be discernable from the raw density map or even the ball-and-stick model.

So while we do not explicitly think of X-ray crystallography as a method that is primarily model building (we call it structure determination), in essence, it is the models at different levels of abstraction that allow us to comprehend and communicate protein function. Likewise, the signals recorded in NMR spectroscopy in themselves do not allow the determination of the 3D structure, but instead they provide initial estimates of amino acid side chains proximity as constraints to feed molecular dynamics algorithms to virtually heat up the molecule and upon "cooling" down to adopt energetically favorable states that satisfy intermolecular distance constraints detected by the NMR spectrum.

As a result, in NMR we often do not obtain one structure, but a number of structures, all of which are compatible with the molecular dynamics calculation and the distance constraints detected in the spectra. The models obtained may differ widely in certain parts of the linear amino acid sequence, whereas other regions appear with much more constraints. This is interpreted as flexible and rigid regions of a protein, respectively, and while this seems a reasonable assumption, nobody knows for sure. In essence, it yields a reasonable model of the macromolecule of interest. Likewise, in X-ray crystallography, regions that are less well defined are interpreted in having a low occupancy state or are called disordered, which means they may vary slightly in their local 3D structure from molecule to molecule. The assignment of temperature factors defines which part of the structure one can be sure about and which parts less certain. All of these are reasonable suggestions and interpretations. But importantly, in any of these approaches, we are dealing with model representations of the actual density maps/3D structures rather than with raw data.

One form of interpretation, often successfully used in cryo-EM single-particle analysis (and even as a starting model for phasing in X-ray crystallography) is the docking of a model (or derived from it an actual density map) of known protein components into a usually larger density map obtained by single particle analysis. When such docking appears reasonable and the fit supports or at least doesn't violate additional (e.g., biochemical) information, and if all other fits appear clearly worse, we declare that this particular portion of the density map "clearly" represents the protein in a certain orientation. Such hybrid models are the accepted standard and, as far as we can tell, represent a reasonable interpretation of the density map. But to be clear, while we tend to think that we determine "THE" 3D structure and organization, what we really do is to interpret the data through model building.

Likewise, in AFM, we use a topological height map to deduce from it changes in the 3D architecture of a protein upon changes in a particular parameter in the experimental matrix. We do not have the information directly in our data, but based on a good model of the protein, the observed changes in topology (height) yield a reasonable model of the molecular changes that accompany the experiment.

In summary, all of the high-resolution approaches of the first section require a substantial amount of model building to obtain the information of interest about the biological system under investigation. What unites all techniques in Section I is that the systems are well defined: we typically know the molecular mass and each component of this purified system. The diversity of the molecules involved is typically low, and nearly all objects are implicitly or explicitly averaged and can be presumed to be identical in composition and often also in conformation and 3D organization.

In Section II, we face a substantial change. The systems become more complex, and although each technique may not capture that compositional complexity, we no longer can assume all or even a significant fraction of the sceneries under investigation to be the same. Indeed, at an atomic resolution level, no cell can reasonably be expected to be identical to the next one, but similarities remain. Imaging cellular sceneries is fundamentally different from macromolecular complexes for two different cells cannot be assumed to be identical. At best, they can be considered similar in terms of composition and organization (and physiological behavior). In reality, all cells are unique and have their own distinct history and status. They share commonalities, but how similar is similar, and how does one measure the similarity between two systems? And what do we mean by similar? Clearly, we need prior knowledge and expert information to determine what aspect of the system we are interested in. Yet while we try to choose our systems carefully, we can never be sure that the particular scenery chosen is representative. The system is inherently complex, underdetermined, and thus ill-defined. Before we carry out extensive analysis, we don't even know whether the system is likely to be in an experimental state that renders a successful observation of the trait of interest.

Furthermore, the techniques in Section II provide information at a somewhat lower resolution, by a factor of one to four orders of magnitude. Hence, they provide complementary but vastly different information than the high-resolution techniques of Section I. Where techniques of Section I relied to a large part on averaging, the approaches in section two typically do not allow averaging of the signals. Instead, each data set must be assumed and treated to be a unique example, with certain constituents common to all data sets collected. Hence, in such cases (i.e., electron tomography and in soft X-ray tomography), we can no longer rely on the averaging of entire data sets. Template matching, the attempt to cross-correlate known template 3D structures with the actual 3D density map and therefore to find the location of proteins, is extremely computationally expensive and works for a handful of selected large complexes of very defined 3D architecture and composition. For most objects, this approach does not work because they are too small and thus their shapes are not sufficiently unique. In some defined instances, we can carry out sub-volume averaging, which may boost the solution enough to fit known candidate molecules. However, for the most part, we can only apply rather standard bioimage-processing tools, such as filtering for noise reduction. But most importantly, we must explicitly build models based on the data obtained and, to a great extent, on our expectations and subjective interpretation of the results. Such models often cannot claim to be a one-to-one replica of the density maps encountered but often a simplified and abstracted version of the data, such as a connectivity between two objects symbolized by a simple stick, with describing the exact shape of the connection. Still, how to craft such models is a challenge: shall we proceed in a manual, computer-assisted, or fully automated manner? The answer may be different for different types of the data sets, and is particularly confusing when very large volumes with complex sceneries need to be analyzed. While a traditional approach with complete manual segmentation may be time-prohibitive, computer-algorithms, including fast marching-based level-set approaches, such as water shedding and boundary segmentation may not be sophisticated enough in cases of complex architecture.

If bioimaging is to have an effect beyond the aesthetically pleasing, at the very minimum, we need to obtain volumetric information, including lengths, widths, volumes, surface area, and distances. Ideally, we seek statistical verification and validation of the maps and the models derived from them. Wherever possible, we must try to fit models into maps, to address the certainty or uncertainty of a particular quantification across the different resolutions, estimate the errors between the maps and the models, and allow for curation of the models based on prior domain knowledge. As soon as possible, we need to employ data abstraction, using the power of simplified models to replace actual density to allow comprehension and avoid drowning in meaningless detail. We need to bring order to chaos by identifying (annotating) objects on the fly, and possibly by building our own local database, filled with relevant information and linked to some more comprehensive public domain databases. A true understanding will require hierarchical and semantic models, leaving the world of voxels altogether and entering the relational world, where a macromolecular complex belongs to a particular position on a particular organelle in a particular cell in a particular tissue, instead of being centered at a particular X, Y, and Z coordinate. Visualization and real-time exploration of such complex volumes are nontrivial, not only due to the terabytes size of modern recorded 3D data sets, but in particular due to its complexity. Should we scroll through the 3D volume slice by slice, and risk missing the 3D relationships that only come by looking at the scenery in 3D, or should we try to display the entire volume in 3D, risking to be overwhelmed by the complexity of the data and missing the forest for the trees? Or would it be more natural and intuitive to use immersive visualization approaches that would allow one to walk around the data, and explore close-up from a ground's prospective a small portion of the 3D volume at the time from the "INSIDE," rather than receiving a large bird's eye view of the entire 3D volume from the "OUTSIDE"? How such 3D data are best served up to the investigator is yet unclear. Should one store and interact with the data locally or globally? What if we could engage the entire world community in the process of segmentation, annotation, and/or interpretation? When building models, one may wonder whether and how such architecture-derived models could/should be subjected to mathematical scrutiny. For example, using computer-assisted design (CAD) models, one might calculate the mechanical properties and the implications of model alterations, or to simulate the ion flow inside the complex 3D volume of a cell.

To go even further, for the sake of the argument, let's pretend we totally understand this complex 3D volume. How can we let the public—the scientific expert community and the lay audience—in on the secret? How can we communicate all the wonderful things we have learned? In the age of CSI TV shows, *National Geographic*, and National Public Broadcasting Service documentary TV shows and Pixar Studios, one ought to think about communicating one's latest finding in a full-blown professionally arranged lengthy video. However, very few scientists have the financial and production-wise depth to prepare a professional video for DVD or the web. Yet with computers getting faster and animation software becoming cheaper and more readily available, such a vision no longer sounds far-fetched. Storytelling in a multimedia format by integrating information from a range of techniques may become the new standard of communication. Before long, we could end up with the biological equivalent of the physics "Grand Unified Theory," maybe the Grand Unified Theory of Life, one movie production at a time.

This section focuses on the computational and mathematical analysis of the various kinds of image data. For high-resolution data, models of the linear protein primary sequence are threaded through the 3D space. Using the amino acid sequence to predict secondary structural elements and constrained by an overall envelope and the knowledge of the 3D density map, some easily recognizable for their unique size and shape (e.g., the aromatic residues), structures these days can often be determined in an automated manner, where data processing and model building are completed autonomously by a computer. In more pathological cases, where data are incomplete or ambiguous, one still requires human intervention and expert knowledge. However, atomic models allow calculation of hydrophobic, hydrophilic, and charged properties of the constituent amino acid side groups, which also allow determination of protein-protein interfaces and hence model prediction/modeling of protein-protein or protein-substrate binding energies. Modeling also can reveal mechanistic insight into protein function. For the more intermediate resolution data (i.e., lower resolution envelopes determined by single-particle cryo-EM), atomic models of known or suspected components can be fitted into the density map outlines, which typically are characterized by higher complexity, therefore allowing the mapping of respective constituent proteins.

Chapter 15, "From Voxel Maps to Models," focuses on the mathematics and computational science behind high-resolution methods such as X-ray crystallography and most prominently single particle cryo-EM analysis. The field of single particle cryo-EM in particular needs all the help that it can get, as the resolution that is typically achievable does not allow direct fitting of the density. First computational approaches help speed up and come to the correct solution by automated particle picking and formation of a starting model that then gets further refined. Once a density map is obtained, one may want to dock known parts into the density map and to segment the overall map into smaller subvolumes, some of which may even be averaged to further increase resolution. Some of these tools (docking, segmentation, and subvolume averaging) are also applicable to electron tomography, which is further complicated by missing information due to limited tilt range that typically result in data anisotropy.

As novel 3D SEM are boldly entering the world of whole cell and large tissue portion imaging, we recognize the need for new computational approaches to support the analysis of the wealth of information that are now and will routinely be obtained in the very near future, easily adding up to over one terabyte per acquired data set. This currently is clearly the cutting edge and we are just making baby steps towards efficient handling of such huge data sets. However, it is clear that the merging of computational science with biological imaging is not only powerful but with the emerging fields absolutely critical for progress.

Chapter 16, titled "Bioinformatics and Modeling Complex Biological Systems," explores the question of how to build 3D atlases, in this case of the brain connectome of the nematode *Caenorhabditis elegans*, as well as fruit fly *Drosophila melanogaster*, with data collected from a variety of different animals. Using 3D optical microscopy such as confocal and two-photon laser scanning microscopy with different color channels, 3D digital atlases have been developed. For such atlases to be derived from different animals, a variety of challenges must be overcome including standardization and registration, segmentation, annotation and data mining, as well as atlas modeling. These steps along with the 3D visualization of complex connectome data are critical to gain biological insight into the inner working of arguably the most complex organ in any animal. Should the brain connectome be sufficiently known in detail and under a variety of different experimental conditions, we might finally have some insight into brain function (and misfunction), and their 3D architectural role.

15

From Voxel Maps to Models

Quantitative Analysis and Visualization of Macromolecular Machines

CHANDRAJIT BAJAJ

1. STRUCTURE DETERMINATION OF MACROMOLECULAR COMPLEXES

The field of structure determination of individual proteins and more recently macromolecular complexes has been dominated over the last half a century by X-ray crystallography, with some notable contributions from NMR spectroscopy and electron crystallography. Crystallography produces models with atomic feature resolution. This accuracy, however, comes with a price. The crystallization process cannot be applied to macromolecules or entire cells; and even for the small complexes where it is applicable, it introduces deformations to the molecules. To understand the complexity of living systems, it is often insufficient to simply determine the structure of a protein or even as a complex. Their interaction in the cellular context, their precise localization, their binding partners, and any conformational changes are crucial for an understanding of the in-vivo function of the protein of interest. For such applications cryo-electron microscopy has become the preferred choice.

1.1. X-Ray Crystallography

The wavelength of electromagnetic radiation used for imaging biological structures must be comparable to the smallest features to be resolved. Given that the size of an atom is around 1 Å, X-rays of that wavelength is the appropriate portion of the electromagnetic spectrum to be used. However, X-ray scattering from a single molecule is extremely weak, and so one resorts to periodically ordered arrays of proteins (crystals) that coherently diffract incoming X-ray beams, and amplify the signal into discrete lattice spots (diffraction pattern). Unlike in electron microscopy where we record magnified images of the biological sample, in X-ray diffraction the data recorded on the detector is a diffraction pattern created by the scattering X-rays by each macromolecular structure with constructive interference at angles defined by the crystal lattice parameters and symmetry.

As a consequence, the intensity patterns of the diffraction patterns, which are indexed as *h, k,* and *l*, define a 3D space, also known as the reciprocal space. The intensity of each spot with the indices *h, k,* and *l*, together with the phase for each spot (that is initially unknown but can be determined) yield a vector, the so-called structure factor $|F(h,k,l)|$ and which leads to an electron density map. Reciprocal space structure factor *F* for each *h, k, l* values and the electron density f^i for each spatial coordinate x, y, and z in a unit cell are related to one another via a Fourier Transformation according to equation (1):

$$F(h,k,l) = \sum_{i=1}^{n} f^{i} \cdot \exp\left(2\pi i\left(hx_i + k_i + lz_i\right)\right) \qquad \text{[Eq. 1]}$$

where f^i denotes the electron density function defined for a single electron and *n* is the number of electrons within a single unit cell of the crystal. Since the entire crystal is a lattice of copies of a single unit cell, it is enough to consider just one unit cell. The recorded intensity value can be simply defined as

$$I(h,k,l) = \left|F(h,k,l)\right|^2 \qquad \text{[Eq. 2]}$$

Equations 1 and 2 build a mapping from the original electron density functions of the crystal to the recorded intensities in the reciprocal space. However, the structure determination problem is an inverse process, that is, from the recorded intensities we map to the original electron density maps.

Due to the duality property of the Fourier Transform, equation 1 can be slightly modified such that we can determine the electron density maps of the original crystals from the function *F (h, k, l)* (known as *structure factors*) defined in the reciprocal space.

$$f(x,y,z) = \frac{1}{V}\sum_{h}\sum_{k}\sum_{l} F(h,k,l) \cdot \exp\left(-2\pi i\left(hx_i + ky_i lz_i\right)\right) \qquad \text{[Eq. 3]}$$

where V is the volume of a single unit cell. We know an *I (h, k, l)* from the detected information and thus we know a $|F(h,k,l)|$, the magnitude of *F (h, k, l)*. However, the phase of *F (h, k, l)* is lost and must be recovered. There are now several techniques to solve this phase recovery problem. The direct method takes advantage of *Patterson maps*, aiming to discover the electron density function directly from the magnitude of the structure factors $|F(h,k,l)|$. This method which uses a computer-based technique is practical to recover the phase for few hundred electrons within each unit cell. However, it suffers from scalability issues when the number of electrons per unit cell increases. Note: A *Patterson map* is the function obtained by equation 2 except that the structure factors *F (h, k, l)* are replaced by $|F(h,k,l)|$, which is already known from the intensity function *I (h, k, l)*. Other methods include the *Molecular replacement (MR)* which is an experimental technique used for homologous structures (144), the *Multiple Anomalous Dispersion (MAD) Phasing* method which uses X-rays with different wavelengths and considers diffraction maps obtained by different X-rays (145), and the *Multiple Isomorphous Replacement (MIR) method* which considers the difference maps with and without the soaked heavy metal atoms (146).

As a consequence of phase recovery, one is rewarded with a high-resolution electron density map of the macromolecule of interest that can be readily inspected and interpreted through model building, often in a semi-automated manner.

1.2. Cryo-Electron Microscopy

While structures of individual proteins or small complexes, such as most of the Protein Data Bank entries, provide important information, they do not necessarily yield the "full picture" of a functional biological complex. The study of large macromolecular complexes, such as viruses, ion channels, the ribosome, and other machines of various types, particularly when imaged in different conformational stages, offer a more progressive structural and functional description of the cellular protein machinery. In contrast to X-ray crystallography, electron cryo-microscopy (cryo-EM) provides a method for imaging large macromolecular complexes without the need for crystallization (1–3). Single particle analysis (SP) has become a powerful tool in revealing the structures of large complexes at subnanometer resolutions (5–10 Å) (4–10). From a map of resolution coarser than about 12 Å, gross morphological and structure arrangements in global symmetry can be detected. As the resolution improves, structural motifs like alpha-helices (8–12 Å), C-alpha chains (5–7 Å), beta-sheets and alpha-helix pitch (4–6 Å), and side chains (2–4 Å) can be discerned via SP from the cryo-EM images.

Alternatively, electron tomography (ET) has the advantage in its ability to image nonrepetitive structures and therefore provide a full picture of cell sceneries in a near natural state. In ET a single sample is imaged over a sequence of orientations, typically a uniform sample over a set of tilt angles between plus/minus 70 degrees. These advantages come at a cost: since missing large tilt angles appear as a wedge in Fourier space, no unique solution for the problem exists and reconstructions typically have resolutions between 20 and 60 Å. While this is too coarse to determining quasi-atomic motifs, it is sufficient to infer large conformations and some super-secondary motifs. Due to the varying trade-offs between resolution, particle size, and experimental challenges, the most accurate structural models must rely on refinement, combining multiple experimental data sources (11, 12). As experimental gains continue to improve towards physical limitations in image quality (13), computational advances are also necessary in order to provide the best quality 3D maps and models. Vitrified biological samples are highly sensitive to radiation damage and data must be collected at very low electron doses, resulting in images with very poor signal-to-noise ratios. Image reconstruction is further complicated by the fact that for thicker specimens the images contain a mix of phase-contrast and amplitude-contrast components and must be corrected using estimates of the microscope contrast transfer function. Images also suffer from effects of blurring owing to specimen charging and specimen movements during exposure to the electron beam and from asymmetries in the imaging optics (e.g., from astigmatism in the objective lens). Finally, in SP the orientations of the particles in the sample are not known precisely and the quality of the reconstruction is limited by the accuracy to which we can estimate these orientations. As a consequence of these and other factors, structures determined today from single particle cryo-EM have resolutions typically in the range of 6–9 Å.

While advancements in SP have propelled structure determination to these sub-nanometer resolutions in many instances, the computational infrastructure for interpretation and visualization of the assemblies still remains relatively undeveloped. For a number of recent cryo-EM structure determination projects structure interpretation from 3D EM has become the rate-limiting step. Advances in automated image acquisition (14) and image reconstruction (15) have made it possible to go from a virus sample to medium resolution (6–9 Å) structures in less than a week and sometimes even within hours, primarily by exploiting virus symmetry. Existing tools (16) can predict the locations of the more obvious secondary features (e.g., long alpha-helices), but are less useful in identifying short helices or small beta-sheets (e.g., with three or fewer beta-strands). There still exist many situations where we have 3D molecular maps but no accompanying X-ray crystallographic data. For example, there are no X-ray structures available for the capsid proteins of any of the dsRNA viruses whose structures have recently been solved (147, 148).

2 CRYO-ELECTRON MICROSCOPY IMAGE PROCESSING

When a high-energy electron wave interacts with a biological specimen, it is scattered due to the electrostatic field of the specimen's constituent atoms. The output wave after passing through the specimen contains information about the electrostatic field of the specimen, and which is captured primarily as image contrast. The main contrast mechanism in EM imaging of unstained biological samples is phase contrast (interference contrast), which results from the quantum superposition (interference) of the crests of the incident wave and the scattered wave. This can often be approximated by projecting the potential of the specimen. Typically, thinner specimens with higher accelerating voltage and lighter atoms will satisfy this projection approximation (1). The emergent electron wave interacts with the microscope optics to generate a 2D projection image. The 2D projection images typically suffer from a number of degrading factors. The images acquired have spatial resolution on the order of Angstroms, but slight mechanical noise during tilting and/or acquisition can cause large misalignment in the images. The projection assumption used above for the electron beam-specimen interaction only holds for weakly scattering specimens. The forward problem tries to approximate the expected errors/noise and apply corrections to reproduce a true projection of the object.

The inverse problem in EM/ET is to reconstruct a 3D map based on 2D images/projections. Mathematically this can be roughly stated as the problem of recovering the best estimate of a signal (a scalar function representing the 3D map) given measured data (the single particle or the tilt series images) obtained from observations of the signal.

2.1. Image Correction

CTF

In general, single particle electron microscopy (SP) and electron tomography (ET) images are degraded by many factors such as lens aberrations, defocusing, and so on. Hence, it is very necessary to correct the recorded images to their true intensities. The restoration functions include Contrast Transfer Function (CTF), Envelope Function, and Background Noise Function. The relationship between the measured image and the true image is described as (21): $M(s)^2 = F(s)^2 C(s)^2 E(s)^2 + N(s)^2$ where M(s), F(s), and N(s) are the Fourier transforms of the measured image, the true image, and the added noise image, respectively. E(s) stands for the envelope function, which can be represented by one parameter as $E(s) = e^{-Bs^2}$. Here C(s) stands for the CTF and is given by:

$$C(s) = A(\sqrt{1 - C_A} \sin \gamma(s) + C_A \cos \gamma(s)) \qquad \text{[Eq. 4]}$$

where A and $C_A (0 \le C_A \le 1)$ are two parameters to be determined and γ is defined by: $\gamma(s) = -2\pi\left(\frac{C_s \lambda^3 S^4}{4} - \frac{\Delta Z \lambda S^2}{2}\right)$ where λ is the electron wavelength and ΔZ stands for the defocusing value. The noise function N(s) is described as: $N(S)^2 = n_1 e^{n_2 + n_3 S^2 + n_4 \sqrt{S}}$ where n_1, n_2, n_3, n_4 are four parameters to be determined.

The above model is reported in EMAN (21), in which there are *eight* parameters in total that need to be estimated: A, C_A, ΔZ, B, n_1, n_2, n_3, n_4. Of course other models can be used for C(s), E(s), and N(s). For example, in (18), the CTF model is given by $2[\sin \gamma(s) - W \cos \gamma(s)]$ where only two parameters require estimation and the noise model is given by $a + b\exp[-(k/c)^2]$ where only three parameters require estimation. The goal of CTF correction is thus to determine the parameters that are seen in the above models such that one can correct and obtain the true image *F(s)* from the recorded image *M(s)*.

Most CTF corrections on recorded EM images *m(x, y)* (a 2D projection image) use the following steps:

1. Compute the Fourier transform of m(x,y): $M(u,v) = \tilde{F}\{m(x,y)\}$ and from that calculate the rotationally averaged power spectrum of *M (u, v)*, denoted by $M(s)^2$.
2. Next, determine each local minima of $M(s)^2$ and their corresponding coordinates (S_m, I_m), *m* = 1, 2, 3, where S_m stands for the special frequency coordinate and I_m stands for the intensity coordinate. From these minima one estimates the four parameters of N(s). After N(s) is determined, it is subtracted from the power spectrum $M(s)_2$. Recall, the amplitude contrast ratio *A* might need further iterative refinement.
3. The three parameters of C(s) can also be estimated from the locations (S_m, I_m), *m* = 1, 2, 3 of minima of $M(s)_2$. Using C(s) one obtains the parameter B of the envelope function E(s) by comparing the shape of $M(s)_2$ after correcting $N(s)_2$ and $C(s)_2$ with the shape of the power spectrum $F(s)_2$ of the true image. Hence, the power spectrum $F(s)_2$ of the true image must be known before the correction. All these corrections are accomplished using data-fitting techniques.

Image Filtering

We know that crystals in X-ray diffraction act like an amplifier where the contribution of each unit cell can add up and largely increase the signal-to-noise ratio. Without crystals, the images obtained from an electron microscope are usually degraded with much noise, which may result in an incorrect reconstruction of the true structures. Although the signal-to-noise level can be improved by increasing the electron doses of the microscopes, high electron dose imaging unfortunately always causes significant radiation damages to the molecular specimen (such as specimen shrinkage as well as mass loss), which in turn reduces the accuracy of the reconstructed structures.

However, even with the help of flash cooling techniques, minimizing radiation damage and therefore damage-induced sample heterogeneity, the collected microscopy images are not good enough to be directly used for accurate reconstruction. Fortunately, one can improve the image quality by explicitly averaging individual particles with identical views, similar to the role that crystals play in crystallography, where averaging occurs implicitly. Note that the "particles" cited here for single-particle analysis (SP) stand for the experimental projection images (2D) of the true 3D biological structure being studied. It is very important to keep in mind that there is an implicit requirement that all the observed particles must be projections along certain directions of identical structures. Without this condition satisfied, averaging becomes erroneous. Therefore, by *single particle reconstruction*, we mean that the samples must contain isolated, non-crystalline and identical biological macromolecular assemblies, which in the real world sometimes can be challenging to accomplish.

2.2. Imaging and Reconstruction Protocols

We describe several reconstruction techniques popular in the past several years for SP and in some cases for ET. In general, different methods of data collection result in different methods of reconstruction. Another way to categorize different reconstruction techniques is by looking at how each technique computes the orientation of each particle, which is recognized as one of the more difficult problems in electron microscopy reconstruction.

- *Direct method for ET*. This method is similar to CT/MRI reconstruction seen in medical imaging. Roughly speaking, reconstruction from a series of 2D projections to 3D volumetric maps becomes possible if the projection images and their relative orientations are available. For this purpose, the sample can be observed using microscopes with different tilt angles. However, the tilt angles cannot exceed a certain degree due to the instrumental difficulty, hence for practical reasons the limit is commonly set at 69^0 (1). Depending on how we change the tilt angle, there are two methods to collect the projection data: *single axis tilt* (Figure 15.1a) and *uniform conical tilt* (Figure 15.1b, shown below). Due to the limit of the tilting angles, both of these two methods have information lost in reciprocal Fourier space. Single axis tilt method has a wedge-shape area missing in Fourier space while the uniform conical tilt method gets a cone-shape area missing. Geometrical considerations indicate that, under the same restriction of the tilting angles, the missing cone seen in uniform conical tilt method is much smaller than the missing wedge seen in single axis tilt method.

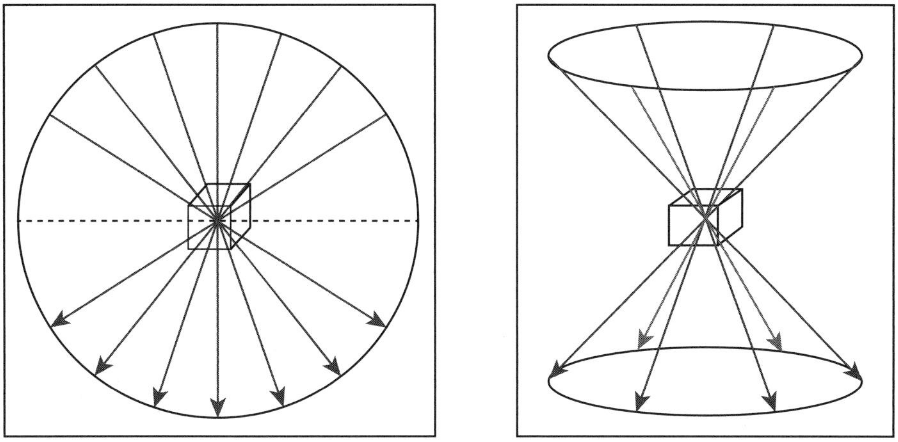

Figure 15.1 Illustration of single axis tilting and uniform conical tilting methods.

The direct method is easy to understand and simple to implement. However, besides the missing wedge or missing cone problems that cause inaccurate reconstruction, this method has several other (and even more serious!) problems. First, we cannot use an averaging technique to improve the signal-to-noise ratio because the reconstruction results are usually not good enough for structure analysis. Second, the molecular specimen have to be exposed under the electron beams for many times to collect enough projections for reconstruction, which imposes significant radiation damages to the specimen and, as a result, largely reduces the accuracy of the reconstruction. Therefore, this direct method is not ideal for molecular structure reconstruction. Strictly speaking, this method does not belong to the category of single particle reconstruction techniques. It was just discussed theoretically and used in an early stage of electron microscopy image reconstruction for a rough structure determination of macromolecular assemblies.

- *Common-line based method for SP*. This method utilizes an averaging technique on particles that have identical appearances to obtain a set of improved projection images before applying a reconstruction algorithm (such as the weighted back projection method). It begins with particle picking, followed by particle classification, particle alignment, and then the technique of averaging, after which the common-line method can be applied to determine the relative orientation of each projection average. Finally, the weighted back projection technique (or any other reconstruction approach) is used to reconstruct the 3D electron density map of the specimen being studied. The overall pipeline is illustrated in Figure 15.2 below. Remember that, in this figure, we ignore the contrast transfer function (CTF) correction that should be implemented before particle picking to determine CTF parameters and correct phases. In practice, the micrographs can be collected at different defocus values. The procedure illustrated in Figure 15.2 is applied to each defocus group and

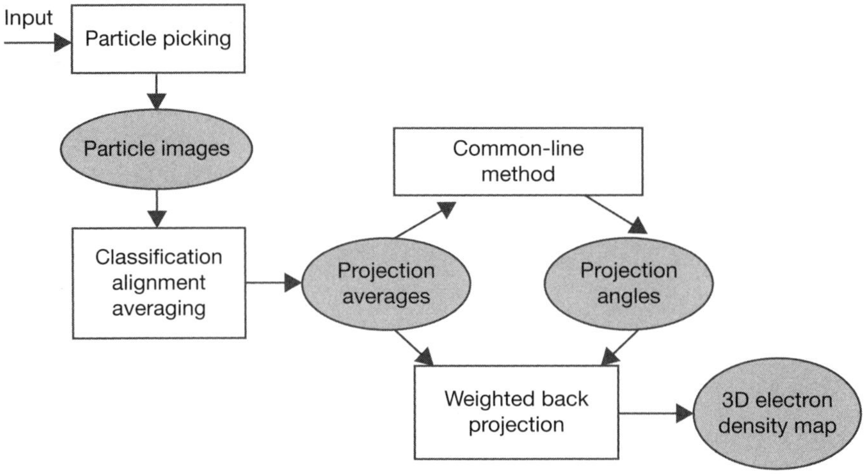

Figure 15.2 Pipeline of reconstruction procedure based on common line and weighted back projection.

then the reconstructed 3D density maps from each defocus group are merged using the CTF parameters previously computed.

The common-line based method was used in one of the popular computer tools, called IMAGIC (19). Compared to the direct method, this method only requires the specimen to be exposed once under the electron beams. Hence, the radiation damage can be reduced to a very low level. In addition, the averaging technique is used to greatly improve the signal-to-noise ratio, which as a result increases the reconstruction accuracy. Despite these improvements, the drawback of this method is that much more computational time is needed for processing images. Additionally, the noise present in the images introduces some errors in particle classification, particle alignment, as well as in the relative orientation determination method, which, to some extent, reduces the resolution of the reconstructed structures.

- *Random conical method for ET*. In this method, the specimen is exposed twice: once in tilt angle and the other in untilt angle. The micrographs in untilt angle are used to pick the particles, classify the selected particles, and align particles within each group. Since there is a one-to-one correspondence between the particles in tilt angle and the particles in untilt angle, the information of particle selection and particle classification obtained from the untilt micrographs can be used to pick and classify the particles in tilt micrographs. Then the reconstruction (e.g., weighted back projection method) can be applied to obtain 3D electron density maps, based on the particle images from the tilt micrographs and the orientation information obtained from the untilt micrographs. Remember that each group from the micrographs (untilt and tilt) independently generates a 3D density map. All of these maps can be merged together using 3D alignment to improve the reconstruction accuracy. The overall pipeline is illustrated in Figure 15.3.

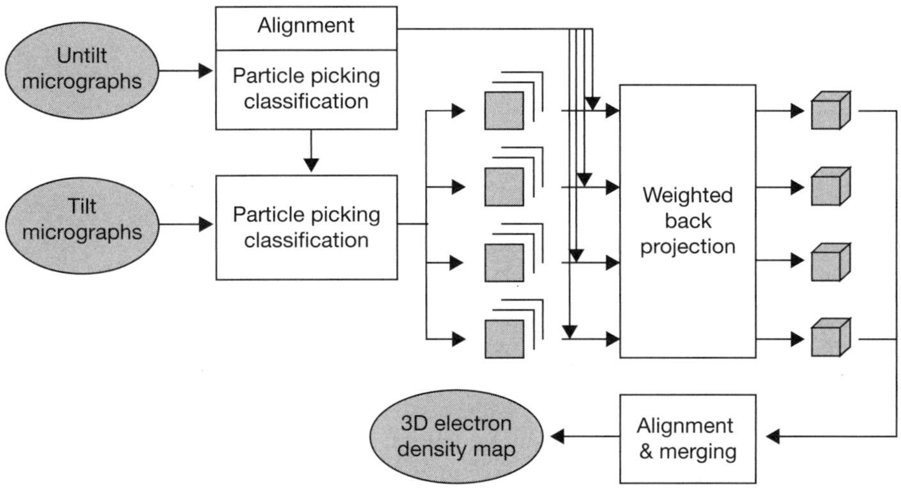

Figure 15.3 Pipeline of reconstruction procedure based on the random conical method.

This method was used in another well-known computer software program called SPIDER (20). It is obvious that, in this method, the radiation damage to the specimen is much less than in the direct method. Although the averaging technique is not used in this method, the merging on the 3D reconstructed structures also improve the accuracy of the electron density map. Compared with common-line based method, the random conical method is much easier to compute the orientation of each particle, making the overall procedure very fast to implement. However, the random conical method is only applicable to macromolecules with one or a few preferred orientations, which, according to (1), is often observed true in experiments.

- *Symmetry-based reconstruction*. All the methods described above require enough number of projection images to "cover" the Fourier space such that interpolation accuracy in Fourier space can be guaranteed at an acceptable level. For some highly symmetric structures, however, a small number of projections are usually enough to interpolate in Fourier space, where one origin-crossing plane (which is also Fourier transform of a projection image) can be copied several times (depending on the degree of symmetry) in Fourier space. An example is helical symmetry, for which only one single projection image is enough for reconstruction because of the high degree of redundancy of views due to the rotational symmetry of the helix. Icosahedral-like symmetry structures such as viruses are another example of biological structures that have been widely studied for advantages gained in 3D reconstruction.

3D Map Refinement

Neither the common-line based method, nor the random conical method nor the symmetry based method can provide a reconstruction at desirable resolution or clarity due to the high noise level seen in electron microscopy images and the finite (small) number of projection images available. In practice, the above-mentioned methods only generate an initial model of the true structure. To achieve improved secondary resolution, a procedure called refinement must be repeatedly employed. As seen in solution of any optimization problems, a good initial model has two advantages: faster convergence and higher possibility of converging to a global optimum. Both IMAGIC and SPIDER provide routines to generate initial models as well as how to refine them. Another well-known computer tool, called EMAN (21) (http://blake.bcm.edu/emanwiki/EMAN/), also provides a complete package for single particle reconstruction, including particle selection, initial modeling, and the refinement loop. EMAN is introduced in this section instead of the last section because EMAN differs from other packages more in refinement than in other reconstruction procedures.

The refinement technique is generally based on what we call a *projection matching method*, which is another technique to determine orientation Euler angles of each selected particle based on a preliminary model or a refined model. Given a 3D reference model, a series of projections can be obtained by summing up the structure along certain directions. Each of the selected particles is then compared to each projection by cross-correlation and the orientation Euler angles of the best-matched projection is assigned to each particle. In this way, the initially selected particles are classified into groups by projection matching. Alignment and averaging techniques may be used for each group and, since the Euler angle for each group is already known, the 3D model can be built from such a set of group averages. The overall pipeline is illustrated in Figure 15.4.

Again, in Figure 15.4, we don't consider CTF correction. In practice, the CTF correction can be applied either in the step of "averaging" (merging of 2D images) as seen in (21) or in the step of "building 3D model" (merging of 3D structures) as seen in (1, 20).

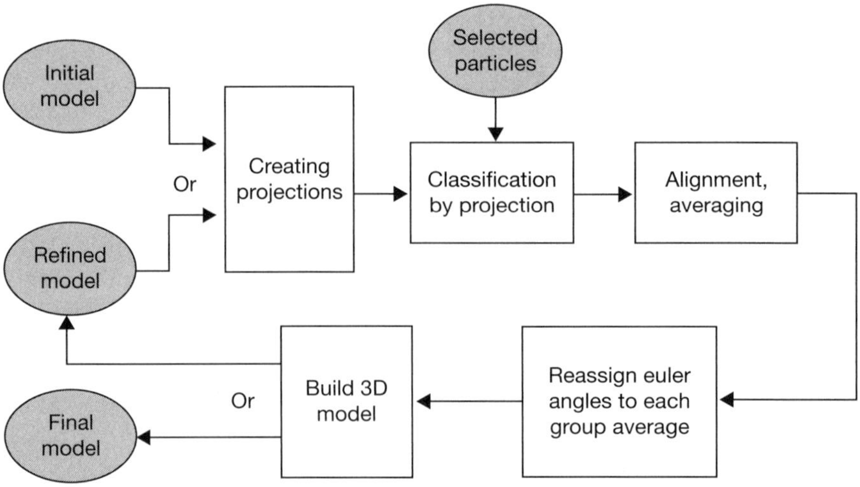

Figure 15.4 Pipeline of refinement based on projection matching and reclassification.

2.3. Advanced 3D Map Reconstruction Methods

Reconstructing 3D maps from the raw images remains a computationally (and theoretically) challenging inverse problem (22, 23). Generically, the reconstruction problem aims to determine the 3D map based on a set of projection images.

Existing reconstruction approaches for cryo-EM can be grouped into two classes: analytical methods and iterative methods. Analytical methods attempt to directly approximate and discretize the inverse operator and thus compute directly. The most common algorithms in this class are filtered/weighted back projection (FBP/WBP) (24, 25) and direct Fourier reconstruction (Fourier) (25). Iterative algorithms generate a sequence of maps, which converge to the least squares solution and include algebraic reconstruction (ART) (26), as well as simultaneous iterative reconstruction (27) along with a many other variants (28–33). Under both approaches, success relies on the particular algorithm's ability to suppress noise, which corrupts the solution to the ill-posed inverse (reconstruction) problem. This has led to a variety of software packages which each provide some subset of the available reconstruction methods: SPIDER (WBP, SIRT), IMAGIC (FBP), EMAN2 (Fourier) (34–42).

Variational and Statistical Regularization Methods for 3D Reconstruction

There are two other classes of methods commonly applied to ill-posed inverse problems that are not available in the standard cryoEM reconstruction packages. Variational methods, which have been successfully applied to numerous problems in image processing (43, 44), but have only recently been used in the context of cryoEM. Inverse problems can also be solved using an approach based on Bayesian statistics, which has been applied successfully for medical imaging (45–48) but has not been thoroughly considered for cryoEM.

The improved pipeline includes variational formalizations and priors tailored specifically to cryoEM, a reconstruction loop that includes elucidated structural information using new regularizers, and superior numerical techniques for the resulting problems. Based on either a classical variational or Bayesian statistical framework, new regularization methods have been devised that target specifically the challenges presented by cryoEM (SP and ET) reconstruction. While traditional 3D reconstruction iteratively refines the results by re-projecting the 3D map for improved particle averaging and alignment, one is now able to use a comprehensive refinement loop including regularizers/priors, based on elucidated 3D ultrastructure (see Figure 15.5).

The cryo-EM reconstruction using variational/statistical regularization seeks the best 3D map f given a set of noisy projection images g_d (tilt series in the case of electron tomography). The variational approach to this problem defines the "best" map as the minimizer of an energy function of the form of $J(f) = J_1(Hf, g_d) + J_2(f)$ where J_1 is a data fidelity term; that is, a penalty when Hf does not match the input data, and J_2 is a regularization term. For well-posed problems, simply minimizing data-fidelity is sufficient (and preferred) but for ill-posed problems (resulting from noise in the data) a regularization term is necessary to suppress instability in the direct inverse operator. The most straightforward approach to regularization is to limit large variations in f caused by noise by penalizing the L^2- or L^1-norm of the gradient of f. This type of regularization

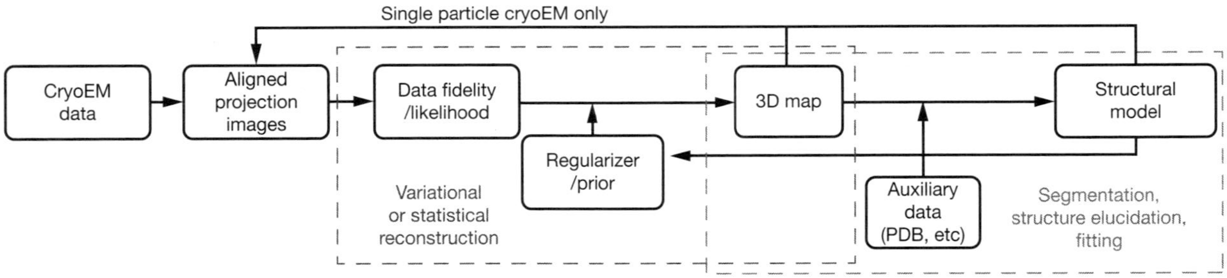

Figure 15.5 General iterative procedure in the variational/statistical regularization methods for cryo-EM reconstruction. While existing methods for single particle cryo-EM use the resulting 3D map to improve the classification and alignment of input images, a more comprehensive framework utilizes resulting structural models both in image alignment and directly in shape regularization to help improve the quality of the final reconstruction.

has been successful in the context of other inverse problems (49) and has been applied to cryo-EM reconstruction in our preliminary work (21, 50, 51). Searching for a 3D map that minimizes the energy functional is performed via the gradient flow:

$$\int \partial_t f \phi + \delta\left(J_1\left(f, g_d\right), \phi\right) + \delta\left(J_2\left(f\right), \phi\right) = 0 \quad \text{[Eq. 5]}$$

where is the variation of functional J_i. For large times, this flow approaches a minimum (at least locally) of the energy. Efficient methods for computing gradient flows have been studied in many contexts (e.g., (62–64)).

Alternatively, reference (3) applies statistical methods to cryo-EM reconstruction using shape-based regularization. In the Bayesian statistical setting, the probability of map f leading to observed images g_d is

$$p\ (f \mid g_d) = p\ (g_d \mid f)\ p\ (f)\ /\ p\ (g) \quad \text{[Eq. 6]}$$

where p (·) and p (· |·) and denote the appropriate probability and conditional probability, respectively.

One then searches for and computes the maximum-a-posteriori (MAP) estimator defined as the 3D map f which maximizes the conditional probability $p\ (f \mid g_d)$; that is, the likelihood that f is the correct map given the data (projection images) g_d. In the Bayesian setting, we are able to maximize $p\ (g_d \mid f)\ p\ (f)\ /\ p\ (g)$ and since the data g_d is fixed, only $p\ (g_d \mid f)\ p\ (f)$ must be considered. $p\ (g_d \mid f)$ can be computed based on the noise model for the imaging process. The term $p\ (f)$ is modeled as a Gibbs distribution $p\ (f) = \exp\ (-S(f))$ where S is an appropriately normalized energy functional; in this initial work, S is a shape-based functional which encodes some local information about the structure (65). Minimizing the negative logarithm of the posteriori gives a form $c\ (f) = log\ (p\ (g_d \mid f)) + \log\ (p\ (f))$; the result is an optimization problem similar to one as seen in the classical variational approach discussed earlier. In certain circumstances, the MAP estimator can be determined using an iterative scheme based on a penalized EM method (56):

$$f^{i+1} := \frac{f^i}{1+\lambda\left|\nabla S\left(f^i\right)\right|} H^T \frac{g_d}{f^i}, \quad \text{[Eq. 7]}$$

where $S(f^i)$ acts as the regularizer, _ a regularization parameter, and H^T is the back-projection operator.

Figure 15.6 shows an example of the variational reconstruction on real SP images. In this case, our variational method outperforms standard methods (as implemented in Xmipp). Statistical reconstruction has been applied to an ET data set of the Simian Immunodeficiency Virus. Shape regularization has been used to generate a prior for the critically important spike protein and applied to spikes observed in the data. This reconstruction significantly improves the measured FSC in the result and (more importantly) can be seen to produce qualitative improvement to the resulting map away from the spikes where the prior is enforced; see Figure 15.7.

Superior stable and efficient numerical methods for solving gradient flows are required to minimize the variational energy functional. These numerical solutions

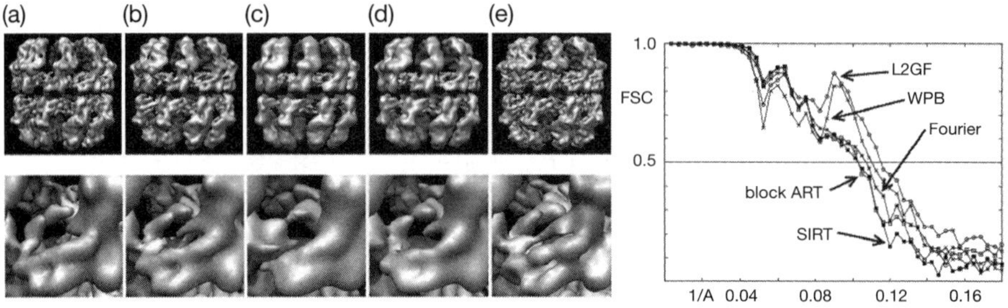

Figure 15.6 Reconstruction of group II chaperonin of the archaea Acidianus tengchongensis strain S5T from 9310 images (53). (top) The reconstructed density maps using WBP (a), Fourier method (b), block ART (138) (c), SIRT (20) (d), and L2GF (50, 51) (e). These maps have been filtered at their resolutions given by the FSC, and normalized in the same way. (bottom) In an enlarged portion, we see a clearer α-helix in the L2GF map. The resolutions computed from the FSC plot (right) for WBP, Fourier, block ART, SIRT, and L2GF (our variational method) are 9.07, 9.35, 9.72, 9.73, and 8.69 Å, respectively. All other methods are applied as implemented in Xmipp.

to optimization naturally lead to a class of operator splitting methods (which are designed to approach each term of the solution in a computationally optimal way) but are subject to numerical instability without carefully analysis. Structural functional and biological unit elucidation provides an alternative to a single composite measure (Fourier Shell Correlation) of the resolution of a map. It additionally allows one to capture the spatial heterogeneity in the reconstructed resolution especially for macromolecular assemblies as multiple instances of a protein/nucleic acid are recoverable from a single map. This information also enables weighted alignment and averaging to obtain the best signal-to-noise model of structure and its functional subunits.

3. MOLECULAR STRUCTURE INTERPRETATION AND VISUALIZATION FROM 3D EM

Identification of key structural features is an essential step in inferring the functional behavior of a biomolecular complex from a 3D map. Since experimental and computational differences in cryo-EM have lead to a wide range of resolutions (4–30^{+}Å), structural modeling encompasses a wide range of computational modeling procedures from coarse segmentation of large functional groups (at poor resolution) to backbone tracing and fitting of atomic models (at the best resolutions). Broadly, the structural interpretation process begins making poor resolution distinctions and progressively refines until the data is insufficient to reliably identify further features. We divide this general procedure into four types of operations: curation, segmentation, ultra-structure elucidation, and fitting.

3.1. Map Filtering and Curation

Filtering (or denoising) stands for the preservation of the most important features (e.g., helices and sheets in proteins) and suppression of unimportant details, and

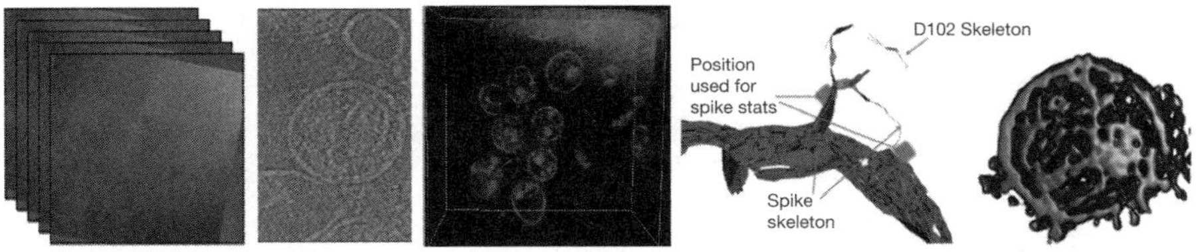

Figure 15.7 Reconstruction of the SIV virus from electron tomography. (left-to-right) Stack of ET images (data courtesy S. Subramaniam) containing 70 four-megapixel images; zoomed image of a single virus; reconstruction using a variational approach (51) (5123 voxels); model of protein spike for identification and alignment based on about 5000 spikes found over 500 virions; shaded rendering of a viral capsid reconstructed using shape-based regularization from the spike model for eight spikes in this particular sample (55).

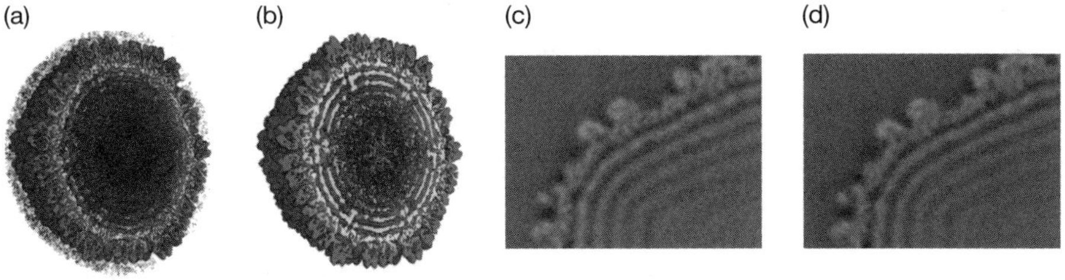

Figure 15.8 (a)–(b) Before and after bilateral filtering. (c)–(d) before and after anisotropic filtering.

hence we encapsulate it here under curation. These features could be described by a voxel or collection of connected voxels forming curves, or surfaces in the 3D map (see Figure 15.8). In the algorithms we describe here the focus is on critical points (local minima, local maxima, and saddle points) of the 3D map density (or reconstructed intensity) function (75). While the local maxima of this 3D function are of relevance in localizing the structures within the 3D map, since higher density values imply the existence of more atoms per unit volume, the saddle and minima prove useful in identifying symmetries and also complementary space regions. As noise is always present in any reconstructed map obtained from experimental data, a pre-filtering process is often advantageous. Filters can be either linear or nonlinear. Linear filters (e.g., Gaussian filters) may suppress weak features, and thus eliminate some critical points, whereas nonlinear filters (57, 58) tend to "flatten" regions, and thus create unwanted critical points.

One approach utilizes a bilateral, pre-filtering, coupled with an evolution-driven, anisotropic, geometric diffusion PDE (partial differential equation) (95). This has been shown to enhance the feature analysis and modeling of macromolecular density maps. The PDE model is: $\partial_t \phi - \|\nabla\phi\| \operatorname{div}\left(D^{\sigma} \frac{\nabla\phi}{\|\nabla\phi\|} \right) = 0$ where Φ is the parameterized image intensity function. The efficacy of this method requires a careful selection of the anisotropic diffusion tensor, $D\sigma$, which is based on estimates of the normal, two principal curvatures, and curvature directions of a feature isosurface (level-set) in

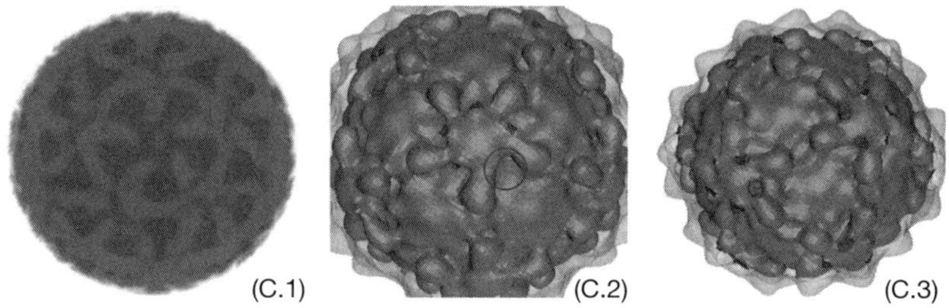

Figure 15.9 An isosurface of the viral capsid of Nodavirus is selected that brings out the inherent symmetry. The isosurface shown in the center is lacking a pocket in one position of the pentametric symmetry (circled), while the isosurface on the right has symmetric geometric and topological features.

3D. Detection of critical points of a reconstructed 3D map can be achieved using anisotropic vector diffusion.

Curation refers to modifications to the 3D map to avoid undesirable characteristics (see Figure 15.9). Solvent flattening (59) is one particular modification: after identifying a boundary of "the particle" in the 3D map, density of the 3D map in the exterior region is flattened under the assumption that this only represents noise in the data. This basic idea is commonly applied in many forms and variations, such as (60–63).

The 3D map curation primarily has been aimed at identifying and correcting surface-based models derived from level sets of the 3D map. Working beyond the fundamental tools of the contour tree (64) and the contour spectrum (65), techniques have been developed for identifying appropriate level sets in a 3D map and ensuring the resulting surfaces and enclosed and complementary volumes have consistent/desirable topological and geometric features (66, 67). Closely related is the work on a higher-order level-set method for molecular surface construction designed to avoid small non-physical pockets or over-inflation which are characteristics of standard Gaussian molecular surfaces (68).

3.2. Symmetry Detection and Segmentation of Structural Units

Symmetry Detection

In general, macromolecular machines can have symmetries of different types, including helical (e.g., tobacco mosaic virus), icosahedral (e.g., rice dwarf virus), and *n*-fold cyclic symmetry (e.g., GroEL). The most direct method for detecting the symmetry axes simply involves correlating the original map with its rotated map (according to the type of symmetry) and searching the resulting correlation map for the peaks (73). However, this method has a very high computational cost, as the time complexity is in the order of $O(NM)$, where N is the number of voxels and M is the number of possible orientations to be searched. Although a number of techniques can be employed to speed up the searching process by reducing the number of the angular bins (e.g., principal axis method (74) or a hierarchical coarse to fine approach), it is still expensive if N is very big. Paper (75) introduced an alternative method for fast detection of rotational symmetries, given that the rotational

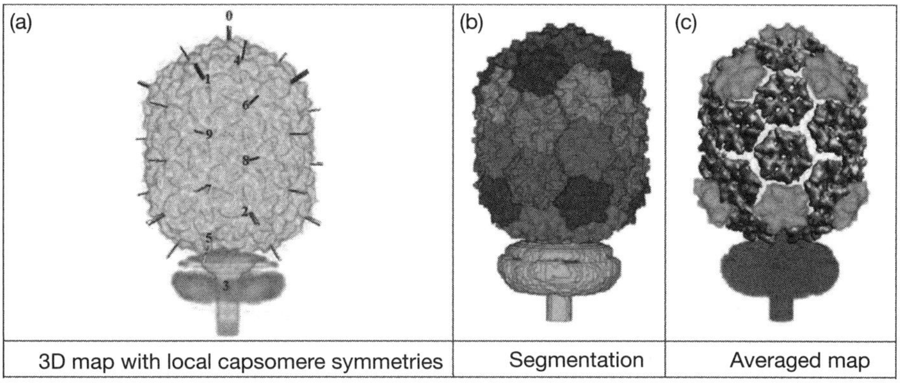

Figure 15.10 Illustrations of segmentation and averaging of Phi 29. (a) The reconstructed cryo-EM 3D map of Phi 29. The symmetry axes are detected using the automatic method (76) and a total of ten independent structure units are labeled. (b) The structure units are segmented based on the segmentation method (76, 77), combined with 3D alignment approach (78). (c) The segmented pentons and hexons can be averaged and the whole structure can be reconstructed from the averaged structure units based on the 3D alignment matrices.

symmetry group is known (76). *N*, the number of voxels to be tested, is reduced by restricting the map search only to a subset of the critical points instead of the entire volume. This strategy tremendously reduces the computational cost in detecting the symmetry of a given volumetric map (both globally and locally).

An algorithm for identification of asymmetric subunits of a density map is as follows (76, 77). Once the symmetry is detected, the critical points are automatically classified based on their symmetrically equivalent positions. As the critical points are assigned, equivalent points are assigned membership in the symmetrically related components.

However, critical points in different subunits are classified with different memberships; the same membership is assigned to critical points only within one subunit. Once all critical points (seeds) are indexed and assigned memberships, the traditional fast marching method (79–81) can be used with the following modifications. First, each object may consist of a number of seeds instead of just one. Secondly, since each seed initiates a marching contour and all contours start to grow simultaneously and independently, each seed (and accordingly, the marching contour) must be attached with a membership index based on the classification of seeds. Once a voxel is conquered by a marching contour, it should be assigned with the same index of the marching contour. Thirdly, two marching contours with the same index should merge into one when they meet, while two marching contours with different indexes should stop on their common boundaries. This idea is known as the multi-seeded fast marching method and has been used elsewhere (82, 83). Figure 15.10 shows an application of symmetry-based segmention for the Phi29 virus.

Segmentation, a classical image processing challenge, aims to separate functional subunits (i.e., the asymmetric subunit of a virus or individual proteins of a complex) typically relying on 3D map intensity to derive boundaries between the subunits (see Figure 15.11). Segmentation often relies on manual work with an assistance of

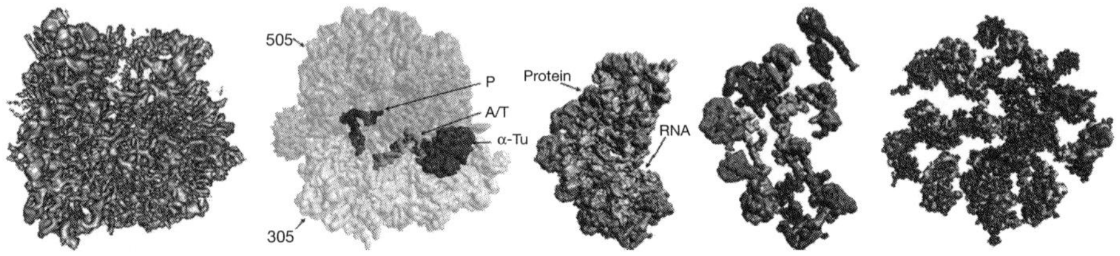

Figure 15.11 Segmentation of a 6.4 Å map for the 70S ribosome (72). (left to right) Original 3D map; segmentation into five primary subunits: 30S, 50S, EF-Tu, P site, and A/T site; further segmentation of the 30S subunit into protein and RNA parts; fine segmentation of the 30S proteins; segmentation of the 50S proteins after fitting with the associated X-ray structure. See insert for color version.

a graphical user interface (84, 85), and the process can still be both tedious and subjective (86, 87). Automated segmentation is still recognized as one of the hardest tasks in the field of image processing although various techniques have been suggested for automated or semi-automated segmentation. Commonly used methods include segmentation based on edge detection, region growing and/or region merging, active curve/surface motion, and model-based segmentation (88–91). The water-shed immersion method (92, 93) and a method based on normalized graph cuts and eigenvector analysis (94) have been applied successfully to data from electron tomography, but processing time can still pose a severe bottleneck (94).

There are two popular categories of segmentation approaches. The first is a pipeline consisting of six steps: (i) bilateral filtering coupled with an evolution-driven, anisotropic, geometric diffusion PDE is used to preserve important molecular features (78, 95), (ii) automatic seed point generation using map critical points, (iii) multi-seeded fast marching methods are applied to segment 3D maps into individual domains of interest (76, 96), (iv) a cloud of boundary/interface points are constructed in the interface between segmented regions from which domain bounding surfaces are constructed (97, 98) and curated (99), (v) geometric PDEs are used to produce smooth B-spline domain surfaces (68), and (vi) similar subunits are identified, aligned, and averaged when appropriate (78). The second segmentation method (100, 101) is particularly promising as applications of variational methods to image processing are relatively recent (87, 102, 103). High computational demands of PDEs to be solved in this formulation make parallelization necessary for use on practical 3D maps and GPU based software has provided significant speed-ups (139).

3.3. Alignment and Averaging of Structural Units

While segmentation algorithms described in the previous subsection can decompose 3D virus maps into individual structure units (also called *subunits* below), it does not tell us how different the segmented subunits are. From the structural point of view, it is important to know the similarity between the segmented subunits and quite often useful to average the subunits of high similarity in order to improve the signal-to-noise ratio. To this end, a fast algorithm (78) has been developed to align the segmented subunits such that the similarity measure and averaging can be conducted between the spatially aligned subunits. Technically, the goal of 3D structure alignment is to find the transformation matrix from one 3D structural unit to another, such that the two 3D maps are best matched according to a similarity scoring function. Given two 3D maps *f* and *g*, a similarity scoring function can be defined in the following way:

$$S_{f,g}(T)=1-\frac{\sum_{m=1}^{M}\left|f(c_m)-g(T(c_m))\right|+\sum_{n=1}^{N}\left|f(T^{-1}(d_n))-g(d_n)\right|}{\sum_{m=1}^{M}\max\{f(c_m),g(T(c_m))\}+\sum_{n=1}^{N}\max\{f(T^{-1}(d_n)),g(d_n)\}}, \qquad \text{[Eq. 8]}$$

where c_m ($m = 1, 2, \ldots, M$) are critical points of *f*, and d_n ($n = 1, 2, \ldots, N$) are critical points of g, and T is a 4×4 matrix using homogeneous coordinates, which is intuitively treated as a function that transforms the coordinate system of *f* to that of *g*. To speed up the correlation search one computes the similarity score on a set of critical points of the 3D maps, instead of the entire collection of voxels that make up the 3D maps of the structure units. The critical points are local maxima, local minima, and saddle points that best capture the features of a molecular density 3D map. If the data is noisy, one initially preprocesses the 3D map using gradient vector diffusion as discussed in (76). The transformation from one structural unit to another (expressed by *T*) has two degrees of freedom: one translation *t* and one rotation *r*, which are determined based on the scoring function. The search space for *t* and *r* can be reduced using a two-level hierarchical method, which uses a coarse sampling of *t* and *r* followed by a dense one, and the search space for *r* can be reduced even further by exploiting the symmetry of the structural units, if any.

If the segmented structure units of a large biomolecular complex (such as viruses) have high similarities, they can be averaged such that the structural analysis of the entire complex can be simplified to the analysis of a single averaged structure unit. Averaging two structure units is done by aligning one of them to the other and taking the average density values of both. In general, the averaged map has a higher signal-to-noise ratio than each individual structure unit, which makes it easier to further analyze the structures (e.g., secondary structure identification). Figure 15.10 shows segmentation and averaging of the structural subunits of Phi 29.

3.4. Ultra-Structure Elucidation

Here we briefly describe three recent algorithms for structural unit identification in quasi-atomic resolution 3D cryo-EM maps: one uses the volumetric Morse graph and the second relies on unstable manifolds of the distance function associated with the molecular surface, and the third skeletonizes the molecule and identifies skeletal components matching familiar structures. These algorithms are more efficient than previously reported methods since they effectively utilize the local topological structure of the map instead of searching exhaustively. Each method has been designed a different target resolution (4–9 Å, >9 Å, and <4 Å, respectively), but precise characterization of the domains of applicability and limitations is still necessary.

The first approach is a volumetric Morse graph method ideally suited for identifying secondary structures (α-helices and β-sheets) in intermediate resolution (5–8Å) 3D maps. Here one performs voxel classification that determines if a voxel belongs to a linear or planar region (78). For every voxel in the 3D map, a 3×3 structure tensor is computed which records the second partial derivatives of the intensity values weighted by a Gaussian kernel. The eigen values l_1, l_2, l_3 (sorted from maximum to minimum) of this structure tensor are then computed and can be used to identify α-helices and β-sheets. For 3D maps of medium-resolution (i.e., finer than 9 Å), this technique identifies most of the α-helices and β -sheets.

While several tests of a structure tensor-based approach (76) have demonstrated its performance for map resolutions of (5–8 Å), the method seemed unable to detect all intertwined helices/sheets for maps of coarser resolution. An alternative algorithm seems to produce more reliable secondary structure elucidation results for maps with resolutions of (>9 Å) as well as demonstrates the ability to identify tertiary folds of even coarser maps (>13 Å) (104, 105). The method exploits the critical point structure of the distance function induced by a suitable molecular surface generated from the 3D map. Such a molecular surface generated from the 3D map, denoted as S, encodes a lot of information about its structural features.

To detect the secondary structural motifs, one first computes a synthetic distance function that assigns to every point in three dimensions its distance to the nearest point on S. The distance function has four types of critical points, namely *maxima, minima, index-1 saddle*, and *index-2 saddle* points. For the purpose of tertiary fold/secondary structure elucidation, we only need index-1 and index-2 saddle points that are interior to the molecular surface. Any noncritical point has a nonzero gradient vector associated with it and if one collects all the points for which the path connecting them end at a critical point, that gives a set which is called an *unstable manifold* of the critical point. A computational approach to compute the unstable manifolds of index-1 and index-2 saddle points from the Voronoi and Delaunay diagram of a point sampling of the molecular surface is described in a recent paper (105). The unstable manifolds of an index 1 saddle point is two-dimensional (medial set of sheets), whereas the unstable manifold of an index 2 saddle point is one-dimensional (axis of helices). Using sizes of known molecular structural elements (e.g., helices (16) and sheets (107–109)), the unstable manifolds can be filtered to produce quasi-atomic structural motif models.

A third method targeted to finer resolution maps relies on skeletonization of the volumetric domain (110, 111). The skeleton is a set of one- and two-dimensional manifolds representing the shape of the domain. The skeleton provides an effective guide when searching for helices (features which are essentially one-dimensional) and sheets (features which are essentially two-dimensional). The primary computational challenge lies in finding the skeleton, which results from a "fire front" evolution. First an initial front is selected as an isocontour of the 3D map. Points on this front are evolved according to the equation $\partial_t p = \alpha \nabla f(p)$. When fronts collide, they merge (which can be handled in a numerically robust fashion) eventually forming a one-dimensional curve in some sections of the map. Figure 15.12 shows examples of the different algorithms explained above.

The simplest robust way to validate results from secondary structure detection is to count the number of helices and sheets detected. If the crystal structure of the

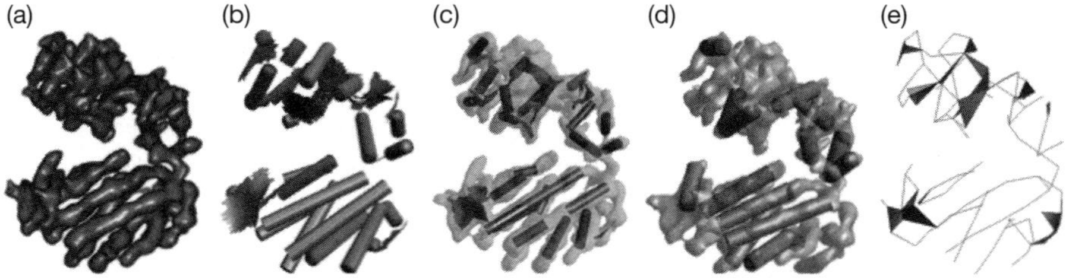

Figure 15.12 Results of the three secondary structure elucidation methods from a 4.2 Å map for GroEL (106). (a) Input 3D map. (b) Structure tensor method. (c) Unstable manifold method. (d) Skeletonization-based method. (e) Skeleton computed yielding the results in (d).

molecule is available, one can count false positives and false negatives, as in (17, 71, 75, 76) comparing to either the secondary structures listed in the PDB file or those automatically computed (104, 108, 109). Errors in the number of identified sheets and helices can be further quantified by constructing a more careful metric that considers the error in the spatial location of the structures and the error in the size (e.g., helix length) of each structure.

3.5. Match and Fit: X-Ray Model with 3D Map

Match and fit methods automatically align X-ray structure models at near atomistic resolution (typically from the PDB) with reconstructed 3D EM maps at varying resolution. The problem of fitting atomic models into 3D maps is a relatively well-studied problem based on both rigid (112–122) and flexible (123–127) models.

Leveraging tools from molecular docking software using the fast Fourier transform to accelerate the translational portion of the search (129, 130), papers (126, 127) have developed for 3D map-atomic model fitting into 3D EM. The most recent improvements utilize non-uniform fast Fourier transforms, which have been developed in both the translational and rotational setting (125, 128). This modification has yielded tangible benefits especially for increasing robustness of low-density fitting and fitting a single subunit into a larger map (127).

Fitting could be validated under several metrics (e.g., 17). One popular metric is the external total ratio which counts the number of atoms of the model which lie outside of a suitably selected isocontour of the 3D map. A second is the mutual information score (126, 127), given by

$$MIS = \sum_{x \in E} \sum_{y \in P} p(x,y) \log \frac{p(x,y)}{p(x)p(y)}. \quad \text{[Eq. 9]}$$

where E is the 3D map, P is the blurred atomic model, $p(x)$ and $p(y)$ are the percentage of voxels in E and P that take on intensities equal to x and y, and $p(x,y)$ is the percentage of voxels in E with intensity x that are aligned with voxels in P with intensity y. Since the range of intensities in a density map is typically very high, a binning scheme with about twenty bins is used. Two additional validation metrics are the Z-score that measures the normalized deviation of each candidate result from the mean, and the normal-vector score (127), a geometric metric that measures the deviation of the surface of the density map from the fitted molecular surface.

3.6. Quantitative Visualization

It is equally important for the macromolecular assemblies to be correctly *reconstructed* as well as being efficiently *visualized*. Generally speaking, the reconstructed volumes represent the electron density maps of the structures being studied. Two commonly seen tools can be used to visualize a volumetric map. One is *surface-based rendering* and the other is *volume-based rendering*. Both are available from *VolumeRover* 2.0 (17). The first one requires surface extraction from the volumetric data (e.g., using segmentation techniques) as well as geometric modeling on the surfaces (such as mesh generation). As a contrary, volume rendering requires classifications of input density data (each voxel is assigned a color and opacity value) and calculations of visibility and shading of each voxel.

Besides visualization of the obtained structures, it is often required to topologically/geometrically and combinatorially quantify features (subdomains) from the structural models and 3D maps (17, 65–69, 75, 82, 150–153). While topological and combinatorial quantification is achievable via methods such as the contour tree and contour spectrum (65, 149, 153), geometric quantification such as feature specific area/volume/curvatures are facilitated by producing finite element surface and volumetric meshed models of molecular domains (140, 152). These finite element meshes are additionally used to derive and quantify biophysical properties of the macromolecules of interest (134, 154). See Figure 15.13 and Figure 15.14 for examples.

Several algorithms have been developed for building quality surface meshes both from atomic (131) and imaging (132–136) data. We briefly describe a couple of algorithms that have been implemented in *VolumeRover* 2.0.

One of them utilizes a tetrahedral mesh generation procedure that generates a simplicial decomposition of the bounded domain of the imaging data (131, 152) (see also Figure 15.13). The boundaries of the volumetric mesh are topologically equivalent to the isocontour (a pair of isocontours in the case of interval volume). More specifically, they are manifold, crack-free, and have the same genus as that of the solid bounded by the isocontour. The meshes are adaptive and geometrically close. The adaptivity is achieved through octree subdivision and geometric fidelity is guided by the user. The resulting mesh is optimal in terms of number of simplices up to the level of octree subdivision since one does not need to add any extra point. We have incorporated feature-sensitive adaptivity in our mesher to control the density of the mesh while preserving sufficient resolution in the regions of interest. The mesh adaptivity is controlled by the user interaction or a feature sensitive error function. The octree cells are either subdivided in the regions of interest to respect the contour features closely, or they are merged to generate a coarser representation. Each octree leaf cell is represented by a point obtained by minimizing a Quadratic Error Function (QEF) (137). However, the simplest star-ring scheme conforming to the triangulated level set produces a tetrahedral mesh

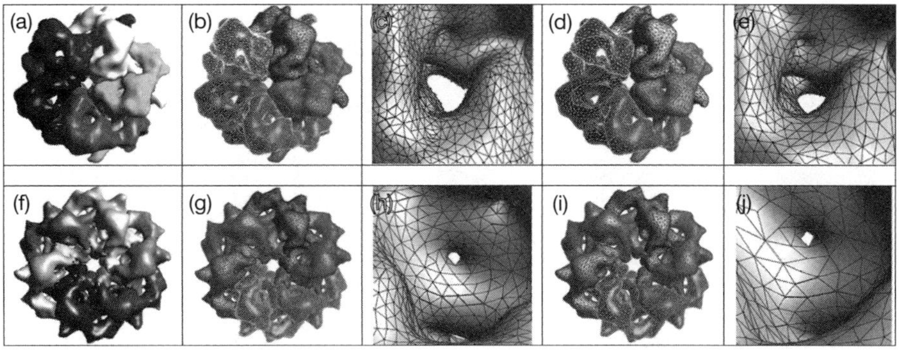

Figure 15.13 Triangular surface meshing of Bacteriophage P22 subunits (pentons and hexons): (a) each penton is composed of five monomers (after decimation and improvement), (b) segmented penton mesh after decimation and quality improvement, (c) zoom of (b), (d) segmented penton mesh after further decimation and improvement, (e) zoom of (d), (f) each hexon is composed of six monomers (after decimation and improvement), (g) segmented hexon mesh after decimation and quality improvement, (h) zoom of (g), (i) segmented penton mesh after further decimation and improvement, (j) zoom of (i).

with cracks and hanging vertices because an octree leaf cell can have neighbors of different subdivision level. To circumvent this problem one needs to star each cell into tetrahedra, taking the subdivision levels of the neighbors into account. The quality of a mesh is measured by the shape of the worst-shaped element. The mesh quality is improved by edge contraction and geometric flow smoothing (140). Figure 15.13 shows triangular surface meshing of the pentameric and hexameric subunits of bacteriophage P22 using this algorithm.

A recent algorithm (132), also implemented in *VolumeRover* 2.0, describes an approach to generate quality triangular/tetrahedral meshes for complicated biomolecular structures directly from the PDB format data, and conforming to a good implicit solvation surface approximation. There are three main steps in our mesh generation process. First, a smooth implicit solvation model is constructed to approximate the molecular surface by using weighted Gaussian isotropic atomic kernel functions and a two-level clustering technique. In the second step, a modified dual contouring method is used to extract triangular and interior/exterior tetrahedral meshes, conforming to the implicit solvation surface. Finally, geometric flows (140) are used to improve the quality of extracted triangular and tetrahedral meshes.

VolumeRover 2.0 encapsulates many of the image processing, reconstruction, segmentation, fitting, secondary and supersecondary elucidation, and macromolecular modeling algorithms described here. *VolumeRover* 2.0 additionally allows the user to easily visualize scalably large raw and processed Electron Microscopy data, including reconstructed volumes, and interactively to adjust the transfer function for combined isosurface and volume visualization (Figure 15.14). Users are able to immediately display results of the encapsulated image and map processing procedures, adjust all the relevant parameters, and then perform subsequent calculations.

The *VolumeRover* 2.0 package has been made freely available to academic users via the website (http://cvcweb.ices.utexas.edu/cvcwp/?page_id=100).

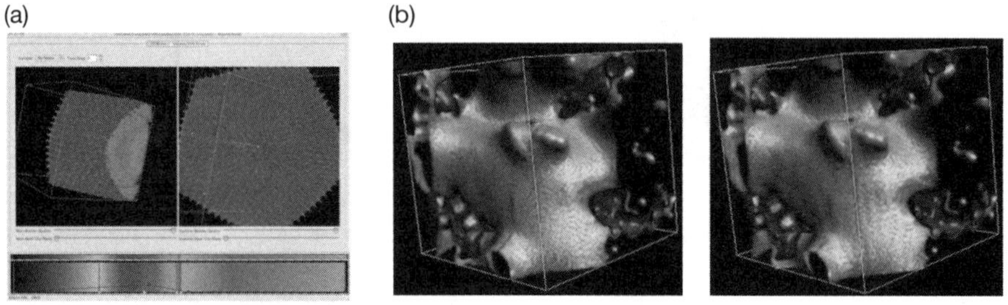

Figure 15.14 High-performance quantitative visualization in *VolumeRover*. (a). *VolumeRover* rendering of the chillo iridescent virus (EMDB 1580) where the right panel shows a low-resolution rendering of the virus and the left panel shows the rendering of the portion of the density map which is inside a moving rover. (b) Comparison of uncompressed (left) and compressed (right) volume rendering shows no visible difference despite 41:1 compression ratio (155). See insert for color version.

4. FUTURE CONSIDERATIONS

Obtaining and interpreting 3D density maps from electron micrographs requires a variety of computational science approaches, some dating back half a century, while some have only become available in recent years, in part due to the increased computational power of even moderate desktop machines, and in part to the increased collaboration between life sciences, imaging scientists, and the computational sciences community. This interface, also known as bioimage informatics (141), increasingly attracts the attention from scientists (typically biologists/ biophysicists) that create three- and multidimensional data sets, as well as from the computational scientists that offer a variety of solutions for the problems encountered in creating meaningful biological information from the wealth of voxels that can be recorded with ever-increasing speed and quality. This chapter mainly focuses on cryo-EM single particle and electron tomography analysis, due to its popularity for structure determination of macromolecular machines that cannot be crystallized and hence are beyond the reach of X-ray diffraction techniques. It also reflects the maturity of cryo-EM single particle analysis, where densities provided by cryo-EM can be interpreted with increasing accuracy and speed, as well as level of automation.

Many approaches can be extended to the much younger field of cryo-electron tomography, where better reconstruction algorithms, and better means for interpretation of the tomograms are urgently needed. Challenges arise from the higher complexity of the 3D data, the lack of precise knowledge of the composition and 3D organization of its macromolecular inventory, as well as the lack of resolution and the anisotropy and the artifacts introduced by the limited tilt angle imaging and reconstruction.

Yet another nascent field is the large volume 3D imaging of entire cells and tissues by focused ion beam scanning electron microscopy (FIB/SEM) (142) and serial block face scanning electron microscopy (SBF/SEM) (143), where we no longer can pretend to recognize proteins and complexes by shape alone due to the resolution limit of the data and the necessity of staining for sufficient contrast. Instead, we must rely on half a century of biological ultrastructural and cell biological knowledge for data annotation and interpretation. Yet, in this nascent discipline we strive for volumetric quantification to describe changes for cells and tissues in health and disease or in development. The complexity, information content, and sheer size of such data sets with tens of thousands of voxels in X, Y, and Z make computational approaches for deriving biological meaning from voxels not a luxury but the key ingredient for a successful analysis, and will require advances in large volume multiresolution, multiscale visualization (potentially through immersive approaches), in segmentation, annotation, volumetric measurements, and data abstraction, as well as in animation, modeling building, and simulation.

It is clear that the ties between the biological imaging community and the computer science community will be become increasingly closer and that we are on the brink of integration of imaging approaches either at the level of experimentation or at the level of analysis and representation, allowing more of a systems biology understanding of complex biological processes.

As stated in (12), current software packages on single particle processing have already been well developed but still have several open problems for solutions from computer science fields (e.g., image processing, pattern recognition, and artificial intelligence). These problems include automated algorithms of particle picking, robust image classification and alignment, and detection of inhomogeneous particles. Besides all of these, detection of substructures also requires new techniques from the fields mentioned above. Current solutions for these problems are either manual, which is always time-consuming and quite tedious, or by semi-automatic methods, which did not guarantee satisfying accuracy especially in the presence of noise.

REFERENCES

(1) J. Frank. Three-dimensional electron microscopy of macromolecular assemblies: visualization of biological molecules in their native state. Oxford University Press, USA, 2006.

(2) J. Frank. Electron tomography: methods for three-dimensional visualization of structures in the cell. Springer Verlag, 2006.

(3) R. M. Glaeser, K. H. Downing, D. DeRosier, W. Chiu, and J. Frank. Electron crystallography of biological macromolecules. Oxford University Press, USA, 2007.

(4) B. Böttcher, S. A. Wynne, and R. A. Crowther. Determination of the fold of the core protein of hepatitis B virus by electron cryomicroscopy. Nature, 386:88–91, 1997.

(5) J. F. Conway, N. Cheng, A. Zlotnick, P. T. Wingfield, S. J. Stahl, and A. C. Steven. Visualization of a 4-helix bundle in the hepatitis B virus capsid by cryo-electron microscopy. Nature, 386:91–94, 1997.

(6) R. Matadeen, A. Patwardhan, B. Gowen, E. V. Orlova, T. Pape, M. Cuff, F. Mueller, R. Brimacombe, and M. van Heel. The Escherichia coli large ribosomal subunit at 7.5 Å resolution. Structure, 7(12): 1575–1583, 1999.

(7) M. van Heel, B. Gowen, R. Matadeen, E.V. Orlova, R. Finn, T. Pape, D. Cohen, H. Stark, R. Schmidt, M. Schatz, and A. Patwardhan. Single-particle electron cryo-microscopy: towards atomic resolution. Quart. Rev. Biophys., 33(4):307–369, 2000.

(8) B. L. Trus, R. B. S. Roden, H. L. Greenstone, M. Vrhel, J. T. Schiller, and F. P. Booy. Novel structural features of bovine papillomavirus capsid revealed by a three-dimensional reconstruction to 9 Å resolution. Nat. Struct. Mol. Bio., 4(5):413–420, 1997.
(9) Z. H. Zhou, M. Dougherty, J. Jakana, J. He, F. J. Rixon, and W. Chiu. Seeing the herpesvirus capsid at 8.5 Å. Science, 288(5467):877, 2000.
(10) R. M. Glaeser. Cryo-electron microscopy of biological nanostructures. Phys. Today, 61(1):48, 2008.
(11) C. V. Robinson, A. Sali, and W. Baumeister. The molecular sociology of the cell. Nature, 450(7172): 973–982, 2007.
(12) J. Zhu, L. Cheng, Q. Fang, Z. H. Zhou, and B. Honig. Building and refining protein models within cryoelectron microscopy density maps based on homology modeling and multiscale structure refinement. J. Mol. Biol., 397(3):835–851, 2010.
(13) R. M. Glaeser and R. J. Hall. Reaching the information limit in cryo-EM of biological macromolecules: experimental aspects. Biophys. J., 100:2331–2337, 2011.
(14) C. Suloway, J. Pulokas, D. Fellmann, A. Cheng, F. Guerra, J. Quispe, S. Stagg, C. S. Potter, and B. Carragher. Automated molecular microscopy: the new Leginon system. J. Struct. Bio., 151(1):41–60, 2005.
(15) X. Yan, R. S. Sinkovits, and T. S. Baker. AUTO3DEM–an automated and high throughput program for image reconstruction of icosahedral particles. J. Struct. Bio., 157(1):73–82, 2007.
(16) W. Jiang, M. L. Baker, S. J. Ludtke, and W. Chiu. Bridging the information gap: computational tools for intermediate resolution structure interpretation. J. Mol. Biol., 308(5):1033–1044, 2001.
(17) Q. Zhang, R. Bettadapura, and C. Bajaj Macromolecular Structure Modeling from 3DEM using VOLROVER 2.0 *Biopolymers, September, 97(9):709–731, 2012*
(18) J. Zhu, P. Penczek, R. Schroder and J. Frank, *Three-dimensional reconstruction with contrast transfer function correction from energy-filtered cryoelectron micrographs: procedure and application to the 70s escherichia coli ribosome*, J. of Structural Biology, Vol.118, pp.197–219, 1997.
(19) M. Van Heel, G. Harauz, E. Orlova, R. Schmidt,and M. Schatz, A new generation of the IMAGIC image processing system, J. Struct. Bio., 116:17–24, 1996.
(20) J. Frank, M. Radermacher, P. Penczek, J. Zhu, Y. Li, M. Ladjadj, and A. Leith, SPIDER and WEB: Processing and visualization of images in 3D electron microscopy and related fields, J. Struct. Bio., 116:190–199, 1996
(21) S. J. Ludtke, P. R. Baldwin, and W. Chiu, EMAN: semiautomated software for high-resolution single-particle reconstructions, J. Struct. Bio., 128:82–97, 1999.
(22) M. Radermacher. Three-dimensional reconstruction from random projections: orientational alignment via Radon transforms. Ultramicroscopy, 53(2):121–136, 1994.
(23) C. O. S. Sorzano, S. Jonic, C. El-Bez, J. M. Carazo, S. De Carlo, P. Thévenaz, and M. Unser, A multiresolution approach to orientation assignment in 3D electron microscopy of single particles. J. Struct. Bio., 146(3):381–392, 2004
(24) M. Radermacher. Weighted back-projection methods. In J. Frank, editor, Electron Tomography: Methods for Three-Dimensional Visualization of Structures in the Cell, pages 245–274. Springer, 2006.
(25) F. Natterer and F. Wübbeling. Mathematical methods in image reconstruction. SIAM, 2001. ISBN: 0898714729.
(26) R. Gordon, R. Bender, and G. T. Herman. Algebraic Reconstruction Techniques (ART) for three-dimensional electron microscopy and X-ray photography. J. Theo. Bio., 29(3):471–481, 1970.

(27) P. Gilbert. Iterative methods for the three-dimensional reconstruction of an object from projections. J. Theo. Bio., 36(1):105–117, 1972.

(28) A H Andersen and A C Kak. Simultaneous algebraic reconstruction technique (SART): a superior implementation of the ART algorithm. Ultrasonic Imaging, 6(1):81–94, 1984.

(29) R. Marabini, G.T. Herman, and J.M. Carazo. 3D reconstruction in electron microscopy using ART with smooth spherically symmetric volume elements (blobs). Ultramicroscopy, 72(1–2):53–65, 1998.

(30) D. C. Youla and H. Webb. Image restoration by the method of convex projections: Part 1-theory. IEEE Trans. Med. Imag., 1(2):81–94, 1982.

(31) M. I. Sezan and H. Stark. Image restoration by the method of convex projections: Part 2-applications and numerical results. IEEE Trans. Med. Imag., 1(2):95–101, 1982.

(32) L. A. Shepp and Y. Vardi. Maximum likelihood reconstruction for emission tomography. IEEE Trans. Med. Imag., 1(2):113–122, 2007.

(33) H. M. Hudson and R. S. Larkin. Accelerated image reconstruction using ordered subsets of projection data. IEEE Trans. Med. Imag., 13(4):601–609, 1994.

(34) J. Frank, B. Shimkin, and H. Dowse. SPIDER–a modular software system for electron image processing. Ultramicroscopy, 6(4):343–357, 1981.

(35) T. R. Shaikh, H. Gao, W. T. Baxter, F. J. Asturias, N. Boisset, A. Leith, and J. Frank. SPIDER image processing for single-particle reconstruction of biological macromolecules from electron micrographs. Nature Protocols, 3(12):1941–1974, 2008.

(36) M. van Heel and W. Keegstra. IMAGIC: A fast, flexible and friendly image analysis software system. Ultramicroscopy, 7(2):113–129, 1981.

(37) G. Tang, L. Peng, P. R. Baldwin, D. S. Mann, W. Jiang, I. Rees, and S. J. Ludtke. EMAN2: an extensible image processing suite for electron microscopy. J. Struct. Bio., 157(1):38–46, 2007.

(38) R. Marabini, I. M. Masegosa, M. C. San Martin, S. Marco, J. J. Fernandez, L. G. De la Fraga, C. Vaquerizo, and J. M. Carazo. Xmipp: an image processing package for electron microscopy. J. Struct. Bio., 116(1): 237, 1996.

(39) C. O. S. Sorzano, R. Marabini, J. Velázquez-Muriel, J. R. Bilbao-Castro, S. H. W. Scheres, J. M. Carazo, and A. Pascual-Montano. XMIPP: a new generation of an open-source image processing package for electron microscopy. J. Struct. Bio., 148(2):194–204, 2004.

(40) D. B. Ress, M. L. Harlow, R. M. Marshall, and U. J. McMahan. Methods for generating high-resolution structural models from electron microscope tomography data. Structure, 12(10):1763–1774, 2004.

(41) S. Nickell, F. Förster, A. Linaroudis, W. D. Net, F. Beck, R. Hegerl, W. Baumeister, and J. M. Plitzko. TOM software toolbox: acquisition and analysis for electron tomography. J. Struct. Bio., 149(3):227–234, 2005.

(42) Y. Liang, E. Y. Ke, and Z. H. Zhou. IMIRS: a high-resolution 3D reconstruction package integrated with a relational image database. J. Struct. Bio., 137(3):292–304, 2002

(43) T. F. Chan and J. Shen. Image processing and analysis: variational, PDE, wavelet, and stochastic methods. SIAM, 2005. ISBN: 089871589X.

(44) T. F. Chan, J. Shen, and L. Vese. Variational PDE models in image processing. Not. AMS, 50(1):14–26, 2005.

(45) G. Gindi, M. Lee, A. Rangarajan, and I. G. Zubal. Bayesian reconstruction of functional images using anatomical information as priors. IEEE/ACM Trans. Med. Imag., 12(4):670–680, 1993.

(46) T. Hsiao, A. Rangarajan, and G. Gindi. Bayesian image reconstruction for transmission tomography using deterministic annealing. J. Elec. Imag., 12:7–16, 2003.
(47) A. Rangarajan, I. T. Hsiao, and G. Gindi. A bayesian joint mixture framework for the integration of anatomical information in functional image reconstruction. J Math. Imag. Vision, 12(3):199–217, 2000.
(48) S. Yoon, A. R. Pineda, and R. Fahrig. Simultaneous segmentation and reconstruction: A level set method approach for limited view computed tomography. Med. Phys., 37:2329, 2010.
(49) L. I. Rudin, S. Osher, and E. Fatemi. Nonlinear total variation based noise removal algorithms. Physica D: Nonlinear Phenomena, 60(1–4):259–268, 1992.
(50) M. Li, G. Xu, C. Sorzano, R. Melero, and C. Bajaj. Electric-potential reconstructions of single particles using L2-gradient flows. In Proc. 3rd Int. Conf. Biomed. Eng. Inf., pages 213–217, 2010.
(51) G. Xu, M. Li, A. Gopinath, and C. Bajaj. Computational inversion of electron tomography images using L^2-gradient flows. J. Comput. Math., 29:501–525, 2011.
(52) A. Caboussat, R. Glowinski, and V. Pons. Numerical methods for non-smooth L1 optimization: applications to free surface flows and image denoising. Int. J Numer. Anal. Model, 6(3):355–374, 2009.
(53) X. Feng and M. Yoon. Finite element approximation of the gradient flow for a class of linear growth energies with applications to color image denoising. Int. J. Numer. Anal. Model, 6(3):389–40, 2009.
(54) D. Boukerroui. Efficient numerical schemes for gradient vector flow. In Proc. 16th IEEE Int. Conf. Image Proc., pages 4057–4060, 2009.
(55) A. Gopinath, G. Xu, D. Ress, O. Oktem, S. Subramaniam, and C. Bajaj. Shape-based regularization of electron tomographic reconstruction. IEEE Transactions on Medical Imaging, 31(12):2241–2252, 2012.
(56) P. J. Green. On use of the EM for penalized likelihood estimation. Journal of the Royal Statistical Society. Series B (Methodological), pages 443–452, 1990.
(57) Perona, P., Malik, J., Scale-space and edge detection using anisotropic diffusion. IEEE Trans. on Pattern Analysis and Machine Intelligence 1990, 12, (7), 629–639.
(58) Weickert, J., Anisotropic Diffusion In Image Processing. ECMI Series, Teubner, Stuttgart, ISBN 3-519-02606-6: 1998
(59) B. C. Wang. Solvent flattening. *Meth. Enzymol*, 115:90–112, 1985.
(60) T. C. Terwilliger. Reciprocal-space solvent flattening. *Acta Crystal. Sect. D: Bio. Crystal.*, 55(11):1863–1871, 1999.
(61) D. E. McRee. Xtalview/xfit–a versatile program for manipulating atomic coordinates and electron density. *J. Struct. Bio.*, 125(2–3):156–165, 1999.
(62) T. C. Terwilliger. Maximum-likelihood density modification. *Acta Crystal. Sect. D: Bio. Crystal.*, 56(8): 965–972, 2000.
(63) K. Cowtan. Recent developments in classical density modification. *Acta Crystal. Sect. D: Bio. Crystal.*, 66(4):470–478, 2010.
(64) M. Van Kreveld, R. Van Oostrum, C., V. Pascucci, and D. Schikore. Contour trees and small seed sets for isosurface traversal. In Proc. 13th Symp. Comput. Geom., pages 212–220, 1997.
(65) C. L. Bajaj, V. Pascucci, and D. R. Schikore. The contour spectrum. In Proc. 8th Conf. Visual., pages 167–173, 1997.
(66) C. Bajaj, A. Gillette, and S. Goswami. Topology based selection and curation of level sets. In A. Wiebel, H. Hege, K. Polthier, and G. Scheuermann, editors, Topology-based Methods in Visualization, pages 45–58. 2009.

(67) C. Bajaj, A. Gillette, S. Goswami, B. J. Kwon, and J. Rivera. Complementary space for enhanced uncertainty and dynamics visualization. In V. Pascucci, X. Tricoche, H. Hagen, and J. Tierny, editors, Topo—logical Data Analysis and Visualization: Theory, Algorithms and Applications, pages 217–228. 2011.

(68) C. L. Bajaj, G. L. Xu, and Q. Zhang. Higher-order level-set method and its application in biomolecular surfaces construction. *J. Comput. Sci. Tech.*, 23(6):1026–1036, 2008.

(69) X. Zhang and C. Bajaj. Extraction, visualization and quantification of protein pockets. In Proc. Int. Conf. Comput. Syst. Bioinf., pages 275–286, 2007.

(70) C. Bajaj and S. Goswami. Secondary and tertiary structural fold elucidation from 3d em maps of macro—molecules. In Proc. Fifth Indian Conf. Comput. Vision Graphics Img. Proc., pages 264–275, 2006.

(71) C. Bajaj, S. Goswami, and Q. Zhang. Detection of secondary and supersecondary structures of proteins for cryo-electron microscopy. *J. Struct. Bio.*, 177(2):367–381, 2012.

(72) J. C. Schuette, F. V. Murphy, A. C. Kelley, J. R. Weir, J. Giesebrecht, S. R. Connell, J. Loerke, T. Mielke, W. Zhang, and P. A. Penczek. Gtpase activation of elongation factor ef-tu by the ribosome during decoding. *EMBO J.*, 28(6):755–765, 2009.

(73) Masuda, T.; Yamamoto, K.; Yamada, H., Detection of partial symmetry using correlation with rotated-reflected images. *Pattern Recognition* 1993, 26, (8), 1245–1253.

(74) Sun, C.; Sherrah, J., 3D symmetry detection using the extended Gaussian image. *IEEE Trans. on Pattern Analysis and Machine Intelligence* 1997, 19, (2), 164–168.

(75) Bajaj, C.; Yu, Z., Geometric and Signal Processing of Reconstructed 3D Maps of Molecular Complexes. *Handbook of Computational Molecular Biology*, Edited by *S. Aluru, Chapman & Hall/CRC Press*, Computer and Information Science Series, ISBN: 1584884061 2005.

(76) Yu, Z.; Bajaj, C., Automatic Ultrastructure Segmentation of Reconstructed CryoEM Maps of Icosahedral Viruses. *IEEE Transactions on Image Processing: Special Issue on Molecular and Cellular Bioimaging* 2005, 14, (9), 1324–1337.

(77) Baker, M.; Yu, Z.; Chiu, W.; Bajaj, C., Automated Segmentation of Molecular Subunits in Electron Cryomicroscopy Density Maps. *Journal of Structural Biology* 2006, 156, 432–441.

(78) Yu, Z.; Bajaj, C., Computational Approaches for Automatic Structural Analysis of Large Bio-molecular Complexes. *IEEE/ACM Transactions on Computational Biology and Bioinformatics* 2008, 5, (4), 568–582.

(79) Sethian, J. A., *Level Set Methods and Fast Marching Methods* (2nd edition). Cambridge University Press: 1999.

(80) Malladi, R.; Sethian, J. A. In A real-time algorithm for medical shape recovery, Proceedings of International Conference on Computer Vision, 1998; 1998; pp. 304–310. 20.

(81) Sethian, J. A., A marching level set method for monotonically advancing fronts. *Proc. Natl. Acad. Sci.* 1996, 93, (4), 1591–1595.

(82) Bajaj, C.; Yu, Z.; Auer, M., Volumetric Feature Extraction and Visualization of Tomographic Molecular Imaging. *Journal of Structural Biology* 2003, 144, 132–143. 22.

(83) Sifakis, E.; Tziritas, G., Moving object localization using a multi-label fast marching algorithm. *Signal Processing: Image Communication* 2001, 16, (10), 963–976.

(84) W. Jiang, Z. Li, Z. Zhang, M.L. Baker, P. E. Prevelige, and W. Chiu. Coat protein fold and maturation transition of bacteriophage P22 seen at subnanometer resolutions. *Nat. Struct. Mol. Bio.*, 10(2):131–135, 2003.

(85) Z. H. Zhou, M. L. Baker, W. Jiang, M. Dougherty, J. Jakana, G. Dong, G. Lu, and W. Chiu. Electron cryomicroscopy and bioinformatics suggest protein fold models for rice dwarf virus. *Nat. Struct. Mol. Bio.*, 8(10):868–873, 2001.

(86) Y. Li, A. Leith, and J. Frank. Tinkerbell-a tool for interactive segmentation of 3d data. *J. Struct. Bio.*, 120 (3):266–275, 1997.

(87) M. Marko and A. D. Leith. Sterecon–three-dimensional reconstructions from stereoscopic contouring. *J. Struct. Bio.*, 116(1):93–98, 1996.

(88) J. A. Sethian. *Level set methods and fast marching methods*. Cambridge University Press, 1999. ISBN 0521645573.

(89) G. Sapiro. *Geometric partial differential equations and image analysis*. Cambridge University Press, 2001. ISBN 0521790751.

(90) S. Osher and R. P. Fedkiw. *Level set methods and dynamic implicit surfaces*. Springer Verlag, 2003.

(91) J. S. Suri and S. Laxminarayan. *PDE and level sets: algorithmic approaches to static and motion imagery*. Springer, 2002. ISBN 0306473534.

(92) N. Volkmann. A novel three-dimensional variant of the watershed transform for segmentation of electron density maps. *J. Struct. Bio.*, 138(1–2):123–129, 2002.

(93) G. D. Pintilie, J. Zhang, T.D. Goddard, W. Chiu, and D.C. Gossard. Quantitative analysis of cryo-em density map segmentation by watershed and scale-space filtering, and fitting of structures by alignment to regions. *J. Struct. Bio.*, 170(3):427–438, 2010.

(94) A. S. Frangakis and R. Hegerl. Segmentation of two-and three-dimensional data from electron microscopy using eigenvector analysis. *J. Struct. Bio.*, 138(1–2):105–113, 2002.

(95) Z.Yu and C.Bajaj. Anisotropic vector diffusion in image smoothing. In Proc. Int. Conf. Image Processing, pages 828–831, 2002.

(96) M. Baker, Z. Yu, W. Chiu, and C. Bajaj. Automated segmentation of molecular subunits in electron cryomicroscopy density maps. *J. Struct. Bio.*, 156(3):432–441, 2006.

(97) C. L. Bajaj, F. Bernardini, and G. Xu. Automatic reconstruction of surfaces and scalar fields from 3D scans. In Proc. Conf. Comput. Graphics Interact. Techniques, pages 109–118, 1995.

(98) T. K. Dey and S. Goswami. Provable surface reconstruction from noisy samples. *Comput. Geom.*, 35 (1–2):124–141, 2006.

(99) C. Bajaj. Geometric modeling and quantitative visualization of virus ultrastructure. *Modeling Biology: Structures, Behaviors, Evolution*, Edited *L. da Fontoura Costa and M. Laublichler*, pages 115–137, 2007.

(100) O. Sharma, Q. Zhang, F. Anton, and C. Bajaj. Multi-domain, higher order level set scheme for 3D image segmentation on the GPU. In IEEE Conf. Comput. Vision Pattern Recognit., pages 2211–2216, 2010.

(101) O. Sharma, Q. Zhang, F. Anton, and C. Bajaj. Fast streaming 3D level set on the GPU for smooth multi-phase segmentation. to appear in Trans. *Comput. Sci.*, 2011.

(102) T. F. Chan and L. A. Vese. A level set algorithm for minimizing the Mumford-Shah functional in image processing. In IEEE Workshop Variational Level Set Meth., pages 161–168, 2001.

(103) L. A. Vese and T. F. Chan. A multiphase level set framework for image segmentation using the Mumford and Shah model. *Int. J. Comput. Vision*, 50(3):271–293, 2002.

(104) Bajaj, C.; Goswami, S. In *Automatic Fold and Structural Motif Elucidation from 3D EM Maps of macromolecules*, ICVGIP, 2006; 2006; pp 264–275. 12.

(105) Goswami, S.; Dey, T. K.; Bajaj, C., Identifying Flat and Tubular Regions of a Shape by Unstable Manifolds. In Proc. 11th Sympos. Solid and Physical Modeling, 2006; pp. 27–37.

(106) Ludtke, S. J.; Baker, M. L.; Chen, D. H.; Song, J. L.; Chuang, D. T.; Chiu, W., De novo backbone trace of GroEL from single particle electron cryomicroscopy. *Structure* 2008, 16, (3), 441–448.

(107) Branden, C.; Tooze, J., *An Introduction to Protein Structure.* Garland Publishing Inc.: New York, 1999.

(108) Kong, Y.; Ma, J., A structural-informatics approach for mining b—sheets: locating sheets in intermediate-resolution density maps. *J. Mol. Bio* 2003, 332, 399–413.

(109) Kong, Y.; Zhang, X.; Baker, T. S.; Ma, J., A structural-informatics approach for tracing b-sheets: building pseudo-Ca traces for b-strands in intermediate—resolution density maps. *J. Mol. Bio.* 2004, 339, 117–130.

(110) Blum, H., A transformation for extracting new descriptors of shape. *Models for the perception of speech and visual form* 1967, 19, (5), 362–380. 32.

(111) Siddiqi, K.; Pizer, S. M., *Medial representations: mathematics, algorithms and applications.* Springer Verlag: 2008.

(112) G.D. Pintilie, J. Zhang, T.D. Goddard, W. Chiu, and D.C. Gossard. Quantitative analysis of cryo-EM density map segmentation by watershed and scale-space filtering, and fitting of structures by alignment to regions. J. Struct. Bio., 170(3):427–438, 2010.

(113) A. S. Frangakis and R. Hegerl. Segmentation of two-and three-dimensional data from electron microscopy using eigenvector analysis. J. Struct. Bio., 138(1–2):105–113, 2002.

(114) Y. Kong and J. Ma. A structural-informatics approach for mining beta-sheets: locating sheets in intermediate-resolution density maps. J. Mol. Biol., 332(2):399–413, 2003.

(115) Y. Kong, X. Zhang, T. S. Baker, and J. Ma. A Structural-informatics approach for tracing beta-sheets: building pseudo-C(alpha) traces for beta-strands in intermediate-resolution density maps. J. Mol. Biol., 339(1): 117–130, 2004.

(116) T. S. Baker and J. E. Johnson. Low resolution meets high: towards a resolution continuum from cells to atoms. Cur. Opin. Struct. Bio., 6(5):585–594, 1996.

(117) F. Fabiola and M. S. Chapman. Fitting of high-resolution structures into electron microscopy reconstruction images. Structure, 13(3):389–400, 2005.

(118) N. Volkmann and D. Hanein. Quantitative fitting of atomic models into observed densities derived by electron microscopy. J. Struct. Bio., 125(2–3):176–184, 1999.

(119) P. Chacón and W. Wriggers. Multi-resolution contour-based fitting of macromolecular structures. J. Mol. Biol., 317(3):375–384, 2002.

(120) W. Wriggers, R. A. Milligan, and J. A. McCammon. Situs: A package for docking crystal structures into low-resolution maps from electron microscopy. J. Struct. Bio., 125(2–3):185–195, 1999.

(121) W. Wriggers, R. A. Milligan, K. Schulten, and J. A. McCammon. Self-organizing neural networks bridge the biomolecular resolution gap. J. Mol. Biol., 284(5):1247–1254, 1998.

(122) W. Wriggers and P. Chacon. Modeling tricks and fitting techniques for multiresolution structures. Structure, 9(9):779–788, 2001.

(123) W. Wriggers, P. Chacón, J. A. Kovacs, F. Tama, and S. Birmanns. Topology representing neural networks reconcile biomolecular shape, structure, and dynamics. Neurocomput., 56:365–379, 2004.

(124) F. Tama, O. Miyashita, and C. L. Brooks III. Normal mode based flexible fitting of high-resolution structure into low-resolution experimental data from cryo-EM. J. Struct. Bio., 147(3):315–326, 2004.

(125) C. Bajaj, B. Bauer, R. Bettadapura, and A. Vollrath. Non-uniform fourier transforms for rigid body and multi-dimensional rotational correlations, 2013. SIAM J. on Scientific Computing, Accepted for Publication.

(126) R. Bettadapura, A. Vollrath, and C. Bajaj. PF2Fit: Polar fast fourier fitting of rigid atomic structures in 3D EM, ICES Technical Report 12-27, The University of Texas at Austin, 2012.

(127) R. Bettadapura, A. Vollrath, and C. Bajaj. PF3Fit: Iterative flexible fitting in 3D EM, ICES Technical Report 12-28, The University of Texas at Austin, 2012.

(128) D. Potts, J. Prestin, and A. Vollrath. A fast algorithm for nonequispaced Fourier transforms on the rotation group. *Numerical Algorithms*, 2009.

(129) C. Bajaj, R. A. Chowdhury, and V. Siddahanavalli. F2Dock: Fast fourier protein-protein docking. IEEE/ACM Trans. Comput. Biol. Bioinf., 8(1):45–58, 2011.

(130) R. Chowdhury, M. Rasheed, D. Keidel, M. Moussalem, M. Rasheed, A. Olson, M. Sanner, and C. Bajaj. Protein-protein docking with F2Dock 2.0 and GB-rerank. PLoS ONE 8(3): e51307. doi:10.1371/journal.pone.0051307

(131) Zhang, Y.; Bajaj, C.; Sohn, B.-S., 3D Finite Element Meshing from Imaging Data. *The special issue of Computer Methods in Applied Mechanics and Engineering (CMAME) on Unstructured Mesh Generation* 2005, 194, (48–49), 5083–5106.

(132) Zhang, Y.; Xu, G.; Bajaj, C., Quality Meshing of Implicit Solvation Models of Biomolecular Structures. *The special issue of Computer Aided Geometric Design (CAGD) on Applications of Geometric Modeling in the Life Sciences* 2006, 23, (6), 510–530.

(133) Bajaj, C.; Pascucci, V.; Holt, R.; Netravali, A., Dynamic Maintenance and Visualization of Molecular Surfaces. Fourth issue in the special series of Discrete Applied Mathematics on Computational Molecular Biology 2001, 127, (1), 23–51.

(134) W. Zhao, G. Xu, C. Bajaj An Algebraic Spline Model of Molecular Surfaces for Energetic Computations *IEEE/ACM Transactions on Computational Biology and Bioinformatics, 8(6):* 1458–1467, 2011

(135) C. Bajaj, R. Chowdhury, M. Rasheed A Dynamic Data Structure for Flexible Molecular Maintenance and Informatics *Bioinformatics* (2011) *27 (1):* 55–62

(136) C. Bajaj, G. Xu, Q. Zhang A Fast Variational Method for the Construction of Adaptive Resolution C2 Smooth Molecular Surfaces *Computer Methods in Applied Mechanics and Engineering*, volume *198*, p1684–1690 (2009)

(137) Garland, M., Heckbert, P. S. In Simplifying surfaces with color and texture using quadric error metrics, VIS '98: Proceedings of the conference on Visualization '98

(138) Censor, Y. On variable block algebraic reconstruction techniques, in *Mathematical Methods in Tomography*, 1999

(139) O. Sharma, Q. Zhang, F. Anton, C. Bajaj, Fast Streaming 3D Level Set Segmentation on the GPU for Smooth Multi-phase Segmentation, *Transactions on Computational Sciences*, 6750:72–91, 2011.

(140) Y. Zhang, C. Bajaj, G. Xu. Surface Smoothing and Quality Improvement of Quadrilateral/Hexahedral Meshes using Geometric Flow, *Comm. in Numerical Methods in Engineering*, Vol. *24*, 2008

(141) H. Peng, Bioimage informatics: a new area of engineering biology, Bioinformations, 24(17), p 1827–1836, (2008)

(142) Hekking LH, Lebbink MN, De Winter DA, Schneijdenberg CT, Brand CM, Humbel BM, Verkleij AJ, Post JA, Focused ion beam-scanning electron microscope: exploring large volumes of atherosclerotic tissue, Jounal of Microscopy, 253(3), p 336–347, (2009)

(143) W. Denk and H. Horstmann, Serial Block-Face Scanning Electron Microscopy to Reconstruct Three-Dimensional Tissue Nanostructure, PLOS Biology, 2004.

(144) P. Evans and A. McCoy, An introduction to molecular replacement, Acta Cryst. (2008), D64, p 1–10.

(145) W. Hendrickson and C. Ogata, Phase determination from multi-wavelength anomalous diffraction measurements. Methods in Enzymology 276: 494–523, 1997

(146) E. de la Fortelle and G. Bricogne, Maximum-Likelihood Heavy-Atom Parameter Refinement for Multiple Isomorphous Replacement and Multiwavelength Anomalous Diffraction Methods. *Methods in Enzymology*. Methods in Enzymology 276: 472–494, 1997

(147) T. S. Baker and R. Henderson, Electron Cryo-Microscopy of Biological Macromolecules, International Tables for Crystallography (2012). Vol. F, Chapter 19.6, pp. 593–614.

(148) J. T. Patton (ed), Segmented Double-stranded Viruses: Structure and Molecular Biology, Caister Academic Press, 2008

(149) X. Zhang, C. Bajaj, B. Kwon, T. Dolinsky, J. Nielsen, N. Baker, "Application of New Multiresolution Methods for the Comparison of Biomolecular Electrostatic Properties in the Absence of Structural Similarity", *Multiscale Modeling and Simulation*, 2006, 5(4): 1196–1213, PMID: 18841247, PMCID: PMC2561295

(150) S. Park, C. Bajaj, "Feature Selection of 3D Volume Data through Multi-Dimensional Transfer Functions", *Pattern Recognition Letters*, 2007, 28(3): 367–374, NIHMSID: 90898.

(151) C. Bajaj, X. Zhang, "Scalable Isosurface Visualization of Massive Datasets on Commodity off-the-shelf Clusters", *Journal of Parallel and Distributed Computing*, Jan 2009, 69(1): 39–53, Available online from http://dx.doi.org/10.1016/j.jpdc.2008.07.006, NIHMSID: 89714, PMC Journal in Process

(152) Q. Zhang, B. Subramanian, G. Xu, C. Bajaj. "Quality Multi-Domain Meshing for Volumetric Data", Proceedings of 2010 3rd International Conference on Biomedical Engineering and Informatics, pp. 472–476.

(153) B. Sohn, C. Bajaj Time-Varying Contour Topology, *IEEE Transactions on Visualization and Computer Graphics (TVCG), 12, 1*, 14–25, 2006

(154) C. Bajaj, S.-C. Chen, and A. Rand, "An Efficient Higher-Order Fast Multipole Boundary Element Solution for Poisson-Boltzmann Based molecular Electrostatics", *SIAM Journal on Scientific Computing*, 33(2) 826–848, 2011.

(155) C. Bajaj, J. Castrillon-Candas, V. Siddavanahalli, Z. Xu Compressed Representations of Macromolecular Structures and Properties *Structure*, Volume *13*, Issue *3*, March 2005, Pages 463–471, doi:10.1016/j.str.2005.02.004,PMID15766547

16

Building and Using 3D Digital Atlases of Complex Model Animals at the Single Cell Resolution

HANCHUAN PENG

1. INTRODUCTION

A systems biology approach to understanding a biological system is to collect data of as many individual elements of this system as possible and then analyze the patterns embedded in these data, with the goal to discover general principles or rules underlying this system. This approach, probably most well known in scientists' exploration of the genomics, proteomics, and other types of -omics data in the last two decades, is highly effective. For high-throughput, quantitative screening of phenotypes of several widely used model animals, such as *C. elegans* (*Caenorhabditis elegans*), fruit fly (*Drosophila melanogaster*), mouse, zebrafish, and so on, it becomes increasingly evident that a similar systems biology approach will also work (1–8). In this chapter, we present a general bioimage informatics framework to build three-dimensional (3D) digital atlases of these model animals at the single cell or single neurite tract resolution. These atlases enable quantitative analysis and mining of the cell distribution patterns. An atlas-based strategy also allows more accurate and more robust analysis of complicated image data sets than previous methods that do not use the prior information in the atlases.

For building and using 3D digital cell atlases of model animals, one prerequisite is the availability of a large amount of 3D image data sets of these systems (9). This becomes possible nowadays, due to the rapid advances of both tissue labeling (e.g., fluorescent proteins) and microscopic imaging techniques (10, 11). One prevailing method is to use 3D light microscopy methods, such as confocal or two-photon laser scanning microscopy, to acquire 3D image stacks of specifically labeled (e.g., genetically) tissues of the model animals (e.g., Figure 16.1) (12). Typically, there are several color channels in an image stack, each channel corresponding to one particular molecule-binding fluorophore that emits light of a certain wavelength when excited.

Computational analysis of these image data is often challenging. In recent years, many research groups started to develop bioimage informatics techniques, which aim at quantitatively analyzing and mining these cellular and molecular image data at a large scale (9, 13, 14). For applications to build and use the digital atlases of model animals, it turns out that a wide range of image analysis, mining, visualization, and atlas-modeling methods are highly desired.

In the following, we first introduce a study to build a quantitative 3D digital nuclear atlas for the model animal *C. elegans* at the single nucleus resolution (2). A general bioimage informatics framework to produce this atlas is discussed. In section 3, we will discuss how to use this *C. elegans* atlas in an important and novel application of the atlas-based cell targeting, which allows precise and robust automated gene expression screening and other studies (15). In section 4, we further extend the general framework, but with much enhanced computational techniques, to build a single neurite tract resolution 3D digital atlas for the fruit fly brain (6). Finally we will conclude in section 5.

2. BUILDING A 3D DIGITAL NUCLEAR ATLAS FOR *C. ELEGANS*

C. elegans (16, 17) is widely used in molecular and cell biology, development biology and aging, neuroscience, and so on. This animal has 959 cells in the adult hermaphrodite and 558 nuclei in L1 larva. The lineage of cells in this animal is invariant. We recently built a 3D digital cell atlas for the L1 larva (2), which contains 351 nuclei in the L1 larva, including 81 body wall muscle cells, 41 hypodermal cells, 20 intestine cells, 22 ventral motoneurons, and so on. This atlas has been refined continuously since its initial publication.

To produce this atlas, we consider a pipeline in Figure 16.2. We first generated transgenic worms with specific types of cells labeled with different reporters, and collected 3D confocal image stacks (Figure 16.2 box ◆) (12). Each image has three color-channels (Figure 16.1) of different reporters. There are five key bioimage informatics steps: registration (or standardization), segmentation, annotation, data mining, and atlas modeling.

- *Standardization and registration* (Figure 16.2 box ❖): Because usually a body of *C. elegans* worm is curved after sample preparation, we developed an automated 3D *C. elegans* body straightening method to standardize and register (align) different images (18). We attained that goal via detecting the medial axis along the anterior-posterior direction of the animal using a deformable curve model, followed by generating a whole sequence of cutting planes, with one-pixel spacing between consecutive planes, orthogonal to the medial anterior-posterior axis. The image data on these cutting planes are restacked to assemble the straightened *C. elegans* body. Typically there is a negligible amount of image resampling error in this step. After the initial standardization step, images of multiple *C. elegans* worms have their anterior and posterior ends matched. These images can be further registered to have a similar size and shape. Additional registration steps can also be used to align the cells as well (after the next cell segmentation step).

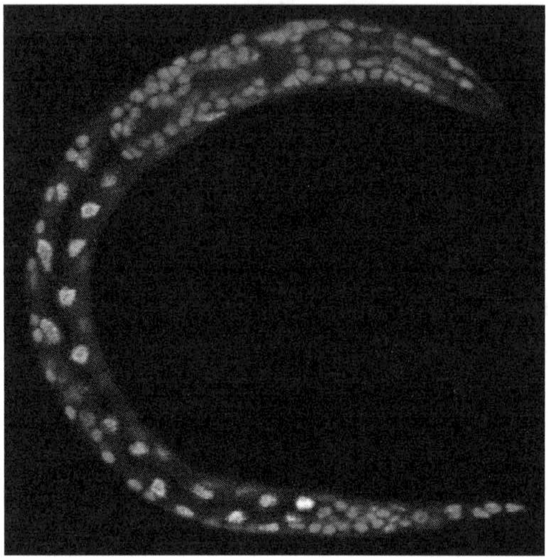

Figure 16.1 A typical confocal laser scanning microscopic image of *C. elegans* (L1 stage), where different nuclei are specifically labeled using fluorescent molecules. Blue: DNA-binding DAPI staining that labels all 558 nuclei of the animal. Green: enhanced Green Fluorescent Protein (eGFP) labeling of 81 body wall muscle cells and 1 depressor muscle cell. Red: mCherry labeling for specific cells for gene expression analysis. See insert for color version.

- *Segmentation* (Figure 16.2 box ◆): The next step in this pipeline is to extract the nuclei from images. For *C. elegans*, the problem is quite challenging, due to the fact that in the head and tail regions there are many clusters of crowded nuclei, which appear in the image as touching image objects, and in the trunk region of a worm there are many donut-shaped nuclei (due to large nucleoli that do not contain DNA and thus not stained by DAPI), which may also touch other smaller nuclei. We developed a method to first automatically fill all the holes in the donut-shaped nuclei, and then used an adaptive thresholding-based watershed method to automatically segment all DNA-labeled nuclei (19). The accuracy of nuclei segmentation is about 95~96 percent. After segmentation, morphological features and spatial relationships of these nuclei are used to characterize them.
- *Annotation* (Figure 16.2 box ☒): It is known that *C. elegans* nuclei/cells have unique identities. How could one incorporate a *C. elegans* biologist's knowledge about cells' unique identities into a database, after these nuclei have been segmented? On the other hand, since the nuclei segmentation is not 100 percent accurate, how might one pinpoint and correct the potential errors in 3D segmented image stacks? We developed an ergonomic 3D manual annotation system, VANO, which allows a biologist to quickly go through the segmentation results in a 3D environment and assign names to any of the segmented cells (20). If the biologist also finds potential segmentation errors, such as over—and under—segmentations, VANO allows rapid correction of these errors using a combination of computer mouse and keyboard operations. After the cell annotation, there is little

segmentation error in the data, which are ready for further pattern mining and atlas modeling.

- *Data mining* (Figure 16.2 box ⍓): The annotation step essentially turns the image data of *C. elegans* into point cloud representations, where each point corresponds to one cell. With many such point cloud data sets, we built data mining tools to detect the invariant spatial layout patterns in *C. elegans* cells (2). This starts by optimally aligning many point cloud data sets using a congealing approach, so that the overall distance of any pair of 3D point cloud data sets is minimized, in terms of the spatial transform (we used the 3D affine transform in this case). Once all representations were aligned, we computed the anterior-posterior, left-right, and dorsal-ventral relationship of all cells, based on their "standardized" 3D spatial locations. Modules of spatially invariant layout patterns of these cells were discovered (2). We also analyzed the relationships among cell locations, types, adjacency to each other, and lineages.
- *Atlas modeling* (Figure 16.2 box ⌘ and Figure 16.3): The last step in this atlas-construction pipeline is to collect statistics of cells' mean locations, standard deviation along three spatial axes (anterior-posterior, left-right, and dorsal-ventral), spatial spacing between adjacent cells, sizes and shape features, etc. (Figure 16.3). This digital atlas can be visualized in 3D. Compared to the widely used 2D cell map, which was originally made by hand thirty years ago (17), this 3D digital atlas (2) does not only provide an additional dimension, but also provides quantitative biology information that the previous 2D version does not have. In addition, for various biological purposes, we can integrate various other pieces of information in this 3D atlas, such as the covariance of cell locations (unpublished data)

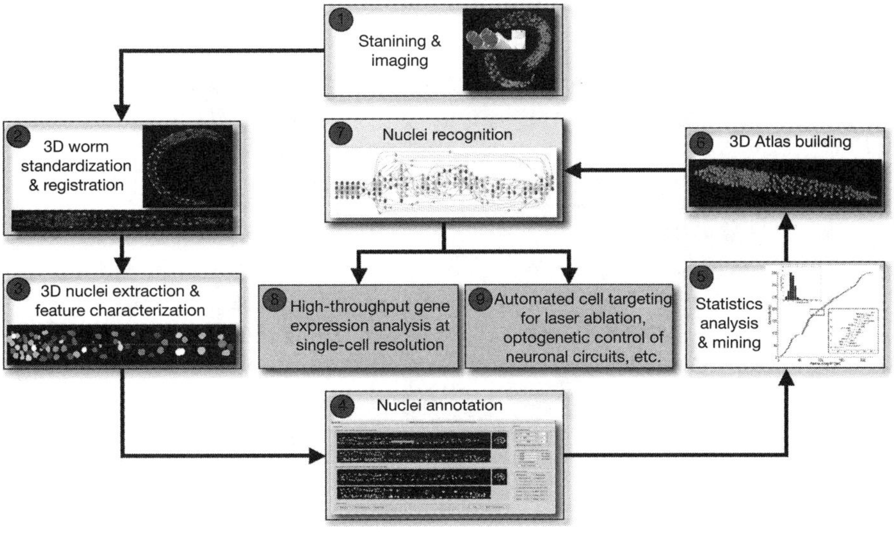

Figure 16.2 The major steps (boxes ♦~⌘) in building a 3D digital cell atlas of L1 larval stage of C. elegans and the applications (boxes ❀~❝) of this atlas. See insert for color version.

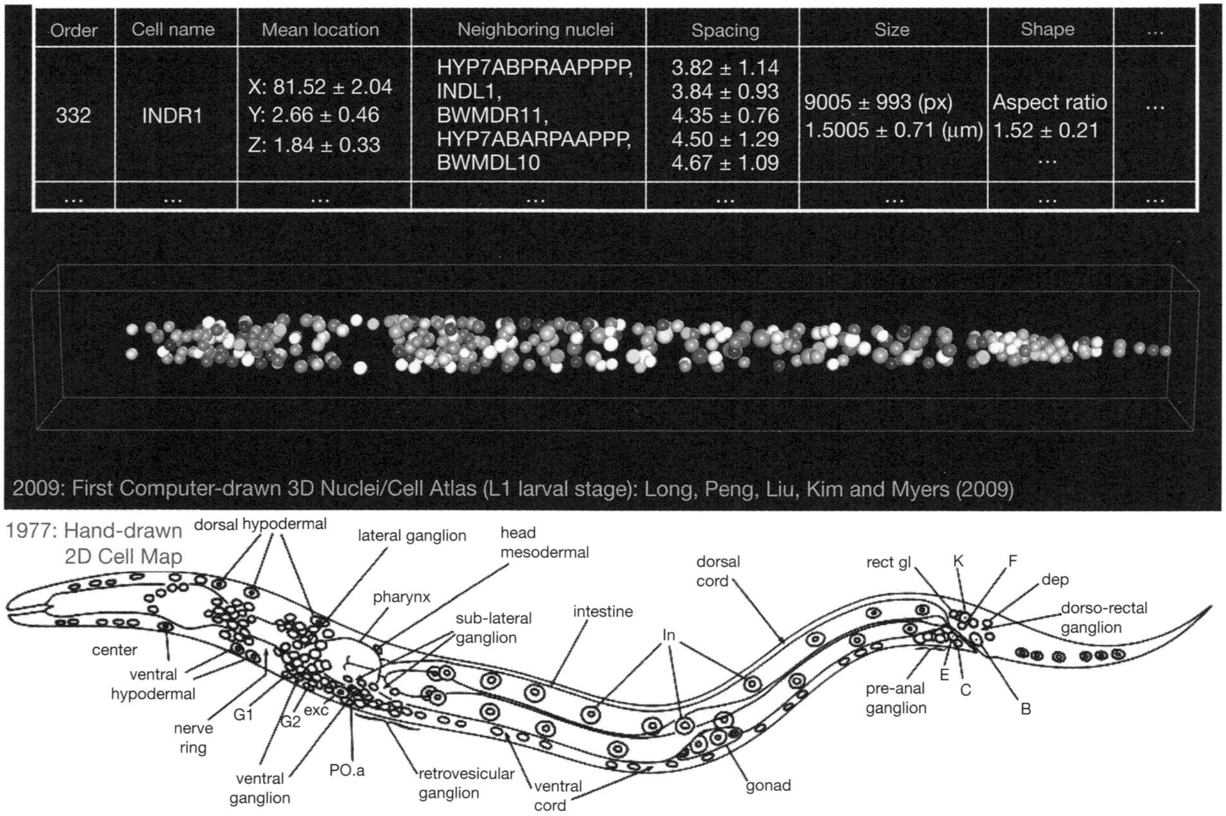

Figure 16.3 Comparison of the 3D digital cell atlas of *C. elegans* (Long, et al, 2009) and an earlier 2D version (Sulston and Horvitz, 1977). In the 3D atlas, each colored sphere corresponds to one nucleus; also for clear visualization, only a subset of nuclei are displayed. See insert for color version.

and gene expression levels (12). Most importantly, this digital atlas allows new ways to study biology (boxes ❁, ✿, ❝).

3. ATLAS-BASED CELL RECOGNITION AND RELATED APPLICATIONS

The single-cell resolution digital atlas has several immediate biological applications. For example, we have developed a cell recognition approach to automatically annotate these cells (Figure 16.2 box ❁) (2). This technique further enables median to high throughput analysis of gene expression in *C. elegans* larvae and has led to the finding of new cell subtypes (Figure 16.2 box ✿, 12). It has been shown to be a powerful tool for studying gene expression and regulation, cell lineages, and cell fates (12). We also proposed to use it for automated manipulation of neuronal circuits based on optogenetic methods and laser ablation (Figure 16.2 box ❝, Peng, et al, unpublished data).

Automatic recognition of cell identities is critical for quantitative measurement, targeting, and manipulation of cells of model animals at single-cell resolution. Indeed, a number of 3D cell and gene expression image segmentation, recognition, and tracking techniques have been developed and applied to widely used model animals, including *C. elegans* (1, 2, 21, 22), fruit fly (4, 5, 23), and zebrafish (8).

However, existing image analysis pipelines for single-cell resolution studies typically start with a segmentation of the cells in a 3D image, followed by an ad hoc recognition or tracking process. It is typically hard to ensure an error-free segmentation of cells in an image sample, especially when (a) the image has an uneven background or low signal-to-noise ratio, or (b) nuclei are so crowded as to touch each other or have irregular morphology. Our experience (2, 21) is that segmentation errors will make the subsequent recognition task even more challenging. In addition, within this previous framework, useful prior information, such as the relative location relationship of cells, and useful statistics from the recognition phase, such as discrepancies between the predicted and a priori cells locations, are hard to incorporate in a way that improves the overall analysis.

In addition, for a variety of experimental settings, it may be difficult to generate some of the currently used fiducial image features that are critical for accurately recognizing cells. For example, in the L1 larval stage of *C. elegans*, there are 81 body wall muscle (BWM) cells. These cells form four nearly symmetrical bundles, each of which has twenty or so cells in one of the ventral-left, ventral-right, dorsal-left, and dorsal-right quadrants. Recognizing these BWM cells is critical because they can serve as additional fiducial points for recognizing other cells in the animal. In experiments using fixed animals (e.g., 2) that also stain all nuclei, one can recognize these four bundles based on the asymmetry of the distribution of other cells (e.g., the 15 ventral motoneurons form an almost linear array along the ventral side of the animal). Nonetheless, for live animal experiments, due to the cost of building transgenic animals and the limited number of fluorescent color-channels, it is desirable to be able to recognize these BWM cells or other cells (e.g., intestinal cells, neurons) directly without additional fiducial patterns.

Figure 16.4 Simultaneous cell recognition and segmentation. In each image, the red spheres show the deformed atlas; the segmented pixels and their corresponding cells in the atlas are connected by lines with different colors. The first row shows the original image overlaid with the initial atlas; other rows show the results of several intermediate steps of the iterative optimization, with which the atlas of cells (body wall muscle cells shown here) deform to the optimal locations and the foreground image voxels are automatically classified to (and thus segmented) each of these cells. The energy values shown are normalized using the total number of image foreground voxels. When we produced this figure, the image intensity was enhanced for better visibility. The surface rendering of the segmented regions can be seen in Figure 16.5. See insert for color version.

Therefore, we proposed a new approach to simultaneously recognize and segment cells in *C. elegans* (15). We essentially eliminated the need for a two-step, segment-then-recognize, process. Instead, our method recognizes cells directly, producing the segmentation as a by-product. As shown in Figure 16.4, we realized this idea as an atlas-guided voxel classification algorithm, which integrates the processes of atlas-to-image mapping and voxel classification under a robust deterministic annealing framework.

The input of our new algorithm consists of a 3D image stack where the target cells have been stained in a single color channel, and a 3D atlas (which is a 3D point cloud representing the stereotyped spatial locations) of the target cells. The goal of the algorithm is to automatically extract the "meaningful" objects in the image and assign one of the cells in the atlas to each of these objects.

Mathematically, suppose we have an image that consists of N voxels $V = \{v_i, i = 1,2,\ldots,N\}$ and an atlas of K cells $C^0 = \{c_j^0, j = 1,2,\ldots,K\}$, where v_i is the ith voxel and c_j^0 is the jth cell, respectively. The goals of our algorithm are (a) to classify (i.e., label) the voxels X into K groups, each of which corresponds to a unique cell, and at the same time (b) to smoothly map each atlas cell c^0_j to a new 3D spatial location in the image that can best represent the corresponding voxel subset. For (a), we use a Euclidean distance and image intensity-based voxel-to-cell assignment matrix, which is continuously updated in our iterative optimization. For (b), to allow local deformation but also preserve cells' relative locations and the global smoothness, we consider the smoothing-thin-plate-spline (Wahba, 1990) transform. The detail of the algorithm can be found in (15).

In our direct atlas-to-image matching (15), since the atlas encodes all target cells' identities and their spatial location relationship, our method recognizes cells via smoothly deforming the atlas to fit the image. This process effectively classifies, or labels, voxels in the image; each group of voxels with the same label represents the extraction of a "meaningful" object and thus the image is also segmented. From a segmentation perspective, the key difference between our new approach and previous ones is that now we can naturally incorporate the relative location relationship of cells that is encoded in the cell atlas when we segment cells, thus reducing the chance of wrong segmentation. More importantly, with the new approach we could directly predict cell identities, without the complication due to problematic segmentation.

We applied this SRS method to a large data set of 161 *C. elegans* image stacks, which contains a total of 12,976 BWM and DEP cells that were carefully annotated by an expert of *C. elegans* anatomy and thus were used as the "ground-truth" identities of cells in our study. SRS correctly predicted the identities of 12,863 cells (99.13 percent recognition rate). SRS was also applied to recognize other types of cells, such as the intestinal cells.

We also compared SRS to previous segmentation and recognition methods. The watershed algorithm is one of most popular methods used in 2D or 3D cell segmentation. We compared SRS to the shape-based watershed algorithm, which is typically thought to be a well-performing method when the image intensity is uneven (such as our test images). To produce the best possible watershed result, we tried multiple different thresholds to define image foreground and multiple Gaussian smoothing kernels to remove image noise. Note that for SRS, we did not do these preprocessing. However, Figure 16.5 shows that compared to SRS, in the best case the watershed result still has substantial over-segmentation and under-segmentation. Of note, previous studies (e.g., 2) may use additional methods to split or merge cells after the initial watershed segmentation. SRS avoids these extra steps.

4. BUILDING A 3D STEREOTYPED NEURITE ATLAS FOR FRUIT FLY

Fruit fly (26) is another widely used model animal. An adult fruit fly brain has about 100,000 or so neurons, many of which are believed to be stereotyped and have

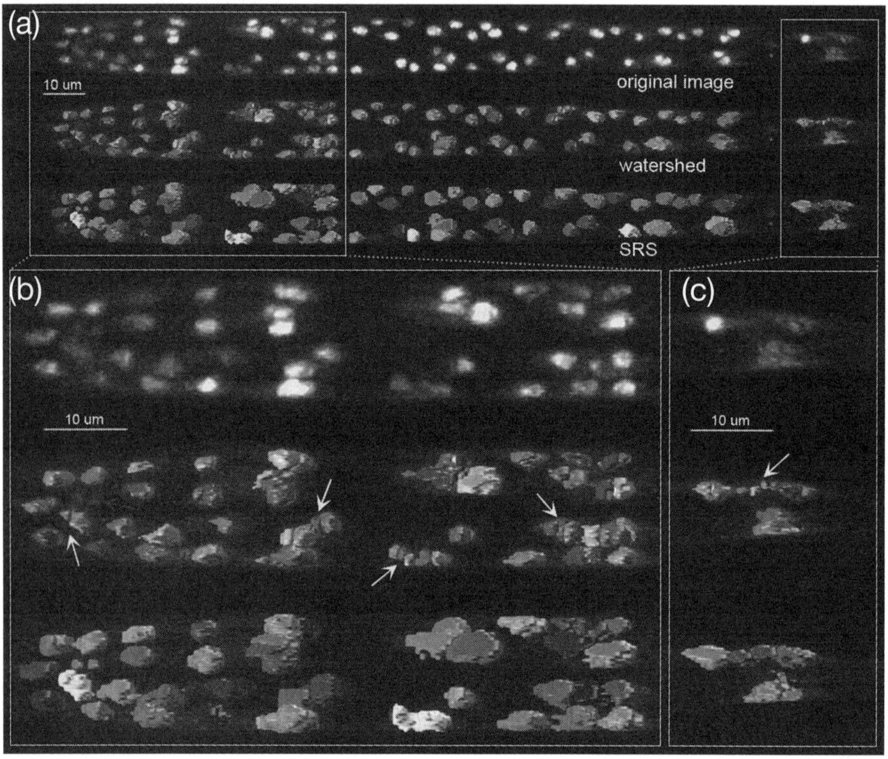

Figure 16.5 Comparison of the segmentation results of watershed and SRS. The same test image in Figure 16.4 was used. The color-surface objects indicate the different segmentation regions. In (c), yellow arrow: over-segmentation; red arrow: under-segmentation. See insert for color version.

invariant lineage, similar to *C. elegans*. However, currently our understanding in the fly brain is still very limited. A digital atlas derived from the statistical analysis of neuron distributions, projections, and connections in a number of individual fly brains can greatly deepen our insight into how a fly brain works.

The informatics pipeline in Figure 16.2 can be generalized. Indeed, from an image-informatics perspective, building the 3D statistical neuronal atlas in a fly brain (Figure 16.7) is conceptually similar to, although technically much more challenging than, building the digital cell atlas of *C. elegans* (Figure 16.2). The common computational components are:

1. 3D image registration, which maps fly neurons or *C. elegans* cells to a standard space;
2. 3D cell segmentation and description, which extracts/characterizes fly neurons or *C. elegans* cells;
3. 3D pattern annotation, which incorporates biologists' domain knowledge of the fly neurons or *C. elegans* cells into databases;
4. 3D image mining, which automatically clusters, annotates, searches fly neurons or *C. elegans* cells, and in some cases detects the statistical associations between cells and functions (or behaviors) of an animal;

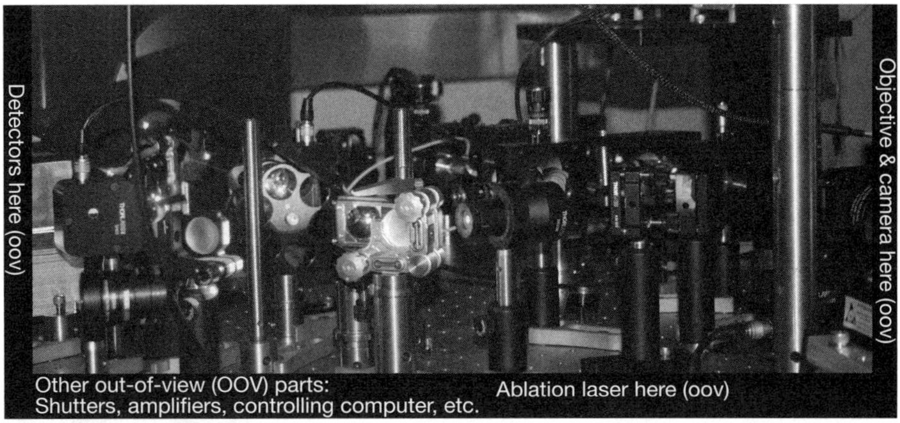

Figure 16.6 A snapshot of a laser scanning microscope driven by V3D (http://vaa3d.org; 25) for systematically perturbing *C. elegans* and other model animals at the single cell resolution (unpublished data).

5. Integration of information into a 3D atlas, which is an abstracted model of the probabilistic distribution of the location and morphological/physiological pattern of fly neurons or *C. elegans* cells.

In a fruit fly brain, cell bodies are located in a rind at the outer surface of the brain. Neurites extend into the interior to form the synaptic neuropil. Specific types of neurons can be labeled with a variety of techniques including antibody detection of proteins restricted to particular neuron types (e.g., 27), or genetic methods such as the UAS-GAL4 system (28), where each GAL4 line drives expression in a different subpopulation of neurons visualized with a membrane-targeted fluorescent reporter protein (e.g., UAS-mCD8-GFP). In Figure 16.7 box ◆, we used many GAL4 lines produced in several research groups (e.g., J. Simpson lab and G. Rubin lab). A large number of confocal images of the fly brains were acquired in these labs through collaboration.

For Figure 16.7 box ❖, automated 3D brain registration needs to be done in an accurate and robust way. Earlier 3D image registration approaches (29–31) use surface- or landmark-based alignment modules of the commercial 3D visualization software AMIRA (Visage Imaging, Inc.) to align sample specimens to a template, which defines the common coordinate system of the resulting atlases. The major disadvantage of these approaches is the huge amount of time it takes a user to manually segment the surfaces or define the landmarks in each subject brain, which makes any of these approaches unsuitable for our large-scale approach of image registration. Previous efforts to automatically register images of the fruit fly nervous system based on image features includes work on adult brains (32, 33), on the adult ventral nerve cord and larval nervous system (34). We conducted comparison tests (Supplementary Note), on several widely used methods for registration (32–37) and all produced unsatisfactory alignments at a rate that make them unsuitable for use in a pipeline that involves thousands of high-resolution 3D laser scanning microscope (LSM) images of Drosophila brains. In addition, all were very time-intensive and none produced quality scores that might alert a user that a particular sample did

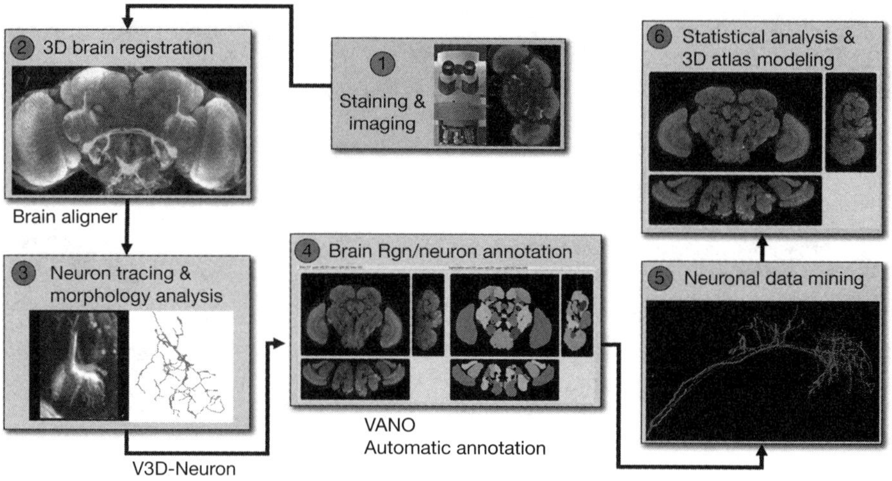

Figure 16.7 A pipeline to build the 3D neuronal atlas for a fruit fly brain.

not align well. We speculate that the failures are because most of the other systems were implicitly tuned for textures with sharp boundaries, which is not true of the neuropil staining pattern. All these issues make it hard to apply them to large-scale high-throughput experiments.

Therefore, we developed a new automatic registration program, BrainAligner (6), for Drosophila brains and have used it to align more than 25,000 large 3D LSM brain images of thousands of different neuronal expression patterns (the typical size of each image has at least 1024×1024×300 voxels). Our algorithm, called reliable landmark matching (RLM) (6), combines several of the best existing approaches into a new strategy based on reliably detecting landmarks in images. Our method is hundreds of times faster than the existing methods for automatically aligning Drosophila brain images, due to the fact that we used a much smaller number of landmarks than previous approaches and optimized image-warping methods. Since our landmarks correspond to the key corners or other high curvature locations in the images, using a smaller set of landmarks indeed makes the algorithm more robust to noise and other imperfectness of the data (e.g., brain damage during dissection).

After brain images have been aligned, the next step is to extract semantically meaningful objects, which are neuronal structures, in these images (Figure 16.7 box ◆). There are a number of studies on neuron reconstruction (39–57, 59, 60) and many semi- or fully automatic neuron tracing methods have been developed. Most of these existing methods have used various Structural Components (SC)—such as 3D spheres, ellipsoids, cylinders, line segments, or irregular compartments—to model a neuron's morphology. The most successful strategy among these algorithms is to build up the reconstruction by incrementally adding more and more such SCs into the morphological modeling of a neuron. Good examples include image voxel scooping (54), ray shooting (46), and template matching (60). These bottom-up local searching methods are suitable for 3D images that have ideally continuous neurite tracts and good signal-to-noise ratio (SNR).

However, for our 3D laser scanning microscopic images, precise digitization of the 3D morphological neuronal structure remains very problematic in practice. It is especially hard when an image has low SNR, and/or broken and fuzzy neurite segments that are due to the intrinsic punctuated neurite structures (e.g., synaptic boutons) or imperfections in sample preparation. Most of the local search methods discussed above cannot easily handle these hard situations, as it is very difficult to cross these gaps (i.e., low signal regions). Therefore, we developed a 3D neuron reconstruction system, called V3D-Neuron, for this goal. V3D-Neuron has several methods to automatically trace the neuron or neurite tract morphology in 3D (6, 25). Our methods tackle this challenging situation by combining both global and local cues. Global prior information, such as the starting and ending locations of neurite structures, will guide the finer-scale optimization using local image content. Such global priors of neurite structure can be supplied easily using a novel and highly effective 3D visualization system V3D (25). The GD algorithm, which is a Graph-augmented Deformable model, can be used to trace the optimal paths from the starting location to each of the ending points automatically (25). Then the entire reconstruction can be assembled automatically by detecting branching points along the merging paths (25). In addition, our neuron reconstruction method is built upon our 3D visualization system V3D (25), therefore a user can optionally proofread and edit the traced neuronal structure right away (25, 60). In this way, the produced neuronal structures are reliable for further annotation (Figure 16.7 box ⊠), mining (Figure 16.7 box ⍓), and atlas modeling (Figure 16.7 box ⌘). This method has also been successfully applied

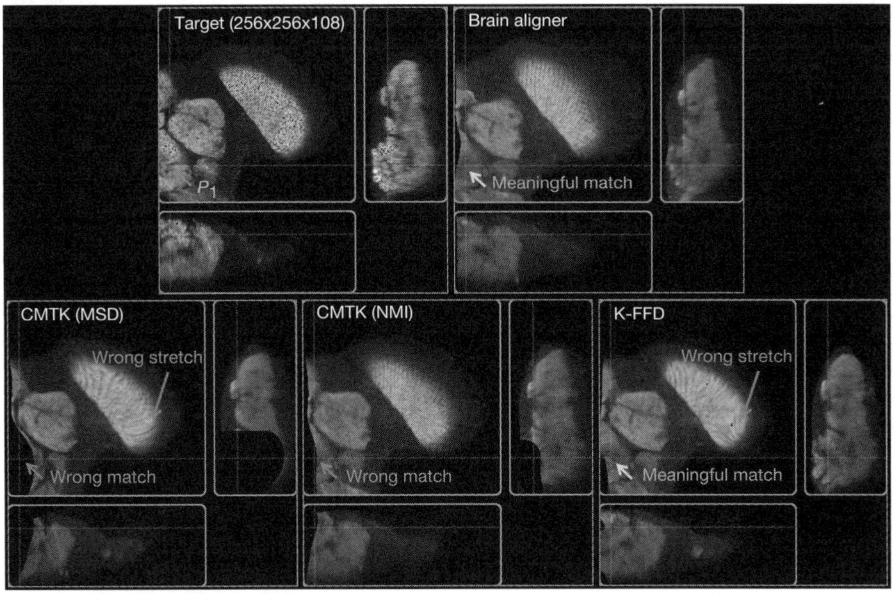

Figure 16.8 Comparison of the aligned images produced by BrainAligner and several previous methods. (a) and (b) show two typical locations we checked the aligned images. Compared to the target image, BrainAligner was able to produce smoothly aligned subject image. The irregular stretching of the warped image is one of the main problems of the previous alignments (38; Qu and Peng, unpublished data). Irregular stretching is due to wrong matching at various locations. See insert for color version.

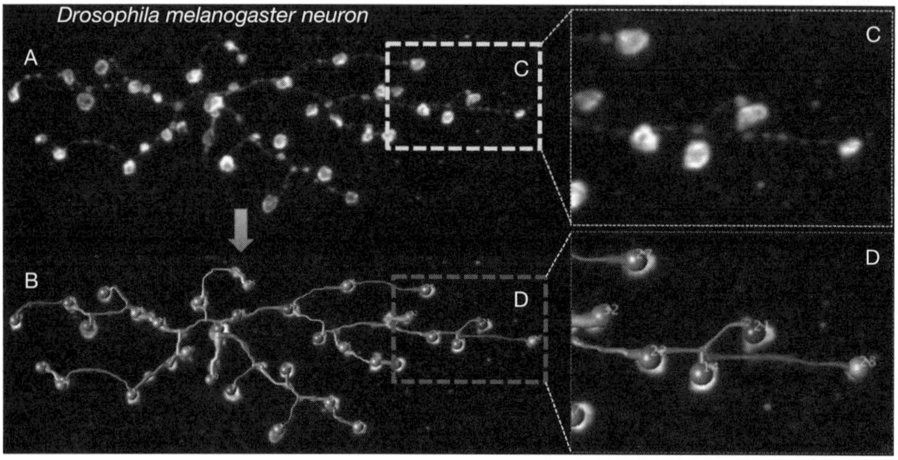

Figure 16.9 A challenging case of fruit fly neuron for tracing. This neuron has broken and often fuzzy structures and low single-to-noise ratio. (A) A single lamina neuron of fruit fly (courtesy of G. Rubin lab) along with the 3D reconstruction. In the input image (A) and (C) the fruit fly neuron has broken and punctuated neurites. We applied the V3D-Neuron to trace the neuron morphology (B) and (D) and visualize as a color-coded neurite model. See insert for color version.

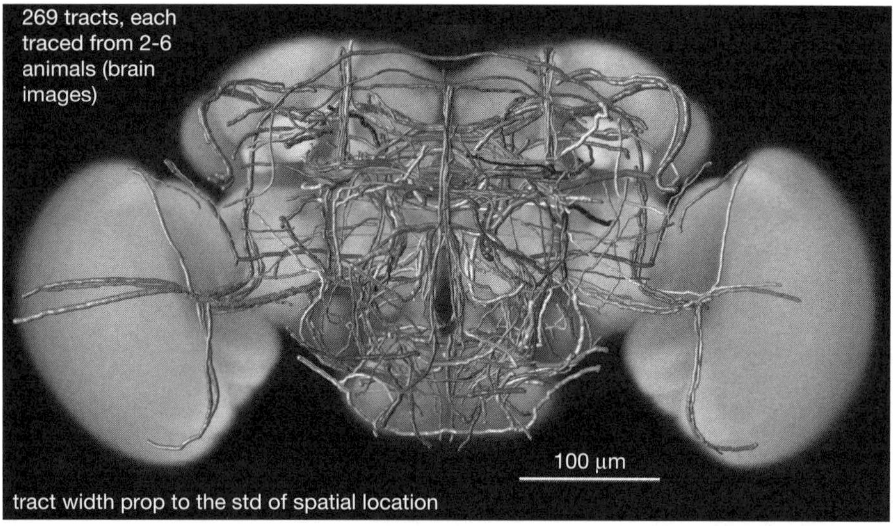

Figure 16.10 A 3D atlas of neurite tracts reconstructed from aligned GAL4 patterns. The neurote tracts were reconstructed from a number of brain images, which were all non-rigidly aligned in 3D to best match a standard target brain (Peng, Chung, et al, 2011). The neurite tracts are randomly color-coded for better visualization. The width of each tract equals the respective spatial deviation. See insert for color version.

to reconstructing neuronal connections in the most detailed mouse brain image atlas to date (61).

Our pipeline in Figure 16.7 is able to produce informative biology. For instance, we built the first stereotyped neurite tract atlas for the fruit fly brain. By aligning thousands of fly brain images and tracing their neurite tracts in 3D, we determined a set of 269 neurite tracts throughout a fly's brain. These tracts together indicate the architecture of the fruit fly brain. We reconstructed each tract from at least two aligned brains of each *GAL4* line. We computed the spatial variations and represented them as the width of each tract for visualization (Figure 16.10). The average variation was 1.98 ± 0.83 μm. This range of variation is within the upper bound of biological stereotypy of the neurite tracts themselves and noise introduced in sample preparation, imaging, and image analysis including registration and tracing. The tracing error was close to zero (25). Compared to the typical size of an adult fly brain (590μm×340μm×120μm), this small variation indicated strong stereotypy of the neurite tracts.

The 3D atlas in Figure 16.10 is significant. First, it quantifies the invariability of the layout of a complicated brain. Second, and more importantly, the stereotypy of these neuron bundles indicate that one will be able to repeatedly find the same neuronal feature at the same location in different individual fruit flies' brains. Therefore, biologists will be able to use this atlas to search patterns and compare them, and thus greatly reduce the amount of time for identifying interesting genetic lines in studying neuronal connectivity and related animal behaviors. It also allows design of a microscope similar to what we developed in Figure 16.6 to potentially manipulate the nervous system and study the functional circuit of a brain.

5. DISCUSSION AND CONCLUSIONS

It can be seen that the framework of atlas building in Figure 16.2 and Figure 16.7 are general. Despite the different image computing techniques in individual steps, the five steps—i.e. of registration, segmentation, annotation, data mining, and atlas modeling—are common. Indeed, the first four steps are also essential for many other bioimage informatics applications, such as modeling the developmental dynamics of animals.

The availability of the 3D atlases at single cell or neuron (or neurite tract) resolution enables new image analysis methods, such as the simultaneous recognition and segmentation of cells discussed in section 3. Combined with the new design of microscopes, these tools may find many biological and biomedical applications.

ACKNOWLEDGEMENT

This chapter is based on a number of collaborations and publications of the author and the following research labs and individuals (in alphabet order): Stuart Kim lab, Fuhui Long, Gene Myers, Lei Qu, Gerry Rubin lab, Julie Simpson lab, Jianyong Tang, and members of the Hanchuan Peng Lab.

REFERENCES

1. Bao,Z. et al. (2006) Automated cell lineage tracing in Caenorhabditis elegans. Proc. Natl. Acad. Sci. USA 103, 2707–2712.
2. Long,F., Peng,H., Liu,X. et al. (2009) A 3D digital atlas of C. elegans and its application to single-cell analyses. Nature Methods, 6(9): 667–672.
3. Murray,J.I. et al. (2008) Automated analysis of embryonic gene expression with cellular resolution in C. elegans. Nature Methods, 5(8): 703–709.
4. Luengo Hendriks CL, et al. (2006) 3D morphology and gene expression in the Drosophila blastoderm at cellular resolution I: data acquisition pipeline. Genome Biol. 7:R123.
5. Fowlkes,C,. et al. (2008) A quantitative spatiotemporal atlas of gene expression in the drosophila blastoderm. Cell, 133(2): 364–374.
6. Peng,H., Chung,P., Long,F., Qu,L., Jenett,A., Seeds,A., Myers,E.W. and Simpson,J.H. (2011) BrainAligner: 3D registration atlases of Drosophila brains. Nature Methods, 8(6): 493–498.
7. Lein, E., et al. (2007) "Genome-wide atlas of gene expression in the adult mouse brain," Nature, 445:168–176.
8. Keller,P.J. et al. (2010) Fast, high-contrast imaging of animal development with scanned light sheet-based structured-illumination microscopy. Nature Methods, 7(8): 637–42.
9. Hanchuan Peng, (2008) "Bioimage informatics: a new area of engineering biology," Bioinformatics, Vol. 24, No. 17, pp. 1827–1836, 2008.
10. Tsien, R.Y. (2003) "Imagining imaging's future," Nature Reviews Molecular Cell Biology, 5:S16–S21.
11. Megason, S.G., and Fraser, S.E. (2007). "Imaging in systems biology," Cell, 130: 784–795.
12. Liu,X., Long,F., Peng,H., Aerni,S.J., Jiang,M., Sanchez-Blanco,A., Murray,J.I., Preston,E., Mericle,B., Batzoglou,S., Myers,E.W. and Kim.S.K. (2009) Analysis of Cell Fate from Single-Cell Gene Expression Profiles in C. elegans. Cell, 139: 623–633.
13. Rittscher, J., Machiraju, R., Wong, S.T.C. (2008) Microscopic Image Analysis for Life Science Applications, Artech House (2008).
14. Swedlow, J.R., Goldberg, I.G., Eliceiri, K.W. and the OME Consortium (2009) "Bioimage informatics for experimental biology," Annual Review of Biophysics, 38: 327–346.
15. Qu, L. Long, F., Liu, X., Kim, S., Myers, E.W., and Peng, H. (2011) "Simultaneous recognition and segmentation of cells: application in C. elegans," Bioinformatics, DOI:10.1093/bioinformatics/btr480, 2011.
16. Brenner, S. (1974). "The Genetics of *Caenorhabditis elegans*". *Genetics*, 77 (1): 71–94
17. Sulston JE, Horvitz HR (1977). "Post-embryonic cell lineages of the nematode, Caenorhabditis elegans". *Dev. Biol.* 56 (1): 110–56.
18. Peng,H., Long,F., Liu,X., Kim,S.K. and Myers,E. (2008) Straightening C. elegans images. Bioinformatics, 24(2): 234–242.
19. Long, F. Peng, H., and Myers, E.W. (2007) "Automatic segmentation of nuclei in 3D microscopy images of C. elegans," Proc. of 2007 IEEE International Symposium on Biomedical Imaging: From Nano to Macro, Washington, D.C., USA, pp.536–539.
20. Peng, H. Long, F. and Myers, E.W. (2009) "VANO: a volume-object image annotation system," Bioinformatics, Vol. 25, No. 5, pp. 695–697.

21. Long,F. et al. (2008) Automatic recognition of cells (ARC) for 3D images of C. elegans. Lecture Notes in Computer Science: Research in Comp. Mol. Biology, 4955: 128–139.
22. Jaensch,S., Decker,M., Hyman,A.A. and Myers,E.W. (2010) Automated tracking and analysis of centrosomes in early Caenorhabditis elegans embryos. Bioinformatics, 26(12): i13–i20.
23. Zhou,J. and Peng,H. (2007) Automatic recognition and annotation of gene expression patterns of fly embryos. Bioinformatics, 23(12): 589–596.
24. Wahba,G. (1990) Spline models for observational data. SIAM, Philadelphia, PA.
25. Peng, H., Ruan, Z. Long, F., Simpson, J.H. & Myers E.W. (2010) V3D enables real-time 3D visualization and quantitative analysis of large-scale biological image data sets. Nature Biotechnology 28, 348–353.
26. Bellen, H.J, Tong, C., and Tsuda, H. (2010) "100 years of Drosophila research and its impact on vertebrate neuroscience: a history lesson for the future," Nature Reviews Neuroscience 11, 514–522.
27. Buchner, E., Bader, R., Buchner, S., Cox, J., Emson, P.C., Flory, E., Heizmann, C.W., Hemm, S., Hofbauer, A., & Oertel, W.H. Cell-specific immuno-probes for the brain of normal and mutant Drosophila melanogaster. I. Wildtype visual system, Cell and Tissue Research 253, 357–370 (1988).
28. Brand, A.H., & Perrimon, N. Targeted gene expression as a means of altering cell fates and generating dominant phenotypes. Development 118, 401–415 (1993).
29. Broughton, S.J., Kitamoto, T., & Greenspan, R.J. Excitatory and Inhibitory Switches for Courtship in the Brain of Drosophila melanogaster. Curr Biol 14, 538–547 (2004).
30. Jenett, A., Schindelin, J.E., & Heisenberg, M. The Virtual Insect Brain protocol: creating and comparing standardized neuroanatomy. BMC Bioinformatics 7, 544 (2006).
31. Boerner, J. & Duch, C. Average shape standard atlas for the adult Drosophila ventral nerve cord. The Journal of Comparative Neurology 518, 2437–2455 (2010).
32. Jefferis, G.S., Potter, C.J., Chan, A.M., Marin, E.C., Rohlfing, T., Maurer, C.R., Jr., and Luo, L. Comprehensive maps of Drosophila higher olfactory centers: spatially segregated fruit and pheromone representation. Cell 128, 1187–1203 (2007).
33. Chiang, A.S., et al., Three-dimensional reconstruction of brain-wide wiring networks in Drosophila at single-cell resolution, Current Biology 21, 1–11 (2011).
34. Qu, L., & Peng, H. A principal skeleton algorithm for standardizing confocal images of fruit fly nervous systems. Bioinformatics 26, 1091–1097 (2010).
35. Thirion, J.P. Image matching as a diffusion process: an analogy with Maxwell's demons, Medical Image Analysis 2, 243–260 (1998).
36. Shen, D. & Davatzikos, C. HAMMER: hierarchical attribute matching mechanism for elastic registration, IEEE Transactions on Medical Imaging 21, 1421–1439 (2002).
37. Vercauteren, T., Pennec, X. Perchant, A. & Ayache, N. Symmetric log-domain diffeomorphic registration: a demons-based approach. Lecture Notes in Computer Science 5241, 754–761 (2008).
38. Rohlfing, T. & Maurer, C.R., (2003) "Nonrigid image registration in shared memory multiprocess environments with application to brains, breasts, and bees," IEEE Transactions on Information Technology in Biomedicine 7, 16–25.
39. Al-Kofahi, K, Lasek, S., Szarowski, D., Pace, C., Nagy, G., Turner, JN, Roysam, B. (2002) "Rapid automated three-dimensional tracing of neurons from confocal image stacks," IEEE Transactions Information Technology in Biomedicine, 6(2).
40. Al-Kofahi, K., Can, A., Lasek, S., Szarowski, D.H., Dowell-Mesfin, N., Shain, W., Turner JN, Roysam, B (2003) "Median-based robust algorithms for tracing

neurons from noisy confocal microscope images," IEEE Transactions on Information Technology in Biomedicine, 7: 302–317.

41. Dima et al. (2002). Automatic segmentation and skeletonization of neurons from confocal microscopy images based on the 3-D wavelet transform. IEEE Transactions on Image Processing 11(7):790–801.
42. Meijering, E., Jacob, M., Sarria, J., Steiner, P., Hirling, H., Unser, M., (2004) "Design and validation of a tool for neurite tracing and analysis in fluorescence microscopy images," Cytometry 58A: 167–176.
43. Schmitt et al. (2004). New methods for the computer-assisted 3-D reconstruction of neurons from confocal image stacks. NeuroImage 23(4):1283–1298.
44. Weaver et al. (2004). Automated algorithms for multiscale morphometry of neuronal dendrites. Neural Computation 16(7):1353–1383.
45. Abdul-Karim et al. (2005). Automatic selection of parameters for vessel/neurite segmentation algorithms. IEEE Transactions on Image Processing 14(9):1338–1350.
46. Wearne, S.L., Rodriguez, A., Ehlenberger, D.B., Rocher, A.B., Hendersion, S.C., and Hof, P.R. (2005) "New Techniques for imaging, digitization and analysis of three-dimensional neural morphology on multiple scales," Neuroscience, 136: 661–680.
47. Evers et al. (2005). Progress in functional neuroanatomy: Precise automatic geometric reconstruction of neuronal morphology from confocal image stacks. Journal of Neurophysiology 93(4):2331–2342.
48. Xiong et al. (2006). Automated neurite labeling and analysis in fluorescence microscopy images. Cytometry Part A 69(6):494–505.
49. Zhang, Y., Zhou, X., Degterev, A., Lipinski, M., Adjeroh, D., Yuan, J., Wong, STC., (2007) "Automated neurite extraction using dynamic programming for high-throughput screening of neuron-based assays," NeuroImage, 35(4): 1502–1515.
50. Narro et al. (2007). NeuronMetrics: Software for semi-automated processing of cultured neuron images. Brain Research 1138:57–75.
51. Zhang et al. (2008). 3D Axon structure extraction and analysis in confocal fluorescence microscopy images. Neural Computation 20(8):1899–1927.
52. Cai et al. (2008). Using nonlinear diffusion and mean shift to detect and connect cross-sections of axons in 3D optical microscopy images. Medical Image Analysis 12(6):666–675.
53. Losavio et al. (2008). Live neuron morphology automatically reconstructed from multiphoton and confocal imaging data. Journal of Neurophysiology, 100(4):2422–2429.
54. Rodriguez et al. (2009). Three-dimensional neuron tracing by voxel scooping. Journal of Neuroscience Methods 184(1):169–175.
55. Sun et al. (2009). Fast linear feature detection using multiple directional non-maximum suppression. Journal of Microscopy 234(2):147–157.
56. Vasilkoski et al. (2009). Detection of the optimal neuron traces in confocal microscopy images. Journal of Neuroscience Methods 178(1):197–204.
57. Yuan et al. (2009). MDL constrained 3-D grayscale skeletonization algorithm for automated extraction of dendrites and spines from fluorescence confocal images. Neuroinformatics 7(4):213–232.
58. Roysam, B., Shain, W., and Ascoli, G.A. (2009) "The central role of neuroinformatics in the national academy of engineering's grandest challenge: reverse engineer the brain," Neuroinformatics, 7:1–5.
59. Xie, J., Zhao, T, Lee, TM, Myers, E, and Peng, H. (2010) Automatic neuron tracing in volumetric microscopy images with anisotropic path searching, Medical

Image Computing And Computer-Assisted Intervention—MICCAI 2010, PT II, 6362: 472–479.

60. Zhao T., et al, (2011) Automated reconstruction of neuronal morphology based on local geometrical and global structural models, Neuroinformatics. (in press)
61. Li A., et al, (2010) Micro-optical sectioning tomography to obtain a high-resolution atlas of the mouse brain, *Science*, 330(6009): 1404–1408.

PART IV

Conclusions

17

Quo Vadis, Imaging?

MANFRED AUER AND GARY C. HOWARD ■

In a relatively short time, our ability to image life has become dramatically more powerful. For millennia, humans could only visualize life macroscopically. Early scientists struggled to imagine what made up the human body, but their own eyes were the limit. In 1268 Roger Bacon noted that a "burning lens"—one that could focus sunlight to start fires—magnified small objects. In the late seventeenth century, early microscopists combined two lens into a rough version of today's microscopy. In just 400 years, we have moved from only what our own eyes can see to an ability to visualize individual atoms. We have learned a great deal about living systems. And the ability to visualize life is accelerating.

While starting out with simple brightfield optical microscopy, the inventory of microscopic imaging techniques in the meantime has expanded to use neutrons and electrons, and a large portion of the electromagnetic spectrum. Each form of microscopy has been optimized over the last fifty years, with ever more refined approaches being invented by physicists and subsequently applied to biological problems. We doubt that any major breakthrough technology in the near future will radically change the way we visualize biological structures.

While it would seem that we are at the end of a long development, we're actually at the beginning of a new era of integrated bioimaging. We will bridge the gaps among the various imaging methods, particularly between the high-resolution structural methods, such as X-ray diffraction and NMR, and those used to define tissue, cellular, and subcellular structure, such as various forms of microscopy, including EM at the higher resolutions and MRI at the lower resolutions.

Not only will this synthesis of methods lead to a holistic visualization of biology, but we contend that integrated bioimaging is the next logical step "beyond" genomics, metagenomics, and proteomics, and many other -omics, that tend to be bulk analytic techniques. Through integrated bioimaging, we will acquire a spatially and temporally distinct understanding of complex systems that goes beyond the parts list and shows how the parts fit together to create biological function.

The large-scale study of proteins of the last twenty years is a good example of how ever-increasing amounts of data acquired by examining individual macromolecules are not sufficient to understand life. Summarized in a seminal article by Alberts (1998), the biochemical community realized that most, if not all, life-sustaining functions require the *coordinated interaction* of several proteins, and their correct spatial and temporal assembly into micromolecular machines. Hence life cannot be adequately understood by knowledge of a single protein structure and function, no matter how detailed its description. Instead, it has become clear that physiological and pathological processes are the result of malfunctions of multiprotein macromolecular complexes and, as a consequence, alterations of their spatial and temporal distribution within their cellular context.

Moreover, such protein-protein interplay could not be easily predicted from the structure of the individual proteins alone, and is not a static process. It requires tools to study the three-dimensional architecture of macromolecular machines, along with a precise description of their dynamic behavior, in vitro and in vivo. The need for 3D architectural information has led to the revival of electron microscopic techniques (i.e., single-particle cryo-electron microscopy, cryo-EM). Cryo-EM and the ever-expanding contribution of X-ray crystallography have revealed the organization and detailed 3D structure for several multiprotein complexes. Indeed, so-called hybrid methods have been developed that aim to use information from both X-ray crystallography, often of partial structures, and somewhat lower-resolution cryo-EM density envelopes for structure fitting and density interpretation. Increasingly, modeling such complexes based on structural constraints contributes to our insight into biological function. It is most powerful in cases where direct structural evidence cannot be obtained but the system of interest is homologous to known model systems. Dynamics and localization are of equal importance.

However, the study of macromolecular machine three-dimensional organization, by itself, cannot explain the pathogenesis of a variety of diseases or the phenomenon of self-sustaining life. A more holistic view of life's inner workings is needed. In its simplest form, many thought that what we must do is to measure pretty much everything we can measure with respect to composition and individual structures, whether it seems to matter or not. "Shoot first, think later" may have become the new motto, and is most prominently visible in the -omics type of research. By collecting data on all kinds of -omics (i.e., genomics, metagenomics, structural genomics, proteomics, metabolomics, lipidomics), we have established an exquisite and ever-expanding parts list.

Yet, simply listing who is likely to be contributing to a particular process is not enough, any more than imagining the content of a movie by looking at the credits. Plant cell walls are an excellent example. More than 1000 proteins are involved in the formation, maintenance, and remodeling of plant cell walls, but no experimentally rooted structural model exists. Likewise, we have exquisite information about which proteins play roles in carcinogenesis, cancer progression, or metastasis, but we still lack a physical or truly mechanistic model of such processes.

In fact, it somewhat seems that we are drowning in data, and it is unclear whether simply collecting more data will accomplish anything. What we propose is that such -omics-data, to become more useful, needs to be put into a 3D spatial and temporal context, that we need wiring diagrams and assembly directions, the architectural plans which allow us to put all the parts together, so we understand their relationships, and from there we can derive function.

What is missing is an understanding of how the parts interact physically, and what effect respective single protein malfunction has (e.g., in the progression of the disease). In other words, what is missing is a blueprint of spatial and temporal protein interactions, distribution, and function that will allow us to predict what cell behavioral consequences such a disruption of a single component implies.

The importance of spatial/temporal organization and the interplay of the respective components does not stop at the subcellular and cellular level: many diseases, such as cancer, cannot be fundamentally understood at a (macro)molecular, subcellular, or cellular level alone. Cancer is a disease of a tissue. Likewise, eukaryotic and even prokaryotic cells alter their behavior when faced with a tissue or microbial community context. The study of the human microbiome (i.e., all microbial communities that colonize human epidermal and epithelial surfaces and vastly outnumber the host cells) is an excellent example. The interactions of these biofilms and the human tissue are likely crucial for health and disease, and hence, a detailed understanding of such interaction and interdependence, while clearly lying far in the future, is imperative.

So, how is one to go about this multiplicity of scale, the dimensions of space and time, and the parameter space of constituents, ranging from metabolic substrates, ligands, and small molecules to protein interaction partners? What is the magic bullet technology that will once and for all answer all such questions? We have already occupied every part of the electromagnetic spectrum, and we have explored many types of radiation and particles. Barring an unforeseen change in physics, the biophysical approaches will not likely change over the coming years. No single approach will measure all the parameters of interest.

But if such biological challenges require a holistic understanding at multiple spatial and temporal scales, the second best option is to mix and match appropriate techniques to measure all necessary experimental aspects. While there is almost always room for incremental improvement in each technique, with ever-lasting quests for larger, faster, and brighter or higher-contrast methods, each technique has reached a level of maturity in terms of instrument development and advancement of experimental protocols. At the same time, each technology has advanced to the point that even scientists from related fields have difficulties in grasping all the nuanced details of a certain expert technique that can make or break the successful outcome of the project.

Our methods have served us well in looking at specific isolated aspects of biology. Now we want to further answer those fundamental questions. We need to find ways to synthesize our findings from multiple levels of organization and aspects. This book was an attempt to do that with the most fundamental of methods: imaging.

Although reasonable explanations for a variety of steps in the rather complex process of evolution have been proposed, how detailed such complexity of spatiotemporal chemical, physical, compositional, and structural organization arose is still hotly debated. We have been extremely successful in understanding in mechanistic chemical detail the function of enzymes and how mutations in their genes affect function, although we are far from predicting the effect of mutations when solely faced with the sequence data or when considering multiple mutations in the same gene simultaneously.

"Seeing is believing," and even today, few things are more powerful for convincing skeptics than an image. Imaging remains a critical component of biomedical research, and imaging methods are improving rapidly and revealing great detail about biology. Nevertheless, those methods, sophisticated and powerful as they are, still offer only snapshots of the living world. Many questions remain. How can we integrate information derived from incomplete methods? How do we combine information from the molecular, cellular, and organismal levels into a coherent whole?

Table 17.1 Critical Biological Questions for Imaging

ARRANGING THE BASIC BUILDING BLOCKS
Does a gene's location in the nucleus matter? Genes on different parts of the chromosome and perhaps even on different chromosomes loop together in domains of active transcription. However, are there levels of organization beyond this?
How does a cell position its proteins? With a typical cell making ~10,000 different types of proteins amounting to over a billion proteins altogether, protein distribution is far from random. How does the cell know where to position these proteins, and while eukaryotic cells employ organelles to compartmentalize metabolic pathways, and proteins often have sorting signals that will guide them to their final destination, "many proteins operate outside organelles and still need to find specific homes or molecular partners within the relatively vast spaces of a cell." mRNA is often non-stochastically distributed, and this observation may explain some of the protein's localization pattern but simply shifts the problem to how mRNA knows where to go.
Are there specialized domains (e.g., lipid rafts) in a cell membranes that allow the specific segregation of membrane proteins (e.g., into organized signaling platforms)?
CONTROLLING CELL DIFFERENTIATION
What changes in cellular organization accompany cell differentiation?
How do cells detect and migrate in response to tiny concentration gradients of chemicals in their environment?
How do groups of cells migrate together (e.g., during development of the organism)? How does a cell or a tissue know its size and maintain it within the normal range? While cells can range in size from small to huge, for each cell type, a narrow range of dimensions is observed. So while obviously small and large volumes are compatible with cellular function, how does each cell know when to stop growing in size? This question can be extended to tissues and organs. How is size so tightly controlled, and do these mechanisms act independently of each other?
MANAGING COMPLEX BIOLOGICAL FUNCTIONS
Can cells communicate over long distances (e.g., via membrane nanotubes)?
How does the malfunction of a single protein lead to complex pathological phenotypes (e.g., cancer, neurodegenerative diseases, psychiatric diseases, organ failure, death)?
Can we mathematically predict the behavior of cells and tissues/microbial communities in response to environmental challenges?
Does the complexity of spatiotemporal organization of cells in the brain and the interplay of the separate components necessarily lead to conscience?
Do microbial cells in a biofilm community coordinate behavior and how do they work together towards a sustainable lifestyle?
OTHER QUESTIONS
How do highly specialized cells and tissues know exactly where they are in the complex multicellular organism to behave appropriately? What is the underlying mechanism of their cellular GPS?
What is the molecular mechanism of evolution? How does the mutation of single traits/protein sequences result in novel species?
What did the first cell look like and how did it function? How was it possible for life as we know it to occur in the first place?

Where do we go from here? One can imagine progress in two directions: improvements to the methodologies and what questions those improved technologies might focus on. Technologies are improving at a rapid rate; for example, multiple variations on STED light and fluorescent microscopy have been developed in recent years and free-electron diffraction might allow imaging of individual biological molecules in aqueous solution. Other methods will allow us to gain information about composition as well as organization. New systems will allow visualization of molecules in living and performing organisms, and better high-throughput systems will allow systems to collect much more data.

Those innovations will be employed to answer new questions about living systems. In 2011 the NewFocus section of the journal *Science* (334:1046–1051) listed a number of fundamental questions remaining in biology. Advances in imaging will surely be important in resolving these questions. Here we note some questions that imaging technologies will help to resolve (Table 17.1).

REFERENCES

Alberts B (1998) The cell as a collection of protein machines: preparing the next generation of molecular biologists. *Cell* 92:291–294

Travis J (2011) Mysteries of the cell. *Science* 334:1046–1051

INDEX